A. B. Prescott

A MANUAL

OF

CHEMISTRY,

ON THE BASIS OF

TURNER'S ELEMENTS OF CHEMISTRY;

CONTAINING, IN A CONDENSED FORM,

ALL THE MOST IMPORTANT FACTS AND PRINCIPLES
OF THE SCIENCE.

DESIGNED AS A TEXT-BOOK

FOR COLLEGES AND OTHER SEMINARIES OF LEARNING.

Sixth Revised Edition.

REWRITTEN AND RESTEREOTYPED, WITH MANY NEW ILLUSTRATIONS.

BY JOHN JOHNSTON, LL. D.,

PROFESSOR OF NATURAL SCIENCE IN THE WESLEYAN UNIVERSITY.

PHILADELPHIA:
CHARLES DESILVER,
No. 714 CHESTNUT STREET,
OPPOSITE THE MASONIC HALL.
1860.

STEREOTYPED BY J. FAGAN.

TO

PARKER CLEAVELAND, LL. D.,

PROFESSOR OF CHEMISTRY, MINERALOGY, AND NATURAL PHILOSOPHY,

IN BOWDOIN COLLEGE, BRUNSWICK, ME.;

DISTINGUISHED NO LESS FOR HIS PERSONAL VIRTUES

THAN

AS THE AUTHOR

OF THE FIRST AMERICAN WORK ON MINERALOGY AND GEOLOGY;

The following Pages are Respectfully Inscribed,

IN TOKEN OF THE DEEP SENSE OF OBLIGATION ENTERTAINED

BY HIS FRIEND AND FORMER PUPIL,

JOHN JOHNSTON.

PREFACE

TO THE PRESENT, OR SIXTH REVISED EDITION.

By the original contract between the publishers and compiler of this work, provision was made for a periodical revision, in order that new and important discoveries might be introduced without delay, and the work be made to conform as much as possible to the rapidly advancing science. These revisions have been carefully attended to, and considerable alterations in the plates from time to time have been required; the whole of the part on Organic Chemistry, in the last preceding revision, having been rewritten and restereotyped.

But the progress of the science is, and has been, still onward; — new and important facts have rapidly been made known, and new views, throwing more or less light on points heretofore considered obscure and doubtful, have been proposed; so that in the present revision an entire recast of the work has been found necessary. Encouraged by the favor heretofore shown the work, the publisher has cheerfully incurred the expense of stereotyping it anew, including the preparation of many new illustrations; and the results of our joint labors are here presented to the public in a book substantially new, though retaining the former title.

The changes which have been made are too great and important to be discussed here; — for a knowledge of them the intelligent reader is referred to the pages of the work itself; they are such only as the new aspects of the science seemed imperatively to' demand. The principles which have been followed in its preparation are indicated in the extracts from the advertisements to former editions, which will be found further on. Many of the new cuts have been derived from the profusely illustrated work of Regnault; others are original, or have been obtained from miscellaneous sources.

The Grouping of the Elements adopted is nearly the same as that of Gmelin; — it is not free from objection, but is considered the best yet proposed on this difficult point. In preparing the remarks introductory to the part on Organic Chemistry, important aid has been derived from Dr. W. Gibbs' "Report on the Recent Progress of Organic Chemistry," prepared for the American Association for the Advancement of Science, and printed in the Proceedings of their ninth Meeting, at Providence, R. I., August, 1855.

Many thanks are due to teachers and other kind friends, for judicious suggestions and encouraging words during the preparation of the work; and it is now offered to the public in the confident expectation that it will be found not less adapted for use in the school or lecture-room than preceding editions.

MIDDLETOWN, CT., *July*, 1856.

EXTRACT FROM THE ADVERTISEMENT TO THE FIRST EDITION. (1840.)

The preparation of the following pages was undertaken by the advice of the late lamented President of the Wesleyan University, with the primary design of providing a suitable Text-book on Chemistry, for the use of the annual classes in that institution.

There are indeed already before the public many excellent works on this branch of science, the great merits of which the subscriber is happy to acknowledge; but he long since became convinced, from his experience in teaching, of the need of a work of a little different character, for the special use of students in our higher seminaries of learning, as a text-book. The object of a great majority of students, even of those who pursue a collegiate course, is, not to make themselves familiar with minute details of facts or processes of manipulation, but to understand the great principles of the science, and the leading facts which serve for its foundation. To facilitate the accomplishment of this purpose is the object of the present work. In preparing it, the excellent "Elements of Chemistry" of the late Dr. Turner has been adopted as the basis, and all of that work incorporated in it which was suited to our purpose. His arrangement has been uniformly followed, with a few unimportant exceptions, which it is not necessary here to particularize. This arrangement, on the whole, is considered the best that has ever been proposed.

The part of Dr. Turner's work omitted is taken up chiefly with details of facts and discussions of opinions and theories, which indeed is important in a work designed for the general student, but which would be out of place in a book prepared expressly to be used as a text-book. Its place, however, has been in part supplied by matter compiled from various other sources, so that the work is thought to be sufficiently large for the ordinary use of students, as the study of this science is usually pursued in this country. It has constantly been an object, while the work should be true to the science, and present in true proportion all its important features, to make it at the same time as practical as possible; to lead the student to apply the principles he learns to the solution of natural phenomena, or processes he may witness in the arts.

EXTRACT FROM THE ADVERTISEMENT TO THE SECOND EDITION.

In the present edition the work has been carefully revised, and indeed re-compiled from the seventh edition of Turner's, and many additions made to adapt it to the advancing state of the science. * * * *

The extracts from other authors are always introduced in their own language, except in cases where it was necessary to make some little change to incorporate the extract the better with the passage with which it comes in connection. In a few instances the names of authors are introduced in the text. To avoid the necessity of constantly introducing quotation marks and references, a list of the authors which have been used will be given.

To facilitate the acquisition of the science, the text is divided into paragraphs, and numbered; and references to important facts and principles introduced as frequently as they seemed necessary. As in many institutions so much time cannot be devoted to this science as would be requisite for a thorough study of the whole work, the less important parts have been printed in smaller type, which may be omitted on the first reading. The intelligent student, however, it is hoped, will not be satisfied without a perusal, at his hours of leisure, of the whole work.

LIST OF WORKS

MADE USE OF, MORE OR LESS, IN THE PREPARATION OF THIS WORK.

Elements of Chemistry, by the late Edward Turner, M.D., F.R.S., &c., edited by J. Liebig, M.D., Ph.D.F.R.S., &c., and Wm. Gregory, M.D., F.R.S.E.
Elements of Chemistry, &c., by Robert Kane, M.D., M.R.I.A., &c. Dublin.
Chemistry of Organic Bodies, by Thomson.
Do. Inorganic Bodies. Two vols.
Ure's Dictionary of Chemistry. Two vols.
Encyclopedia Metropolitana. Articles, Electro-magnetism, Electricity, Galvanism, Heat, Light, and Chemistry.
Library of Useful Knowledge. Articles, Electricity, Galvanism, Magnetism, Electro-magnetism, Chemistry, &c.
Thomson's Outlines of the Sciences of Heat and Electricity.
Traité de Chimie Appliquée aux Arts, par M. Dumas. Six tomes.
Traité de Chimie, par J. J. Berzelius; traduit par Me. Esslinger. Huit tomes.
Abrégé Elémentaire de Chimie, par J. L. Lassaigne. Deux tomes.
Organic Chemistry in its applications to Agriculture and Physiology, by Liebig, edited by Webster.
Animal Chemistry, or Organic Chemistry in its applications to Physiology and Pathology, by Liebig.
Lectures on Agricultural Chemistry and Geology, by J. F. W. Johnston.
Elements of do. do.
Thomson's "First Principles." Two vols.
Prof. Silliman's Chemistry. Two vols.
Prof. Hare's Compendium of Chemistry.
Faraday's Chemical Manipulation, edited by Dr. J. K. Mitchell.
Thomson's History of Chemistry. Two vols.
A Treatise on Chemistry by Michael Donovan, Esq.; Lardner's Cabinet Cyclopedia.
Prof. John W. Webster's Manual of Chemistry, on the basis of Prof. Brande's.
United States' Dispensatory, by Drs. Wood and Bache.
American Journal of Science and the Arts, conducted by Prof. Silliman.
Henry's Elements of Chemistry. Three vols.
Cleaveland's Mineralogy and Geology.
Dana's Mineralogy.
Shepherd's Mineralogy. Three vols.
Griffin's Chemical Recreations.
Journal of the Franklin Institute.
Parke's Chemical Catechism.
Chaptal's Chemistry applied to Agriculture.
Elements of Chemistry, by M. Lavoisier, translated from the French by R. Kerr, F.R.S.
Watson's Chemical Essays. Five vols.
Noad's Chemical Manipulation and Analysis.
Do. Lectures on Electricity.
Knapp's Chemical Technology. Vols. I., II.
Gibbs' Report on the Recent Progress of Organic Chemistry.
Gmelin's (L.) Hand-book of Chemistry, translated by Henry Watts. Vols. I. to IX.
Traité de Chimie Elémentaire, Theorique et Pratique, par L. J. Thenard. Cinq tomes.
Cours de Chimie Elémentaire, par A. Bouchardat. Deux tomes.
Leçons sur la Philosophie Chimie, professées au Collége de France, par M. Dumas.
Théorie des Proportions Chimiques, et Table Synoptique des Poids Antomiques, etc., par J. J. Berzelius.
Traité de Mineralogie, par M. L'Abbé Haüy. Quatre tomes.
Eléments de Physique, etc., par M. Pouillet. do.
Lehrbuch der Chimie, von E. Mitscherlich, Berlin, 1844.
Grundriss der Chimie, von Professor Dr. F. F. Runge.
Cours de Chimie Generale, par J. Pelouze et E. Fremy. Trois tomes, accompagné d'un Atlas de 46 Planches.
Gerhardt (Ch.), Traité Chimie Organique.
Regnault, Cours Elémentaire de Chimie.
The same, translated into English by Dr. T. F. Betton, M.D.

Besides the above, reference has often been made to various other works, as Le Dictionnaire des Sciences Naturelles, Annales de Chimie et de Physique, the various Encyclopedias, Philosophical Transactions, &c.

CONTENTS.

PART I

THE IMPONDERABLE AGENTS.

PART II.

GENERAL CHEMISTRY.

PART III.

SPECIAL CHEMISTRY—INORGANIC.

PART IV.

SPECIAL CHEMISTRY—ORGANIC.

MANUAL OF CHEMISTRY.

PART I.

THE IMPONDERABLE AGENTS.

INTRODUCTION.

1. WE recognize as matter or substance whatever possesses the four properties of extension, impenetrability, inertia, and gravity, or weight. By the first of these properties every body occupies a portion of space; by the second, it refuses to allow another body to occupy this space at the same time with itself; by the third, it is incapable, of itself, of changing its state, whether of rest or motion; and by the fourth, if unsupported, it falls to the earth. Whatever does not possess all these properties is not recognized as matter.

2. Natural science embraces the whole range of material things: their properties, the changes they are capable of undergoing, and the laws of their changes.

3. As has been suggested by Gmelin, all the changes of which any portion of matter is capable may be referred to the three causes or forces of *Repulsion*, *Attraction*, and *Vitality*.

4. **Repulsion** is manifest in the property of matter denominated impenetrability, and in the expansion of bodies, especially by the influence of heat, as will be shown hereafter.

QUESTIONS. — 1. What is matter or substance? Define what is meant by the four properties mentioned. — 2. What does *Natural Science* embrace? — 3. To what three causes may all changes of matter be referred? — 4. In what is repulsion manifest?

5. **Attraction** manifests itself in a variety of forms: 1. As *Gravitation*, or that force which acts at all distances, however great, and between the largest masses. 2. *Cohesion*, or that force which, acting only at distances immeasurably small, unites the parts of the same mass. 3. *Electrical* and *Magnetic Attraction*. 4. *Chemical Attraction* or *Affinity*, which acts only at insensible distances, and between the ultimate particles of bodies, and produces homogeneous compounds.

6. **Vitality** is that peculiar force or power, possessed both by animals and plants, by which the simple affinities of the various substances contained in their bodies are so modified and controlled in their action, as to produce the complex, and almost innumerable organic compounds, such as sugar, woody-fibre, albumen, &c.

Changes produced by all the varieties of attraction above mentioned, except the fourth, or last, pertain properly to Physics or Natural Philosophy; while those produced by *Affinity*, either alone, or as it is controlled by vitality in the bodies of plants and animals, belong to Chemistry.

The changes produced by the action of affinity consist in the combination of dissimilar substances into a homogeneous mass, or, occasionally, the separation of dissimilar substances from a homogeneous mass. We may, therefore, define Chemistry as the science which treats of the combination of dissimilar substances into homogeneous compounds, and of the separation of dissimilar bodies from homogeneous compounds.

7. **Molecules or Atoms.**—All bodies, it is believed, are made up of infinite numbers of indefinitely small particles—too small to be detected by the eye, even when aided by the most powerful microscopes—which are called *molecules* or *atoms* (from *a*, privative, and *temno*, I cut), indicating their supposed indivisibility. Our knowledge of them is obtained indirectly, as we shall see hereafter; but it is believed that all the molecules of the same substance are precisely alike in weight, size, and form, as well as other properties.

8. **Simple and Compound Bodies.**—From what has been said above, the distinction between simple and compound bodies is obvious. Simple substances are such as are believed to be com-

---

QUESTIONS.—5. What are the different varieties of attraction? Define the several varieties.—6. What is vitality? What changes pertain to *Natural Philosophy* or *Physics?* What to *Chemistry?* The changes produced by the action of affinity consist in what?—7. Of what are all bodies composed? Do we have any direct knowledge of these atoms?—8. What are *simple bodies?*

posed of only one kind of particles, as carbon, sulphur, copper, and gold; compound substances are composed of two or more kinds of particles, which are held in union more or less intimate by their affinity. The separation of the elements of a compound is called its decomposition.

The composition of a body may be determined in two ways, analytically or synthetically. By *analysis*, the elements of a compound are separated from one another, as when water is resolved by the agency of galvanism into oxygen and hydrogen; by *synthesis* they are made to combine, as when oxygen and hydrogen unite by the electric spark, and generate a portion of water. Each of these kinds of proof is satisfactory; but when they are conjoined—when we first resolve a particle of water into its elements, and then reproduce it by causing them to unite—the evidence is in the highest degree conclusive.

9. **Matter is Indestructible**; that is, it cannot be made to cease to exist. This statement seems at first view contrary to fact. Water and other volatile substances are dissipated by heat; and coals and wood are consumed in the fire, and disappear. But in these and other similar phenomena, not a particle of matter is annihilated: the apparent destruction is owing merely to a change of form or of composition. The power of the chemist is, therefore, limited to the production of these changes.

10. **Different Forms of Matter.**—Matter exists in three forms or states: the solid, liquid, and gaseous. Besides these, there are the three imponderable agents, Heat, Light, and Electricity, which, if they are ever proved to be material, will constitute a fourth form of matter.

It is believed that the particles of a substance, even the most solid, are never in actual contact, but are held in close proximity by the two opposite forces of attraction and repulsion; and that the particular state, whether solid, liquid, or gaseous, in which a body is seen, depends upon the relative intensity, for the time, of these forces.

If the force of attraction altogether preponderates in a body, it is *solid*, and the particles, in general, are held firmly in their

---

QUESTIONS.—What are *compound bodies?* Give an illustration. In what two modes may the composition of a body be determined? Explain analysis and synthesis.—9. Can matter be destroyed? To what is the power of the chemist limited?—10. What different *forms* of matter are there? What is said of the imponderable agents? Are the particles of matter ever in contact? Upon what will the state of matter in any particular case depend? Are not the particles of solids in contact? What reasons are given for this opinion?

places, and are incapable of motion among themselves. But the particles are not in actual contact, for, by cooling, or by great pressure, the dimensions of any body may be contracted, and, therefore, its particles brought nearer to each other. This will appear more fully hereafter.

In *liquids*, there is a degree of cohesion among the particles which, however, are capable of perfectly free motion among them selves. That there is a degree of cohesion existing between the particles is shown by the drop, which is composed of particles held together by a slight force; but this slight force does not interfere with the freedom of their movements.

Gases are distinguished by their tendency to expand, or enlarge their volume, when external pressure is removed. In them cohesion is entirely wanting. The term *fluid* is applied to both liquids and gases.

Some substances are found naturally existing in one of these states, and some in another; and many can be made to pass from one state or form to another, simply by varying their temperature, or the pressure to which they are exposed. Thus, water at a moderate temperature is liquid, but in the cold weather of winter it freezes, that is, becomes solid; and if it be heated sufficiently, it is changed into steam, or becomes gaseous. The metal, platinum, is found always in the solid state, though it may be melted by very great heat; but carbon is known only as a solid. Several substances, found naturally in the gaseous state, may be changed to liquids by great pressure, or by extreme cold; and, by a still greater cold, some of them may be frozen. Others, as atmospheric air, have hitherto resisted all attempts to reduce them to the liquid or solid form.

Heat, light, and electricity are said to be imponderable, because they possess no appreciable weight; but they certainly exhibit some of the ordinary properties of matter. They may be accumulated in bodies, are capable of being attracted and repelled, and often produce various chemical and mechanical effects. But because they possess no weight, so far as we can determine, many choose to consider them, not as matter, but only properties of matter.

QUESTIONS.—Is there any cohesion among the particles of liquids? How is this shown? How are *gases* distinguished? How is the word *fluid* used? What is said of the natural state of substances? What are the imponderables? Why are they so called?

I. HEAT.

NATURE AND SOURCES OF HEAT.

11. The word *Heat* is used indiscriminately to indicate the sensation we experience by placing the hand in contact with a heated body, or the cause of the sensation. To indicate the latter, the word *caloric* has sometimes been used.

The discussion of this subject properly pertains to Physics, or Natural Philosophy, (6,) but the agency of heat is so intimately connected with nearly all chemical changes, that a treatise upon Chemistry would be imperfect without a previous development of some of its more important laws and phenomena.

12. Nature of Heat. — Heat cannot be obtained separate from matter; it is invisible, and, so far as we are able to determine, entirely destitute of weight. It is not, therefore, (10,) believed to be material; but in describing its effects, and its relations to matter in general, we speak of it as an exceedingly subtile fluid, the particles of which constantly repel each other, but are attracted by other substances — as capable of being transmitted through space, and the interior of bodies, and of being accumulated in quantities in them. It is present in all bodies, and cannot be wholly separated from them. For if a substance, however cold, be transferred into an atmosphere which is still colder, a thermometer placed in the body will indicate the escape of heat.

Heat appears to be attracted by all bodies, but is self-repellent, as is shown by the fact that two bodies easily movable, when heated in a vacuum, repel each other.

13. Sources of Heat.—The chief sources of heat are: the Sun, Combustion, and other chemical changes, Friction, Electricity, and Vital Action.

The *Sun* is the great source of heat to our system. The intensity of the solar heat appears to be directly in proportion to the number of rays that can be collected upon a given surface; and at one time philosophers were able to produce a greater heat by

QUESTIONS. — 11. How is the word *Heat* used? *Caloric?* Is the agency of heat connected with chemical changes? — 12. Can heat be obtained separate from matter? Do we speak of heat as being material? Is heat present in all bodies? Is it attracted by matter? — 13. What *sources* of heat are mentioned? How may the sun's rays be concentrated so as to produce a great heat?

collecting the sun's rays by means of the convex lens or concave mirror, than by any other mode.

But although the sun's rays are not made use of in the arts when great heat is required, yet their momentous importance to all the inhabitants of the earth cannot be over-estimated. Without them all the water upon the face of the globe would soon be congealed, and animal and vegetable life cease to exist.

Combustion is the great source of artificial heat, as the sun is the source of natural heat. Besides wood, nature has provided immense deposits of combustible material, in the form of mineral coal, in the bosom of the earth. These are found in almost every country, and seem to be provided by the Creator as an unfailing resource for man, when, from the increase of the species, or from his own negligence or extravagance, the supply from the vegetable world should fail or become deficient.

Friction is a well-known source of heat. By the friction of the parts of heavy machinery, especially when not well oiled, heat has often been evolved sufficient to ignite wood; and the same effect is said to have been produced in ships by the rapid descent of the cable. Some tribes of the aborigines of this country were accustomed to kindle their fires by rubbing smartly one piece of wood against another. In the boring of cannon, heat enough has been evolved to raise the temperature of a considerable quantity of water so as to boil.

The heating effects of *electricity* will be considered hereafter.

The influence of *vital action* in developing heat is seen in all warm-blooded animals, which are maintained at a temperature often much above that of the air and other surrounding bodies, though heat must constantly be escaping from them.

EXPANSION OF BODIES BY HEAT.—THERMOMETERS.

14. All bodies, with a very few exceptions, expand when their temperature is increased, and contract when it is reduced. How this effect is produced we really do not know, but appearances indicate that the particles of heat entering among the particles of the body, partially overcome their cohesion, and cause them to sepa-

---

QUESTIONS.—What is the great source of artificial heat? What is said of friction as a source of heat?—14. Are all bodies expanded by heat? How are bodies affected by a reduction of their temperature?

rate farther from each other. On the other hand, when the particles of heat are withdrawn, the molecules of the body are allowed to approximate each other more closely. A substance is therefore less dense when heated, than when cold.

15. Expansion of Solids.—The expansion of solids by heat is not very considerable, but may easily be made very sensible. Let a bar of brass be accurately fitted into a gauge, when cold, and then let it be slightly heated; it will be found to have increased so much in length as not to fit the gauge. If the gauge be also made of brass, and the experiment performed in the warm weather of summer, the same result will be produced by cooling the gauge in ice-water, because of its contraction by the cold. This experiment indicates a change only in length, but a corresponding change is at the same time produced, both in breadth and thickness, as may be demonstrated in various modes, which the ingenious student will readily devise.

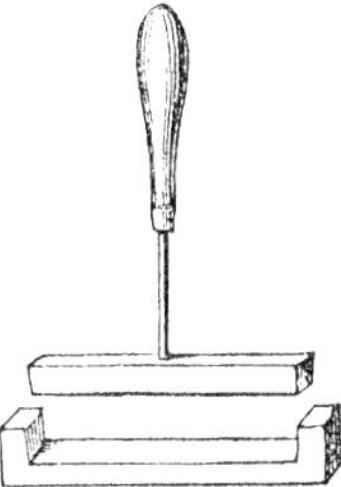
Expansion of Solids.

16. Different Solids, when equally heated, do not expand equally; every substance possesses an expansibility peculiar to itself. But a body expanded by heat, and again cooled to the same temperature it had at first, suffers no change in its dimensions.

Nor does the same substance expand equally at all temperatures with an equal increase of heat; in general, the expansibility increases with the temperature. Thus, a body heated ten degrees at a high temperature, expands more than when the same amount of heat is added at a low temperature.

The different expansibility of the two metals, copper and platinum, may be shown by soldering together a thin slip of each, and applying a moderate heat to the compound bar. Both plates will be equally heated, but the copper being the most expansible, the bar will be curved, the cop-

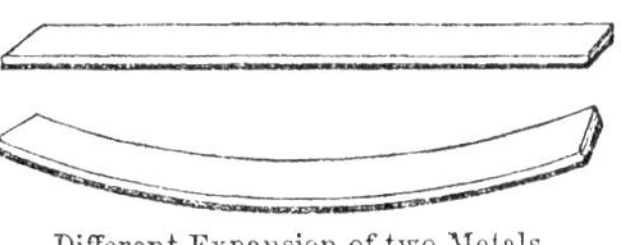
Different Expansion of two Metals.

QUESTIONS.—15. How may the expansion of a solid by heat be shown 16. Do all bodies expand equally when equally heated? Does the same substance at different temperatures expand equally for equal increases of temperature? How may the different expansibilities of two metals, as copper and platinum, be shown?

per being on the convex side. See figure, in which the copper is supposed to be on the lower, and the platinum on the upper side. Other metals, used in pairs in a similar manner, would show the same result, but with many of them the effect would be less decided.

An instrument like the following, at the same time that it shows the different expansibilities of two metals, serves as an excellent thermometer for many practical purposes. A and B are pieces of iron wire $\frac{2}{10}$ths of an inch in diameter, and a foot long; and C a piece of brass wire of the same size and length. At the bottom they are all fastened together by brazing or otherwise; at the top, a piece of brass is fixed to the two pieces of iron, and through it, near the centre, is a hole in which the brass wire, C, plays freely. Now, by immersing the thin wires in boiling water, hot oil, or melted lead, they are all expanded; but the brass expanding more than the iron, its upper end is pushed upward against the lever, D, which in turn acts upon E, producing considerable motion at its extremity, where may be placed a graduated scale, as S. Such an instrument will be sensibly effected by even moderate changes of temperature.

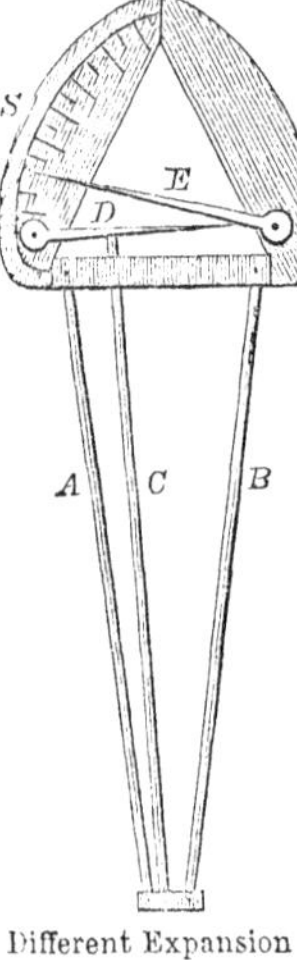

Different Expansion of Metals.

The following table shows the expansion in length of rods of several substances, when transferred from the freezing to the boiling point of water:

Substances.	Expansion in Fraction of Length.	Substances.	Expansion in Fraction of Length.
Flint Glass..........	$\frac{1}{1248}$	Copper..............	$\frac{1}{588}$
Wood..............	$\frac{1}{1289}$	Brass...............	$\frac{1}{530}$
Platinum............	$\frac{1}{1111}$	Zinc................	$\frac{1}{339}$
Gold................	$\frac{1}{660}$	Tin.................	$\frac{1}{430}$
Silver..............	$\frac{1}{505}$	Bismuth.............	$\frac{1}{718}$
Iron................	$\frac{1}{899}$	Lead................	$\frac{1}{351}$
Steel...............	$\frac{1}{927}$	Antimony...........	$\frac{1}{923}$

QUESTIONS.—Describe the instrument represented by the second figure of this paragraph. What is the design of the instrument? What are some of the most expansible of the metals, as indicated in the table?

17. Practical Applications.—This property of bodies, and particularly of the metals, has been applied to various useful purposes in the arts. The iron band or *tire* of a carriage-wheel is made a little smaller than the circumference of the wheel, but, being expanded, is sufficiently enlarged to be slipped on; and the immediate application of water prevents it from burning the wood, and brings the iron to its original dimensions, causing it to grasp the wheel with great firmness. Other examples are of frequent occurrence in the arts.

The expansions and contractions of bodies by change of temperature also occasion some inconveniences. The accurate movement of clocks depends upon the length of their pendulums, which being sensibly affected by changes of temperature, they are made to go faster in cold, and slower in warm weather.

Brittle substances, when unequally heated, are often broken by the unequal expansions and contractions to which they are liable. The danger is greater if the substance is a bad conductor of heat, as is the case with glass, and particularly if it is thick. Hence, glass vessels that are to be used about the fire, or with hot water, should be made as thin as is consistent with the requisite strength.

Metallic or other instruments used for measuring length or capacity vary with change of temperature—a circumstance that sometimes occasions serious difficulty where very great accuracy of measurement is required.

It has been found by very accurate examination, that the Bunker Hill Monument, which is built of granite, is daily made to change its position slightly, by the heat of the sun, which expands the sides upon which the rays fall.

18. Expansion of Liquids.—In solids, the expansive force of heat is opposed by the cohesion of their particles, and is therefore less effective than in liquids, in which there is only a very slight cohesion of the particles. A liquid, therefore, will expand on being heated, much more than a solid.

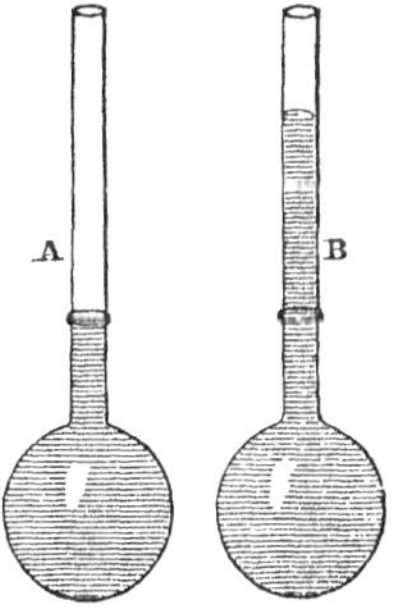

Expansion of Liquids.

The expansion of a liquid may be shown in the following manner. Take a glass flask (called a mattrass or bolt-head), of the form represented in the figure, and partly fill it with some liquid, as water, and tie a thread around the stem, as on A, to indicate the height of the water in it; and then apply for a few minutes the heat of a spirit-lamp. Both the glass and the water will

QUESTIONS.—17. What is said of the *tire* of wheels? How are brittle substances affected by sudden changes of temperature? What is said of the Bunker Hill Monument? 18. What is said of the expansion of liquids by heat? How may the expansion of a liquid be shown?

be expanded; but the water will expand more than the glass, and will then rise in the stem, as shown in B.

But all liquids when equally heated do not expand alike,—every one possesses an expansibility peculiar to itself. Thus, it has been found by making the experiment, that 1000 parts of water, at the freezing point, when heated so as to boil, are expanded to 1046 parts; but 1000 parts of mercury, heated in like manner, expand only to 1008 parts. Ether is more expansible than alcohol, and alcohol more expansible than water.

Liquids, as well as solids (16), are expanded more at high than at low temperatures, by a given addition of heat.

19. Expansion of Gases.—All gases expand equally when equally heated, and the expansion is proportional to the increase of temperature. When 1000 parts of any gas are heated from 32° to 212° of Fahrenheit's thermometer (an instrument soon to be described), they expand to 1365 parts, or $\frac{1}{493}$ part of the volume at 32° for each degree.

In the case of gases that are capable of becoming liquid by pressure, this law does not hold strictly true when they are about to assume the liquid form.

Expansion of Gases.

To show the expansion of air by heat, let a glass flask, filled with air, be placed as in the figure, with its mouth immersed in water; then warm it slightly, by grasping the bulb in the hands, or breathing upon it, when the air will escape in bubbles, in consequence of its expansion by the heat. On cooling, the air within contracts, and the water rises in the stem to supply the place of the air which was expelled.

THERMOMETERS.

20. Thermometers are instruments for ascertaining and measuring changes of the temperature of bodies, of which there are several kinds. The name is derived from the two Greek words, *thermos*, heat, and *metron*, a measure.

The first instrument of the kind, so far as we know, was constructed but little more than two hundred and fifty years ago, by Sanctorio, an Italian philosopher.

Sanctorio's thermometer was made in the following manner. A glass tube of small diameter, having a bulb blown at one end, was

---

QUESTIONS.—Do all liquids expand equally when equally heated? How much do 1000 parts of water expand when heated from the freezing to the boiling point? 19. Do gases expand by heat? What is the amount of their expansion for each degree of heat? How may the expansion of air by heat be shown? 20. What are *thermometers?*

partly filled with a colored liquid, and the stem passed through a cork, and inverted in a vessel containing the same kind of liquid, and having a wide bottom, as in the figure, so as to stand upright firmly. Through the cork a small perforation was made, so as to allow the air to pass freely, and to the stem a graduated scale was attached, to mark the rising and falling of the liquid in it.

Now, in an instrument of this kind, it is plain that when the bulb is heated, the air within will be expanded, as before explained, and the liquid in the stem will fall; and a motion of the liquid in the opposite direction will take place when the bulb is cooled. The rise and fall of the liquid will also be proportional to the change of temperature in the bulb.

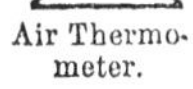

Air Thermometer.

This thermometer will very well answer some specific purposes, but as it will be affected by changes of atmospheric pressure as well as by changes of temperature, it cannot be applied to general use.

The *differential thermometer* may be considered as a modification of the preceding. It consists of a glass tube, bent twice at right angles, with a bulb at each end, and is supported on a stand, as shown in the figure. In the tube is contained a portion of colored oil of vitriol, or other liquid; but both bulbs are left filled with air, and to one of the arms is attached a graduated scale. When both bulbs are equally heated or cooled, this instrument indicates no change: but if one is heated or cooled more than the other, a motion is at once occasioned in the liquid in the stem, the *direction* of which will be readily understood from the explanations already given. This thermometer therefore indicates the *difference* of temperature at any time existing between the bulbs, and hence its name. It is exceedingly delicate, and is especially adapted for some particular purposes.

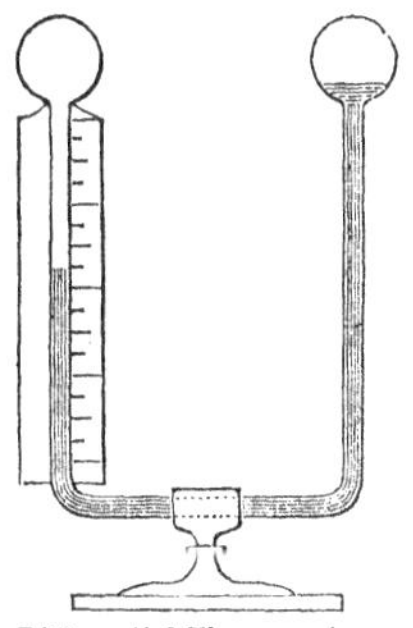

Differential Thermomter.

QUESTIONS.—Describe Sanctorio's *air thermometer*. What objection is there to its use? Describe the *differential thermometer*.

21. The Common Thermometer.—The thermometer in common use consists simply of a glass tube of an exceedingly small bore, with a bulb blown at one extremity, and filled with mercury to about one-third the height of the stem. The air being expelled, the tube is hermetically* sealed, and the *freezing point* ascertained by holding it a short time in water containing ice, and the *boiling point* by holding it in the same manner in boiling water. Both points are marked on the stem by a file. It is necessary that these two points should be accurately determined, in order that the indications of different instruments may be compared with each other.

By the term freezing point here, is meant the temperature at which water freezes or ice melts, which, with certain exceptions, is always the same, as will be fully explained hereafter; so, also, pure water always boils at the same temperature, provided attention is paid to certain circumstances to be discussed further on in the work. This temperature is called its boiling point.

It will be unnecessary here to give a *minute* description of the method of making thermometers, as, at the present day, they can be everywhere obtained at a very moderate price. "Besides, the construction, though simple in theory, is difficult in practice. It requires great tact and dexterity to produce one of very moderate goodness; and without steadily watching the process as performed by another, or previously possessing much practical knowledge in glass-blowing, &c., it would be a vain attempt."—*Faraday's Chemical Manipulation*, p. 144.

The *graduation* of the scale of the thermometer is a matter of great importance; and it would be fortunate for us if we had but one, instead of three or more, as is the fact. We have seen that in all thermometers there are two fixed points; and the question now before us is, into how many parts or degrees, shall the space between them be divided? Unfortunately, this question has been answered differently by different artists, and in a manner entirely arbitrary.

Fahrenheit, a German artist, whose thermometer is generally used in this country and in England, divided it into 180 parts or degrees, and placed the zero, or the beginning of the scale, 32 degrees below the freezing point; so that the temperature of melting ice or freezing water is 32 degrees, and that of boiling water ($32 + 180 =$) 212 degrees.

Celsius of Sweden proposed to divide the space into 100 parts, and placed the zero at the freezing point. His thermometer is called the *centigrade* thermometer, and is used in France and Sweden, and some other parts of Europe.

* A glass tube is sealed *hermetically* by melting the end by means of the blow-pipe, and thus perfectly closing it. For this purpose the end is usually drawn out into a fine point.

QUESTIONS.—21. Describe the *common thermometer*. What are the *freezing* and *boiling points?* How are these determined? Describe the scale adopted in Fahrenheit's thermometer. Where is the zero or beginning of this scale? Describe the scale of the centigrade thermometer.

Reaumur divided it into only 80 parts, placing the zero, or beginning of the scale, like Celsius, at the freezing point; of course the boiling point is at 80.

Below zero of each of the scales, and above the boiling point, degrees are marked, of precisely equal magnitude with those of the other part of the scale. Temperatures below zero are usually indicated by placing a horizontal line before the figures representing the degrees. Thus, —12° means 12 degrees below zero on the scale used.

The numbers 180, 100, and 80, which severally represent the number of degrees on the above scales, are to each other as 9, 5, and 4. Recollecting, therefore, that the zero of Fahrenheit is 32 degrees below that of the other scales, the expert arithmetician will find no difficulty in reducing the degrees of one scale to those of another.

Thus, to convert the degree of temperature indicated by Fahrenheit's scale into its centigrade equivalent, we multiply the degrees above or below 32° by 5, and divide by 9. Suppose the temperature by Farenheit's thermometer is 140°, what is the corresponding degree in the centigrade? *Ex.* $140 - 32 = 108$, and $108 \times 5 = 540$, and $540 \div 9 = 60$. On Fahrenheit's scale, therefore, 140° are equivalent to 60° of the centigrade thermometer.

Let us suppose again that the temperature by the centigrade thermometer is 60°; it is required to find the corresponding degree by Fahrenheit's instrument. *Ex.* $60 \times 9 = 540$, and $540 \div 5 = 108$. To this (108) we must now add 32, because the beginning of Fahrenheit's scale is 32° below that of the centigrade. Thus $108° + 32° = 140°$.

In this work, and in most works in the English language, if nothing is said to the contrary, it is always to be understood that temperatures are expressed in degrees of Fahrenheit's scale; but, to avoid confusion, we often place F., C., or R., after the figures expressing the degrees, to indicate what thermometer has been used.

The relation between the three scales above described is indicated in the accompanying figure.

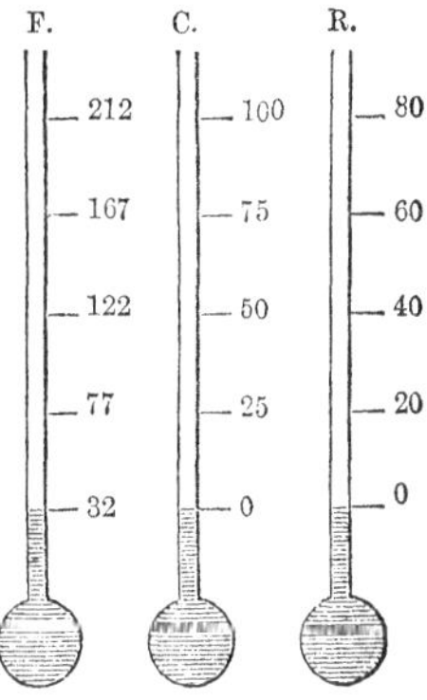

Different Thermometers.

Though mercury is chiefly used in filling thermometers, yet other liquids are also sometimes employed. At very low temperatures mercury is frozen, so that it ceases to answer the purpose designed; in such cases, therefore, alcohol thermometers alone can be used.

22. The Register Thermometer, while it answers the same purpose as another thermometer, at the same time indicates or *registers*

QUESTIONS.—Describe the scale of Reaumur's thermometer. How are the degrees determined below the freezing point and above the boiling point? How may we convert temperatures as indicated by Fahrenheit's scale into its centigrade equivalent? How may we convert centigrade into Farenheit degrees? Is mercury always used in constructing thermometers?

the extremes of temperature that may occur during the absence of the observer. It consists of two thermometers, with the stems bent near the bulb, and placed in a horizontal position, attached to the same frame, as shown in the following figure:

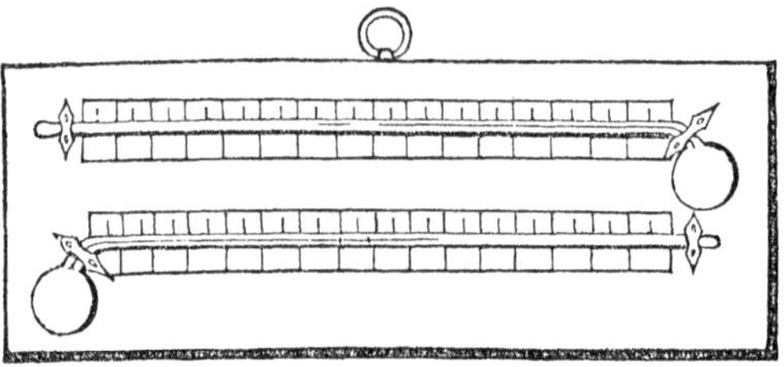

Register Thermometer.

The one usually placed uppermost is a mercurial thermometer, having in the tube a small piece of iron or steel wire, which is pushed forward by the mercury as it expands, but does not recede with it when it contracts. The point at which the iron is left, of course shows the maximum temperature attained. The other thermometer is filled with colored alcohol, and contains in the liquid in the stem a similar piece of iron, inclosed in glass to prevent oxidation, around which the alcohol flows while that in the bulb is expanding, so as not to be moved, but which is drawn along with it by capillary attraction, when it contracts so as to be kept at its surface. It is therefore left at the lowest point to which the spirit has contracted, and of course shows the *minimum* temperature. Both pieces of iron or steel, which thus serve as indices, may be brought to any position in their respective tubes by means of a magnet applied on the outside.

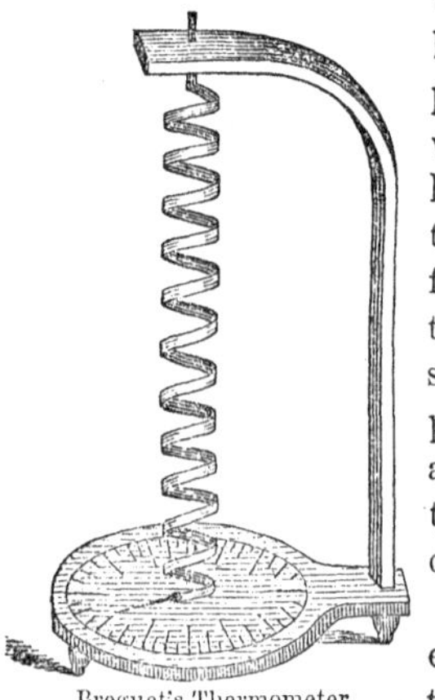

Breguet's Thermometer.

23. Breguet's Thermometer is made entirely of solids. It consists of a very thin strip of platinum, soldered to a

---

QUESTIONS.—22. Describe the *register thermometer*. *Breguet's thermometer.*

similar strip of silver, and coiled in a spiral, as shown in the figure (page 26). The upper end of the coil is then attached to a firm support, and to the other extremity is fixed a pointer or index, which is made to revolve by any change of temperature, by reason of the unequal expansions and contractions of the two metals. Beneath the pointer is placed a circle which may be graduated to any scale desired. It is a very delicate instrument.

A modification of this instrument is used in the United States' Coast Survey, for determining the temperature of the water in deep soundings, at sea.

The *Pyrometer* (from *pur*, fire, and *metron*, a measure,) is an instrument for measuring temperatures too high to admit of the use of the thermometer. The only one now in use is Daniel's pyrometer, which is not of sufficient importance to require description here. By it the melting point of cast iron has been shown to be about 2786° F., that of gold to be about 2016° F., copper 1996°, silver 1860°, and zinc 713°.

24. Exceptions to the general Law of Expansion.—There is a remarkable exception to the general law (14) concerning the expansion of bodies by heat, as above stated. Water is most dense at the temperature of about 40°, and expands, whether it is heated above this point or cooled below it.

To show this, fill an ounce vial with water at a temperature of 65° or 70°, and adapt to it a cork, through which passes a glass tube of small bore. Then insert the cork and tube, and fill the latter with water one or two inches above the neck of the vial, and expose the whole to the cold atmosphere of winter, or immerse the vial in a freezing mixture of snow and salt; the contraction of the water in the vial will very soon be made evident by the fall of that in the tube; but the falling will shortly cease, and an upward motion commence, indicating an expansion of the water in the vial, although its temperature must be all the time falling. The volume of the water has therefore first been diminished by reduction of its heat, and again expanded; and by making use of the thermometer, it is found that the change takes place at about 39° or 40°.

Water Expansion when Freezing.

A large thermometer tube, nearly filled with water, may be used for the same purpose.

25. The most important effects result from this remarkable property of water. If the density of water continued to increase until it arrived at the freezing point, as is the case with mercury

QUESTIONS.—What is the design of the *Pyrometer?* 24. What exceptions are there to the general law of expansion of bodies by heat? Describe the method of showing the expansion of water by reduction of temperature.

and other liquids, ice would be heavier than water, and as soon as formed would subside to the bottom in successive flakes, until the whole of the water, however deep, would become solid. The effects of such an arrangement can be easily conceived. Countries which, in the present state of things, are the delightful abodes of innumerable animated beings, would be rendered uninhabitable, and must inevitably become dreary and desolate wastes. But, since water expands previously to its freezing, as well as during this change, ice is lighter than water, and floats upon its surface, protecting the water, to some extent, from the further influence of frost.

The cause of the expansion of water at the moment of freezing is attributed to a new and peculiar arrangement of its particles. Ice is in reality crystallized water, and during its formation the particles arrange themselves in ranks and lines, which cross each other at angles of 60° and 120°, and consequently occupy more space than when liquid. This may be seen by examining the surface of water while freezing, and still better by receiving particles of snow as they fall upon a piece of black cloth. They will often be found to be small but beautiful crystals or collections of crystals, presenting a great variety of forms. Some of the more common forms are shown in the figure.

Snow Crystals.

QUESTIONS.—25. What is said of the importance of this remarkable property of water? What is suggested as the cause of this expansion of water in freezing?

26. The view just taken of the cause of the expansion of water when freezing is sustained by the facts observed in the formation of *anchor ice*, or *ground ice*, as it is often called. This ice is found in certain circumstances at the *bottom* of bodies of water, and not at the *top*, as with ordinary ice. It has little tenacity, and may be supposed to consist of the primary crystals of water. Separately, they are believed to possess a higher specific gravity than water, but, when aggregated according to the law stated above, at angles of 60° and 120° to form common ice, on account of the insterstices necessarily left among them, the volume is so increased as to diminish the specific gravity to the point we usually witness.—*Manuscript Notes of Professor Cleaveland's Lectures in Bowdoin College*, 1832.

DISTRIBUTION OF HEAT.

Heat constantly tends to diffuse itself, and its distribution is effected by *Conduction*, *Convection*, *Radiation*, *Reflection*, and *Transmission*.

27. Conduction of Heat.—Heat is said to be conducted, when it is transmitted from particle to particle through a body, as when one end of a metallic bar is held in the fire until the whole becomes heated.

The passage of the heat in such cases is evidently progressive, as may be shown in the following manner. Take a small bar of copper, 18 or 20 inches in length, and cement to it several small bullets, or marbles, about two inches from each other, by means of wax, as shown in the figure, and then apply the heat of a lamp to one end. As the heat progresses along the bar, it will melt the wax, and the balls will drop off in succession,

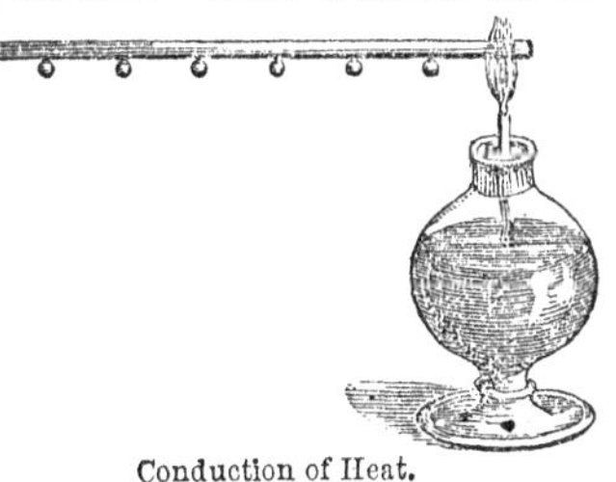
Conduction of Heat.

QUESTIONS.—26. What is said of *anchor ice* in this connection? What are some of the modes by which heat tends to diffuse itself? 27. When is heat said to be conducted? How may the conduction of heat along a bar of copper be shown?

the one nearest the lamp falling first, and the one farthest from it last.

Substances differ greatly in their power of conducting heat; a rod of glass, or a piece of charcoal, an inch long, may be heated to redness at one extremity, and yet be held in the fingers by the other extremity; but it cannot be done with a similar piece of metal, because, on account of its better conducting power, the whole very soon becomes too much heated.

The apparent temperature of a body, as determined by the hand, will often depend upon its conducting power. Thus, if on a cold morning of winter, the hand is placed upon a piece of metal, and then upon a piece of woollen cloth, the former will feel much colder than the latter, because the metal in equal times conveys away from the hand more heat than the cloth.

28. Substances are divided into two classes in reference to their ability to conduct heat, called *conductors* and *non-conductors.* There are, however, no absolute non-conductors; heat penetrates the substance of all bodies; the only difference in them, in this respect, is in the rapidity with which the process takes place. Gold is usually considered the best conducting substance known; and very porous solids, the interstices of which are filled with air, as cotton, or sheep's wool, and fur, are the poorest conductors.

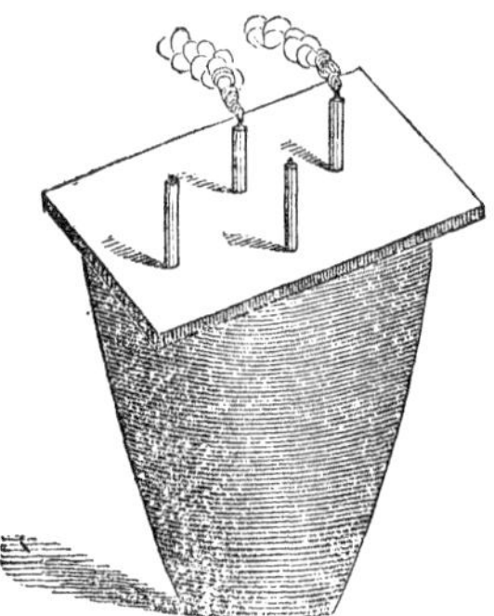
Conduction of Heat.

A convenient method to determine the relative conducting power of different substances, is, to have them made into cylinders of equal diameter, and set in a thin piece of wood at sufficient distances from each other, both extremities of each piece projecting a little from the wood. If the board be held in

QUESTIONS.—Do substances differ from each other in their power to conduct heat? Why will some substances feel colder than others, when it is known that all must be at the same temperature? 28. Into what two classes are substances divided in reference to their conducting powers? What is usually considered as the best conductor known? What method for determining the relative conducting power of several substances is pointed out?

a horizontal position, a small piece of phosphorus may be placed upon the upper extremity of each of the substances experimented upon, and the lower ends exposed to the same temperature by plunging them in heated oil or sand: and the times that elapse before the ignition of the phosphorus upon the several substances, will indicate with some accuracy their relative conducting powers.

The following table exhibits the relative conducting power of several metals and other substances:

Gold	1000	Tin	304
Silver	973	Lead	180
Copper	898	Marble	23
Platinum	381	Porcelain	12
Iron	374	Fine Clay	11
Zinc	363		

In the arts, advantage is taken of the imperfect conducting powers of bodies, to prevent the passage of heat in any direction, particularly in confining it. Hence furnaces are generally lined with "fire-brick," or a thick coating of clay and sand. Wooden handles are fitted to metallic vessels, or a stratum of wood or ivory is interposed between the hot vessel and the metal handle. Ice-houses are constructed with double walls, which have their interstices filled with fine charcoal, saw-dust, or some other non-conducting substance, to prevent the influx of heat from without.

The design of clothing is to retain the heat produced by the system; and hence the warmest clothing will be that which possesses the least conducting power. In winter, the poorest conductors are selected, and in summer the best, as it is then desirable that the superfluous heat may be permitted at once to escape. If, in summer, the temperature of the atmosphere should rise considerably above that of the system, it would be found advantageous to use the same clothing as in cold weather.

Snow, in consequence of its imperfect conducting power, serves as clothing to the earth, and prevents its surface from being cooled down as low as it would otherwise be.

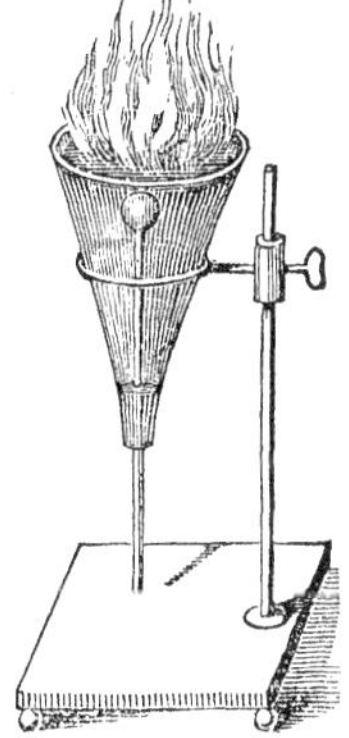

Ether burns on the surface of Water.

Liquids of all kinds, except mercury, are poor conductors of heat. This may be shown by cementing a thermometer tube in a glass funnel, inverting it, and filling it with water, so as to cover the bulb about a quarter of an inch, or even less, as shown in the figure. Then pour upon the surface

QUESTIONS.—What is the use of "*fire-brick*" in coal-stoves? How are ice-houses constructed? What is the design of clothing? What is said of the benefits of snow in winter? How is the poor conducting power of liquids shown by the burning of ether?

of the water a little sulphuric ether, and inflame it; the ether will burn brilliantly, but without affecting the thermometer for some time, although the flame is so very near the bulb.

Water boils in vessel with Ice.

In like manner, heated oil, poured upon the surface of water in a tumbler, can scarcely be made to affect a small thermometer placed at the bottom.

If a tube ten or twelve inches long be nearly filled with water and placed in an inclined position, so that the heat of a spirit-lamp can be applied near the centre, the water in the upper part of the tube may be made to boil, while the lower portion will remain perfectly cold. If, before applying the heat, a piece of ice be confined to the bottom, it will remain unmelted while the water above is boiling. Mercury, though liquid, is a very good conductor of heat.

Gases are even poorer conductors than liquids; and it is for this reason that very hot or very cold air can be endured in contact with a person, though exposure to a liquid of the same temperature would produce intense pain, or perhaps even worse effects. Double windows and double doors, with air between them, are sometimes used to insure the greater warmth of dwellings.

29. Convection of Heat.—Though fluids are poor conductors of heat, yet, if the heat be applied to the bottom of the vessel containing them, in consequence of the mobility of their particles, it is rapidly diffused through the whole mass. The heated portions are expanded, and becoming, in consequence, specifically lighter than the rest, they rise through the centre of the vessel, the colder portions around the sides at the same time descending to take their place. Thus an upward and a downward current will be at the same time established, which will continue until the whole is heated to the boiling point. This mode of distribution is called the *convection* of heat.

QUESTIONS.—How is the poor conducting power of liquids shown by the boiling of water in a vessel containing ice? 29. How is the heat distributed through a liquid when it is applied to the bottom of the vessel containing it? What name is given to this mode of the distribution of heat?

These currents may readily be shown by filling a flask with water containing some insoluble powder, as pulverized gum copal, and applying the heat of a small lamp, as represented in the figure.

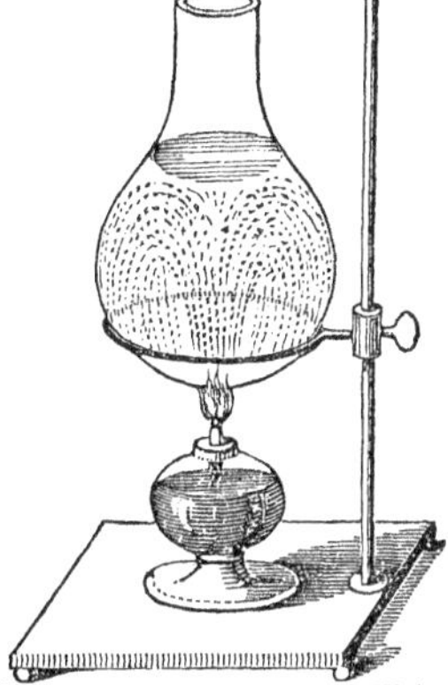

Currents formed in vessel of Water when heated.

When large quantities of water are slowly heated, the upper portions will frequently be found quite warm, while that in the lower part of the vessel will remain comparatively cold; and this though the fire is applied beneath. Hence it is not unfrequent, in bathing establishments, to draw both warm and cold water from the same reservoir.

Similar currents are produced in gases when heated; and it is on this account that the heated air, with the smoke and other gases from a fire, ascend in a chimney, or the pipe from a stove.

30. Radiation of Heat.—A hot body suspended in the air emits heat in all directions in right lines, like radii drawn from the centre to the surface of a sphere. This mode of distribution is termed the *radiation* of heat.

The radiation of heat from hot bodies is singularly influenced by the *nature and condition of their surfaces*, which is perhaps the most important circumstance connected with the subject. It is probable that every substance in nature has a radiating power peculiar to itself, but, in any case, very much will depend upon the nature of the surface of the body. By many experiments, it has been proved that bodies with bright polished surfaces retain their heat much longer than when their surfaces are rough and unpolished. Adding even a thin coat of whiting or lampblack to a bright tin vessel greatly increases the radiating power of its surface, so that boiling water or other hot liquid contained in it will be cooled more rapidly in consequence. The same effect will be produced by scratching its surface with coarse sand-paper.

Some important practical considerations will naturally suggest themselves in connection with this subject. Whenever it is desired that the heat of a fluid or other substance should be retained, vessels with bright and polished metallic surfaces should be used, but the reverse if the heat is to be distributed. Thus tea and coffee pots are usually made of some bright metal, while stoves and stove-pipes, for the diffusion of heat, are made with dark and rough surfaces. Pipes to convey steam from the boilers in steam-engines to the cylinders, and pipes to convey heated air from furnaces to the different apartments of a building, should be bright, or else they should be protected by some non-conducting covering.

QUESTIONS.—30. When is heat said to be *radiated?* How is the radiation of heat affected by the nature of the surface of the heated body? What surfaces radiate best? What practical considerations suggest themselves in view of these principles?

31. Reflection of Heat.—That heat may be reflected, may be shown by standing at the side of a fire in such a position that the heat cannot reach the face directly, and then placing a plate of tinned iron opposite the grate, and at such an inclination as permits the observer to see in it the reflection of the fire; as soon as it is brought to this inclination, a distinct impression of heat will be produced upon the face.

If a line be drawn from a radiating substance to the point of a plane surface by which its rays are reflected, and a second line from that point to the spot where its heating power is exerted, the angles which these lines form with a line perpendicular to the reflecting plane are called the angles of *incidence* and *reflection*, and are invariably equal to each other.

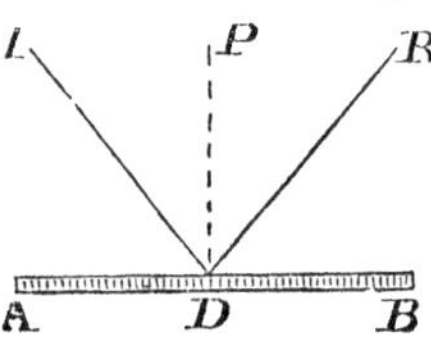

Reflection of Heat.

Thus, let AB (see figure) be the reflecting surface, and R a ray of heat, which strikes this surface at D, in the direction RD; it will be thrown off or reflected in the direction DI. If a perpendicular PD be erected at the point D, the angle RDP will be the angle of incidence, and IDP the angle of reflection.

These principles, which have just been developed concerning heat, apply as well to the invisible rays emitted from a moderately heated substance, as to those accompanied by light from an incandescent body, or the rays of the sun.

32. *The Absorption of Heat* by bodies sustains an intimate relation both to its radiation and reflection. Bright and polished surfaces, it is well known, are the best reflectors; and these are just the ones, we have seen (30), which radiate least. And rough, unpolished surfaces, which radiate heat best, are found to be the best absorbers.

Surfaces may therefore be divided into two classes, those which afford an easy passage to heat, and those which do not. The former will be good radiators and absorbers, and the latter good reflectors and retainers.

The color of a body influences considerably its absorbing, but not its radiating power; surfaces that are black, other things being equal, absorbing heat more readily than those of a lighter color.

QUESTIONS.—31. How may the *reflection* of heat be shown? Define the angles of incidence and reflection. Do these principles apply to rays of heat unaccompanied by light? 32. What surfaces are the best absorbers of heat? Into what two classes may surfaces be divided in reference to their power to transmit heat?

Both the radiation and the reflection of heat are well shown by placing a heated cannon-ball in the focus of a concave reflector, having another similar reflector facing it at a distance, in the focus of which is placed one

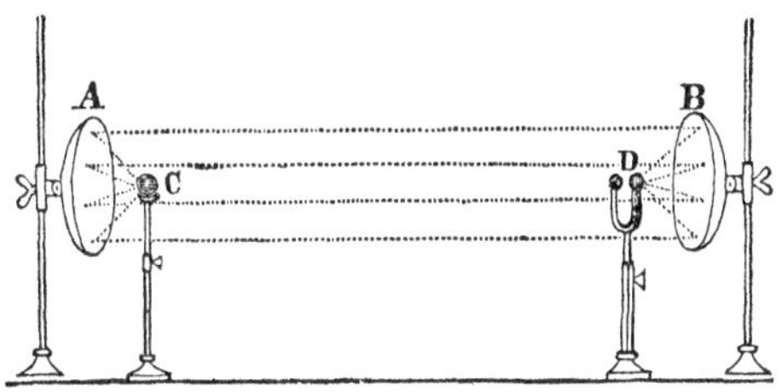

Parabolic Reflectors.

of the bulbs of a differential thermometer. The rays from the ball C are reflected in parallel lines from the reflector A (see figure), and are again concentrated on the thermometer D, by reflection from the second concave mirror B.

If a piece of phosphorus be substituted for the thermometer at D, it may often be inflamed, even when the reflectors are ten or twenty feet distant from each other.

If a lump of ice is made use of, instead of the heated ball, the thermometer in the focus of the other reflector will fall; in which case the bulb of the thermometer is the radiating body, and its heat is received by the ice.

33. Transmission of Heat.—When a ray of heat is thrown upon a body, it must either be *reflected*, *absorbed*, or *transmitted* by the body. We have already seen the conditions upon which reflection and absorption depend, and it remains only to consider the circumstances of transmission.

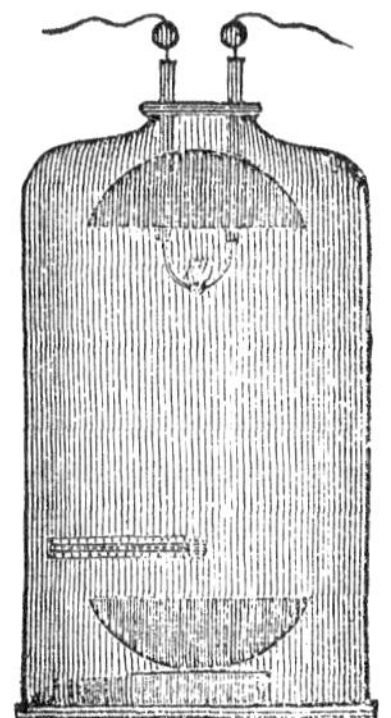

Transmission of Heat in a Vacuum.

In general, transparent substances afford the most ready transmission of heat, but there is a great difference among them in this respect. Even atmospheric air transmits heat but imperfectly. This is shown conclusively by an experiment of Davy. He contrived to heat a platinum wire by means of galvanism, within a receiver containing two concave reflectors, with a thermometer in the focus of one of them, the heated wire being in the focus of the other. Now, when the air was exhausted to $\frac{1}{120}$th part of its ordinary density, the thermometer, it was

QUESTIONS.—Explain the effect of parabolic reflectors in reflecting heat. How is a lump of ice to be used instead of a heated ball? 33. When a ray of heat falls upon a body, in what three ways may it be disposed of? How does the presence of the air affect the transmission of heat?

found, would be raised, by means of the ignited wire, three times as high as when the air in the receiver was at its natural pressure.

Bodies that transmit heat freely are said to be *diathermanous* (from the two Greek words, *dia*, through, and *thermos*, heat), as those which afford a free passage to light are said to be *transparent.*

By experiments made with a very delicate piece of apparatus, called the *thermo-multiplier*, it has been shown that the most diathermanous substance known is rock-salt, in pure transparent crystals. Of different specimens of glass, some are much more diathermanous than others, though all are equally transparent; and some colored glasses, and other bodies only partially transparent, afford a ready passage to heat, or are highly diathermanous.

It appears, also, that the ray of heat, like a ray of light, is really compound, or composed of rays of heat of different kinds, some of which have a greater penetrating power as regards most diathermanous media, than others. In this respect heat from an oil-lamp will differ from that of a spirit-lamp, though both are equally intense; and the heat of both will differ from that of heated metal.

The rays of heat from the sun possess this penetrating power, as I have called it, in a greater degree than any kind of artificial heat. Thus, a pane of window-glass, held between the face and a coal-fire, is found at once to intercept most of the heat; but no such effect is produced by holding it before the face when exposed to the direct solar ray.

The rays of heat, like those of light, may be refracted; and some of them being more refrangible than others, like the different colors of light, they may be separated from each other by means of the prism.

The ray of heat, like a ray of light, may also be polarized, and in a similar manner. See *Decomposition and Polarization of Light.*

RELATION OF HEAT TO CHANGES IN THE STATE OF BODIES.

34. Relation of the Three Forms of Matter to each other.—We have seen above (10), that, omitting the imponderable agents, which are not known to be material, every substance must be in one of the three forms, or states, solid, liquid, or gaseous; and that the particular form a body assumes will depend upon the relative intensity of the cohesive and repulsive forces existing among its particles.

If the repulsive force be comparatively feeble, the particles will adhere so firmly together, that they cannot move freely upon one another, thus

---

QUESTIONS.—What are diathermanous bodies? Will heat from all sources be transmitted alike? What is said of the rays of the sun in this connection? May the rays of heat be refracted? Polarized? 34. Upon what will the particular form or state of a body depend?

constituting a solid. If cohesion is so far counteracted by repulsion that the particles move on each other freely, a liquid is formed; and, should the cohesive attraction be entirely overcome, so that the particles not only move freely on each other, but would, unless restrained by external pressure, separate or expand to an indefinite extent, an aeriform substance will be produced.

Now, the property of repulsion is manifestly owing to heat; and as it is easy, within certain limits, to increase or diminish the quantity of this principle in any substance, it follows that the forms of bodies may be made to vary at pleasure: that is, by heat sufficiently intense, we have reason to believe, every solid, provided decomposition does not take place, may be converted into a liquid, and every liquid into vapor. The converse ought also to be true; and, accordingly, several of the gases have been condensed into liquids by means of cold, or cold and pressure combined, and the liquids have been solidified. The temperature at which liquefaction takes place is called the *melting point*, or point of fusion; and that at which liquids solidify, their *freezing point*, or point of congelation. Both these points are different for different substances, but usually the same, under similar circumstances, in the same body.

35. Liquefaction.—By the liquefaction of a substance, we mean its reduction from either the solid or gaseous to the liquid state; but generally it is the former change which is intended.

If, when the temperature of the air is at zero, as is often the case in some parts of our country, a quantity of ice be brought into a room, and placed near a fire, it will be gradually heated, like any other solid, as a thermometer placed in it will indicate, until the temperature reaches 32°; but it will stand at this point until the whole is melted. The thermometer will then begin again to rise, as it did before. Now, it is plain that it must have been receiving heat as rapidly while the thermometer was stationary, as before and afterwards; but the heat thus communicated did not affect the thermometer, because it was all absorbed by the ice, and was expended in changing the ice into water. It has therefore become insensible to the thermometer, and is properly called *latent heat;* and if it was known to be material, we might perhaps, with some propriety, consider water as a compound of ice and heat.

The quantity of heat which is thus lost or becomes insensible, during the melting of a mass of ice, is sufficient to raise the tem-

---

QUESTIONS.—To what is the property of repulsion owing? What is the *melting point* of a body? The *freezing point?* 35. What is meant by the *liquefaction* of a body? Is heat always required to produce liquefaction? Why cannot ice be heated above 32°?

perature of an equal weight of water about 140°, as may be shown in the following manner: Let a pound of water at 32° be mixed with a pound of water at 172°, and the temperature of the mixture will be intermediate between them, or 102°. But if a pound of water at 172° be added to a pound of ice at 32°, the ice will quickly dissolve, and on placing a thermometer in the mixture, it will be found to stand, not at 102°, but at 32°. In this experiment, the pound of hot water which was originally at 172°, actually loses 140° of heat, all of which enters into the ice, and causes its liquefaction, without affecting its temperature.

36. Heat of Fluidity.—The heat thus required for the liquefaction of solids is often called their *heat of fluidity;* and the quantity necessary for the purpose is not the same in any two substances. While the heat of fluidity of water is, as we have just seen, 140°, that of spermaceti is 145°, that of lead 162°, that of tin 500°, and that of bismuth 550°. That is, to melt any given weight of one of these substances, an amount of heat is required that would heat the same weight of the substance the number of degrees indicated, provided no change of state should take place during the process.

The *melting point* of nearly all substances is the same as their freezing point; but this point varies greatly in different substances. Thus, solid mercury melts (and liquid mercury freezes) at —39°; ice at 32°; spermaceti at 132°; sulphur at 226°; tin at 442°; lead at 612°; zinc at 773°; silver at 1873°, and gold at 2016°.

37. *Freezing Mixtures* are made of various salts and liquids, which have such an affinity for each other, that rapid liquefaction is produced, without the direct application of heat. But as this agent is always required when this change takes place, it must be absorbed from surrounding objects, which therefore lose their heat, or become cold. A good mixture of this kind is made of snow, or finely broken ice, and common salt, both of which, when mixed together, become rapidly liquid; and the process is attended with great cold, so that a thermometer immersed in it will fall to zero,

---

QUESTIONS.—What is the quantity of heat absorbed by ice when it is melted? How is this shown? 36. What is the *heat of fluidity* of a substance? Will it be the same in all substances? What are the *melting points* of the several substances mentioned? 37. What are *freezing mixtures?*

or below. Of course, if a vessel of water be immersed in it, the water will in a short time be frozen.

Saltpetre, dissolved in cold spring water, will often reduce the temperature to 32°, or lower, so that water may be frozen by it; but the greatest cold is produced in this mode by mixtures of certain of the salts and acids. Thus, powdered sulphate of soda three parts, and diluted nitric acid two parts, mixed at 50°, will sink the temperature to —3°; and phosphate of soda nine parts, and the same diluted acid four parts, will produce a cold of —12°. The greatest cold that can be produced in this way is found to be about —100°, but by other means still lower temperatures have been obtained. But it is not possible, in the present state of our knowledge, entirely to deprive a body of heat.

Since solids, on becoming liquid, absorb heat, as we have seen, it necessarily follows, that when liquids become solid, heat must be given out. The freezing point of water is usually said to be 32°; but if it be contained in a close vessel, and cooled very slowly without agitation, its temperature may be reduced, without freezing, to 20°, or lower. Slight agitation will now cause it to freeze suddenly, and the temperature will rise at once to 32°, the ordinary freezing point. The portion that has frozen, therefore, has given out sufficient heat to raise the temperature of the whole mass some 12°. Saturated solutions of several of the salts, made at elevated temperatures, upon being slowly cooled, exhibit the same phenomenon.

A beautiful experiment may be performed by making a saturated solution of Glauber's salt in warm water, and setting it aside in a closely corked vial till it cools. If now the cork be removed, or the vessel violently agitated, the salt will immediately crystallize, and a thermometer placed in it will rise several degrees.

38. Provision of Nature.—We cannot but notice here the beautiful and unexpected manner by which nature, to some extent at least, checks the cold of winter, which might otherwise be destructive. The cold atmosphere causes large quantities of water to congeal, but at the same time heat is given out, which prevents so great a reduction of temperature as might, but for this circumstance, be experienced.

QUESTIONS.—What is the greatest cold that can be produced by freezing mixtures? Is heat given out when liquids become solid? May water have its temperature reduced below the ordinary freezing point? What is the effect on the thermometer when freezing at length is produced? Describe the experiment with Glauber's salt. 38. How is the excessive cold of winter to some extent checked?

The peculiar mode provided by the Creator to check the heat of summer, which might otherwise become excessive, will be noticed hereafter.

39. Vaporization.—By the term vaporization is meant the conversion of solids or liquids into gases.

Aeriform bodies are often divided into *vapors* and *gases*, according to the relative force with which they resist condensation; but the distinction is of little consequence.

Heat is always required to convert a solid or liquid into a gas; usually, it is communicated directly, as when water is made to boil over a fire, but if not applied directly, it will always be absorbed from surrounding bodies.

In most cases, when heat is applied to solids, they first melt, or become liquid, and afterwards, by a continuance of the heat, are converted into vapor; but some, as metallic arsenic, and certain salts, pass at once, when heated, from the solid to the gaseous state.

Gases occupy considerably more space than the liquids from which they are formed. Water, when converted into steam, expands about 1700 times, so that a cubic inch of water forms nearly a cubic foot of steam; but most liquids expand much less than this. Alcohol, for instance, is expanded, when converted into vapor, only 659 times its original volume, and sulphuric ether 443 times.

Volatile substances are such as are readily converted into vapor by heat or at ordinary temperatures, while those that are incapable of this change are often called *fixed* substances.

Two or more gases, whatever may be their density, when brought in contact, readily intermix with each other, and become equally diffused through the vessel that may contain them. This is seen in the atmosphere, which is composed of gases differing in density, but they remain uniformly diffused.

If we fill two bottles with gases of different densities, as hydrogen and carbonic acid, and connect them by a narrow tube, as shown in the figure on p. 41, the lower containing the most dense gas, in a short time the two

---

QUESTIONS.—What is meant by *vaporization?* Is heat always required when a vapor is formed? Do gases occupy more space than the solids or liquids from which they are formed? How many times does water expand when it takes the form of steam? Alcohol? Ether?

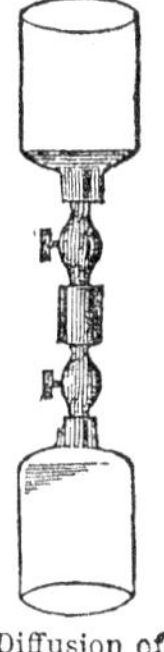
Diffusion of Gases.

gases will diffuse themselves equally through the whole space. The mixture of the gases will even take place through thin membranes, whether animal or vegetable; the least dense of the gases passing the most rapidly.

A good method to show this is to fill an ox bladder with carbonic acid, or even atmospheric air, and suspend it with the neck tied firmly in a large vessel filled with hydrogen. A transfer of the gases through the substance of the bladder will take place, but the hydrogen entering more rapidly than the denser gas escapes, the bladder will after a time be burst.

40. Ebullition—Boiling Point.—When a liquid in an open vessel is exposed to any source of heat, the temperature gradually rises, like that of any other substance in similar circumstances, until a certain point is attained, when a violent motion commences in it, called *ebullition* or *boiling;* and no heat can then cause any further increase of temperature. If the heat be continued, the quantity of liquid gradually diminishes, or, as we familiarly say, is boiled away, until the whole is gone. The commotion in the liquid is occasioned by portions of it at the bottom, where the heat is applied, being converted into vapor, and rising in bubbles to the surface.

Ordinarily it will be found that water boils at 212°, which is therefore called its *boiling point;* but the temperature at which other liquids boil is not necessarily the same, every liquid having a boiling point peculiar to itself. Thus, the boiling point of alcohol is only 173°, and that of sulphuric ether 96°, while that of sulphuric acid is 620°, and that of mercury 662°.

41. Boiling Point dependent upon Circumstances.—But the boiling point of a liquid is not to be considered as perfectly constant; it depends upon several circumstances, the most important of which is the pressure of the atmosphere upon the surface of the liquid. By heating a small vessel of water to about 200°, and placing it under the receiver of an air-pump, it begins to boil when the air is very moderately exhausted. So, on ascending a mountain, by

---

QUESTIONS.—What is said of the diffusion of gases of different densities? Describe the experiment with the ox bladder. 40. What is *ebullition* or boiling? What is the effect of continuing the heat? What is the *boiling point* of water? Is this point in other liquids the same? 41. What are the circumstances which affect the boiling point of a liquid? Describe the experiment with the air-pump.

which a part of the atmospheric pressure is avoided, the boiling point falls in proportion to the ascent. At the hospital of St. Bernard, situated upon a point on the Alps, about 8400 feet above the sea, water boils at 196°; and on the top of Mount Blanc it was observed by Sausure to boil at 184°.

This is just as we should expect, for the expansion of the vapoı has to take place directly against the pressure of the atmosphere on the surface of the liquid; and the degree of heat necessary to produce the expansion will be to some extent proportional to th expansive force required.

The pressure of the atmosphere at the surface of the sea is usually about 15 pounds to each square inch, but it is subject to some variation; and the boiling point of any liquid will of course vary at the same time with the atmospheric pressure.

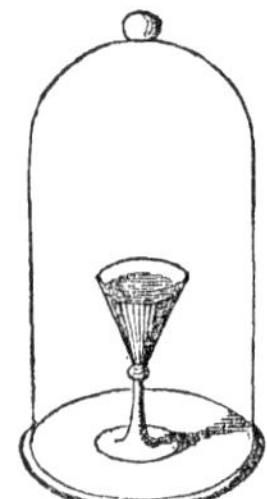
Boiling of Ether.

In a perfect vacuum, water boils at 72°, and sulphuric ether at —46°, or about 140° lower than in the open air.

As might be expected, sulphuric ether may easily be made to boil under an exhausted receiver, without heat, even in the coldest weather. For this purpose let a little good ether, in a wine-glass, be placed under an air-pump receiver, as represented in the figure; upon working the pump, it will boil violently. The experiment will usually succeed best if some small pieces of metal are dropped into the ether, before placing it under the receiver.

While the boiling is in progress considerable reduction of temperature takes place, and water contained in a small vial placed in the ether will be frozen.

42. *Other circumstances* affecting the boiling point are, the nature of the containing vessel, and the presence of soluble substances in the liquid. Thus water boils in metallic vessels at 212°, but in a clear glass vessel one or two degrees more are

---

QUESTIONS.—What is said of the boiling point upon high mountains? What is the boiling point of water in a vacuum? Describe the experiment with ether under the exhausted receiver of the air-pump? While the ether boils how is the thermometer in it affected? 42. What other circumstances are mentioned as affecting the boiling point?

required; so any substance, as a salt, held in solution in the water, causes a rise of the boiling point. Water saturated with common salt boils at 224°; saturated with saltpetre at 238°, and with chloride of calcium at 264°.

43. Effect of increased Pressure.—If water or other liquids boil at a lower temperature by diminishing the pressure upon the surface, so a higher temperature is required for this purpose when the pressure is increased, as in a steam boiler. Water cannot be heated above 212° in the open air, because any additional heat is expended in converting a portion of it into steam, which at once makes its escape into the air; but if it be confined in a strong vessel, it may be heated to any temperature even to redness.

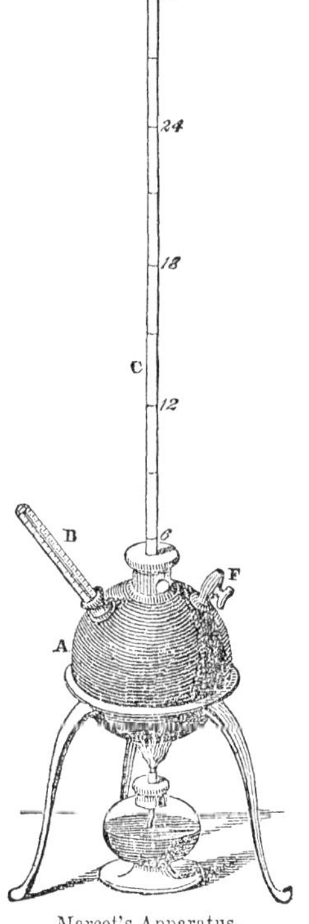

Marcet's Apparatus.

The rise of the boiling point under increasing pressure is well illustrated and proved by *Marcet's steam apparatus*, which is represented in the accompanying figure. A is a hollow brass globe, supported on a stand, and in it is contained a little mercury, and a small quantity of water. Through an air-tight collar, a graduated glass tube, C, is inserted, so as to reach very nearly to the bottom, both ends of it being open. B is a thermometer, having its bulb in the water or mercury. Now, by applying a lamp the water is heated, and when the temperature has risen to 212°, the steam will begin to issue freely through the faucet, F; but, by closing the faucet, the escape of the steam will be prevented, and the temperature will rise; the mercury at the same time by the pressure of the steam within, being forced up the tube C. And the

QUESTIONS.—43. Why cannot water be heated above 212° in the open air? What, if the steam be confined? Describe Marcet's steam apparatus.

height to which it may rise will always show the exact amount of the pressure.

By this means it has been determined that, at a heat of 250°, the tension of steam, thus confined in a boiler, is equal to two atmospheres, or 30 pounds to the square inch; at 275° its tension is equal to three atmospheres; and four atmospheres, or 60 pounds to the square inch, at 294°.

The expansive force of steam confined in this manner is the propelling power in the *steam engine.* (See author's *Nat. Philosophy*, p. 173.)

44. Evaporation.—But it is not only when a liquid is heated, and made to boil, that it is changed into vapor; this change, in most, and probably in all liquids, and many solids, is ever taking place, whatever may be their temperature, when they are contained in open vessels. This slow formation of vapor is termed *evaporation.* It is seen in the drying of clothes, when wet with water or alcohol, and in the gradual diminution of a quantity of either of these liquids, when left in an open vessel. Even in the forms of ice and snow water gradually evaporates.

Evaporation is much more rapid in some liquids than in others; and it is always found that those which have the lowest boiling point evaporate with the greatest rapidity. Thus, alcohol, which boils at a lower temperature than water, evaporates also more freely; and ether, whose point of ebullition is yet lower than that of alcohol, evaporates with still greater rapidity.

The chief circumstances that influence the process of evaporation, are extent of surface, and the state of the air as to temperature, *dryness, stillness,* and *density.*

The same quantity of liquid, exposed in a shallow vessel, wil evaporate more rapidly than in one of a different form, because of the large amount of surface in contact with the air; so, also currents in the air increase evaporation by removing the vapor as fast as it is formed. Increased pressure of the atmosphere diminishes evaporation.

QUESTIONS.—What is the tension of steam at 250°? 44. Do water and other liquids take the form of vapor without ebullition? Do all liquids evaporate with the same facility? What are the chief circumstances which influence evaporation?

45. Heat is absorbed by the Formation of Vapor.—During the slow evaporation of water, or other liquids, as well as when they are evaporated by boiling, a large amount of heat is absorbed, and becomes latent in the vapor produced. It is on this account that ether, alcohol, or even water, though at the same temperature as the air, always feels cold when a little is dropped upon the hand. The natural heat of the hand is absorbed and carried off in the vapor that is formed.

The evaporation of good sulphuric ether may easily be made to freeze water, even in the warmest weather. For this purpose let a very small glass vial, covered with muslin, be filled with water, and suspended by the neck from some convenient support; then drop slowly upon the muslin good sulphuric ether, from the mouth of a vial, or by means of a dropping tube. In a few minutes, ice will begin to form; and if the operation be continued, the whole of the water will be frozen, perhaps breaking the vial containing it.

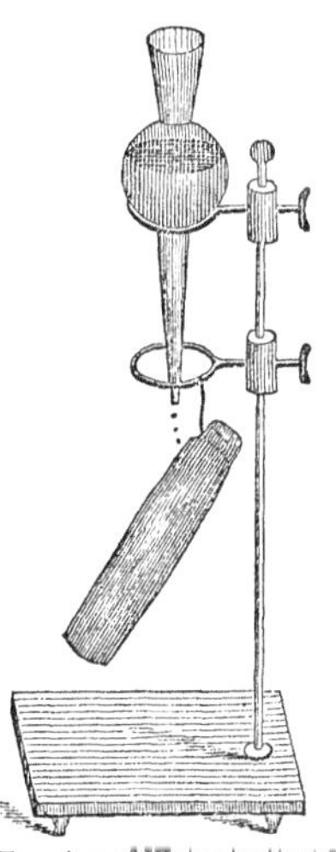
Freezing of Water by Evaporation of Ether.

Porous earthen vessels are often used in hotels and other places, in warm weather, to contain water for drinking. A portion of the water gradually exudes through the vessels, and evaporates from the surface, by which that within is kept several degrees colder than the temperature of the atmosphere. Such vessels are said to be much used in Spain, where they are called *alcarrazas*. People crossing the deserts of Arabia in caravans, are said sometimes to load camels with earthenware bottles filled with water, which is kept cool by wrapping the jars with linen cloths, and keeping them moist with water.

46. The freezing of water by its own evaporation under the receiver of an air-pump, is a common experiment. A shallow dish containing strong oil of vitriol is first placed upon the plate of the machine, and over it, supported by a tripod of wire, is placed a small capsule filled with water. The receiver being now put in its place, covering the whole, by working the pump the

QUESTIONS.—45. Is heat absorbed during the slow evaporation of a liquid? Describe the mode of freezing of water by the evaporation of ether. 46. Describe the experiment with water and sulphuric acid under the receiver of the air-pump.

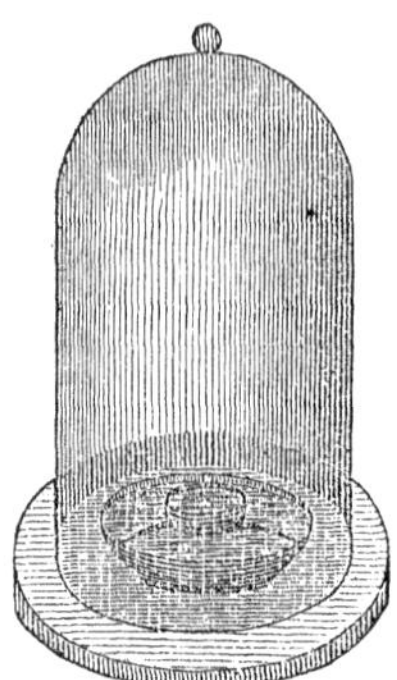
Freezing Water under Exhausted Receiver with Oil of Vitriol.

air is exhausted, rapid evaporation from the surface of the water commences, which is continued because of the absorption of the vapor by the acid beneath, until the water is frozen. Without the acid, or other substance to produce the same effect, the receiver would soon be filled with vapor, and no further evaporation take place; the vapor of water, at ordinary temperatures, not having sufficient tension to lift the valves of the pump, as it is worked. Indeed, a small drop of water may be frozen under the receiver of an air-pump by its own evaporation, without the use of any substance to absorb the vapor. Let a single drop of water, on a piece of charred cork, hollowed a little on its upper surface, be placed under the air-pump receiver, and by working the pump a few seconds, it will be frozen by the rapid evaporation which takes place from its surface. The burnt cork capsule is preferable to one of glass or metal, since, as the water does not adhere to its surface, not so much heat is received from it.

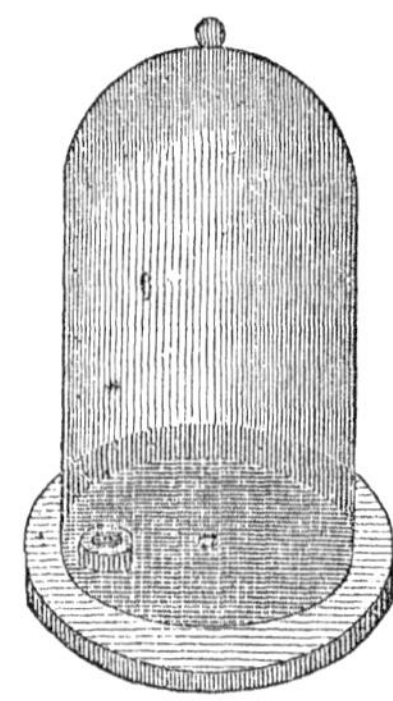
Freezing Water by its own Evaporation.

47. Latent Heat of Vapors.—It is not easy to determine with precision the amount of latent heat in vapors, or the relative quantity of heat absorbed as they are formed. The results obtained by different experimenters, therefore, are not uniform. It is believed that water, in taking the form of vapor, absorbs nearly 1000° of heat, or heat enough to raise the temperature of an equal weight of water 1000°, if it could be confined. The amount of heat in different vapors varies

---

QUESTIONS.—What purpose does the acid serve? Explain the method of freezing a drop of water upon a piece of burnt cork under an exhausted receiver. 47. What is the amount of latent heat in steam?

with their nature; in no two vapors is it the same. Thus, while the latent heat of watery vapor is, as we have seen, about 1000°, that of vapor of alcohol is only 373°, that of vapor of ether 163°, vapor of oil of turpentine 138°, and of sulphide of carbon 144°.

The heat which is absorbed when water or other liquid is converted into vapor, will, as a matter of course, be given out again when this vapor is condensed into the liquid form. On this principle, steam is often used for warming buildings, being conveyed in pipes through the different apartments. As it passes along the pipes, it is condensed, giving out its heat; and the water that is formed runs back again into the boiler.

48. Distillation.—The process of distillation consists simply in evaporating a substance, and again condensing the vapor, by causing it to come in contact with a cold surface. This is usually accomplished by having a tube of considerable length, leading from the top of a close boiler, and passing in the form of a spiral through a vessel which is kept filled with cold water.

In the laboratory, the apparatus figured below answers well for distilling small quantities of any liquid. A retort, R, contains the liquid to be distilled, and the vapor is received into a flask, F, the mouth of which is slipped on the neck of the retort, but the joint not made perfectly air-tight. The flask should be kept cold by being immersed in cold water, or by having a small stream of water constantly falling upon it from a vessel above.

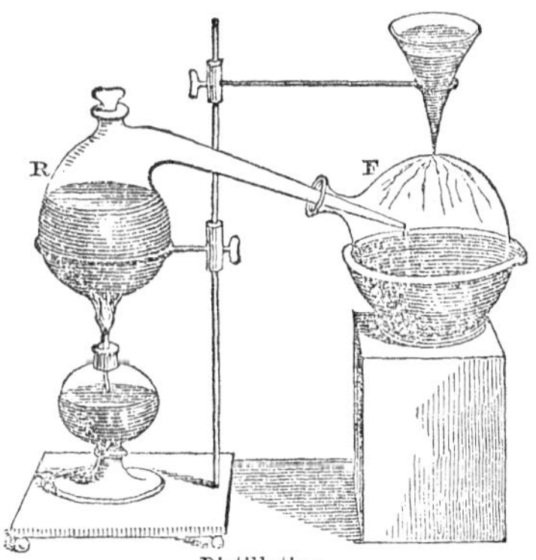

Distillation.

For larger operations, a Leibig's condenser is indispensable. It consists of a glass flask, for a boiler, which may be heated in a small furnace, as represented in the figure on page 48, or by means of a spirit-lamp, and a metallic case, *a*, in which is inserted, through perforated corks, a glass tube, *d d*, designed

---

QUESTIONS.—Describe the mode of warming buildings by the use of steam. 48. In what does the process of *distillation* consist?

to be kept constantly surrounded with cold water. From the vessel, *i*, a stream of cold water enters the funnel, *c*, and, as it is heated, escapes at the highest part by the tube, *h*, and is collected in a basin, *b*. The glass tube, *d d*, is connected at one end, by means of a smaller tube, with the boiler, and at the other end, with the receiving-vessel, *e*, in which is col

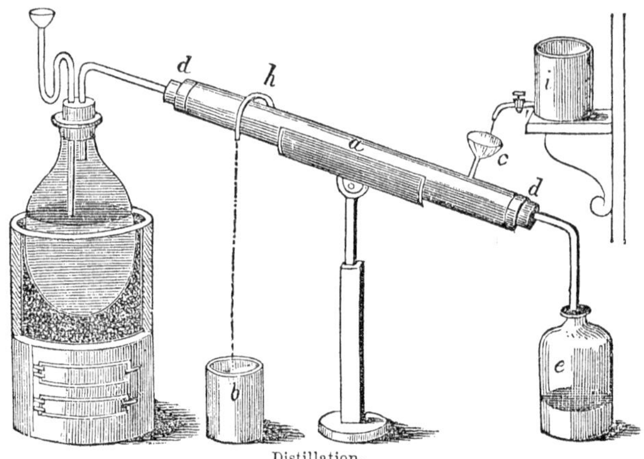

Distillation.

lected the distilled liquid condensed in passing through the tube, *d d*. The crooked funnel in the boiler serves to introduce the liquid to be distilled.

By the process of distillation volatile substances, whether liquid or solid, may be separated from those that are fixed, or even from such as are less volatile than themselves Water is distilled to purify it from salts or other substances it may contain in solution or suspension; and alcohol, by distillation, is separated from water, which is less volatile than itself, as well as from fixed substances.

The application of this process to solids is usually termed their *sublimation*.

49. Boiling produced by Cold.—We have seen above (41) the effects of diminishing or removing the atmospheric pressure in promoting ebullition, and we are now prepared to understand another ingenious method of accomplishing this object. Let a flask, with a cork well fitted to its mouth, be partly filled with water, and made to boil briskly by means of a spirit-lamp; then suddenly insert the cork and remove the lamp: the water will

---

QUESTIONS.—Describe Leibig's *condenser*. How is it that substances may be separated from one another by distillation? 49. Describe the experiment of boiling water by the application of cold.

continue to boil, and by immersing it in cold water, as shown in the figure, the boiling will become violent. The same effect will be produced by inverting the flask and applying snow or even cold water to the bottom. But if the flask be held again a moment over the lamp, the boiling will instantly cease. The reason of this is, because the upper part of the flask, when the cork is inserted, is filled with steam, which is condensed by the application of cold to the outside, and a vacuum produced. The warm water within then boils, as in a vacuum produced by any other means; but if heat be applied, steam will be again formed, and fill the upper part of the flask, and, by its pressure upon the surface of the water, prevent further boiling.

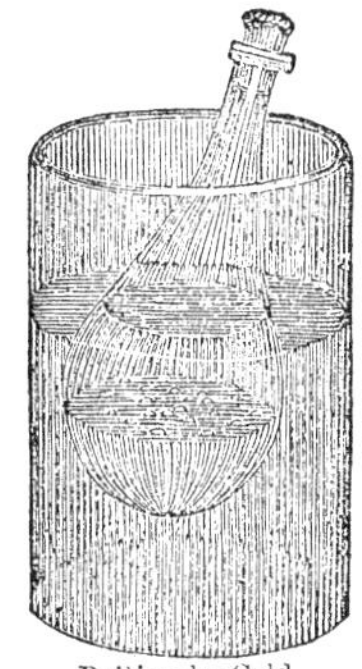
Boiling by Cold.

If the flask is firmly corked, so as to exclude the air perfectly, when it has become cold the water within, as the flask is handled, will fall from side to side, almost like masses of ice, and producing a similar sound to the ear. This is because there being no air within to break up the water as it is thrown in any direction, it falls in a mass and strikes against the sides of the glass with much the same effect as a solid. A small toy of this kind, made of glass tube, and hermetically sealed, is called a *water-hammer.*

50. The Cryophorus. — The cryophorus, or *frost-bearer* (from the two Greek words, *cruos*, frost, and *phero*, I bear), is an instrument for freezing water by its own evaporation, which beautifully illustrates some of the foregoing principles. It consists of a tube, half an inch or more in diameter, with a bulb blown at each end, one of them having a small aperture, A, by which a small quantity of water is introduced, sufficient only partly to fill one of the bulbs. This water is first all collected in the lower bulb, and the heat of a lamp applied, so as to cause it to boil briskly; and while the interior is filled with steam, the aperture at A is quickly

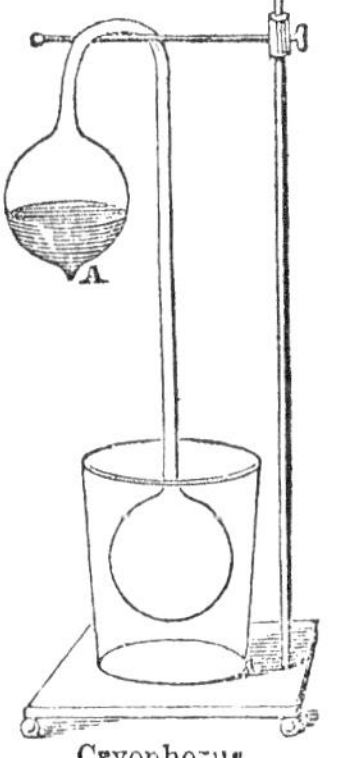

Cryophorus.

QUESTIONS.—What will be the effect of holding the flask over a lamp? What is the effect if, when cold, the flask is shaken? 50. Describe the *cryophorus.*

sealed hermetically, and the lamp removed. When it has become cold, the water is passed to the upper bulb, as represented in the figure (p. 49), and the instrument supported on a stand, with the lower bulb in a beaker glass. All the interior is now filled with vapor of water, except a part of the upper bulb, but no evaporation of the water can take place, because of the presence of this vapor. But by removing the vapor, which is accomplished by surrounding the lower bulb with a freezing mixture of salt and snow, to condense it rapidly, evaporation of the water is produced, attended with cold sufficient to freeze the most of it, even in the warmest weather.

The *pulse-glass*, as it is called, is a very similar instrument, and is made in the same manner, except that ether is used in it, instead of water. By grasping one of the bulbs firmly in the hand, the vapor, by its expansion, will immediately force all the liquid into the other; and the moment it has all passed through the stem, an appearance of violent ebullition is produced, attended by a distinct sensation of cold in the hand which grasps the bulb. This is occasioned by the rapid evaporation of the film of liquid lining the inside of the bulb.

Pulse-glass.

51. Effect of Perspiration upon the Animal System.—The effect of evaporation in withdrawing heat is admirably illustrated by the process of perspiration. The natural temperature of the human body is about 98°, but when we take active exercise, or are exposed to a great degree of heat, there is a tendency to a rise of temperature above that which is conducive to health; and the most injurious effects would ensue, if they were not prevented by the rapid evaporation which takes place from all parts of the surface of the system.

Examples of the power of the human body to sustain great and apparently even dangerous elevations of temperature, are on record. It is well known that individuals have voluntarily exposed them-

---

QUESTIONS.—Describe the *pulse-glass*. 51. What is the effect of *perspiration* upon the animal system? Will the human body sustain high temperatures for a time without injury?

selves for several minutes, in ovens, to temperatures even a hundred degrees above that of boiling water, without suffering any injury. The very rapid perspiration that takes place in such circumstances, prevents the destructive elevation of temperature in the system which would otherwise take place.

52. Temperature of the Seasons Modified.—The heat of summer is always greatly modified by the evaporation which take place from the surface of the earth, and the stalks and leaves of plants and trees. When a stalk of Indian corn (*zea maize*), or other plant, is cut down in midsummer, or a branch removed from a tree, the leaves soon begin to wither, because of the evaporation of the moisture in them. But the evaporation is no more rapid from them after being cut than before, but now the supply of water from the earth, received by the roots, ceases, and the withering we notice is a necessary consequence. We see therefore that vegetation in warm weather is sending forth into the atmosphere immense quantities of water by evaporation, besides that which rises from the surface of the earth itself; and as a result, the temperature of the atmosphere is more or less cooled. In other words, the heat is thus prevented from becoming as excessive as it would be but for this arrangement.

We have seen above (38) that heat given out by the freezing of water in winter, prevents the low reduction of temperature that would otherwise be experienced; and we cannot here less admire the wonderful provision of Providence, by which, on the other hand, the excessive heat of summer is, to some extent, limited.

53. Liquids upon very Hot Surfaces.—Liquids, as water, thrown upon metallic surfaces, heated nearly to redness, instead of adhering to the surface and rapidly evaporating, will sometimes be seen to roll around in globules, apparently without touching, until at length they gradually disappear. This is occasioned by an atmosphere of vapor that is formed around the globules of liquid, by its rapid formation preventing the temperature from rising as high as the boiling point, and also by its elasticity preventing the liquid from coming in contact with the metallic plate. Alcohol dropped upon the surface of heated oil of vitriol, exhibits a like phenomenon. This has been called the *spheroidal state* of liquids.

QUESTIONS.—52. How is the heat of summer modified? Why does a plant-stalk, separated from its root, so rapidly wilt in warm weather? Is water constantly evaporating from the leaves and stalks of plants? 53. Describe the action of a drop of water upon a very hot surface.

54. Dew.—Dew is a deposit of moisture from the atmosphere upon a cold surface in contact with it. If, in the summer, a vessel is left but a few minutes filled with ice-water, or even cold spring-water, dew soon collects upon it, and after a time, the water thus condensed trickles down the surface in drops. A surface upon which dew is seen to form will always be found colder than the surrounding air; and the particular temperature at which it begins to form is called the *dew-point.* When the air is very dry, this point will always be considerably below the temperature of the air; but when there is much moisture present this will not be the case.

In fair weather, during the summer season, there is usually seen, in the morning, a copious deposit of dew upon the leaves of plants, and upon other substances exposed to the open air. This is occasioned by the radiation of heat from bodies at the surface of the earth, which takes place rapidly during the night, cooling them down considerably below the temperature of the air. Substances, therefore, which radiate slowly (30), as polished metallic surfaces, seldom have any dew upon them, while good radiating surfaces near them will be abundantly covered with it.

In cloudy weather (without rain), there is generally little dew, because the heat radiated from the earth is reflected back by the clouds; and by suspending even a small handkerchief by the four corners, a few inches from the earth, the deposition of dew on substances under it is, for the same reason, prevented.

In some warm countries, water is said to be frozen during the night by the rapid radiation which takes place from its surface. The water for this purpose is poured into shallow pans, so situated as to receive as little heat as possible from the earth.

55. Hygrometers.—Hygrometers are instruments for determining the relative quantity of watery vapor present at any time in the atmosphere. Daniel's hygrometer (represented in the figure on p. 53) is much in use. It consists of a tube, A, with a bulb at each end, and is formed in the same manner as the cryophorus

---

QUESTIONS.—54. What is *dew?* Under what circumstances is it deposited? How do the leaves of plants and other bodies at the earth's surface become colder than the air? Why is there usually little dew in cloudy weather? 55. What are *hygrometers?*

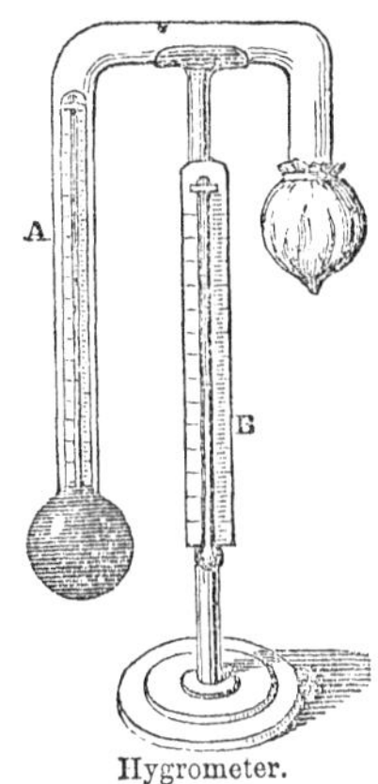

Hygrometer.

(50), except that it contains ether instead of water. The tube is supported by a stand; and the lower bulb, which is usually made of colored glass, is about half filled with the ether, having in it the bulb of a very delicate thermometer, with its stem extending upward in the tube. The other bulb is empty, or contains only the vapor of ether, and is covered with muslin. To the stand B is attached a small thermometer, to indicate the temperature of the air. By pouring a little ether upon the muslin, the bulb is cooled, and the vapor of ether within condensed, and a rapid evaporation of the ether in the bulb produced, as in the cryophorus. This occasions a cooling of the colored bulb, and a deposition of dew upon its surface, the small thermometer within showing the exact temperature at which the process commences, which is taken as the dew-point. Properly, however, it is the difference between the temperature thus obtained, and the temperature of the air, which shows the real state of the air as to moisture.

A decided objection to the use of this instrument is found in the fact, that it is extremely difficult to determine accurately the moment when the formation of dew upon the bulb actually commences. Its indications, therefore, cannot always be fully relied on.

The *wet-bulb thermometer* is now mostly used to determine the hygrometric state of the atmosphere. Two thermometers are attached to the same support, as shown in the figure on p. 54, one of them having a piece of muslin wrapped around its bulb, which is kept wet by a string leading to it from a small fountain of water, attached also to the support between the thermometers. Now, as the evaporation from the muslin necessarily reduces the temperature, this thermometer will always stand a little lower than the other, the bulb of which is dry; and, moreover, as the

QUESTIONS. — Describe Daniel's *hygrometer*. Describe the *wet-bulb thermometer*.

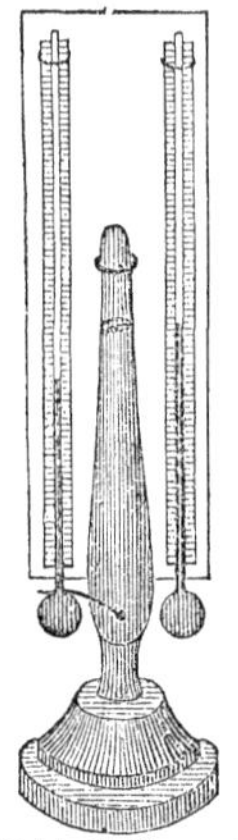
Wet-bulb Thermometer.

rapidity of the evaporation from the muslin will depend upon the dryness of the air, the difference between the readings of the thermometers will indicate its true hygrometric condition.

56. Watery vapor exists in three different states: 1. As transparent, invisible steam, as it rises from boiling water, and before it comes in contact with the air; 2. As it appears partially condensed, after escaping into the air; and 3. As it exists in the atmosphere at all temperatures, but invisible to the eye. That steam, before coming in contact with the atmosphere, is perfectly transparent and invisible, is shown by partly filling a glass vessel with water and causing it to boil rapidly. The steam within, above the water, cannot be seen until it escapes into the air, and becomes partially condensed, as stated above.

Clouds are collections of watery vapor, in this partially condensed state, in the upper regions of the atmosphere, and differ from *fog* only in their more elevated position.

The moisture that constitutes clouds, when fully condensed, falls in *rain* upon the earth, or is solidified and falls in beautiful crystals (25), as *snow*. If the drops of rain are frozen after they are formed, *hail* is produced.

If in warm weather a quantity of air be forced into a large glass receiver, so as to produce a pressure of at least two atmospheres, a slight mistiness will usually be seen within, occasioned by a partial condensation of the watery vapor forced in with the air. If the process is several times repeated, drops of water may be obtained, forming a kind of artificial rain.

In the manner stated above, all the water upon the surface of the earth is subjected to a constant natural distillation; pure water, in the form of vapor, rises in the air from the leaves of plants, from the earth, and from the surface of the ocean, rivers, and lakes, to be again diffused, in rain and snow, over the earth, producing everywhere vigor and life, both in the vegetable and animal world.

57. Liquefaction and Congelation of Gases.—By great pressure, or by pressure and a low reduction of temperature, many of

QUESTIONS.—56. In what three states does watery vapor exist? What are *clouds?* What is *rain?* 57. How may many of the gases be reduced to the liquid state?

the gases may be reduced to the liquid state, and the liquids thus formed solidified or frozen. Indeed, all gases may be considered as the vapors of extremely volatile liquids. Some of them, however, have never yet been reduced to the liquid state.

The usual method to liquefy a gas, is to put the materials from which it is to be formed into a strong glass tube, bent in the middle, as represented in the figure, and hermetically sealing it. As the gas is evolved the pressure of course increases, but at length a point is attained, depending upon the temperature and the nature of the gas, when it begins to condense as a liquid, the quantity of which is increased by the further evolution of gas from the materials, without any increase of pressure, if the temperature is kept uniform. The bent tube is particularly adapted for the liquefaction of cyanogen gas. To form this gas, dry cyanide of mercury is used, a portion of it being placed in the closed end of the tube, and the other end hermetically sealed. Moderate heat is then applied to the end containing the cyanide, the other end being cooled by a freezing mixture of snow and salt. As the cyanide is decomposed by the heat, the cyanogen first takes the gaseous form, but is subsequently condensed by the pressure and cold, and collects in the empty end of the tube.

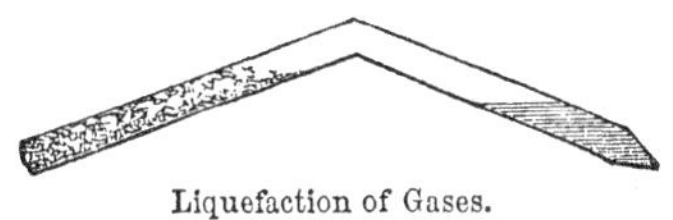

Liquefaction of Gases.

Of the different gases, some require a much greater pressure than others to condense them to the liquid state. At 0° sulphurous acid gas becomes liquid under the ordinary atmospheric pressure, but at 32° it requires a pressure of 2 atmospheres to produce the effect. Carbonic acid gas at 0° has a tension of 23 atmospheres, and at 32° a tension of 36 atmospheres; at higher temperatures the tension is still more increased.

The liquids formed from the gases, in the manner described, may be frozen by the great cold produced by their own evaporation, or by exposing them in tubes to intense cold. In the former case, the solids formed will appear like snow, and in the latter, like clear, transparent ice.

QUESTIONS.—What is the usual method to liquefy a gas? Do all gases require a like pressure to reduce them to the liquid state?

58. For preparing small quantities of solid carbonic acid, the following apparatus answers well, and is much less expensive than such as are usually purchased of the manufacturers of philosophical instruments.

The generator, A, is made of a common mercury flask, having the aperture at the neck a little enlarged, so as to be about an inch and a quarter in diameter. A plug of cast-steel, B, is then made of a bar two inches at least in diameter, and turned with a wide and smooth shoulder so as to fit accurately upon a collar of block-tin, when screwed into its place, as represented in the figure. The valves, which are the most difficult part to construct, on account of the great pressure that is to be overcome, are inserted in the plugs, a second one of which, precisely like the preceding, is made to screw into the receiver, C. Into the upper end of each plug, a hole an inch in diameter is bored about one inch deep, and terminates in a conical point; from which an aperture, a tenth of an inch in diameter, is bored quite through the plug. E H is composed of two parts, so constructed that when screwed firmly into the cast-steel plug, and the part H which terminates in a conical point screwed down, all escape of the gas from the generator is effectually prevented. When the part H is screwed upward, the escape of the gas around E is prevented by the firm pressure of the shoulder of E upon the washer I, and a shoulder upon the lower part H, which presses against the bottom of E, and produces the same effect with regard to the escape of the gas around the thread of the screw H.

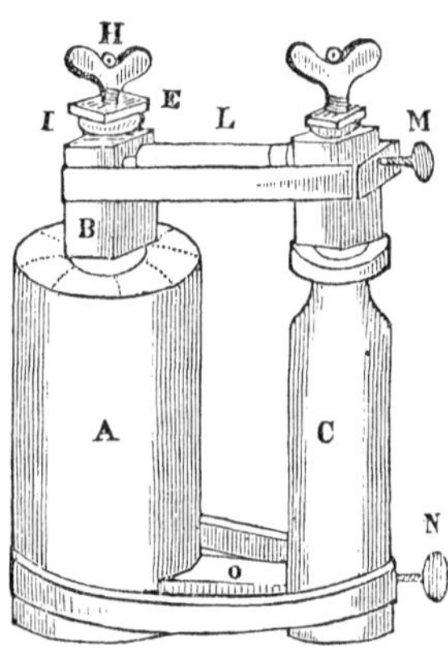

Solidifying Carbonic Acid.

Instead of the valve described above, the following answers better for the generator, as the passage at the bottom of the plug is not liable, as in the other construction, to be closed by the sulphate of soda which is formed. The part H extends quite through the plug, having at the lower extremity a nut, P, attached firmly by a screw and soldered. Now when the screw H is turned upward, the thread on which extends from I downward about an inch, the nut P perfectly closes the passage below, but by turning the screw down, the passage through the plug is opened at P, and closed at I, allowing the gas to escape laterally, as in the other construction.

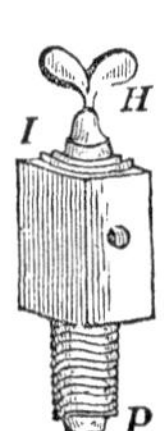

The receiver, C, is made of common boiler iron, and should be about two inches internal diameter, and of the same height as the generator, which will make it of the capacity of

Questions.—58. Describe the apparatus for preparing solid carbonic acid.

about a pint. The tube L should screw into the plug connected with the receiver, having its other extremity terminate in a conical point to fit into a cavity prepared for it in the other plug. By means of the stirrup-screws M and N, and the block of wood O, the receiver may then be firmly screwed in its place; and when both the valves are open, there will be a free passage between it and the generator, but no communication of either with the open air.

To make use of this apparatus, the generator and receiver are separated, and the plug B being removed, 2½ pounds of bicarbonate of soda, made into a paste with the same weight of water, are introduced into A, and 21½ ounces of strong sulphuric acid are poured into several copper vessels, made a little shorter than the length internally of the generator, and of such a diameter that they will just pass the aperture. These being nearly filled with acid are dropped into the generator, which, after the plug B is inserted, is allowed to lie on one side for fifteen or twenty minutes, and several times rolled over, to mix the acid with the soda. The receiver is then attached to it as seen in the figure, by means of the stirrup-screws M and N; and, if kept sufficiently cool by means of ice, the liquid carbonic acid formed in A will shortly be distilled over into C, the passage between them being of course previously opened.

The valves are now to be closed, and the receiver, which contains the liquid carbonic acid, separated from the generator. A small tin cup (not represented in the figure) is then to be attached to the tube L, to receive the jet of acid from the receiver. It is essential that the *liquid* acid should escape into this cup, which is effected by having a small tube pass from the steel plug nearly to the bottom of the receiver, or by inverting the receiver before opening the valve.

The apparatus should be well tested, at least three times, before running any risk by venturing to handle it while charged. This is best done by means of a hydraulic press; but the same object may be accomplished very effectually by standing the apparatus when charged in a tub of water heated to about 150°, so that when the apparatus and water have attained the same temperature, it shall not be lower than 130°. If a more severe test is desired, the water may be made still warmer.

In constructing an apparatus, care should always be taken to make the receiver of not more than one-fifth the capacity of the generator. The quantity of materials used should also be just sufficient *very nearly* to fill the generator.

In using this apparatus, when the liquid is received in the cup, it hisses and boils with the greatest violence; and the cold produced by the evaporation of a part of it is so great as to freeze the rest, which is retained in the cup as a fine white snow. By rolling this in balls, and wrapping it in cotton, it may be kept some time; but in the open air it evaporates rapidly, and intense cold is produced, equal, it is said, to —148°. By moistening the solid with ether, and placing it in an exhausted receiver, it is claimed that a temperature as low as 175° or 180° below zero has been produced.

The solid does not mix with water when immersed in it; a ball of it thrown upon the surface of water floats about lightly, and at

QUESTIONS.—How is the freezing of the liquid accomplished? At what temperature is the solid formed?

length a portion of water in contact with it is frozen by the intense cold. With sulphuric ether or chloroform it mixes readily, and the pasty mass rapidly evaporates, producing intense cold.

Mercury which congeals at about —40° is readily frozen by being kept a short time in contact with the solid, surrounded by some cotton, or by immersing it in a mixture of the solid and ether or chloroform.

SPECIFIC HEAT.—CAPACITY OF BODIES FOR HEAT.

59. When a body is exposed to any source of heat, its temperature rises, and the substance of heat is supposed to accumulate in it; but the same quantity of heat, imparted to different bodies, will not raise their temperature alike. Thus, if a pound of water and a pound of mercury, in similar vessels, and at the same temperature, be exposed to the same source of heat, the temperature of the mercury will rise about 30°, while that of the water rises only 1°. It appears, therefore, that it requires 30 times as much heat to raise the temperature of water any given amount, as it does to produce the same effect upon mercury. This idea is expressed by saying that the *specific heats* of these substances are as 30 to 1; or we say (as some prefer) that the *capacity for heat* of water is to that of mercury as 30 to 1.

If, instead of comparing equal weights of the two substances, we take equal volumes—as a pint of each—and expose them to the same uniform source of heat, we shall find that while the water gains 1° of heat, the mercury will gain 2°, or a little more. To express this relation we use the term *relative heat;* and we say therefore that water has more than twice the relative heat of mercury.

Other methods of determining the specific heat of bodies have been devised, one or two examples of which will be given. If a

---

QUESTIONS.—What is said of the solution of the solid in chloroform and ether? How may mercury be frozen by use of the solid acid? 59. Will the same quantity of heat imparted to different bodies heat them alike? If like quantities of water and mercury are exposed to the same source of heat will they in the same time be heated alike?

pound of olive-oil and a pound of water be heated to some given temperature, say 80°, and then placed in a cold room, and the number of minutes noted which is required for each to cool an equal number of degrees, say to 50°, it will be found that the oil will cool in less than half the time required by the water; but as both substances must be supposed to lose equal quantities of heat in equal times, it follows that the water must have contained more than twice as much as the oil; or the capacity of water for heat is more than twice that of this oil.

If a piece of copper, of a pound weight, be heated to 300°, by holding it a few minutes in mercury at this temperature, and then immersed in a pound of water at 50°, the copper will give out heat to the water until the temperature of both will be at 72°. Now, the copper has lost 228° of heat, and the water has acquired 22°. The specific heat of water, therefore, is to that of copper as 228 to 22.

It is usual to make water the standard in comparing the specific heats of bodies, considering its specific heat as 1·000; we shall then have the specific heat of mercury $\frac{1\cdot000}{30}=\cdot033$, and that of copper $\frac{22}{228}=\cdot096$.

No two substances have the same specific heat, but every substance has a specific heat peculiar to itself, and which is to be considered as one of its own peculiar properties.

The following table exhibits the specific heats of several well-known substances:—

Water	1·000	Gold	0·032
Iron	0·114	Silver	0·057
Copper	0·096	Alcohol	0·645
Lead	0·031	Tin	0·056
Sulphur	0·203	Platinum	0·032
Phosphorus	0·188	Zinc	0·095
Diamond	0·119	Mercury	0·033
Graphite	0·202	Oil of Turpentine	0·467
Charcoal	0·201	Ether	0·503

The specific heat of a body depends to some extent upon its temperature, being greater as the temperature is higher.

Change of density in a body is usually attended by a corresponding change in its specific heat, which is increased as the density is

---

QUESTIONS.—What is the ratio of their *capacities for heat*, or their *specific heats?* What is taken as the standard for the specific heat of bodies?

diminished, and diminished as the density is increased. This is seen in solids, as when a piece of iron is heated by hammering, which increases its density and causes a portion of its latent heat to be given out, thus raising its temperature; but is best illustrated by the gases, the densities of which are more easily made to vary at pleasure. Thus, if we suddenly compress a portion of a gas, its temperature is sensibly raised; but, on the other hand, if we diminish its density, it becomes colder by the absorption of heat occasioned by its increased capacity for heat. A delicate thermometer placed under the receiver of an air-pump falls as the air is exhausted.

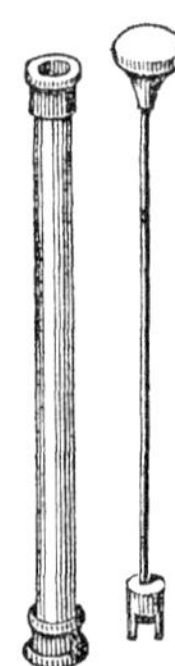
Fire Syringe.

60. The *Fire Syringe* is an instrument by which dry tinder or spunk may be ignited by the heat produced by the sudden compression of a portion of atmospheric air. It consists of a hollow cylinder, closed at one end, and a solid piston fitting it accurately, and having in its under side a cavity to receive the substance to be ignited. To use it, the tinder or spunk is put in the place fitted for it, and the piston is then plunged forcibly into the cylinder; on removing it, the combustible substance will generally be found ignited. If the cylinder is of glass, a flash of light will often be seen when the air is compressed.

II. LIGHT.

NATURE AND SOURCES OF LIGHT.

61. Nature of Light.—Although innumerable observations and experiments have been made upon light, yet it must be admitted that some doubt and obscurity still remain concerning its real nature. But, in the absence of positive knowledge, two theories of light have been proposed, by each of which nearly all

---

QUESTIONS.—How does change of density in a body affect its specific heat? 60. Describe the *fire syringe*. 61. Do we understand fully the real nature of light?

the phenomena attending it may be satisfactorily explained; and it is admitted that each is also attended with its peculiar difficulties. These are called the *Newtonian*, or *corpuscular*, and the *undulatory* theories.

62. The *Newtonian theory* supposes light to be material, and to consist of inconceivably minute particles, which, however, are too subtile to exhibit the common properties of matter. These particles, emanating from luminous bodies, such as the sun, the fixed stars, and incandescent substances, and traveling with immense velocity, excite the sensation of light, it is supposed, by passing bodily through the substance of the eye, and striking against the expanded nerve of vision, the retina. The whole language of optics is founded on this theory.

63. The *undulatory theory*, or *theory of Huygens*, which is now generally adopted, denies to light a separate material existence, and ascribes its effects to the vibrations or undulations of a subtile ethereal medium, supposed to be universally present in nature, the pulses of which, in some way excited by luminous objects, pass through space and transparent bodies, and give rise to vision by impressing the retina, in the same way as pulsations of air impress the nerve of hearing, to produce the sensation of sound.—(See *Natural Philosophy*, p. 188.)

64. Light is not a homogeneous substance, as might be supposed, but the white light of the sun is made up of rays of several different colors, as will be shown when we come to speak of its decomposition, or analysis. So, also, it is capable of producing several distinct classes of effects, which have been attributed to the action of distinct agents; as the *colorific* rays, or the rays which produce the phenomena of color, the *heating* rays, and the *chemical* rays, or those which are capable of producing chemical changes. Thus it is possible, by causing the solar ray to pass through certain substances, to separate the heat entirely from it; or its illuminating power may be destroyed, and a distinct, invisible ray of heat be obtained. So, also, chemical effects may be produced by rays which seem to be destitute of any heating or illuminating power.

QUESTIONS.—62. Describe the *Newtonian*, or *corpuscular* theory of light. 63. Describe the *undulatory theory*. 64. What colored rays are contained in the white light of the sun? What other rays?

65. Sources of Light.—The sun is the great source of light to the earth, and all things upon its surface. As rays of heat always accompany the light of the sun, it is natural to suppose that the sun is an intensely heated mass, which is constantly throwing off both light and heat in every direction, like a red-hot cannon-ball suspended in the air; but this cannot be proved. At the present day, it is generally believed that the body of the sun is a dark, opaque substance, surrounded by luminous clouds, unlike anything, perhaps, with which we are acquainted upon the earth, but which are the real source of the sun's rays. These clouds are supposed to be of great thickness; but occasionally they break away in places, showing the body of the sun beneath them, which constitute the spots often seen upon his surface.

The great distance of the sun from the earth—95,000,000 of miles—very probably will ever prevent us from knowing more with certainty of his real nature.

Artificial light is produced by various modes, but chiefly by combustion, by the burning of a lamp or candle, or a mass of charcoal; but it may also be produced by galvanism,—in a manner to be hereafter explained,—by decaying animal and vegetable substances, called *phosphori*, and by every means which produce great heat.

All bodies begin to emit light when heat is accumulated within them in great quantity; and the appearance of glowing or shining, which they then assume, is called *incandescence*. The temperature at which solids in general begin to shine in the dark, is between 600° and 700°; but they do not appear luminous in broad daylight till they are heated to about 1000°. The color of incandescent bodies varies with the intensity of the heat. The first degree of luminousness is an obscure red. As the heat augments, the redness becomes more and more vivid, till at last it acquires a full red glow. If the temperature still increase, the character of the glow changes, and by degrees it becomes white, shining with increasing brilliancy as the heat augments. Liquids and gases become incandescent when strongly heated; but a very

---

QUESTIONS.—65. What is the great source of light to the earth? How is artificial light produced? At what temperature do bodies begin to emit light?

high temperature is required to render a gas luminous, more than is sufficient for heating a solid body even to whiteness. The different kinds of flame, as that of a wood-fire, candles, and gas-lights, are instances of incandescent gaseous matter.

Artificial lights differ greatly in color, some being of a brilliant white, and others being red, blue, yellow, or green. The chemical agency of artificial light is in general analogous to that from the sun; but in most cases it is too feeble to produce very decided effects.

66. Many substances have the power of emitting a feeble light, unattended by sensible heat, and are called *phosphori* (from two Greek words, *phos*, light, and *phero*, I bear). Certain living animals also possess the same property, as the glow-worm, and the common fire-fly. This property of bodies is termed their *phosphorescence*.

Some phosphori, as that prepared by mixing sulphur and oyster-shells, and exposing the mixture for a time to a strong heat, the diamond, fluor-spar, &c., shine only after having been heated, or exposed for a few moments to a strong light; while others, as moist, decaying wood, and decaying fish, shine without such preparation, even at ordinary temperatures.

Light sometimes appears during the process of crystallization. This is exemplified by a tepid solution of sulphate of potassa in the act of crystallizing; and it has been likewise witnessed under similar circumstances in a solution of fluoride of sodium and nitrate of strontia. Another instance of the kind is afforded by the sublimation of benzoic acid. Allied to this phenomenon is the phosphorescence which attends the sudden contraction of porous substances. Thus, on decomposing by heat the hydrates of zirconia, peroxide of iron, and green oxide of chromium, the dissipation of the water is followed by a sudden increase of density suited to the changed state of the oxide, and a vivid glow appears at the same instant. The essential conditions are that a substance should be naturally denser after decomposition than it was previously, and that the transition from one mechanical state to the other should be abrupt.

67. *Photometers* are instruments for measuring the comparative intensities of different lights, of which several kinds have been proposed; but it does not enter into our present purpose to describe or discuss their comparative merits.

To determine the comparative intensities of two lights, as that from different candles or lamps, the following method, originally proposed by Count Rumford, is perhaps as reliable as any; and has the advantage of requiring little and very cheap apparatus. Let L be a lamp, and C a candle, the lights of which we desire tc

---

QUESTIONS.—Do artificial lights differ in their colors? 66. What are *phosphori?* 67. What are *photometers?*

compare. Provide a screen, S, of white paper, which is to be put in a frame and properly supported by a stand, as shown in the figure, and also a small cylinder, C, of wood or some opaque substance. Then place this cylinder in an upright position in front of the screen, in such a position that its shadow from both of the lights shall be thrown side by side upon the screen, but not overlapping each other, and removing the lights to different distances, until the shadows appear of perfectly equal intensities. The comparative intensities of the two lights will then be as the squares of the distances of the lights from the screen. In the present case, L′ will be the shadow from the light of the lamp, and C′ that from the light of the candle; and the intensity of the lamp light will be to that of the candle light as LL'^2 is to CC'^2. If LL′

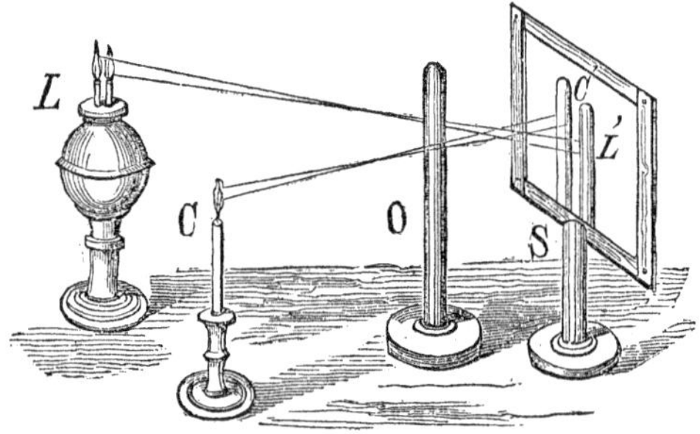

Photometer.

is 50 inches and CC′ 45 inches, then will the light of the lamp be to that of the candle as 2500 is to 2025, or as 1·234 is to 1.

This method is founded upon the fact to be illustrated in the next paragraph, that the intensity of the light from any luminous body, at different distances, will be inversely as the squares of those distances.

The experiment must, of course, be conducted in a dark room.

QUESTIONS.—Describe the method of determining the comparative intensities of two lights by a comparison of the shadows they produce

DISTRIBUTION OF LIGHT.

68. *Light is distributed*, or diffused abroad, by several modes; as by *radiation*, *reflection*, *refraction*, &c.

Light emanates from every point in the surface of a luminous body, and is equally distributed on all sides, if not intercepted, diverging like radii drawn from the centre to the surface of a sphere. Thus, if a single luminous point were placed in the centre of a hollow sphere, every point of its concavity would be illuminated, and equal areas would receive equal quantities of light. Each ray, when not interrupted in its course, and while it remains in the same medium, moves in a straight line, as is obvious by the appearance of shadows cast by the side of a house, or of a sun-beam admitted through a small aperture into a dark room. Owing to these modes of distribution, it follows that the quantity of light which falls upon a given surface decreases as the square of its distance from the luminous object increases—the same law which regulates the heating power of a hot body.

69. The passage of light is progressive, time being required for its motion from one place to another. It comes to the earth from the sun in about 8½ minutes,—a distance of 95,000,000 miles,—which shows its rate of progress to be about 195,000 miles a second. Owing to this prodigious velocity, the light caused by the firing of a cannon or a sky-rocket is seen by different spectators at the same instant, whatever may be their respective distances from the rocket, the time required for light to travel 100 or 1000 miles being inappreciable to our senses.

70. Reflection of Light.—Light is reflected in the same manner as heat (31), obeying precisely the same laws. This always takes place when it passes from one medium into another of different nature or density, whether the media be solid, liquid, or gaseous. Different media, however, differ much in their power of reflection.

Bright metallic surfaces, as polished silver or clean mercury,

---

QUESTIONS.—68. How is light distributed? Does light emanate from every point of an object? What is said of the course of a ray when not interrupted? 69. Is the passage of light from point to point instantaneous? What is its velocity? 70. When is light said to be *reflected?*

reflect nearly all the rays which fall upon them; while those which are dull and rough reflect but a few. The reflection of light, like that of heat, takes place at the surface of bodies, and appears to be influenced rather by the condition of the surface than by the internal nature or structure of the reflecting body.

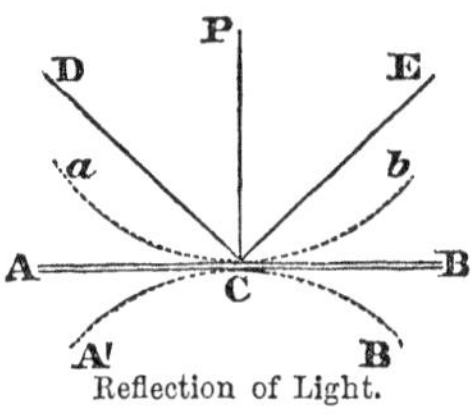

Reflection of Light.

Let A B be the reflecting surface, E C the ray incident at C, and C D the reflected ray, and let P C be perpendicular to A B. E C P will then be the angle of incidence, and P C D the angle of reflection, both of which will be equal. It is not necessary that the reflecting surface should be a plane, as might be supposed;—it may be either concave, as *a b*, or convex, as A′ B′, and the same results would follow.

Light has precisely the same characters after being reflected as before, but is less intense, because of the absorption of a part of the rays.

71. Refraction of Light.—When a ray of light passes through the same medium, as glass or water, or when it passes perpendicularly from one transparent medium to another, it moves in perfectly straight lines; but when it passes *obliquely* from one medium to another of different density, it is thrown more or less out of its first direction, and is said to be *refracted.*

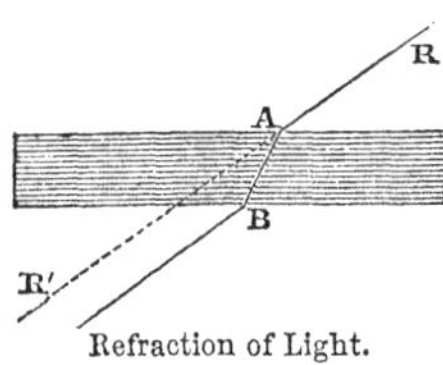

Refraction of Light.

Thus, a ray of light, R, passing through the air, when it comes in contact with a piece of polished glass at A, does not move on in a straight line to R′, but is bent downward or refracted, and emerges from the glass at B, where it is again refracted in the opposite direction, and takes the same course, though not precisely the same path, as it had at first.

Refraction always takes place when a ray of light passes obliquely from one medium to another of different density, but not always to the same amount; this will depend upon the *refracting power* of

---

QUESTIONS.—Describe the angles of incidence and reflection. 71. When is light said to be *refracted?*

the two media, and also upon the obliquity of the ray to the surfaces of the media in contact.

When the ray passes from a rare to a dense medium, it is always refracted or bent *towards* a line perpendicular to the surface at the point of contact, and *from* this line when it passes in the opposite direction, from a dense to a rare medium.

To understand the relative positions of the incident and th refracted ray, in the case of any two media, the following law needs to be well studied. Let A B in the figure be the surface of some transparent medium more dense than air, as water, and let I C be a ray of light incident upon it at C; it will not pass on in a straight line, but on entering the water will be bent downward, to E; I C is then called the incident ray, and C E the refracted ray. Let P P′ be a line perpendicular to the surface at C, then angle I C P will be the *angle of incidence*, and E C P′ the *angle of refraction.*

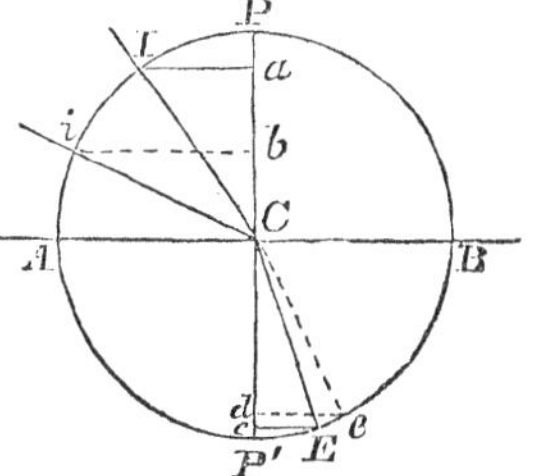

Index of Refraction.

If now from C as a centre we draw the circle A P B P′, and also the lines I *a* and E *c* both at right angles to P P′, then will I *a* be the sine of the angle of incidence and E *c* the sine of the angle of refraction. And for the same two media these lines will always have the same ratio to each other, whatever may be the angle of incidence. Thus, if a second ray, *i* C be incident at C, it will emerge at *e*; and the sines of the angles of incidence and refraction, that is, the lines *i b* and *e d* will have to each other the same ratio as the lines I *a* and E *c*; and the same will be true for the same media, whatever may be the angle of the incident ray.

The quotient obtained by dividing the sine of the angle of incidence by that of the angle of refraction is called the *index of refraction* for the media used; for the same media it is always

QUESTIONS.—What is the course of the ray in passing from a rare to a dense medium? When from a dense to a rare medium? Describe the angles of *incidence* and of *refraction.* Describe the *index of refraction.*

the same, but varies with different media. Usually the air is taken as one of the media, so that the index of refraction for any substance is the quotient thus obtained in the case of a ray of light in passing from air into that substance. The index of refraction for water is thus found to be 1·33, for common flint glass 1·56, for oil of cassia 1·64, for phosphorus 2·22, and for the diamond 2·43.

(For the Polarization of Light, see the author's "Natural Philosophy.")

DECOMPOSITION OF LIGHT.

72. The Solar Spectrum.—The white light of the sun is not a homogeneous substance, but is capable of being separated into several rays of entirely different colors. This was first effected by Newton, by passing it through a triangular piece of clear, solid glass, called a *prism*.

In the figure following, let S be a ray of light from the sun, admitted into a darkened room through the window-shutter, D E; it will pass downward to the floor, at a little distance from the wall, producing a circular spot of clear white light, W. Then let the prism A B C be held in the ray, and at once the spot at W will disappear, and, in its stead, an elongated and beautifully colored image of the sun will be seen upon a screen hung up in front of the window, or on the wall at the opposite side of the room, if no screen be used. The several colors will appear in the order indicated, the violet being uppermost and the red lowest.

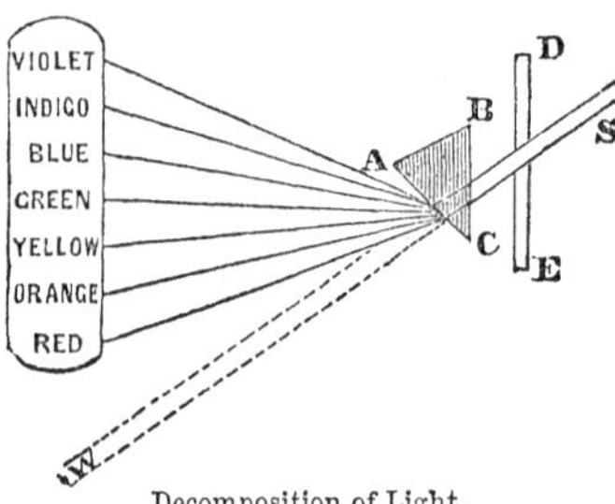

Decomposition of Light.

QUESTIONS.—What is the index of refraction for water? Flint-glass? Diamond? 72. How may light be decomposed? Describe the experiment with the prism.

It will be seen that the light, in passing the prism, has been twice refracted, or bent upward, first as it entered the glass, and again as it issued from it; and that the separation of the several colors has been in consequence of their different refrangibilities. The violet being most refrangible, is found uppermost in the picture, and the red is lowest, because least refrangible. The other colors occupy intermediate positions, depending upon their respective refrangibilities.

The colored image thus produced is called the *solar spectrum;* and, according to Newton, it is composed of the seven colors named in the figure, which are therefore called *primary colors.*

73. More recent investigations by Brewster render it probable that there are in the spectrum really only three colors, *red, yellow,* and *blue,* and that the other shades are produced by mixtures of these in different proportions; a mixture of the blue and the yellow, for instance, producing the green, and a like mixture of the red and yellow producing the orange. Indeed, it is believed that each of these three colors extends over the whole spectrum, but each is much more intense at one part of the spectrum than elsewhere, the blue being most intense near the top, and the red near the bottom, with the most intense portion of the yellow between them. The solar spectrum, therefore, as produced by the prism, may be considered as composed of three simple spectra superimposed upon each other.

The distribution of the rays in each of these simple spectra is represented by the shading of the annexed figures, in which B represents the blue, Y the yellow, and R the red, each color being supposed to be separated from the others. If the three spectra be thrown one upon another on the same screen, the ordinary solar spectrum will be produced.

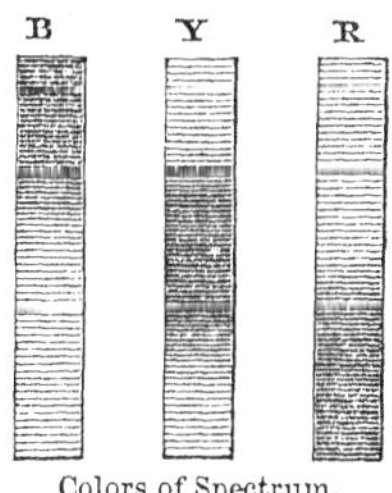

Colors of Spectrum.

74. Heating and Chemical Rays.—It has been stated above that light is capable of producing several distinct classes of

QUESTIONS.—What are Newton's *primary colors?* Why are these separated by the prism? 73. What colors only are contained in the spectrum, according to Brewster? How are the other colors produced? 74. In what part of the spectrum are the greatest heating effects produced?

effects, as those of color, those of heat, and besides these, others which may strictly be called chemical effects. Now, these several effects are not produced in every part of the spectrum with equal facility; the greatest illuminating power is found to be in the yellow, while the greatest heat is in the red, or a little below it, and the greatest chemical effects are produced in the extreme violet.

The *chemical effects* of light are various and important; a mixture of chlorine and hydrogen gases may be kept together in the dark for any length of time, without combining, but unite with an explosion when placed in the direct sunlight. On the other hand, many compound substances are decomposed by light, as certain preparations of gold and silver. If a piece of white paper be coated over with a thin film of white chloride of silver, carefully prepared in the dark, and then placed in the solar spectrum, the part in the violet ray will soon become black, while that in the red will scarcely be affected. Between these extremes there will be produced various shades of gray and purple.

It appears, therefore, that the sun's light is made up of three kinds of rays, viz., the *colorific rays*, or rays of light proper, the *heating rays*, and the *chemical rays*, the last of which are most, and the heating rays least, refrangible.

The light of the sun produces most important effects in the vegetable world; many plants will not grow in the dark, and others growing in the shade have their nature entirely changed. But a discussion of these topics does not belong to our present subject.

75. Photography.—By this name we designate the various modes of producing pictures by the action of light. If a piece of white paper is moistened with a dilute solution of common salt, and then one side of it washed with a solution of nitrate of silver, the surface becomes coated with chloride of silver, which readily turns black or dark chestnut by exposure to the direct rays of the sun. If, now, before exposing paper thus prepared to the light, any small flat object, as a flower, or piece of lace, be placed upon it, an image of the object will remain upon the paper, and may be

---

QUESTIONS.—In what part of the spectrum are the greatest chemical effects produced? Does the light of the sun produce any important effects upon vegetable bodies? 75. What is *photography?* What method is mentioned for preparing a photographic paper?

rendered permanent by soaking it *immediately* in a saturated solution of common salt, or of iodide of potassium.

The unaltered chloride of silver in the paper is by this soaking dissolved out, while the part that has become colored resists the action of the solvent and therefore remains in the paper.

Talbot's Calotype Process, invented by a gentleman of this name, is conducted as follows: A sheet of writing paper of a firm texture is brushed over on one side with a solution of 50 grains of nitrate of silver in an ounce of water, and then dried in a dark room, and subsequently soaked two or three minutes in a solution of iodide of potassium, containing about an ounce of the iodide to a pint of distilled water. It is then to be again soaked for some minutes in water, and thoroughly dried, and preserved for use.

When required for use, the paper is to be washed on the side previously iodized by *gallo-nitrate of silver*, prepared for the occasion in the following manner: Dissolve 100 grains of nitrate of silver in 2 ounces of distilled water, and add to it an equal volume of strong acetic acid, and then mix with it several volumes of saturated solution of gallic acid in cold distilled water. This last preparation should be made only in small quantity for the particular occasion, as it spoils by keeping. The last washing should be made in the dark, or with only a feeble candle light; and the paper dried carefully, excluding the light of day, to the action of which it is exceedingly sensitive.

Paper prepared in this way may be used in the manner first described, or in the camera obscura, for the taking of portraits. If the picture at first is not sufficiently distinct, it may be improved by washing it again in the gallo-nitrate of silver. Finally, it is to be rendered permanent by washing it with solution of bromide of potassium, or of common salt.

The picture thus formed is what is called a *negative picture*—that is, the light and shade are reversed, as compared with an ordinary engraving; but a positive one may be formed from the first by using it as an object for forming a second picture upon another sheet of the same prepared paper. For this purpose it is

---

QUESTIONS.—Describe Talbot's *Calotype process*. What are *negative* and what *positive* pictures?

placed with its face downward upon the prepared paper, and then exposed to the direct rays of the sun, as directed above (75). The new picture is to be rendered permanent in the same manner as before.

Numerous other preparations are in use for producing pictures upon paper, but none of them equal in sensitiveness the one just described. Though the materials to be used are different, as well as the processes, yet the essential principles are the same in all. The light produces a chemical change in the parts of the picture exposed to its influence, and the picture is *fixed* by soaking the paper in a solution capable of dissolving out the sensitive substance contained in the parts which have not undergone this change, in consequence of being in the shade.

Paper prepared for the taking of pictures is called *photogenic paper*, and generally cannot be long kept, even in the dark.

76. The *Daguerreotype* process, so called from the name of the inventor, is applied only to plates of silver, which are usually spread upon plates of copper. The silver surface is first very thoroughly cleaned by washing with dilute nitric acid, and rubbing with leather or cotton and some polishing substance, as very fine colcothar. It is then to be subjected for a few minutes to the action of vapor of iodine, by placing it in a box which has some crystals of iodine spread upon the bottom; by this means, an exceedingly thin coating of iodide of silver is formed upon the surface of a straw-yellow color, which is very sensitive to the action of light. It is then placed in a camera obscura, and the image of any object in front is made to fall upon it for a few moments, by which such a chemical change is produced in the thin coating of iodide of silver, that subsequent exposure to the vapor of mercury, at about 165° F., brings out a beautiful picture of the object. By close inspection, it will be found that in the parts of the picture where the most light has fallen such a change has been produced that the mercury is capable of acting upon it, but in other parts the bright surface of the silver remains unaffected; and further, the action of the mercury upon the silver plate will be in proportion to the intensity of the light upon the different parts.

Instead of pure iodine, the bromide or chloride of iodine may be used for preparing the plates; but the last compound is said to be, on the whole, much the best.

QUESTIONS.—How are pictures prepared in this way fixed or rendered permanent? 76. Describe the Daguerreotype process.

The picture, when taken from the mercurial process, is rendered permanent by removing the coating of iodide of silver, which is readily done by merely pouring over it a warm solution of hyposulphite of soda or of common salt.

It is further improved, and the shades rendered deeper, by heating upon it a solution of chloride of gold and hyposulphite of soda.

The Daguerreotype process is very simple, but to insure success, close attention must be paid to various minute particulars, which cannot here be discussed.

77. *Thermography* is a name which has been given to certain modes, dependent, it is believed, upon heat, by which one body is made to depict itself more or less minutely upon another, either in contact with it or in its vicinity. Thus, if we write with some soft substance upon glass, and then breathe upon it, the writing becomes visible. So if we allow a piece of coin to lie for a time on a plate of metal or glass, and then breathe upon it, an image of the coin will be produced. If while the piece of coin lies upon the metallic plate it is gently heated by a spirit-lamp, and when cold exposed to the vapor of mercury, a very distinct image of the coin will be formed. In some cases, we are told, this effect will be produced when the coin has not touched the plate, but only remained for a time near it.

78. Double Refraction of light takes place when a ray is passed through certain transparent crystals, and some organized substances, so that objects seen through them in particular directions appear double; and the rays emerging from them are found to have undergone a further change, by which they have acquired peculiar properties on different sides, and are said to be *polarized.* Light is also polarized by other means, as by reflection at particular angles from most non-metallic substances, and by refraction. For a very full discussion of the subject, see *Natural Philosophy*, page 259.

Rays of heat may be polarized in the same manner and by the same means as those of light.

QUESTIONS.—77. What is *thermography?* 78. When is light said to be *doubly refracted?*

III. ELECTRICITY.

NATURE OF ELECTRICITY.—ELECTRICAL THEORIES.

79. Nature of Electricity.—As in the cases of heat and light, we know nothing of the real nature of electricity, all our knowledge on the subject being limited to its effects.

Like heat and light, it is imponderable; no accumulation of it in any substance adds to the weight of that substance, even when tried by the most delicate balances; but many of its effects are so like those of a mechanical agent, that it is usually considered a separate material substance.

When certain substances, such as amber, glass, sealing-wax, and sulphur, are rubbed with dry silk or cloth, they are found to have acquired a property, not observable in their ordinary state, of causing contiguous light bodies to move towards them; or, if the substances so rubbed be light and freely suspended, they will move towards contiguous bodies. After a while this curious phénomenon ceases; but it may be renewed an indefinite number of times by friction. This property was first noticed in amber; and therefore the principle thus developed was called *electricity* (from the Greek, *electron*, amber).

When a substance, by friction or other means, has acquired the property just stated, it is said to be *electrified*, or to be *electrically excited;* and its motion towards other bodies, or of other bodies towards it, is ascribed to a force called *electric attraction*. But its influence, on examination, will be found to be not merely attractive; on the contrary, light substances, after touching the electrified body, will be disposed to *recede* from it just as actively as they approached it before contact. This is termed *electric repulsion*.

80. Theories of Electricity.—In the absence of positive knowledge in regard to the nature of this agent, two *theories* have been proposed, to account for and connect together the established facts.

QUESTIONS.—79. Do we understand the real nature of electricity? Is it imponderable? What is the derivation of the term *electricity?* When is a substance said to be *excited* or *electrified?*

Dufay's theory (from the name of its proposer) supposes that every substance, in its *natural state*, contains in itself two highly subtile and elastic fluids, in such a state of combination that their presence is entirely disguised; but that the various phenomena of electrical excitement are produced by one or the other of them, accumulated in a body in excess. The particles of each fluid are supposed to have a strong attraction for those of the opposite kind, and for other matter, but are highly repulsive of each other.

These fluids are supposed to be separated by the various modes of producing electrical excitement, to be hereafter described; and one of them being collected in excess in a body, as just stated, produces the phenomena witnessed.

In most cases, when glass or any other vitreous substance is rubbed, the electricity which is collected is the reverse of that obtained when sealing-wax is subjected to friction; and hence the former is called *vitreous*, and the latter *resinous* electricity.

Franklin's theory of electricity supposes that all bodies, in their natural state, contain in their substance a certain quantity, called their *natural share*, of a single, subtile, elastic fluid, which produces no sensible effects; but that the phenomena of electrical excitement are produced when the body is made to contain either less or more than its natural share. It supposes that the particles of this fluid repel each other strongly, but are attracted by all other matter. When a body contains more than its natural share, it is said to be *positively* electrified; and *negatively* electrified, when it contains less.

Glass and other vitreous substances, when rubbed, are supposed to take more than their natural share of the fluid, or become *positive*; while resinous substances, in the same circumstances, lose a portion of their natural electricity, or become *negative*. These states are often indicated by the algebraic signs + and —.

The terms *vitreous* and *positive*, of the two theories, are there-synonymous, as are also the terms *resinous* and *negative*.

Either of these theories is found to answer well in explaining most of the phenomena of electricity, but that of Dufay is gene-

QUESTIONS. — 80. Describe Dufay's theory of electricity. Describe F[ra]nklin's theory. What terms were proposed by him? What terms of t[he] two theories are synonymous?

rally preferred; though the terms positive and negative, of Franklin's theory, are almost universally used.

From the above it will readily be seen, that when two bodies are either positively or negatively electrified, they repel each other, but attract each other when one is positive and the other negative.

81. **Electrometers.**—Electrometers are instruments for indicating the presence of electricity, or its intensity. A pith-ball, suspended by a dry silk thread from any convenient support, answers the purpose quite well; but the following, called the gold-leaf electrometer, is a more sensitive instrument. It consists of two slips of gold-leaf, suspended in a cylindrical glass vessel, from a metallic plate at the top. If the bottom is also made of metal, its sensitiveness will be increased. When an excited body is brought near the metallic plate, the leaves at once diverge, in consequence of their being brought into the same state, whether positive or negative, by the inductive influence of the excited body, in a manner to be hereafter explained.

Electrometer.

DISTRIBUTION OF ELECTRICITY.

82. **Conduction of Electricity.**—Some substances allow the electric fluids to pass over them freely, and are therefore called *conductors;* while others, that do not possess this property, or only imperfectly, are called *non-conductors.* If electricity be imparted to one end of a conductor, such as a copper wire, the other extremity of which touches the ground, or is held by a person standing on the ground, the electricity will pass along its whole length and escape in an instant, though the wire were several miles long; whereas excited glass and resin, which are non-conductors, may be freely handled without losing any electricity except at the parts actually touched.

QUESTIONS.—Which of these theories is now generally preferred? When do bodies attract and when repel each other? 81. What are *electrometers?* Describe the gold-leaf electrometer. 82. What is said of the conduction of electricity? What are *conductors* and *non-conductors?*

To the class of conductors belong the metals, charcoal, plumbago, water, and aqueous solutions, and substances generally which are moist, or contain water in its liquid state, such as animals and plants, and the surface of the earth. These, however, differ in their conducting power. Of the metals, silver and copper are found to be the best conductors; and after these follow gold, zinc, platinum, iron, tin, lead, antimony, and bismuth Aqueous solutions of acids and salts conduct much better than pure water.

To the list of non-conductors belong glass, resins, sulphur, diamond, dried wood, precious stones, earth, and most rocks when quite dry, silk, hair, and wool. Air and gases in general are non-conductors if dry, but act as conductors when saturated with moisture.

It is not, however, to be understood that any very definite line can be drawn between the two classes of conductors and non-conductors; but there seems to be a very regular gradation from the most perfect conductor to the most imperfect, or most perfect non-conductor. This division of substances is, however, found very convenient, though in some instances individuals might differ with regard to the class to which a particular substance should be assigned.

83. Insulation.—When a conductor is supported upon a non-conducting substance, it is said to be *insulated*, and electricity may be retained upon it for a time; but even then it will be gradually diffused and disappear. This is occasioned in part by the conducting power of the air, which is considerable, except when it is very dry. In damp weather, many electrical experiments cannot well be performed, because of the rapid diffusion of the fluid through the air, and the deposition of moisture upon the surfaces of insulators.

When two substances are rubbed together, both electricities are always developed, one of them going to one of the substances, and the other to the other substance; and both electricities may be retained, if the two substances rubbed together are insulated.

QUESTIONS.—What substances are classed with conductors, and what with non-conductors? Can any definite line be drawn between the conductors and non-conductors? 83. When is a body said to be insulated? Why do electrical experiments often fail in damp weather? Are both electricities always developed by friction?

84. Induction of Electricity.—An electrified body always exerts a peculiar influence on the natural electricity of other bodies in its vicinity, called *induction*, the nature of which will be seen from the following explanation: Let A be a positively excited glass tube, held near one end of an insulated conductor, B, supposed to be in its natural state; the natural electricity in B will instantly be disturbed, and, on examination, it will be found that the end next the excited glass is negatively electrified, and the other end positively, as shown by the algebraic signs. If, instead of the glass tube, some other substance, negatively electrified, had been used, the electricities of the two ends of the conductor B would have been reversed. In every case, the part of the conductor next to the excited body will be in the opposite state of excitement, while the other end will be in the same state as the excited body.

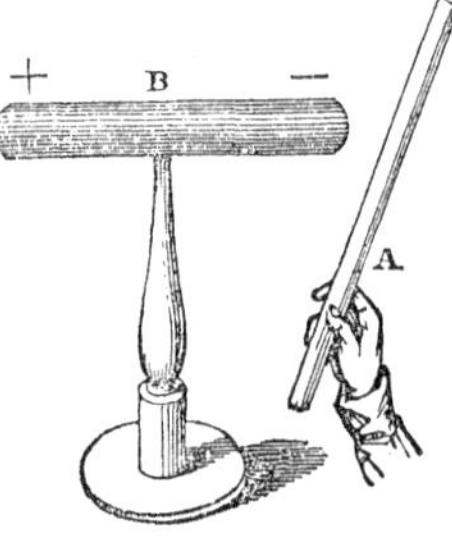

Induction of Electricity.

In the experiment nothing but air is supposed to be between the excited body A, and the conductor B, but the inductive influence is exerted through all non-conductors. Thus, if a clean and dry pane of glass be held between A and B, the result will be the same.

Let A and B, in the next figure, be two metallic discs, supported upon pillars of glass, their edges being towards the eye, and let a spark of positive electricity be communicated to one of them, as A;—it will immediately act by induction upon the natural electricity of B, causing the side next to A to be negative and the other to be positive. The effect is precisely the same as in the preceding experiment, but the form of the conductors different. If now we touch the

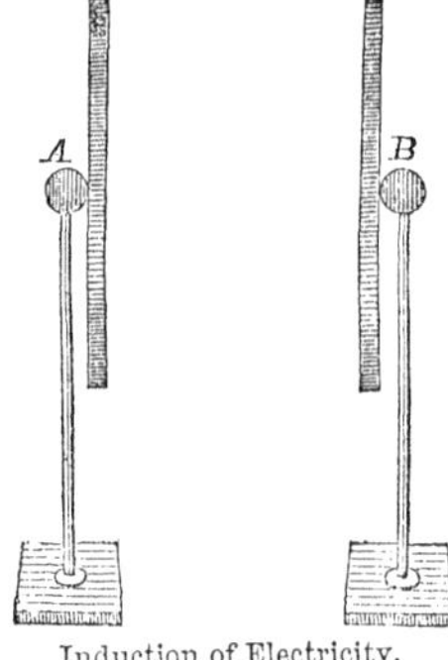

Induction of Electricity.

QUESTIONS.—84. What is meant by *induction?* Explain the experiment described in this paragraph. Explain the experiment described in connection with the next figure.

back of B with the finger, the positive fluid escapes, and the whole disc becomes negative. The action of the positive body, A, has taken place through the stratum of intervening air; and any other non-conductor may be substituted for it. If, for instance, a plate of glass be interposed, the two plates may then be brought much nearer together, and the same results will follow. Instead of the metallic discs, we may simply apply a metallic coating to the two sides of a pane of glass; which, if the coating do not reach within one or two inches of the edge, serves as a sufficient insulator.

85. The Leyden Jar.—The Leyden jar receives its name from the city of Leyden, in Holland, where it was invented. It is essentially the same thing as just described, except that a glass jar is substituted in the place of the pane. It consists of a glass jar, coated both inside and outside with tin-foil, except a part around the top, as shown in the figure. Through a varnished wooden cover, A, a wire, having a knob at top, is passed, and a chain, B, extends to the inside coating. Now, when either positive or negative electricity is communicated to the knob at the top, it is immediately diffused over the whole inside coating; and by its inductive influence, the outside coating takes on the opposite kind. When in this state,—the two coatings being oppositely electrified,—the jar is said to be *charged;* and a *discharge* takes place when a communication is established between the knob and the outside coating, the equilibrium being restored with a bright flash of light and a sharp report. As the human system is a good conductor, this discharge may take place through it, by grasping the outside coating with one hand, and touching the knob at the top with the other; or several persons may form a line by grasping hands, the one at one extreme touching the outside coating, while the one at the other extreme touches the knob. All will feel the shock, as it is called, at the same instant.

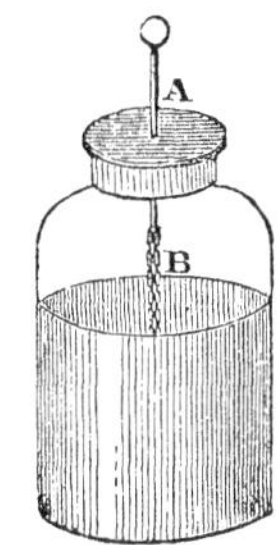

Leyden Jar.

While the jar is receiving the charge, it must not be insulated, that is, the outside must communicate with the earth. As the

QUESTIONS.—85. Why is the Leyden jar so called? Describe its construction and use. How is the shock produced in the system?

positive fluid collects on the inside, the outside becomes negative by the expulsion of the positive fluid naturally in it, and the accumulation of the negative fluid in its stead, drawn from the earth. But if the outside is insulated these transfers to and from it cannot take place, and therefore the jar cannot become charged.

86. Free Electricity resides in the Surface of Bodies.—It has been demonstrated that the electricity of an excited body resides entirely upon its surface. Let A be a sphere of metal, suspended by a silk thread, and excited by receiving a spark of electricity; and let B B be two covers of paper, gilt outside and inside, and held by glass handles. Let them now be applied to the excited globe, and then instantly removed; it will be found that the electricity has been entirely removed from the ball to the covers.

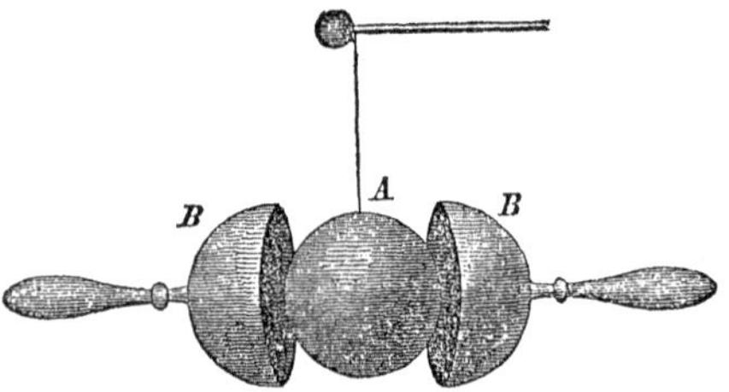

Resides upon the Surface.

The free electricity therefore was entirely accumulated upon the surface of the ball.

87. Distribution over Surface.—The fluid will be distributed over the surface of an excited conductor, in a mode dependent upon its form;—if it be a perfect sphere, the fluid will be distributed equally over every part, but if it be more or less elongated, as in the prolate spheroid, the fluid accumulates in the ends, where the intensity is greatly increased if the spheroid happens to be very considerably elongated.

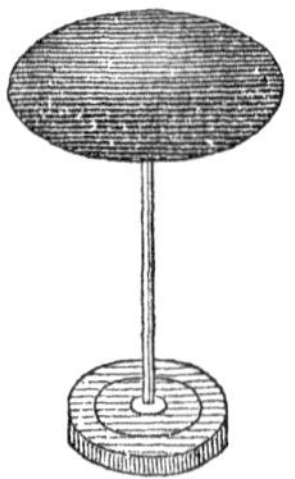

Distribution on Ellipsoid.

QUESTIONS.—While receiving the charge must the jar be insulated? Why? 86. In what part of an excited conductor does the electricity reside? 87. In what manner is electricity diffused over the surface of a sphere? How is it diffused over the prolate spheroid?

If the extremity of the conductor is carried out to a point, the fluid at once escapes from it, and all excitement disappears, even though it remains insulated. In the same manner, a sharp point projecting from a conductor receives the fluid silently upon it, and the body becomes excited. The effect of points in discharging or receiving either of the fluids is therefore apparent; and the circumstance must always be particularly regarded in the construction of electrical apparatus.

The escape of positive electricity from a point in a dark room is always attended by the appearance of a faint blue light in the form of a brush, as represented in A, but the escape of the negative fluid, in the same circumstances, presents the appearance of a star, as shown in B.

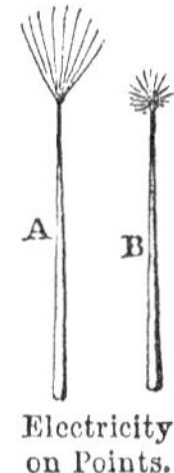

Electricity on Points.

It is to be noticed, that in such experiments the escape of either fluid is to be considered as precisely equivalent to the entrance of the other.

SOURCES OF ELECTRICITY.

88. As we have seen above, electricity is believed to be contained in all bodies, which are therefore properly its sources; but the earth, as being by far the largest mass to which we have access, is its chief source. We propose, however, under this head, to speak of the different modes of exciting or collecting it, which are *friction*, *change of temperature*, and *chemical action*.

89. Friction.—It is believed that electricity is always developed when one substance is rubbed against another, one of the fluids passing to one of the substances, and the other to the other substance, as before stated; but, in most cases, neither of the fluids is retained, because the rubbing substances are not insulated. If

---

QUESTIONS.—What is the effect if one part of an insulated conductor is extended out to a point? What is said of the influence of points in receiving the fluid? 88. What is the great source of electricity? What are the different modes of exciting electricity? 89. Is electricity always developed when one substance is rubbed against another? How may both be collected and retained?

both be insulated, both the positive and negative fluids may be retained (83).

90. *The electrical machine* is an instrument for developing electricity by friction more abundantly than it can be done by the simple means heretofore pointed out, though most of the great principles of the science, as we have seen, may be demonstrated without it. The figure in the margin represents the cylinder machine in its usual form. A is a cylinder of glass, firmly supported, and capable of being turned on its axis by a handle; and R is a conductor, supported on a pillar, having the rubber attached to it, with a flap of silk, S, extending nearly over the cylinder. C is made of sheet-brass, and is called the prime conductor, because it receives the electricity from the cylinder as it is turned, by means of several pointed wires (87), extending inwards towards the cylinder. It is supported upon a pillar of glass.

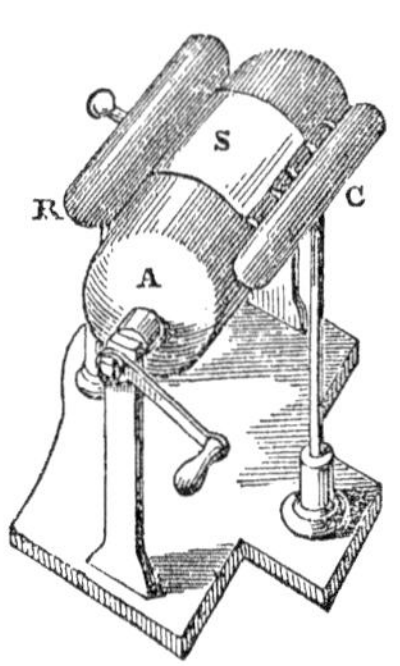

Electrical Machine.

Now, when the cylinder is turned, electricity is abundantly developed by the friction of the rubber against its surface, and is received by the prime conductor, in which it accumulates. The use of the flap of silk, S, is to prevent the fluid from escaping in the air, as the cylinder is turned.

From the principles heretofore discussed, the learner will readily perceive that it is the positive electricity that will be accumulated in the prime conductor; but the negative (83) will also at the same time accumulate in the rubber, if it be insulated. But no considerable quantity of electricity can usually be collected, unless the rubber communicates with the earth, or, which is the same thing, with the floor of the room.

An elegant plate electrical machine is represented in the next figure (p. 83). A B is a firm base of wood, mounted on castors, so as to allow the machine to be moved around easily upon the floor; C C C the prime conductor, supported upon pillars of glass; P P two circular plates of

---

QUESTIONS.—90. Describe the *electrical machine*. Describe the large plate machine.

glass upon the same axis, and turned by the handle H; R R R R the rubbers, of which there are eight, and F F flaps of silk to prevent the

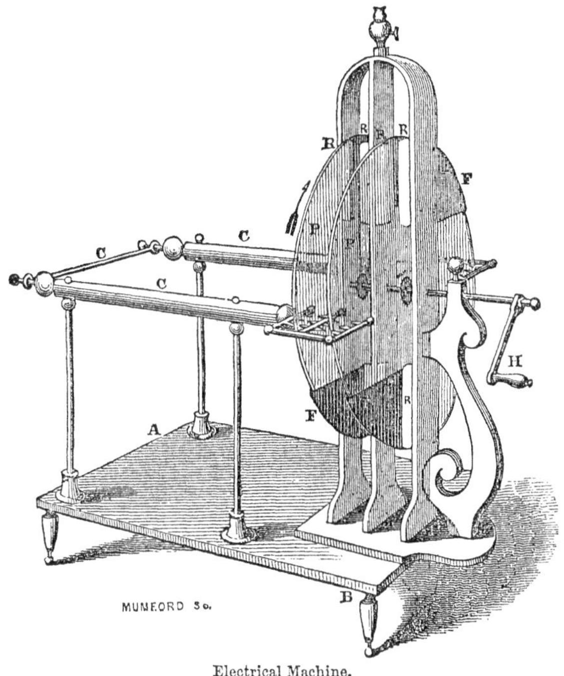

Electrical Machine.

escape of the fluid before reaching the point from the prime conductor. This machine, when put in proper order, developes electricity rapidly, and is decidedly preferable to the cylinder machine.

When used, the machine should be dry and warm, and perfectly clean and free from dust. Its action is also greatly increased by spreading the surface of the rubber, where it presses against the cylinder, with a soft amalgam of zinc, tin, and mercury, or with the yellow sulphide of tin, called *aurum musivum*, the latter, on the whole, being preferable.

QUESTION.—What is the substance spread over the rubber?

To prepare the amalgam, melt in a crucible three parts of zinc and one of tin, and, after removing it from the fire, add four or five parts of mercury. Stir the mass with a stick a few seconds, and pour it out upon a clear marble slab, or plate of metal, and allow it to remain several hours before breaking it up. When wanted for use, grind it as fine as possible in a mortar, and mix it with sufficient tallow to cause it to adhere well to the rubber. If the aurum musivum is used, it is to be mixed with tallow and spread upon the rubber in the same manner.

When the machine operates properly, if the knuckle be presented near the prime conductor, a vivid spark passes between them, and a slight stinging sensation is felt; the same thing also takes place on presenting the knuckle to the rubber, provided it be insulated. In the first case the effect is produced by accumulated positive electricity; in the second, by the negative.

91. *The Electrophorus* (from *electron* and *phero*, I bear) is an instrument for readily obtaining small quantities of electricity. It consists of a plate of resin, A, about 12 inches in diameter, contained in a shallow dish of metal, and a metallic disc, D, a little smaller than the plate of resin, provided with a glass handle, for removing it from the resin at pleasure. To operate well, the surface of the resin should be perfectly smooth.

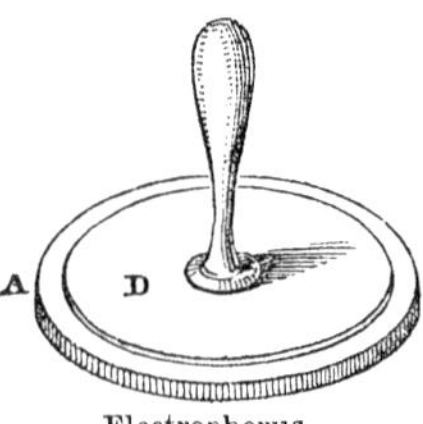

Electrophorus.

To charge the electrophorus, the disc is removed, and the surface of the resin rubbed briskly with a piece of warm, dry flannel, or struck several times with a dry silk handkerchief, folded up for the purpose, by which a negative electricity is excited. If, now, the disc of metal be restored by means of its insulating handle, its lower surface will become positive by induction (84), and its upper surface negative. By touching the upper surface of the disc, when in this position, with the finger, the negative electricity will be discharged; and if it be then removed carefully by its handle, it will be found highly charged with positive electricity, so that a considerable spark may be obtained from it. As the cake of resin has lost nothing of its electricity by the operation, the process may be repeated any number of times, with the same result.

QUESTIONS.—How may a spark be obtained from the machine? 91. Describe the *electrophorus*.

The Hydro-Electric Machine is an instrument for exciting electricity by means of high-pressure steam. The excitement is attributed to the friction of the steam, carrying with it drops of water, against the pipes from which it issues.

92. Atmospheric Electricity.—The general phenomena of thunder and lightning are well known. They are occasioned, as Franklin first demonstrated, about a century ago, by immense accumulations of electricity in the clouds, between which and objects upon the earth violent discharges are frequently taking place. The discharge is believed to differ in nothing from the discharge of a spark from the conductor of an electrical machine, except what necessarily results from the quantity and intensity of the fluid accumulated.

Lightning-rods, which are so common at the present day, are rods of metal erected upon buildings, extending a distance above them at the top, and at the bottom connecting with the moist earth. Being made of metal, which is a good conducting material, any discharge that may happen upon the building will be conveyed by them without danger to the ground.

Often several of them are attached to the same building, at different points, but they should always be connected together, and also make two connections, at least, with the moist earth. Any attempts to insulate the rod from the building are, to say the least, useless. Buildings with tin or copper roofs, and metallic water-conductors extending downward, require special electrical conductors only for the chimneys or other projecting parts, and also from the water-conductors to the moist earth below.

Electricity excited by friction is frequently called *statical electricity*, to distinguish it from *dynamic electricity*, which will be hereafter described, under the head of Galvanism.

93. Thermo-Electricity.—If a crystal of tourmaline, the extremities of which are dissimilar, is slightly heated in the flame of a spirit-lamp, one end will be found, on examination by a delicate electrometer, to be positive, and the other negative;

QUESTIONS.—What is the hydro-electric machine? 92. What is said of lightning and thunder? Who first explained their cause? What are *lightning-rods?* What is said of buildings with metallic roofs? 93. How may crystals sometimes be electrically excited?

but the excitement is very feeble. Crystals of some other substances may be excited in the same manner.

But the more common method of exciting electricity by change of temperature, is to heat slightly the ends of two or more small rods of different metals at their junction, as represented in the figure. Let A be a small rod of antimony, and B another of bismuth, soldered together at one end; and then let the heat of a spirit-lamp be applied, for a moment, at the point where they are soldered; while the bars are warming, the bismuth will be negative, and the antimony positive. The bismuth is called the positive, and the antimony the negative metal, because, while heating, the positive fluid appears to originate in the former, and flow to the latter; but an electrometer will show the different states of the metals, as above indicated.

Thermo-Electricity.

Other metals—and even non-metallic substances—may be used, with similar results. German silver and brass answer very well, the former corresponding in its action with the antimony, and the latter with the bismuth.

The effect will be considerably increased, if several pairs of the metals, arranged as above, are associated together, as shown in the accompanying figure, the alternate rods, A, being of German silver, and the intermediate ones, B, of brass. When the metals are gently heated at the points of junction at one extremity of the bars, and kept cool at the other, the terminal bars become excited, and a constant current flows over any conducting substance, as a copper wire, connecting the extremities of the series. If the upper ends of the rods be heated, the direction of the current over the wire will be as shown by the arrow. If the lower ends be heated, or the upper ends cooled, its direction will be reversed.

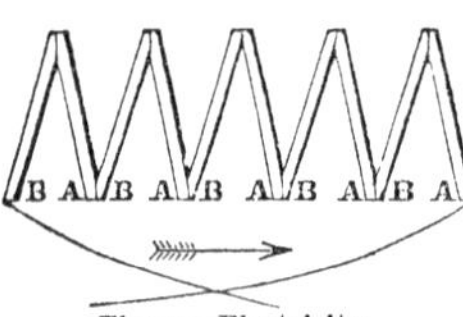

Thermo-Electricity.

QUESTIONS.—Describe the mode with two metals. How may the effect be increased?

To render the instrument more compact, the metallic bars may be placed side by side, with only a slip of silk between them, the ends being bent a little, so as to admit of being soldered as before. Such an instrument constitutes the *thermo-electric pile*, which is figured in the margin, P and N being the positive and negative poles.

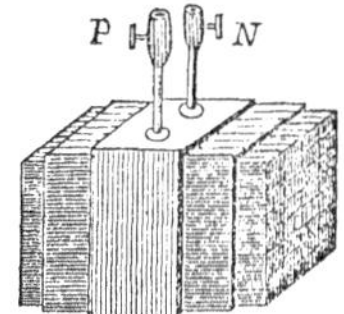

Thermo-Electric Pile.

The existence and direction of these currents are best shown by a delicate *galvanometer*, an instrument to be hereafter described.

By reversing the experiment, and passing a current of electricity through the series of bars, they will be heated or cooled, according to the direction in which the current is made to pass.

94. Chemical Action.—Chemical action, as the solution of a metal in an acid, and the combustion of charcoal in an ordinary fire, it is believed, is always attended by the development of electricity. In the combustion of charcoal, the gas arising from the coal is positive, while the coal itself, if insulated, is negative; and when a metal is dissolved in an acid, a current of positive electricity always passes from the metal to the liquid, and any conducting substance, as a plate of copper, contained in it.

Let Z be a zinc plate immersed in water, acidulated with a little sulphuric acid, contained in a glass vessel, and C a plate of copper, also immersed in the same liquid; the zinc will be gradually corroded, and a current of positive electricity pass from it through the liquid to the copper; and if the plates are connected by a wire, the current will pass over it in the direction indicated by the arrows.

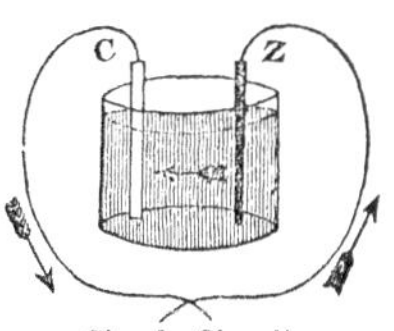

Simple Circuit.

But, although we have, in these and other cases of chemical action, such decided and even powerful developments of electricity, it is admitted, that in very many cases where chemical action really takes place, no indications of electricity have as yet been observed. The action of one salt upon another, of one metal upon another, or

---

QUESTIONS.—Describe the *thermo-electric pile*. 94. Is chemical action always attended by the development of electricity? Describe the simple galvanic circle. Are indications of electrical excitement always observed during chemical action?

of a simple element, as oxygen or sulphur, upon a metal, may be mentioned, as instances in which no electrical excitement is actually known to take place.

The electricity of chemical action, sometimes called *dynamical electricity* (92), properly constitutes the subdivision of the general subject of electricity called GALVANISM; and under this title it will be discussed more at length, as it is this branch of the general subject which more especially concerns the student of chemistry.

GALVANISM.

95. The science of Galvanism owes its name and origin to the experiments on animal irritability made by Galvani, professor of anatomy at Bologna, Italy, in the year 1790. In the course of some of his investigations, he discovered the fact that muscular contractions are excited in the leg of a frog recently killed, when two metals, such as zinc and silver, one of which touches the crural nerve, and the other the muscles to which it is distributed, are brought into contact with one another.

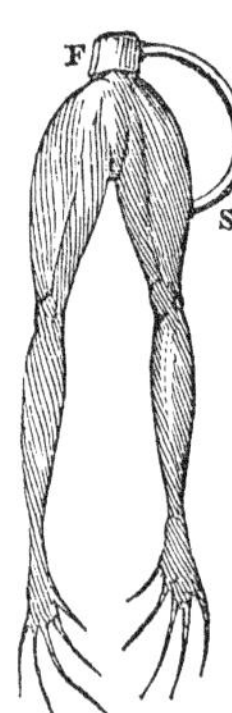

Experiment with Frog.

The experiment with the legs of a recently killed frog is easily repeated, in the following manner:— After killing the frog, immediately separate the hind-legs, with a small portion of the spine, and remove the skin; then bind around the part of the spine removed with the legs a piece of tin-foil, F, and, holding it up with the left hand, apply a piece of silver coin, or a rod of silver, S, bent, if necessary, so that it shall touch the tin-foil and the flesh of one of the legs at the same time. At each contact of the metals, the muscles of the leg will be violently contracted, and jerking of the legs produced. The experiment succeeds best when the whole is kept wet with clean water. The irrita-

QUESTIONS.—By what name is the electricity of chemical action generally known? 95. What was the discovery of Galvani? Describe the experiment with the legs of the frog.

bility of the muscles will gradually subside, but sometimes it will continue more than an hour after the death of the animal.

The large legs of some insects, especially those of the grasshopper, may be used for the same purpose. It is necessary only to remove with a sharp penknife a portion of the skin from each side of the thick part of one of the leaping legs, so as to expose the flesh; then by laying the under side of the leg upon a small piece of moistened zinc, Z, and bringing a piece of copper, C, in contact with the flesh exposed on the upper side, no motions will be observed until the copper also touches the zinc, when quick movements or jerks of the lower part of the leg, A B, will be seen, each time the contact is made.

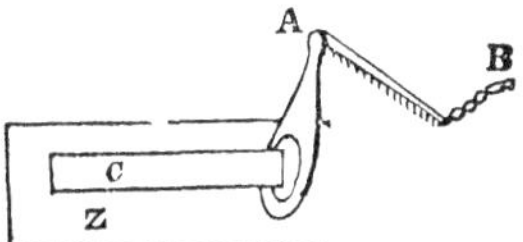

Experiment with Grasshopper.

96. Simple Galvanic Circles.—A simple galvanic circle is formed of three substances, two of which are usually metals, and the third a liquid, as a dilute acid. The arrangement described in paragraph 94 constitutes such a circle. The zinc is acted upon by the acid, and the electrical disturbance takes place over all that part of its surface covered with the liquid; and a current of positive electricity flows to the liquid. If, now, a plate of copper, or other metal not capable of being acted upon by the liquid, be introduced, it will become positive by receiving electricity from the liquid; and, by connecting the two plates by wires, a constant current is established over these wires, as shown by the arrows. It matters not, so far as galvanic action is concerned, at what part the plates touch each other. A current is formed, whether contact between the plates is made below, where covered with liquid, above, where uncovered, or along the whole length of the plates, provided both plates are immersed in the same vessel of diluted acid. But in every case a circuit must be formed, around which the electricity may traverse, either in a single current, or in many partial currents, into which it ma divide itself, as will be the case when the metals are in contact along their whole surfaces. This last result it is desirable to avoid; and therefore the metals are always to be kept separate below the liquid, and above it also, except at the part where the

---

QUESTIONS.—Describe the experiment with the leg of a grasshopper. 96. What constitutes a *simple galvanic circle?* When is the electricity excited? What is the use of the plate of copper? In what direction does the positive current flow? Must the circuit always be complete?

current is desired to pass. Usually, a wire is connected with each plate, which may be brought in contact or separated at pleasure. When they are in contact, the circuit is said to be *closed;* when they are separated, it is said to be *broken*, or *open*.

97. As the electricity is developed entirely by the chemical action between the zinc plate and the acid, it is only upon the surface of the zinc covered by the acid that the electric disturbance takes place; and, other things being equal, the quantity of electricity set in motion will be proportional to the extent of zinc surface thus exposed to the acid. And in every case, in order to establish the current, the circuit must be made complete. Thus, if we take two cups of dilute acid, immersing in one a copper, C, and in the other a zinc, Z, plate, and connect the two by a wire, as shown in the figure, though some chemical action will take place, no current will be established, for the reason that no circuit has been formed. Chemical action does indeed take place in the cup containing the zinc plate, and its electricity no doubt is disturbed; but, still no current can be established. For this it is necessary, further, to add the conducting wire A B, as shown in the next figure; the direction of the current will then be as shown by the arrows.

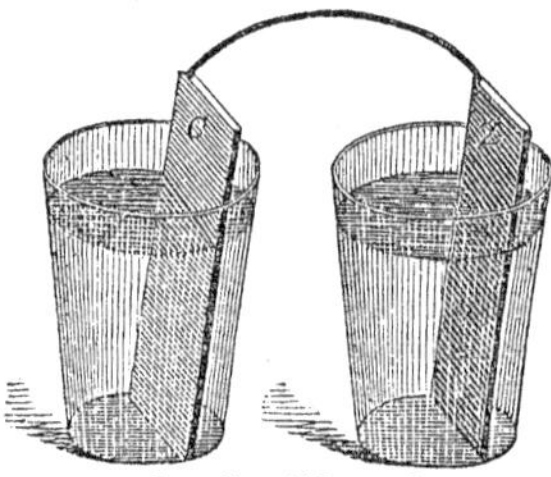

Circuit not Formed.

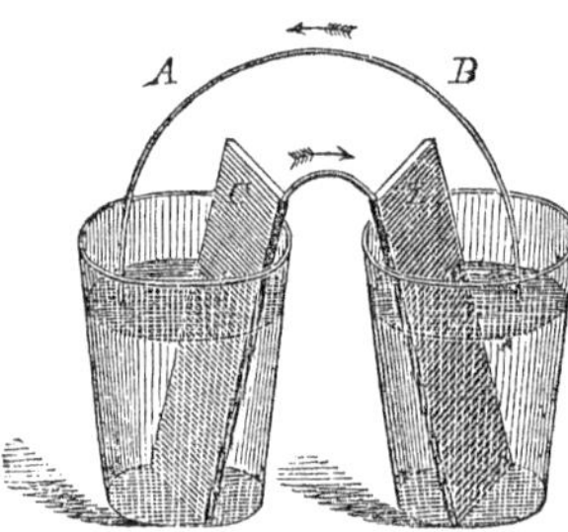

Circuit Complete.

It is to be understood that here, as elsewhere, in using this language, we have reference to the positive fluid; but in reality there is just as much reason to believe there is also at the same time a

QUESTIONS.—When is the circuit said to be closed? When open? 97. To what will the quantity of fluid put in motion be proportional? Explain the necessity of completing the circuit, as illustrated by the figures in this paragraph.

negative current established in the opposite direction. In every case, the direction of the positive current will be *from the metal acted upon to the liquid*, and that of the negative, of course, the reverse. The direction of the positive current, therefore, in the apparatus last figured, is from the zinc cell to the copper cell, over the wire connecting the cells, and from the copper to the zinc over the wire connecting the plates, as shown in both cases by the arrows. To break the circuit it matters not which of these wires is interrupted; both are equally necessary to complete the circuit.

98. A simple galvanic circle may be formed of one metal and two liquids, provided the liquids are such that a stronger chemical action is induced on one side than on the other. Nay, even a plate of metal, with two portions of the same liquid, may be made to constitute the simple circuit, provided only the conditions are such that one side of the metal shall be acted upon by the liquid more readily than the other. This will be effected, if one portion of the liquid is warmer or stronger than the other, or if one surface of the metal is rough and the other polished.

We have above represented the positive current as passing from the zinc, through the liquid, to the copper, and in the opposite direction over the wires connecting the plates above the liquid; this will always take place when a diluted acid is used, which attacks the zinc more violently than it does the copper; but if a solution of ammonia be substituted for the acid, the copper will be most acted upon, and the current will move in the opposite direction.

It is not necessary that copper and zinc alone should always be used in these experiments; other metals may be adopted, with equal, and, in some cases, with even more decisive results. Nor is it required that the liquid should always contain an acid; other substances, as solutions of the salts, are often very efficacious in exciting this subtile fluid. The conditions required are, that the metals and liquid used should be such that chemical action will take place more readily between one of them and the liquid than between the other and the liquid; and that metal is always found positive (below the surface of the liquid) which is most acted upon by it. Other things being equal, the galvanic action will be more intense, the greater the difference between the the two metals

---

QUESTIONS.—Is there a current of the negative fluid? What is its direction as compared with that of the positive? 98. How may a simple circle be formed of a single metal and two liquids? Will the positive current always pass from the zinc to the copper? Why is the direction reversed when aqua ammonia is used? May other metals besides copper and zinc be used? What are the conditions required?

used, as regards their tendency to be acted on by the particular menstruum in which they are immersed.

Besides the above arrangement, there are several modifications of the simple galvanic circle, each possessing its own peculiar advantages, which will be described hereafter.

99. Compound Galvanic Circles.—Galvanic Batteries.—The compound galvanic circle, or galvanic battery, consists of a number of simple circles, so arranged in a series, that the copper of each simple circle is connected with the zinc of the one adjacent. One extreme of the series, it will be evident, will be copper, and the other zinc; they are often called the *poles* of the battery, the former being positive and the latter negative.

The *voltaic pile* (from the name of its inventor) deserves here to be noticed, as the earliest and simplest instrument of this kind, though by no means the most efficient. It is formed of pieces of copper, *c*, zinc, *z*, and cloth, the latter being moistened with a solution of salt or acidulated water. Commencing with either of the metals, upon this is placed a piece of cloth, and then a piece of the other metal; the three, of course, constituting a simple galvanic circle. Upon this circle other simple circles are then formed in the same manner, care being taken to place the metals throughout the series in the same order.

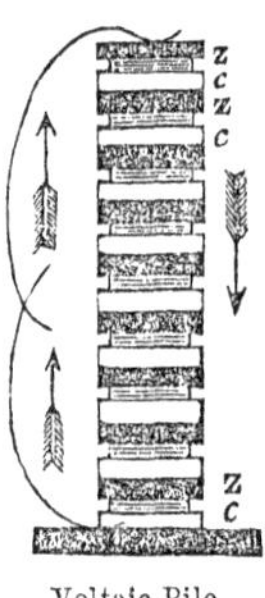

Voltaic Pile.

The series may be extended indefinitely; but, usually, from fifty to one hundred pairs of metallic plates will be found as many as can be employed advantageously. When in action, the extreme zinc plate, which is represented as uppermost in the figure, will be negative, and the extreme copper plate positive; and if they be connected by wires, the current will flow in the direction of the arrows, both through the series, and over the wires. The extremes of the series are called its *poles*, or *electrodes* (from *electron*, and *odos*, a way).

QUESTIONS.—99. What constitutes the *galvanic battery*, or *compound circle?* What constitutes the *poles* of the arrangement? Describe the *voltaic pile*. How many pairs of plates are needed? What are the poles, or *electrodes?*

The voltaic pile is now rarely employed, because we possess other modes of forming galvanic combinations which are far more powerful and convenient. Cruickshank's battery, one of the earliest invented, consists of a trough of baked wood, about thirty inches long, in which are placed, at equal distances, fifty pairs of zinc and copper plates, previously soldered together, and so arranged that the same metal shall always be on the same side. Each pair is fixed in a groove cut in the sides and bottom of the box, the points of junction being made water-tight by cement. The apparatus thus constructed is always ready for use, and is brought into action by filling the cells left between the pairs of plates with some convenient solution, which serves the same purpose as the moistened cloth in the pile of Volta.

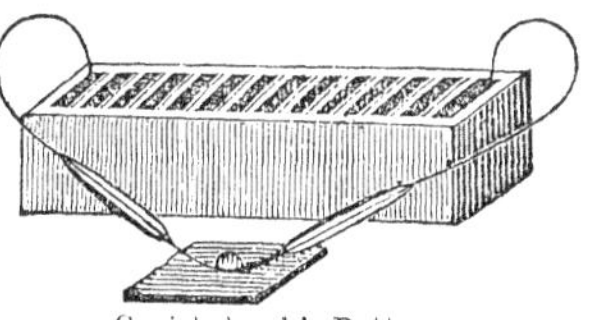

Cruickshank's Battery.

An excellent compound circle, or battery, is formed by combining a number of cups like that represented in paragraph 94. Each cup contains a zinc, Z, and copper plate, C, the zinc of one cup being connected with the copper of the next, through the whole series, leaving the two extreme plates free; and the cups are filled with diluted acid, or a solution of salt. The two free plates, constituting the extremities of the series, will be the poles, or electrodes; and when they are connected by wires, the current will be established in the direction of the arrows.

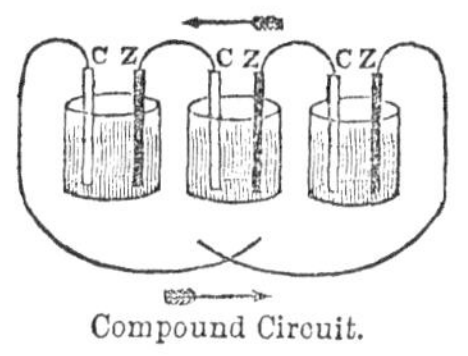

Compound Circuit.

By studying closely this arrangement, it will be seen that the actual quantity of electricity flowing over the wires connecting the electrodes, is no more than when a single cup only, with a single pair of plates, is used. The same electrical disturbance takes place in each cup, over the whole surface of the zinc plate which is covered with the liquid, the part of this plate above the liquid

---

QUESTIONS.—Describe the trough battery. What is said of the quantity of electricity flowing in the compound circle, or battery? How does this appear?

becoming negative, and the liquid becoming equally positive; the copper plate then serves as a conductor to take up this positive electricity, and convey it to the negative zinc of the next cup, by which it will be exactly neutralized. However extensive the series may be, this will take place with every alternate zinc and copper plate except the extreme ones; so that the quantity passing over the wires connecting the electrodes, will be only that of the single pair of plates constituting the extremes of the series.

100. But although the *quantity* of electricity developed at the electrodes of the compound circle is no more than we should obtain by using a single pair of plates, yet it will be found to have acquired a very important property, called *intensity*. By this term is meant its power to overcome resistances which may impede the passage of the current. The current from a single pair of plates, however large they may be, will always be exceedingly feeble, and will not flow unless the wires connected with them are in actual contact; but, if the polar wires of the compound circle are once brought in contact, they may be separated at a little distance, and the current will continue to pass between them, with a brilliant flame. The reason is, because the current from the compound circle possesses sufficient *intensity* to overcome the resistance of the thin stratum of air between the wires, which is not the case with the current of the simple circle.

So, if the polar wires of the simple circle are grasped, one in each hand, by the operator, not the least sensation is felt, because there is not sufficient intensity to impel the current through the system, which is comparatively a poor conductor; but if the same is done with the polar wires of the compound circle, especially if the series be extensive, the moment the circuit is formed, a powerful shock is experienced, similar to that received from the Leyden jar. This is owing to the increased intensity of the current from the compound circle, enabling it to overcome the resistance which is interposed.

QUESTIONS.—100. What benefit then is derived from increasing the series? What is meant by intensity? Why will not any sensation be produced in the system by a single cell?

101. From the above facts we deduce this important practical principle: that, to produce a current of *quantity*, a single pair only of large plates is wanted; but to give *intensity*, a number of simple circles must be combined, in the manner described.

Compared with frictional or statical electricity, that of the most powerful galvanic battery has always only a feeble intensity, but the quantity is often immense.

The energy of any battery, whether composed of one or many simple circles, will depend very much upon the nature of the liquid used. A solution of common salt, sulphate of soda, nitrate of potassa, alum, or other salt, will answer the purpose, but acids are better. Generally, one of the stronger acids is used, diluted with 15 or 20 times its weight of water. For ordinary purposes, a mixture of equal parts of nitric and sulphuric acids, diluted with 20 times their weight of water, will be found to answer well.

102. Nature of the Chemical Action in the Battery.—The development of electricity by the usual galvanic arrangements is attributed chiefly, if not entirely, to the decomposition of water in contact with the zinc plates. Water is a compound of oxygen and hydrogen, and is incapable of acting upon zinc, unless some acid be also present. But when a piece of this metal, in its usual impure state, is immersed in diluted acid, the water is decomposed, its oxygen combining with the zinc, forming oxide of zinc, while the hydrogen rises in bubbles and escapes in the air, and at the same time the oxide of zinc is taken up by the acid. If, now, a plate of copper is placed in the diluted acid at a little distance from the zinc, and the two connected by a wire, constituting a simple galvanic circle (97), the bubbles of hydrogen will not rise around the zinc as before, but around the copper; showing that the gas has in some way been transmitted through the liquid between the plates, though not the least appearance of any motion can be observed by the eye. If the connecting wire is removed, the evolution of hydrogen at the copper plate at once ceases with the cessation of the electrical current, but continues to rise slowly

---

QUESTIONS.—101. What is said of the intensity of the excitement produced by the galvanic battery, as compared with that of statical electricity? What acids are recommended for use in the battery? 102. What is said of the chemical action that takes place when a piece of zinc is immersed in a dilute acid? What is the occasion of the bubbles upon the zinc? When a copper plate is connected with the zinc where do the bubbles of hydrogen make their appearance?

around the zinc. This latter effect, however, is occasioned by the impurity of the zinc. Thus it would seem that the development of the current is occasioned entirely by the decomposition of the water, and the formation of oxide of zinc.

When the battery is in active operation, though the hydrogen is rapidly evolved at the copper plates, yet the whole surfaces of these plates will be all the time covered with a film of this gas, which interferes with its conducting power, and prevents the free passage of the current; at the same time, as a matter of course, diminishing very considerably the energy of the battery. To remedy this difficulty, several new forms of the battery have recently been invented, which act with great energy, and will here be briefly described.

103. Daniel's Constant Battery.—This excellent instrument consists of a cylindrical vessel of copper, C, in which is placed another smaller one, L, made of unoiled leather, or unglazed porcelain, through which water will gradually percolate; and in the latter is contained a rod of zinc, Z, about an inch in diameter. To charge it, the inner porous vessel, which contains the rod of zinc, is filled with diluted sulphuric acid (acid 1 part, and water 8 parts), and the space around the inner vessel with a saturated solution of blue vitriol, acidulated with sulphuric acid. To the side of the copper vessel, and also to the zinc rod, wires are soldered, with binding screws for holding the polar wires; the one connected with the copper being positive, and the other negative, as shown by the algebraic signs.

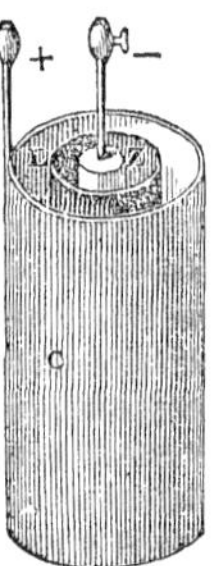

Daniel's Cell.

Usually, the zinc rod is amalgamated with mercury, which is done by rubbing the surface with mercury, while covered with some weak acid. This renders the chemical action over the whole surface more uniform, and the action of the battery more constant; but the same thing may be accomplished with less trouble, by using the zinc in its ordinary state, and substituting, in the porous

QUESTIONS.—103. Describe *Daniel's battery.* How is the zinc amalgamated?

cell, a saturated solution of sulphate of soda (Glauber's salt), instead of the diluted acid.

The above arrangement, it is plain, constitutes a simple galvanic circle; and its action is particularly energetic and constant, in consequence of the accumulation of hydrogen gas upon the copper plate (102) being completely avoided. This will appear from the following explanation.

In this instrument, as in the more common one first described, the electricity is developed at the surface of the zinc, by the decomposition of the water; the oxygen combining with the zinc, and the hydrogen passing through the porous vessel into the vitriol solution, and thence to the sides of the copper vessel, which constitutes the copper plate of the series (98). Here the hydrogen does not make its appearance in bubbles upon the surface of the copper, as in the common arrangement, but enters into a new combination with the oxygen of the oxide of copper deposited from the vitriol solution. Blue vitriol is a compound of sulphuric acid and oxide of copper; and while the battery is in operation, it is all the time decomposing, its acid passing into the porous cup, to act upon the zinc, while the oxide of copper is itself also decomposed, its oxygen combining with the hydrogen at the surface of the outer vessel, which receives a new coating of metallic copper.

Any number of these arrangements may be united, by connecting the zinc of one with the copper of the next, as heretofore described (99); and a battery so constructed has this great advantage, that no action takes place when the circuit is not closed. It is also very constant in its action, affording a uniform current for several hours in succession.

The figure in the margin represents a battery of this kind, of six series.

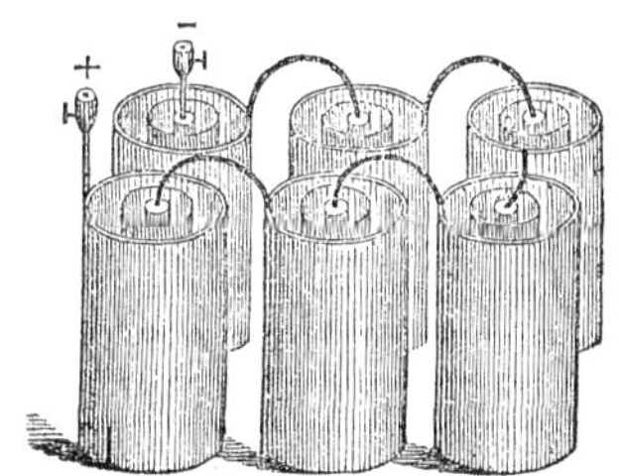

Daniel's Battery.

When a battery of this kind is to be used some time, a quantity of undissolved blue vitriol should be kept in the upper part of the copper cell, either upon a shelf provided for the purpose, or in a muslin bag, to keep the solution constantly saturated

---

QUESTIONS.—How is the hydrogen disposed of in this arrangement? Why is it called a *constant* battery? How are the separate cells to be united to form the compound arrangement?

104. Grove's Cell.—This battery, invented by Professor Grove, of London, is remarkable, not only for the constancy of its action, but also for its great intensity.

The construction of this battery is shown in the accompanying figure. A glass or porcelain cup, of a proper size, is used, and in it is placed a hollow cylinder of zinc, Z, (usually having a slit in one side, to allow a free passage to the liquid), and inside of this, a small cylindrical cup, C, of porous or unglazed porcelain. The glass cup is then filled with diluted sulphuric acid (of the same strength as is used in Daniel's battery), and the porous cup with strong aquafortis (nitric acid). Lastly, a thin slip of platinum, P, is suspended in the aquafortis, and supported by a piece of wood attached to the side of the glass cup, through which a wire passes, and is bent in the form represented in the figure. A projection upward from the zinc supports a binding screw, shown on the right, and another is soldered to the wire to which the platinum is attached, shown at the left. These, of course, serve as the positive and negative poles of the circle; and to them the polar wires may be attached, when required.

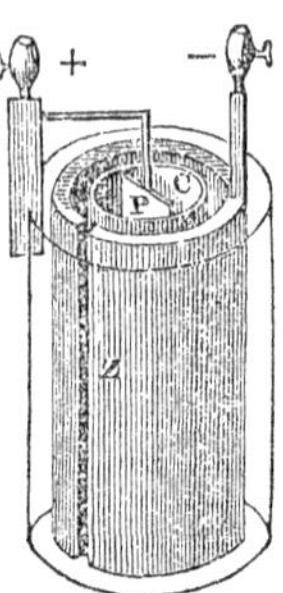

Grove's Cell.

The zinc should be well amalgamated, as heretofore described.

The action of this battery is essentially the same as that of Daniel's, except that the hydrogen from the decomposition of the water, passing into the porous cell, is expended in decomposing the nitric acid, which, when the instrument is in action, gives off copious nitrous fumes. To understand this, it is necessary to recollect that nitric acid is a compound of nitrogen and oxygen; and the hydrogen entering the porous cell takes away a part of its oxygen, forming water; and the binoxide of nitrogen which is liberated rises, and uniting with more oxygen from the air, forms the nitrous fumes.

A battery of twelve series of Grove's arrangement is represented in the following figure. The cups are contained in a box of wood, for the sake of convenience in handling, and between the cups are partitions to

---

QUESTIONS.—104. Describe *Grove's cell.* What in this case becomes of the hydrogen?

hold them more steadily. The hollow cylinders of zinc, when made for this purpose, are cast with arms rising from one side, and then project-

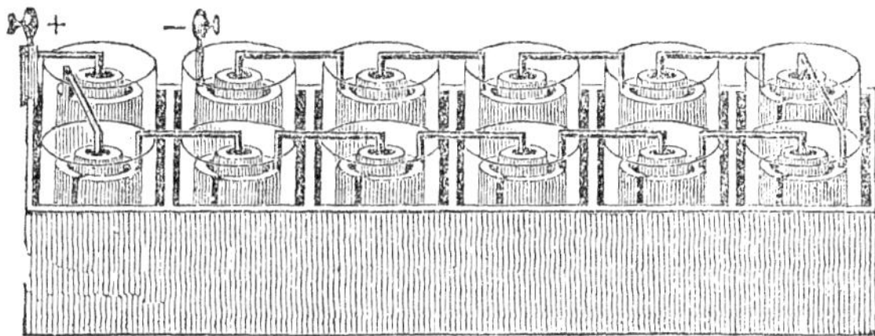

Grove's Battery.

ing a distance horizontally; and to the end the platinum plate is firmly soldered, so as to hang directly in the nitric acid cup of the next adjacent series. The whole, it will be observed, forms a connected series; and the terminal zinc at one extremity, and the platinum at the other, with each of which a binding screw is connected, constitute the negative and positive poles. The terminal platinum plate is supported in its place by a wire passing through a piece of wood attached to the side of the box or the cup, as represented in the figure. When a greater power is required than is afforded by a battery of this size, it is found more convenient to connect two or more batteries together than to have a larger number united in one series. The batteries are connected by extending a wire from the positive pole of one to the negative of the next; they then in reality become a single series.

This is the kind of battery generally used in working the magnetic telegraph, as it excels all others in the constancy and intensity of its action.

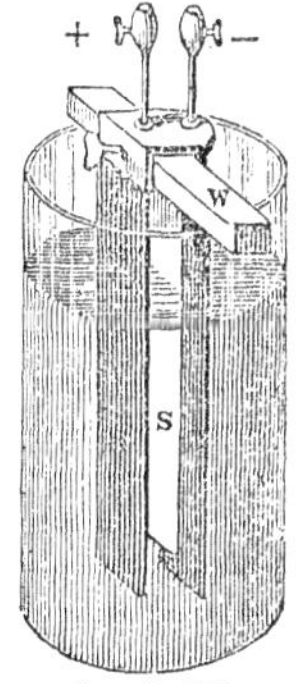

Smee's Cell.

105. Smee's Cell.—This very efficient arrangement is represented in the figure in the margin. Two plates of zinc, well amalgamated, are firmly held together against a piece of wood, W, by means of a brass clamp and screw; and between them is a plate of silver, S, the surface of which has been coated over with metallic platinum, in a state of fine powder, called platinum black. The zinc and silver plates are then suspended in a glass vessel, the piece of wood resting upon the top. A binding screw, constituting the + pole, is connected by a wire

---

QUESTIONS.—What is said of the character of batteries of this kind? 105. Describe *Smee's arrangement.*

(which is insulated from the brass clamp through which it passes) with the silver plate, and another, constituting the — pole, is soldered to the clamp which holds the zinc plates in their places. The liquid used in the glass cup is sulphuric acid, diluted with 10 or 15 times its weight of water.

The peculiar advantage possessed by this form of the battery depends upon the rapid evolution of hydrogen from the platinized silver plate. We have seen above (102) that the accumulation of this gas upon the smooth surface of the copper plate, in the old arrangement, considerably retards the passage of the electric current, by preventing the liquid in a measure from coming in contact with the plate; but this is avoided, in Smee's arrangement, by the roughened surface of the silver plate, produced by the platinum deposit.

Compound circles or batteries are readily formed by combining the simple series of Daniel with each other, or those of Smee with each other; but the proper mode of doing it will be easily understood without further illustrations. For many purposes, they answer well; but neither of them possesses the energy of that of Grove above described.

As batteries are sometimes constructed, and as they are usually represented in books, the direction of the current appears to be the reverse of that in the simple circuit, but this is in consequence of there being a superfluous plate at each extremity which serves only as a conductor. In the figures in this book, however, these superfluous plates are not represented, and the remarks in the text are made with reference to instruments of this construction.

106. *Bunsen's arrangement* is constructed on the same principle as Grove's, except that hollow cylinders of carbon surrounding the zinc, are substituted instead of the platinum plates. The carbon cylinders are made of carbon from gas-retorts, by grinding it to a powder, and then kneading it with flour dough, and afterwards baking it in a strong heat.

107. *Ohm's Formula.*—The effective force of a battery of any construction will depend upon a number of conditions, or circumstances, which are well expressed in a formula devised by Prof. Ohm, and therefore generally known by his name.

To estimate the effective force of any galvanic arrangement, two things are to be considered, viz: 1. The electromotive force, or absolute tension of the electric fluid, and 2. The resistance to be overcome. As there must always be some resistance to the passage

---

QUESTION.—What is the peculiar advantage of Smee's arrangement? 106. Describe *Bunsen's arrangement*. 107. What is the design of *Ohm's formula?* To estimate the effective force of a battery what two things must be considered?

of the fluid, the effective force must always be less than the full electromotive power, if such resistance did not exist. The resistance will be occasioned chiefly by the wires connecting the poles, but something is due also to the liquid between the plates.

In the case of a single cell, let the absolute tension be represented by t, and let r be the resistance of the wire, and r' the resistance of the liquid between the plates, and A the effective force; then will $A = \frac{t}{r+r'}$. That is, the effective power of the cell will be directly proportional to the absolute tension of the electric fluid, which will in general depend upon the activity of the chemical action taking place, and inversely as the sum of the resistances of the liquid between the plates, and that of the conducting wire.

To apply the same law in the case of a compound series or battery, we have the following formula: $A = \frac{nt}{r+nr'}$, in which n is used to represent the whole number of cells. We see, therefore, that the effective power will be directly proportional to the electromotive force of each cell, multiplied by the number of cells, and inversely as the resistance, which is made up of the resistance of the wire connecting the poles, and the resistance of the fluid of the cells.

The resistance r will be proportional to the length of the wire and inversely as the area of its section. The resistance r' will be proportional to the thickness of the stratum of liquid, and inversely as its conducting power.

108. Animal Electricity.—Several animals, as the *gymnotus electricus*, which is found in the fresh waters of some parts of South America, the *silurus electricus*, found in some African rivers, and the *torpedo*, a species of which has been taken, though rarely, upon the sea-coast of Massachusetts, possess the power of giving strong shocks of electricity, by means of peculiar galvanic arrangements with which nature has provided them.

The electrical apparatus of these animals is plainly connected with the nervous system, and is perfectly under the animal's con-

QUESTIONS.—What resistance will there always be to the passage of the current? To what is the effective force of a battery proportional? 108. What animals are mentioned as possessing the power of giving shocks of electricity? Is their electrical apparatus under control of the will?

trol. It is made use of as a defence against enemies, and in paralyzing other small fishes, which are immediately seized for food. The shock given by some of these animals is very powerful.

EFFECTS OF GALVANIC ELECTRICITY.

109. Galvanic electricity, as we have seen, differs essentially from ordinary, or statical electricity, in several important particulars, as its feeble intensity, its immense quantity, and its flowing in a continuous current; and, as we should expect, there is a corresponding difference in the effects which these two kinds of electricity are capable of producing.

In consequence of the feeble intensity of the most powerful galvanic battery, it is quite incapable of producing many of the more brilliant effects of the common electrical machine, but others are produced not less important. In fact, in the galvanic battery the ordinary signs of electrical excitement are almost wholly wanting, plainly for the reason that these indications depend chiefly upon intensity. Thus, when the circuit of a powerful battery is broken, that is, when the poles or electrodes are disconnected, both of them give signs of electrical excitement, if examined by the ordinary tests; one of them being positive, and the other negative, as before explained; but the indications are exceedingly feeble.

A Leyden vial may also be charged by establishing a communication between one of its surfaces and one of the electrodes, while the other surface is connected with the other electrode; but the charge will always be slight.

Either of the electrodes will give a spark to a conductor presented to it, but it is shown best by bringing the two polar wires in close proximity; and on establishing the communication between the electrodes by the hands, previously moistened, a powerful shock is felt, precisely like that produced by the

QUESTIONS.—109. In what respect does galvanic differ from statical electricity? In the charged galvanic battery, are the ordinary signs of electrical excitement apparent? May the Leyden vial be charged by an active battery? How may a spark be obtained from one of the electrodes? How is the shock produced?

discharge of the Leyden jar. But the shock is felt only at the moment the current is established; if afterwards the hands be held firmly in their places a peculiar feeling of numbness in the muscles in the line of the current succeeds, and continues until the current is again broken.

110. Heating Effects—Deflagration.—When the communication between the electrodes of an active battery is established by various substances capable of conducting the current, they often become intensely heated, and sometimes even consumed or dissipated in vapor by the heat. Thus a small wire interposed in the circuit will become red-hot in an instant, and gold and silver leaf, in the same circumstances, will take fire and burn with brilliant scintillations.

This heating effect of the galvanic current may be produced at a distance from the battery, and gun-powder or gun-cotton and other combustibles ignited. Let W be two copper wires, coated with tarred twine, so as to insulate them perfectly, having their extremities connected by a very small platinum wire soldered to them, as shown through the glass cup C. Let the cup be now filled with gunpowder or gun-cotton, and the other ends of the wires brought in contact with the poles of the battery; immediately on the passage of the current, the platinum wire will be heated, and the powder exploded. The wires, if well coated, may be extended under water, and a submarine magazine exploded. This method has been used for exploding the powder in the blasting of rocks.

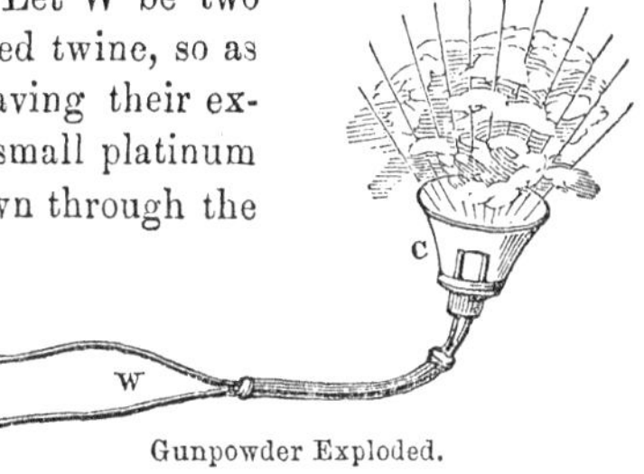

Gunpowder Exploded.

A mixture of oxygen and hydrogen may be inflamed in the same manner; for this purpose the mixed gases are contained in a pistol, into which the wires are inserted, as shown in the figure on p. 104.

QUESTIONS.—Does the shock continue while the current flows? 110. How may the heating effects of the current be shown? How may gunpowder be inflamed? Describe the method of exploding a mixture of oxygen and hydrogen.

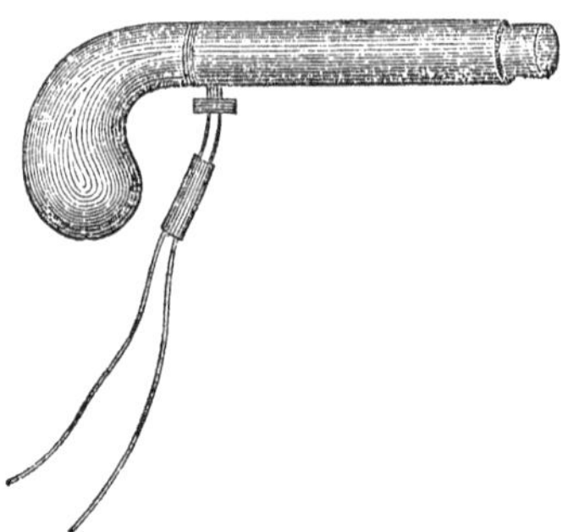

Pistol.

When pieces of well-burnt charcoal are attached to the polar wires, on bringing them near each other, a most brilliant arc of flame appears between them, and the points appear to be vividly ignited, as if heated by any other means. But the heating effect does not depend upon the combustion of the coal, since it is equally as great when the charcoal points are excluded from the air. To produce the effect, the points must first be brought into contact, but they may afterwards be separated some distance, and the arc of flame will continue. At the same time a portion of the carbon point in connection with the positive pole disappears, and the other point is elongated as if by matter transferred from the other side. If the charcoal connected with the positive pole be hollowed out so as to receive pieces of metal, as silver, gold, or platinum, when the current passes the metals will not only be fused, but appear to be even volatilized, and pass off in fumes.

Charcoal Points.

Metal Volatalized.

111. Chemical Effects, Decomposition.—The chemical effects of the galvanic current are seen chiefly in the decomposition of compound substances, and in the operations of electro-metallurgy, soon to be explained. Thus, when two gold or platinum wires

QUESTIONS.—Describe the experiment with the charcoal points. How may the metals, as gold and platinum, be even volatilized? 111. In what are the chemical effects of galvanism chiefly seen?

are connected with the opposite ends of a battery, and their free extremities are plunged into the same cup of water, but without touching each other, hydrogen gas is disengaged at the negative, and oxygen at the positive wire, these being, as we have seen (102), the elements of water. If the wires are brought into actual contact in the water the current is conducted directly through without acting upon the water, and if any other metal except gold or platinum is used for the positive pole, it will be rapidly attacked by the liberated oxygen.

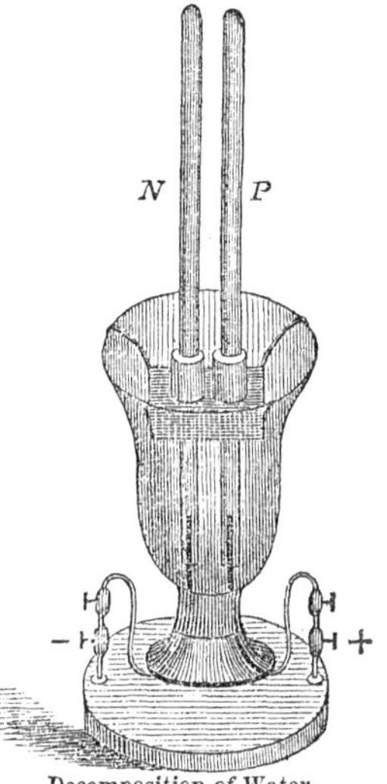

Decomposition of Water.

By using one or the other of the two pieces of apparatus figured in the margin, the oxygen and hydrogen may be collected separately or together, as may be desired. The first piece consists of an open vessel, with a shelf and support for two tubes, N and P, which are closed at the top, and binding screws marked + and — which connect with wires passing underneath the edge of the glass, and terminating in the mouths of the tubes. After filling the tubes and inserting them in their places,—the vessel being previously nearly filled with water, the current from the battery is made to pass through in the ordinary way. Very soon the gases will be seen to collect in the tubes; and as water contains by volume twice as much hydrogen as oxygen, the tube N con-

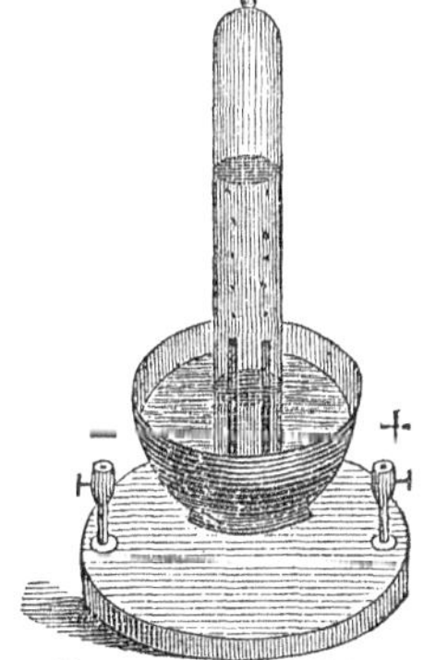

Decomposition of Water.

Questions.—How may water be decomposed by the current? Describe the apparatus for decomposing water and collecting the gases formed in separate tubes. Describe that for collecting the gases together.

taining it will be found to fill proportionably faster than the other.

When the two wires terminate in the same tube, the two gases will be collected together; and by passing afterwards a spark of common electricity through the mixture they will be again united, with an explosion.

In decomposing water, it is well always to add a few drops of oil of vitriol, to increase its conducting power.

If other compound bodies, such as some acids and solutions of salts, are exposed to the action of galvanism, they are also decomposed, one of their elements appearing at one electrode, and the other at the other. An exact uniformity in the circumstances attending the decomposition is also remarked. Thus, in decomposing water or other compounds, the same kind of body is always disengaged at the same side of the battery. The metals, inflammable substances in general, the alkalies, earths, and the oxides of the common metals, are found at the negative electrode; while oxygen, chlorine, and the acids, go over to the positive electrode.

112. Those substances which appear at the positive side have been called *electro-negative* bodies, while those that are separated at the negative wire are called *electro-positive* bodies.

The decomposition of a salt, which must be in solution, may be shown in the following manner:—Let two wine-glasses be filled with a solution of sulphate of soda, (which is a compound of sulphuric acid and soda,) and let some fibres of moistened cotton be extended between them, as shown in the figure. If the current is then transmitted through the cups, the salt will soon be decomposed, and the cup in connection with the positive electrode will be found to contain weak sulphuric acid, and that in connection with the negative electrode a solution of soda. If, now, a little red cabbage-water be poured into each, the acid liquid will become red, and the soda solution green.

Decomposition of Salt.

QUESTIONS.—May other compounds also be decomposed? In decomposing a compound, will the same element always appear at the same electrode? 112. What are *electro-negative* and *electro-positive* substances?

A compound to be decomposed by the current must be in the liquid state, and must of course be a conductor; but all compounds answering these conditions are not by this means capable of direct decomposition. Yet compounds, not directly decomposable by the current, are often decomposed indirectly by the hydrogen derived from the decomposition of water that is present. Thus, sulphuric acid is not directly decomposed by the current, but hydrogen separated from the water present, making its appearance at the negative pole, unites with the oxygen of the acid, and causes the evolution of fumes of sulphuric acid.

Following the suggestions of Faraday, the term *electrolysis* (from *electron*, and *luo*, I unloose) is now very generally used to express this electro-chemical decomposition; and the term *electrolyte* to indicate a compound capable of being thus decomposed. The positive pole or electrode is also called the *anode*, and the negative pole the *cathode* (from *ana*, upward, *kata*, downward, and *odos*, a way).

But instead of the terms *anode* and *cathode*, most writers prefer to say positive or negative electrode, as the case may be.

The elements of a compound capable of separation by this mode, are termed *ions* (from the Greek participle *ion*, going); the *anion* being the element which appears at the anode, and the *cation* the element which goes to the cathode. It will at once be seen that the anions are *electro-negatives*, and the cations *electro-positives*.

It is not to be inferred from the above remarks, that every substance will always make its appearance at the same electrode, whatever may be the other substance from which it is separated by the electrolytic action. Oxygen does indeed always appear at the positive electrode, and potassium at the negative; but in the electrolyses of the compounds of other substances, that element will appear at the positive electrode which is *most* electro-negative, and that at the negative which is *most* positive.

113. The following table exhibits the electrical relations of several of the more important elements. It is to be understood that each sub-

QUESTIONS.—Are all compounds directly decomposable by the current? Define the terms, *electrolyte*, *anode* and *cathode*. What are *ions?*—what *anions* and *cations?* Will every substance always make its appearance at the same electrode, whatever may be the other from which it is separated? What is the law in this respect?

stance in the first column is electro-negative as compared with each one below it, but in the second column each substance is negative as compared with all above it:

—	+
Oxygen.	Potassium.
Fluorine.	Sodium.
Chlorine.	Calcium.
Iodine.	Carbon.
Sulphur.	Hydrogen.
Nitrogen.	Zinc.
Phosphorus.	Iron.
Arsenic.	Bismuth.
Antimony.	Tin.
Gold.	Lead.
Mercury.	Copper.
Silver.	Silver.

114. We have seen that the decomposition of a compound takes place only when the current is made to pass through it—then one of the ingredients is collected at one electrode, and the other at the other electrode; thus in the electrolysis of water, the oxygen is collected at the positive and the hydrogen at the negative electrode. But how is it that these effects are produced? and at what precise point or points? Are both gases liberated at each electrode, but one only, hydrogen (the electro-positive element), *collecting* at the negative electrode, and the other, oxygen (the electro-negative element), *collecting* at the positive pole? This would require that opposite currents of oxygen and hydrogen should be passing by each other in the liquid between the poles while the battery is in operation, of which we have no evidence. Or, does the decomposition take place along the whole line between the poles traversed by the current? Other similar inquiries may be made, which cannot be answered positively, but the following is believed to be a correct representation of the phenomena witnessed, so far as we are able to determine. Water is a compound of hydrogen, which we will represent by H, and oxygen, which we will indicate by O;—and the compound by $\substack{H\\O}$. A tier or row of particles between the poles we may then represent thus:

$$\begin{array}{cccccccc} & H, & H, & H, & H, & H, & H & - \\ + & O & O & O & O & O & O & \end{array}$$

QUESTIONS.—114. When only does the decomposition of a compound take place? Describe the manner in which the decomposition of water is supposed to be effected by the passage of the electric current.

the sign + indicating the position of the positive pole, and — that of the negative.

Instantly as the current begins to pass, oxygen from the decomposition of the water appears on the positive side, and hydrogen from the same decomposition at the negative side, a state of things which may be represented as follows:

$$\begin{array}{l} \quad , H, H, H, H, H, H \; - \\ + \; O \;\; O \;\; O \;\; O \;\; O \;\; O \end{array}$$

We have now one less particle of water than before, and one of the elements of the decomposed particle is found at one electrode, and the other at the other electrode. To produce this effect, it is only necessary that, by the action of the current, all the particles of hydrogen should be removed one place to the right, or all the particles of oxygen one place to the left. That is, the electrolysis of every particle of water in the track of the current has taken place, but the immediate reunion of all the particles has followed, except the extreme particle of oxygen, on the one hand, and the extreme hydrogen on the other.

The oxygen and hydrogen thus liberated, at once make their escape; and the continued action of the current produces a like effect on other particles of water until the action ceases, or all the water is decomposed.

115. The quantity of electricity required to effect the decomposition of any compound is probably always the same, as would be shown if we had means to measure it accurately; and the relative quantities of several electrolites decomposed by a given quantity of electricity will be represented by the combining numbers of these compounds. We shall hereafter have occasion to allude to this point again, in connection with the subject of combining proportions or equivalents.

116. *Electro-Metallurgy* is the name applied to the deposition of the metals from their compounds by the chemical agency of the galvanic current. We have seen above (103), in describing Daniel's battery, that during its action there is a constant deposition of metallic copper upon the negative plate. Now, if, for the copper plate in this arrangement, a medal, or coin, or other conducting body be substituted, the deposition of the copper upon it will take place in the same manner, and all its minute pecu-

Questions.—115. Is the quantity of electricity required to effect the same decomposition always the same? 116. What is *electro-metallurgy*?

liarities will be copied. An apparatus of this kind, called the *electrotype*, is figured below.

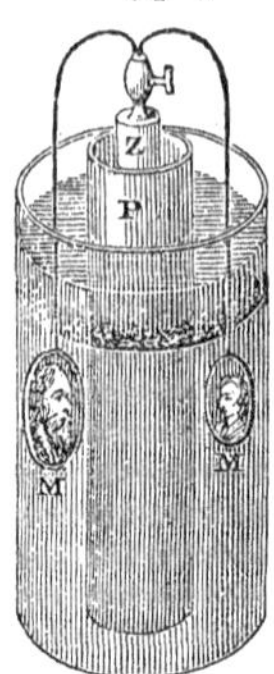

Electrotype.

A glass vessel is partly filled with a saturated solution of blue vitriol, and in this is placed a porous vessel, containing dilute sulphuric acid, and a rod of zinc, Z, having a binding screw at top. The medal or coin to be copied is then suspended in the vitriol solution by means of a wire inserted in the binding screw. The surface of the medal on which the copper is to be deposited should be perfectly clean, and the other surface should be protected by a coating of wax or varnish. In the figure, two medals, M, M, are supposed to be connected at the same time with the zinc.

A better method than the above is to use a regular Smee's battery, and to have the blue vitriol solution in a separate vessel, as in the annexed figure. Then let the article to be copied, A, be connected with the zinc of the battery, and a plate of copper, C, with the silver, both being suspended, at a little distance from each other, in the vitriol solution. By the action of the battery, the piece of copper will be gradually dissolved, and a corresponding deposit of metallic copper made upon the medal.

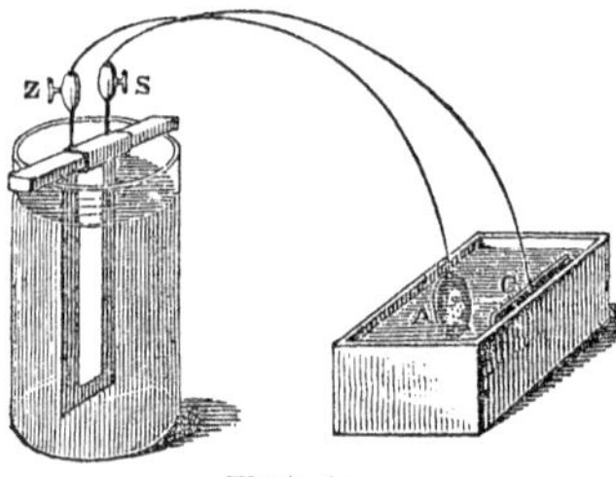

Electrotype.

In the battery here used, one of the zinc plates is supposed to be removed, presenting clearly to view the plate of silver.

117. Other metals besides copper may be deposited in this manner, but a battery of several cells is in most cases required. The most important application that has been made of this discovery is in depositing silver and gold in thin laminæ upon other metals, called plating or gilding. For this purpose a Daniel's or Smee's battery of about four cells answers well, and a fifth, which is called the depositing cell. This is

QUESTION.—Describe the apparatus called the *electrotype*.

filled with a solution of cyanide of potassium, and used in the same manner as just described for obtaining a deposit of copper, except that silver or gold must be used instead of the copper, C, according as one or the other of these metals is to be precipitated.

ELECTRO-MAGNETISM.

118. Natural Magnet, or Loadstone.—Among the ores of iron, pieces are often found which possess the property of attracting and retaining pieces of iron or steel with more or less force, and are called *magnets* or *loadstone.* Each magnet always has two points in which the attractive force appears to be concentrated, which are called the *poles.* They are always readily found by rolling the magnet in iron filings, which will be collected more at these points than in other places. Generally they are nearly opposite to each other. (See figure.) If a magnet is broken into several pieces, each always retains the same magnetic properties as the whole mass, but in less degree. Sometimes a magnet has more than two poles.

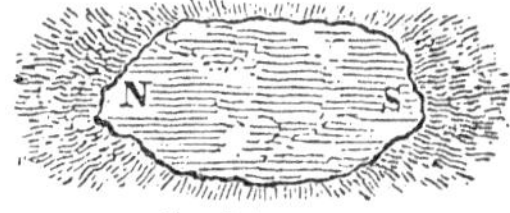

Loadstone.

119. If a magnet be suspended horizontally by a thread, or placed upon a piece of cork floating in a vessel of water, one of the poles will turn towards the north, and is hence called the *north pole,* and the other towards the south, and is called the *south pole.*

If two magnets thus suspended are brought near each other, it will be found that like poles repel each other, but unlike poles attract. These attractions and repulsions extend to some distance, and are not affected by the interposition of other bodies, not capable of becoming magnetic. This may easily be shown by interposing a pane of glass, or plate of copper, or a thin piece

QUESTIONS.—118. What is the natural magnet, or *loadstone?* What are the poles? What is the effect if the magnet be broken in several pieces? 119. What is the north pole? What the south pole? When two magnets freely suspended are brought near each other, what is observed of like, and what of unlike poles?

of wood between the two magnets, when it will be seen that the action of the magnets upon each other is the same as before, and does not even suffer diminution by the interposed substance, except as the distance between them is increased.

120. Magnets and Diamagnets.—It has of late been very satisfactorily determined that all substances may be divided into two classes,—the *magnetic* and the *diamagnetic.* To the first class belong all substances which, like iron, nickel and cobalt, are attracted by either pole of a magnet when presented near them; and when shaped into bars and free to move (as when suspended by a thread in the centre) in the vicinity of a magnet, they arrange themselves in the direction of a line uniting its poles. Two bar magnets best illustrate this property; when in the vicinity of each other, and free to move, they take the position indicated in the figure.

Magnets.

The substances already named are the only ones that possess this property in any considerable degree; but others have recently been added to the list, as manganese, chromium, titanium, palladium, platinum, oxygen gas, &c., in which the magnetic property is manifested only when a powerful magnet is used.

Diamagnetic substances are such as are repelled by either pole of a magnet, and when made into the form of bars and free to move, in the vicinity of the poles of a magnet, arrange themselves at right angles to a line uniting its poles. This relative position of the magnet and diamagnetic bar is shown in the accompanying figure.

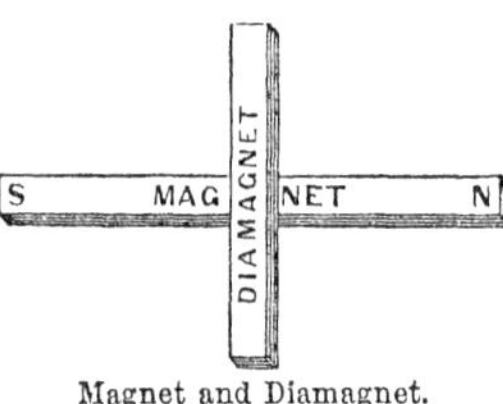

Magnet and Diamagnet.

This diamagnetic property is manifested more powerfully by bismuth than by any other substance, but phosphorus, antimony, zinc, tin, sodium, mercury, copper, gold, glass, ether, alcohol, and many other bodies, are similarly affected. The experiment can only be shown by using very powerful magnets.

QUESTION.—120. Into what two classes may all substances be divided? How do magnetic substances arrange themselves when brought into the vicinity of a magnet? How do diamagnetic substances?

121. Magnetic Induction.—When either pole of a magnet is brought in contact with a piece of soft iron, or only very near it, the iron itself becomes magnetic, and remains so until the magnet is removed. The figure following represents several pieces of iron, placed in different positions near the poles of a magnet; on examination, they will all be found to have the magnetic property, their poles being developed as indicated by the letters N and S.

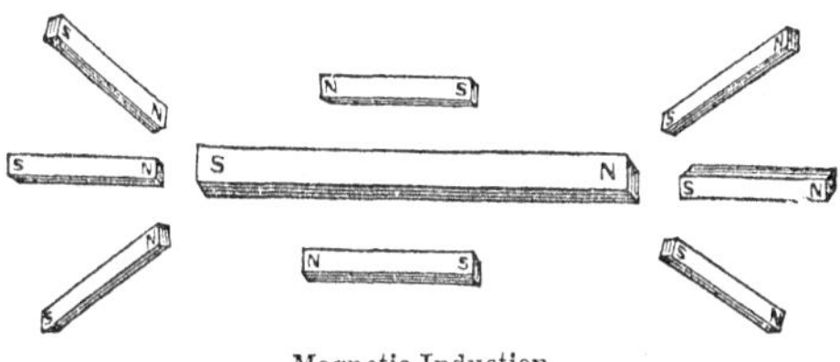

Magnetic Induction.

This influence of a magnet upon pieces of iron, which extends to a distance around its poles, is called its *inductive influence.*

This influence is exerted by a powerful magnet to a considerable distance from either pole, or even through other bodies (119) which are not themselves capable of becoming magnetic.

A piece of iron or other substance which is attracted by either pole of a magnet, is evidently first rendered magnetic by induction, and then the attraction follows as a necessary consequence. But the first piece, as a nail, being rendered magnetic by induction, will act in like manner upon a second, and this upon a third, and so on; so that several pieces may be lifted one after another, as represented in the figure. But it will be found, when the connection with the first magnet is broken, the pieces of iron instantly lose their magnetism, and fall asunder.

122. Pieces of hardened steel will also be affected in a similar manner, but much less readily, and, unlike

Magnetic Induction.

QUESTIONS.—121. What is the effect when either pole of a magnet is brought near a piece of soft iron? What is meant by magnetic induction? Why is a piece of soft iron attracted by a magnet?

iron, they retain the magnetism that has been induced. They therefore become permanently magnetic, and for nearly every purpose are superior to the natural magnet, and may be denominated *artificial magnets.* The same magnet may be used successively to magnetize any number of steel bars, without losing any of its virtue; from which it follows that the magnet communicates nothing to them, but only by its influence developes some hidden principle already there. Artificial magnets are frequently made in the form of the horse-shoe, and are called horse-shoe magnets. To the poles a short piece of soft iron is usually accurately fitted, called the *armature* or *keeper.*

123. The Magnetic Needle—Dipping Needle.—A slender bar of magnetized steel, suspended upon a pivot, so as to revolve freely, constitutes the *magnetic needle.* Sometimes it is attached to a circular card, and suspended upon a pivot, as in the mariner's compass.

124. If a steel bar be suspended by its centre of gravity, and afterwards carefully magnetized, it will be found not only to place itself in the magnetic meridian, but to assume a position inclined to the horizon. In northern latitudes, the north pole will be depressed and the south pole elevated, while in southern latitudes the south pole will be depressed. The angle of inclination is generally nearly the same in the same place, and is called the *dip of the needle;* and a needle nicely balanced and adjusted for showing the dip is called a *dipping needle.*

The dip of the needle is subject to considerable variation, but at the present time it is, at Baltimore, about 71° 30′; at Philadelphia, 72° 15′; at New York, 73°; at Middletown, Conn., 73° 30′; and at Boston, 74° 24.

The magnetic needle does not always point to the true north and south, but deviates more or less from this position at different times and places. This is called its *variation.* At Philadelphia, in 1840, the variation was 3° 52′ W., and at Middletown, Conn., about 6° 40′.

125. Terrestrial Magnetism.—The earth may be considered as a great natural magnet, which, by its action on the needle, in the same manner as any other magnet, causes it to place itself in

---

QUESTIONS.—122. May magnetism be induced in pieces of tempered steel? What are *artificial magnets*? 123. What constitutes the *magnetic needle?* 124. What is a *dipping needle?* 125. What may the earth be considered?

the position of north and south. Indeed, it is by the inductive influence of the earth that magnetism is developed in bodies upon its surface. This is shown in bars of iron or steel that have stood long in a vertical position, which are always found to be magnetic. Tongs and pokers, from their being usually kept in an upright position, are almost always found to be magnetic.

As it has been agreed to call that pole of the needle which points northward, the north pole, it is evident that the pole of the earth situated *north* must be a *south pole;* that is, it must possess *southern polarity*. So, also, the *south pole* of the earth must possess *northern polarity*.

126. Relation between the Electric Current and Magnetism.—It has been long known that a discharge of lightning will often affect seriously the magnetic needle, sometimes reversing its poles; but it was not until 1819 that Œrsted made his famous discovery, which has served as the basis of the beautiful science of Electro-Magnetism.

He first observed that when the wire, connecting the electrodes of an active galvanic battery, is brought near a magnetic needle, it is made to deviate from its ordinary position, and assume a new one, depending upon the direction of the current and the position of the wire in regard to the needle. Thus, the needle being in its natural position,—1st, if the connecting wire be above the needle and parallel to it, the pole next the negative electrode will move westward; 2d, if the wire be beneath the needle, it will move eastward; 3d, if the wire is on the west side, this pole will be depressed; and, 4th, if it be on the east side, it will be elevated. The figure in the margin indicates the motion that will be produced in the first of the above cases.

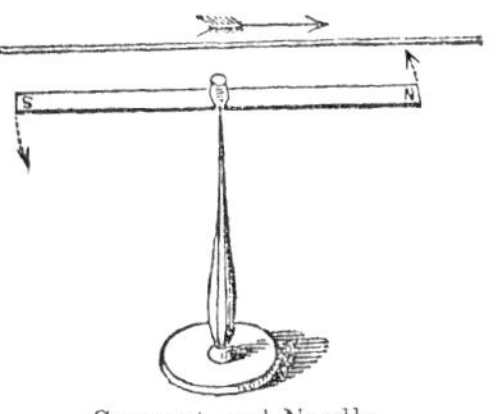

Current and Needle.

If the wire be placed under the needle, and the current made

QUESTIONS.—Why do tongs and pokers become magnetic by standing in a vertical position? What is said of the north and south poles of the earth? 126. What was the fact first observed by Œrsted as regards the wire conducting a galvanic current and a magnet?

to pass from north to south, the motion of the needle will be the same as indicated in the figure.

It follows, therefore, if the wire be bent around the needle, in the form of a rectangle, so as to convey the current in one direction above the needle, and back again, in the opposite direction beneath it, both parts will conspire to produce the same effect, and the motion of the needle will be much increased. Such an arrangement is represented in the annexed figure.

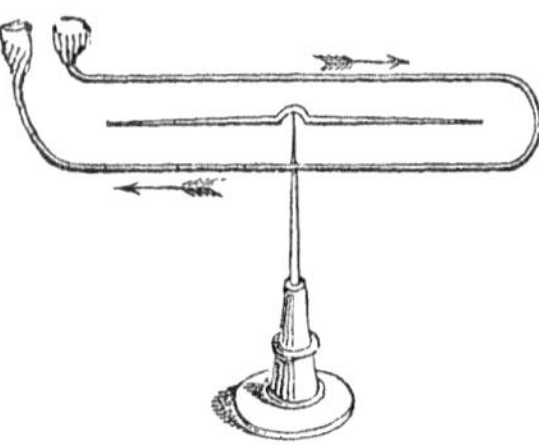
Current and Needle.

In all these cases, the tendency of the needle is to settle directly across the wire, or at right angles to the direction of the current, while the influence of the earth is exerted to bring it in its first position, parallel with the wire, supposing the experiment to be commenced with the needle in its natural position. The position it will ultimately take will therefore be intermediate between these two.

127. To avoid the directive influence of the earth upon the needle, the *astatic* (from the Greek *astatos*, unstable) needle has been contrived. It consists of two needles, of *nearly* equal strength, fixed to the same axis, with their poles reversed in reference to each other, and suspended by a thread, as shown in the figure. One of the needles being a little more highly magnetized than the other, or a little larger, they will have a slight tendency to settle in the meridian. If, now, a wire is bent several times around the lower needle, each turn or coil being coated with some insulating substance, when the current is passed around it, its influence on both needles will be to turn them in the same direction; and the

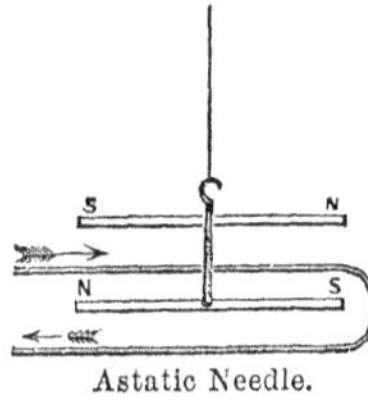

Astatic Needle.

QUESTIONS.—What is the effect if the wire is bent around so as to pass several times around the needle? What is the real tendency of the needle? 127. What is the *astatic needle?* Describe its construction.

arrangement becomes a most delicate instrument for indicating the passage of the feeblest currents. Such an instrument is called a *galvanometer*, or *galvanoscope*. To protect it from currents of air, the whole is usually enclosed in a glass case; and beneath the upper needle, and above the coil of wire, a graduated circle is placed, to indicate exactly the comparative deflections of the needle.

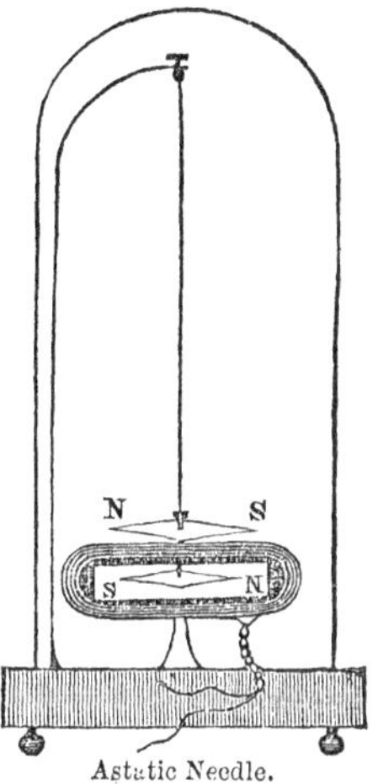

Astatic Needle.

Such an instrument, in connection with a thermo-electric pile (93), becomes a most delicate thermometer, which is capable of indicating a change of temperature of only a very small fraction of a degree.

128. Tangential Force.—By a careful inspection of the motions produced in the needle by the current in the several positions of the wire, as described in paragraph 126, it will be evident that the real tendency of each pole is to revolve around the connecting wire in a circle, the plane of which is perpendicular to the wire;—around the north pole in one direction and around the south pole in the other. The force which causes this motion is exerted in lines which are tangents to the circumference of these circles, and is therefore called a *tangential force*.

The motion of the needle *actually produced* will of course depend upon the mode in which it is supported, as well as the position of the wire in reference to it. It is to be remarked, too, that the pole is a mere point situated near the extremity of the needle;—between this point and the wire the force is exerted. This point,—the magnetic pole,—cannot exist independent of the needle itself, which must therefore always move with it; and therefore in order to determine the real motion in any particular

QUESTIONS.—Describe the galvanometer, or galvanoscope. 128. What is the real tendency of each pole of a magnet when brought near the wire? What is the force called which produces this motion? Upon what will the motion actually produced depend?

case we are to consider the two circumstances,—First, the motion the pole tends to make, and Secondly, the motion of which the needle is susceptible. Upon these two conditions plainly will depend the actual motion that will result.

129. In order easily to remember the particular motion a pole will tend to make in any given case, take the following example. Let us suppose the conducting wire to be placed in a vertical position, and the current of positive electricity to be *descending* through it, the tendency of a north pole in the vicinity of the wire will be to move around it in a horizontal circle, in the direction indicated by the arrows in the figure, or in the direction of the hands of a watch with the dial upward. The tendency of the south pole would be to revolve in the opposite direction. If the direction of the electrical current is reversed, and it is made to pass upward, both poles would tend to revolve in the opposite direction from that described above. Whatever may be the position of the conducting wire with reference to the needle, the motions produced will always be in accordance with these statements; and reference being had to this particular position of the conducting wire and the needle, we can always determine by inspection in what direction the needle will move, whatever may be the position of the wire in regard to it.

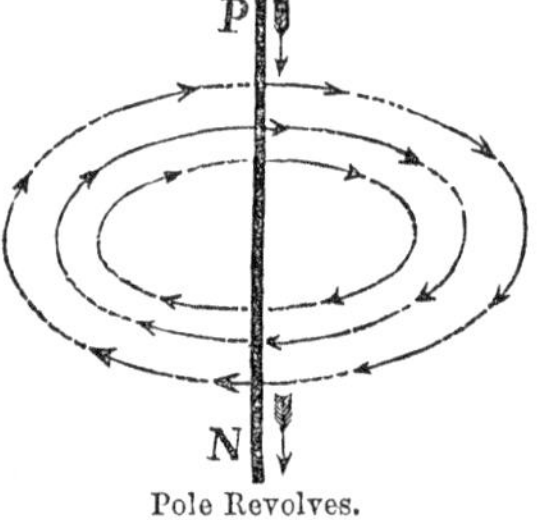

Pole Revolves.

We have here been speaking only of motions produced in the needle by the conducting wire; but it is plain that if the pole of a magnetic needle tends to revolve around a fixed conducting wire, a free wire will have a tendency to revolve around the pole of a fixed needle. The fact is, the influence is mutual, and both the wire and the pole tend to revolve, as we will proceed to show.

QUESTIONS.—To determine the real motion that will be produced in a given case, what two things are to be considered? 129. When the current is descending perpendicularly, what will be the tendency of the north pole of a needle in the vicinity of the wire?

130. Both the revolution of a pole around a fixed wire, and the revolution of a wire around a fixed pole, may be shown by the apparatus, a section of which is seen in the cut in the margin. A and B are two glass vessels, filled nearly to the top with mercury. A wire, supported by a pillar between the vessels, has one end bent down so as to dip into the mercury in A; and the other end is bent into a hook, on which a short piece of wire is suspended, so that its lower end shall also dip into the mercury in the vessel B. Conducting wires pass through the bottoms of both vessels, with binding screws, C and D, to connect them with the poles of the battery; and to that in A, a small magnet is attached by a thread, so that one of its poles may be a little above the surface of the mercury; and in the vessel B another small magnet is firmly fixed, with one pole a little above the mercury. When the current is passed through this apparatus, as indicated by the arrows, the upper pole of the magnet in A will revolve slowly around the conducting wire, and the free conducting wire suspended in B will revolve around the fixed pole near it.

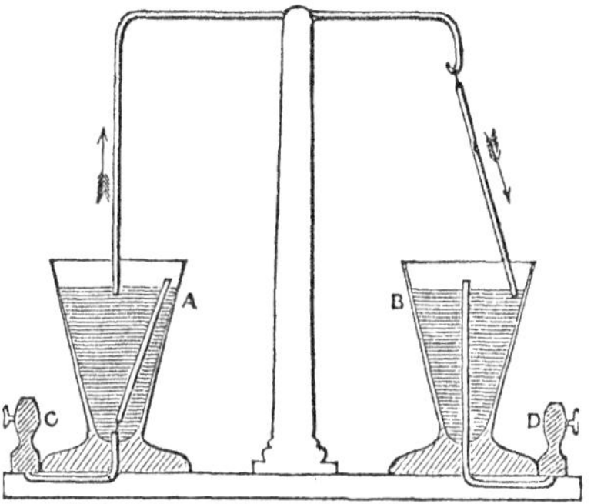

Both Pole and Connecting Wire revolve.

131. Further Experiments illustrating the Relation of the Magnet and a Galvanic Current.—Experiments to illustrate the relation of a current and the magnet may be multiplied almost indefinitely.

De la Rive's Ring is a very beautiful contrivance to show the influence of a magnetic pole upon a conducting wire capable of motion. A copper wire is bent into a circle about an inch in

QUESTIONS.—130. Describe the apparatus figured in paragraph 130. What is the point illustrated?

diameter, and the two ends passed through a piece of wood or cork, and soldered, one to a slip of zinc, Z, and the other to a slip of copper, C. If, now, it be placed in a bowl of water containing a little acid, a current of electricity will circulate along the wire, in the direction of the arrows; and if a bar magnet be brought near it, it will be attracted or repelled, according as one pole or the other of the magnet is presented to it. As represented in the figure, if the north pole of the magnet is presented to it, it will be attracted; the south pole will first repel it, but in a little time it will turn around, and will then be attracted. We may, therefore, properly regard it as a flat magnet, having its two poles in the centre of the circle, the one on one side, and the other on the other; the *south pole* being in that surface on which, when held before you, the positive current flows in the direction of the hands of a watch. The apparatus will be more powerful if the conducting wire, covered with silk, to prevent lateral communication, be formed into several circles of the same diameter, on the principle of the multiplier.

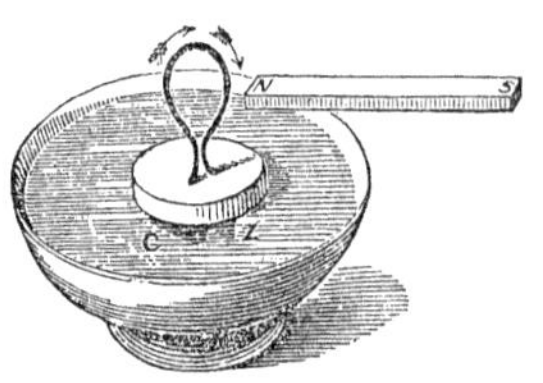

De la Rive's Ring.

132. But what are the forces that have produced these special movements? Giving our attention for the present only to the part of the current upward on one side of the ring and downward on the other, we have seen (129) that the downward current tends to revolve around the north pole of a needle in the direction of the hands of a watch, and the upward current in the opposite direction; let these facts be kept in mind while the eye rests upon the figure, and it is plain that the effect must be to cause the ring to approach the pole, or apparent attraction ensues. If, now, the magnet be suddenly removed, and the south pole presented, in accordance with the same laws, the ring recedes, turns around, and again approaches the pole!

QUESTIONS.—131. Describe De la Rive's Ring, and the mode of using it. What may we regard it? 132. Give the reasons for these movements.

133. *The galvanoscope* is represented by the next figure N and S are the north and south poles of a permanent horse-shoe magnet, supported upon a stand; and between them, in a small glass tube, is a strip of gold leaf, connected at top and bottom with binding screws, as shown in the figure. Now when the current is passed through the gold leaf, it is made to curve to one side or the other, according as the motion of the current is upward or downward. As the poles are situated in the figure, if the direction of the current is downward, the curving or convexity of the gold leaf will be upward from the plane of the paper; but if the current is upward the strip of gold leaf will be convex in the opposite direction, or backward from the plane of the paper.

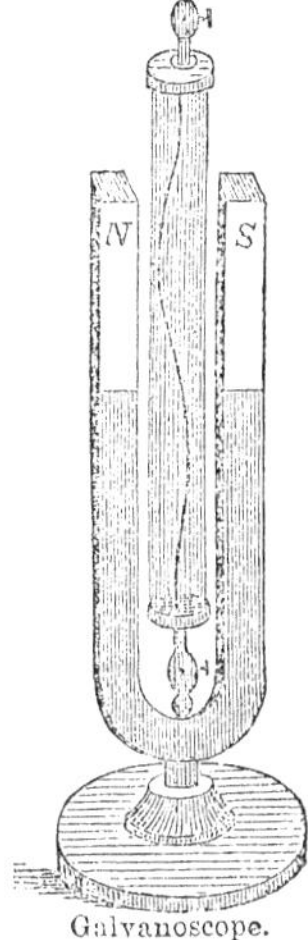

Galvanoscope.

134. *The Revolving Spur Wheel*, figured in the margin, is made to revolve by the action of a magnet upon the galvanic current. A wheel W made of sheet brass, is supported in such a manner that its rays touch in a globule of mercury in the base beneath, from which a concealed wire extends to the binding screw A, the other binding screw B being connected, in like manner, with the wire R which supports the wheel. N S is a magnet placed so

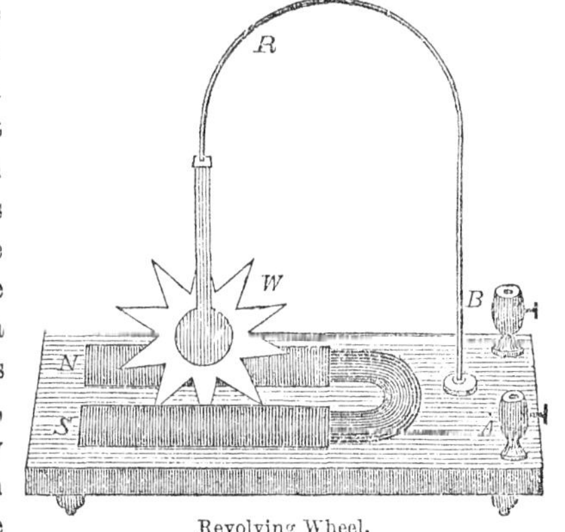

Revolving Wheel.

QUESTIONS.—133. Describe the *galvanoscope*. 134. Describe the revolving spur wheel, and give the reasons for the movement produced.

11

that its poles shall be as nearly as possible one on each side of the ray of the wheel which for the time is in contact with the mercury. When the poles of the galvanic battery are connected with the binding screws, a ray of the wheel being in contact with the globule of mercury, the current passes through it, and the wheel is made to revolve on its axis. A permanent magnet may be used, or an electro-magnet, as represented in the figure (p. 121).

135. Relation of Current to the Earth's Magnetism.—The ring described in paragraph 131 is acted upon by the earth's magnetism precisely in the same manner as by the magnet, only that the influence is less in degree. In order that it may be made to move by the magnetism of the earth, it is only necessary that it should be enlarged sufficiently, and a powerful current passed through it. The following modification of it answers a good purpose.

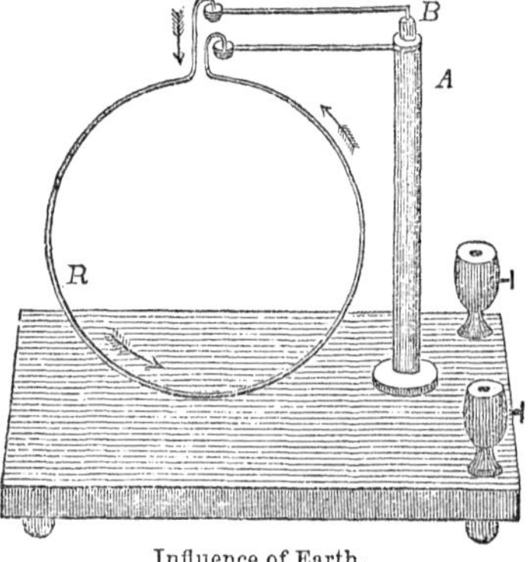

Influence of Earth.

Let A and B be two metallic pillars, insulated from each other, provided each with a horizontal arm terminating in a small cup filled with mercury, and connected at the bottom with binding screws. Let a ring R of copper wire, 8 or 10 inches in diameter, be suspended by the ends of the wire from the cups of mercury, and the poles of the battery connected with the binding screws;—the current will pass around the circle of wire, which, by the influence of the earth's magnetism, will be made to move until at length it will take a position with its plane at right angles to the meridian. Supposing the direction of the current to be as indicated by the arrows, and recollecting that the

---

QUESTIONS.—135. Does the earth's magnetism act upon a current in the same manner as a magnet? Describe the movements produced in the ring here figured when conveying the current by virtue of the earth's magnetism.

north pole of the earth really possesses southern (125) polarity, let the intelligent student determine which side of the circle of wire will settle to the east and which to the west!

It may afford some aid to reflect that the action on the circle of wire will be the same as if it were square, as shown in the margin, in which it appears there are four parts conveying the current, two on which its motion is horizontal in opposite directions and on the same side of the point of support. The influence of these is lost and may not therefore be further considered. On the other two parts the motion of the current is in opposite directions,—being on one upward and on the other downward—and on opposite sides of the point of support. Now the tendency of the part conveying the upward current will be to revolve round the north pole of the earth in one direction, and of the part conveying the downward current to revolve around this pole in the opposite direction; considering then the mode in which the wire is suspended, it is plain that the tendency of the circle of wire will be to settle with its plane due east and west, or at right angles to the meridian. If, then, before passing the current, the plane of the circle of wire is placed in the meridian when the current is made to pass it will turn on its points of support until it takes a position at right angles to the meridian. In every case, the side of the circle conveying the downward current will move towards the east, and the other side of course towards the west.

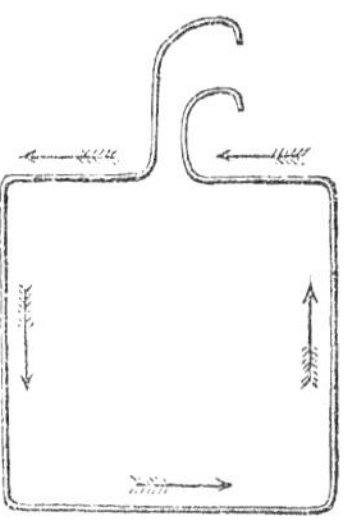
Upward and Downward Current.

The same movement will be produced if the wire conveying the current is bent several times around, as represented in the first figure on the next page, and supported by its two ends resting in cups of mercury, with which the two poles of the battery are to be connected by binding screws not represented in the figure. The influence of each coil conspires to produce the same effect, and therefore the motions described will be more readily produced.

QUESTIONS.—Explain the reasons for the motions which are produced by the earth's magnetism, as illustrated by the preceding figure, and also the two figures on the next page.

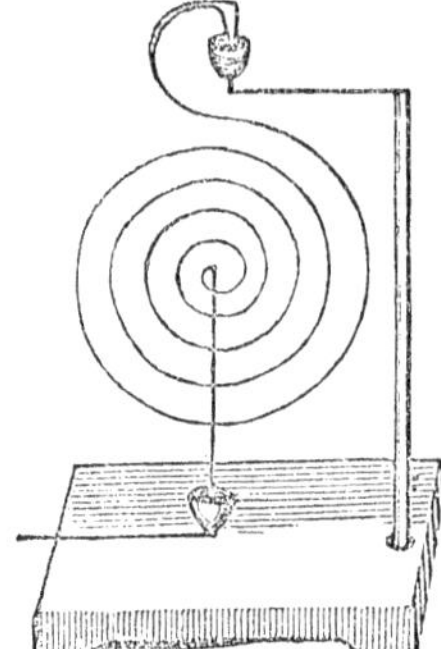

But it is not necessary that the different coils of the wire should lie in the same plane, the result being the same if it is wound in the form of a helix,

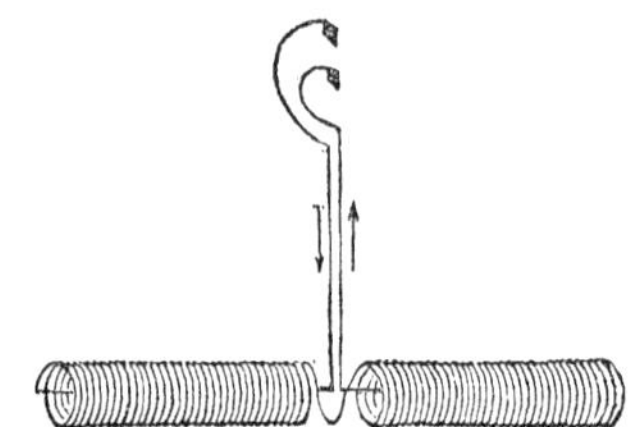

Influence of Earth's Magnetism.

as represented in the next figure; the helix then truly representing a bar magnet, as will be found by holding either pole of a magnet near either end.

By examining the direction of the current in this helix, considering it as a bar magnet, it will be found that is the south pole in which (when it is held up before the eye) the current is moving in the direction of the hands of a watch. It is scarcely necessary to remark that, in the other pole, the direction of the current is the reverse of this.

136. Induction of Magnetism by a Current.—Magnetism is induced in a bar of soft iron by the simple passage of a current near it, in a direction at right angles to the bar. Thus, if W I be the conducting wire, and N S a small piece of iron lying under the wire, while the current is passing, magnetic polarity will be induced in the iron, N being a north pole, and S a south pole, when the current passes from W to I; but if the current pass from I to W, the poles will be reversed.

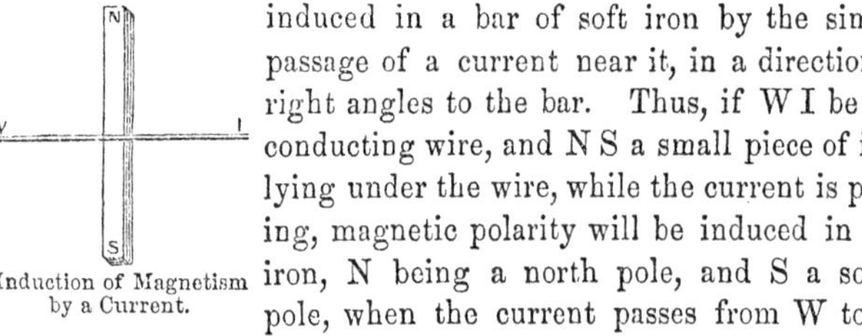

Induction of Magnetism by a Current.

But, if the wire is made to pass around the iron many times, the effect will be greatly increased. This is accomplished by

QUESTIONS.—When the south pole of the helix is held up before the face, in what direction is the current found to be moving? 136. How is magnetism induced in a piece of iron by the galvanic current?

placing the iron in a helix of wire, as shown in the figure. The current being then made to pass in the direction of the arrows, the iron becomes strongly magnetic, with its poles as shown by the letters N and S. The cups P and N serve to connect it with the battery. When the current is broken, the iron ceases at once to be magnetic; but if a piece of hardened steel be substituted for the iron, it retains its magnetism permanently.

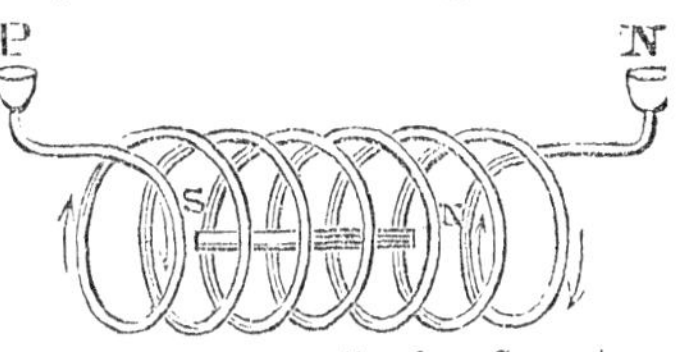

Induction of Magnetism by a Current.

A magnet formed in this manner, by the passage of a current of electricity around it, is very properly termed an *electro-magnet*. Magnets of this kind have sometimes been made of sufficient strength to sustain a weight of 2000 or even 3000 pounds. For this purpose, a bar of iron of considerable size is bent in the form of a horse-shoe, and the conducting wire made to pass many times around it, as represented in the figure. An armature of soft iron (122) is attached to the poles, as with the common magnet, from which the weights are suspended. As stated above, when the current of electricity is interrupted, the magnetism will be immediately destroyed, and the weights drop off.

Electro-magnet.

137. *The Magic Circle*, as it has been termed, possesses too much interest to be here omitted, though no new principle is developed by its action. Two semi-circles, A and B, are made of a stout bar of soft iron, and well fitted together, so as to form a circle, C, and include a small helix of wire, H, the two ends of which are to be connected with the electrodes of a small but active battery. While the current is passing, the semi-circles are made powerfully magnetic, and will adhere with a force which is capable of sustaining a weight of many pounds.

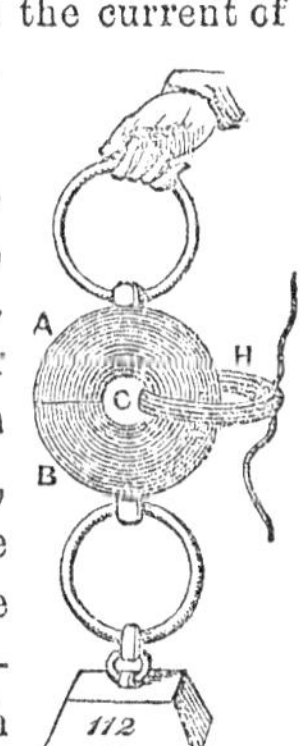

Magic Circle.

QUESTIONS.—What is the effect when the current is made to pass many times around the piece of iron? What constitutes the *electro-magnet?* 137. Describe the *magic circle.*

The wire used in all these arrangements is supposed to be wound with some insulating substance, as silk or cotton, to prevent the contact of the separate coils, which would permit a lateral discharge of the current, so that it would not pass around the whole length of the wire.

138. To form an effective electro-magnet, the total length of copper wire intended to be used is cut into several portions, each of which, covered with thread, is coiled separately on the iron. The ends of all the wires are then collected into separate parcels, and are made to communicate with the same battery, taking care that the positive current shall pass along each wire in the same direction. The current is thus divided into a number of branches, and has only a short passage from one end of the battery to the other, though it gives energy to a multitude of coils. This was first made known by Prof. Henry, now Secretary of the Smithsonian Institution.

139. **Magnetism of the Earth.**—We have seen above (125) that the earth may be considered as a great magnet, having its south pole somewhere near its north geographical pole, and its north pole near its south geographical pole; and it has been suggested that this may be occasioned by currents of electricity flowing around it, beneath its surface, from east to west, or in the direction opposite to that of its motion on its axis. Indeed, it has even been suggested that these currents may have their origin in the heating influence of the sun's rays, as successive portions of the earth's surface come under their influence by its daily motion; but this is to be regarded as only plausible conjecture. That the needle would be influenced by such a current, precisely as it really is, at different points of the earth's surface, may be shown by winding a covered wire several times around the equator of a common globe, and placing a small magnetic needle upon it while the current is passing, as shown in the

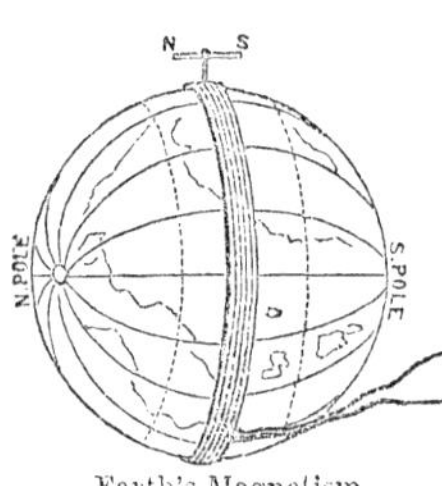

Earth's Magnetism.

QUESTIONS.—Why must the wire used for this purpose be wound with some non-conducting substance? 138. What is said of the total length of wire used for this purpose? 139. What is said of the earth in reference to this subject? How may the magnetism of the earth be occasioned? Describe the figure in this paragraph.

figure (p. 126). The small needle will uniformly settle in the meridian.

140. Electro-magnetic Engines.—All the above motions, it will be observed, whether of the magnetic needle or of the conducting wire, have been produced by the tangential force; but various machines have been devised, called *electro-magnetic engines*, for producing motion by direct influence upon each other of th. unlike and like poles of two or more magnets; one of which at least must be an electro-magnet.

The next figure represents a very simple electro-magnetic machine, which by its motion rings a bell with considerable force. N and S are the north and south poles of a common horse-shoe magnet, which stands in a vertical position, with its poles downward. A is a piece of soft iron wound with copper wire and fixed upon a vertical axis, so as to revolve very accurately between the poles of a magnet; and C C are binding screws for connecting the wires from the galvanic battery. The extremities of the wire around the magnet A connect with C C, by means of parts not represented in the figure, in such a manner that the current flows in the proper direction to develope north polarity in the further extremity A as it approaches the south pole of the magnet S, and south polarity in the other extremity as it approaches the north pole N: but the moment the revolving magnet has arrived at a position between N and S, the two poles of the stationary magnet, where it would be held if the same polarity remained, the direction of the current is reversed, and, as a necessary consequence, the polarity of the revolving magnet. If its momentum has now carried it a little beyond the point of greatest attraction, it will be urged on by repulsion until it has made a quarter of a revolution, when it will be attracted as before. At S on the vertical axis is a perpetual screw which acts upon the teeth of a ratchet wheel, and causes the hammer to strike the bell B.

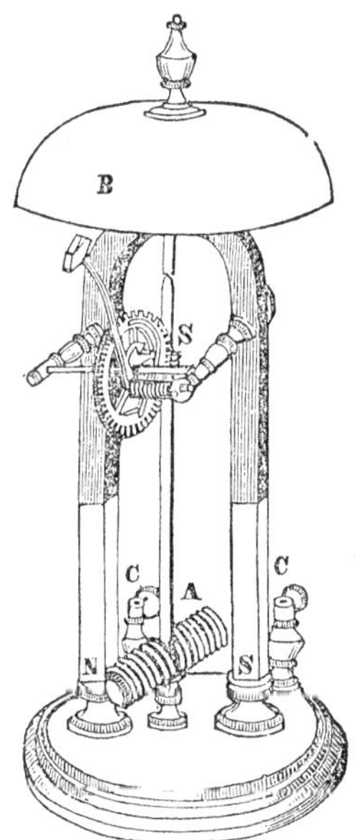

Electro-magnetic Engine.

The machine may be constructed with more than two magnets, and in a variety of forms, but whether it will ever be made practically useful remains to be determined.

QUESTIONS. — 140. What are *electro-magnetic ngines?* Describe the engine represented in the figure.

141. Mutual Influence of Two Currents.—Two currents moving in the same direction in the vicinity of each other are mutually attracted, but if moving in opposite directions they repel each other.

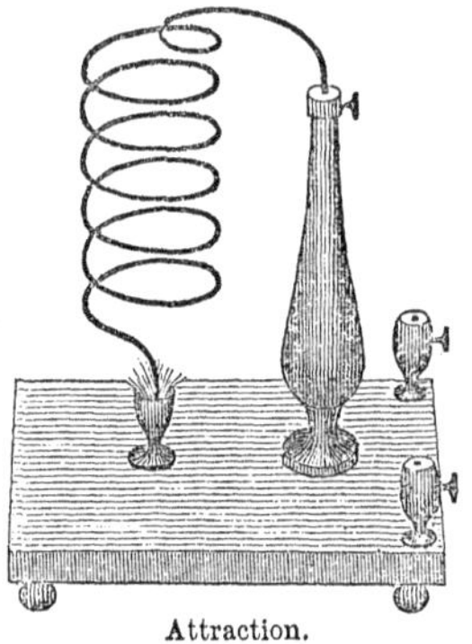
Attraction.

Let a helix of slender wire be suspended by the top so that the bottom may just touch in a cup of mercury; and let a current from a battery be passed through it. In consequence of the attractions between the coils, the spiral will be shortened, and the lower end raised from the mercury, thus breaking the circuit. But when the current ceases the end of the wire again falls so as to touch the mercury, and the current is renewed with the same effects as before. Every time the wire touches the mercury a brilliant spark is produced. Two currents therefore in the same direction attract each other.

142. Electro-Dynamic Induction.—A current of electricity passing through a conductor in a given direction, induces a current in the opposite direction in a second conductor situated parallel to the first. Thus, let A B be a portion of a wire connecting the poles of a galvanic battery, and M N a portion of a second wire parallel, and near to the first;—at the moment the circuit is formed and the current passes through the first wire in the direction from + to — a secondary current is induced in the second wire in the opposite direction. This current is but for an instant, but is renewed in the opposite direction when the battery current is again broken.

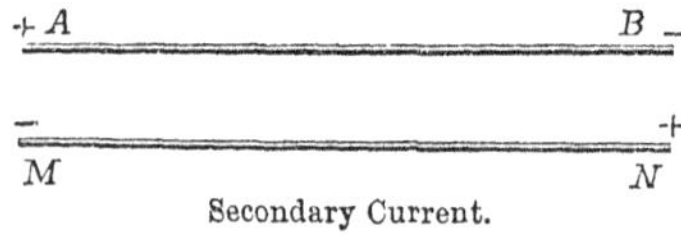

Secondary Current.

QUESTIONS.—141. How do two currents moving in the same direction affect each other? Describe the experiment illustrated in the annexed figure. 142. How may a current be made to induce another current called a secondary current?

A better method to demonstrate the existence of this secondary current is as follows: Let A be a coil of large copper wire, or, better, of copper ribbon, covered with some non-conducting substance, and having binding screws attached to each end, to receive the polar wires of the battery. Above this is placed a coil, W, of fine covered wire, with a metallic handle at each end; and when the battery current is made to pass through the lower coil, the secondary current will be induced in the upper coil, as will be plainly indicated by grasping the handles with the moist hands. In this case, as before, the secondary current is induced at the moment the battery current is established, in the direction opposite that of the battery current, and again when the battery current ceases in the same direction as this latter current.

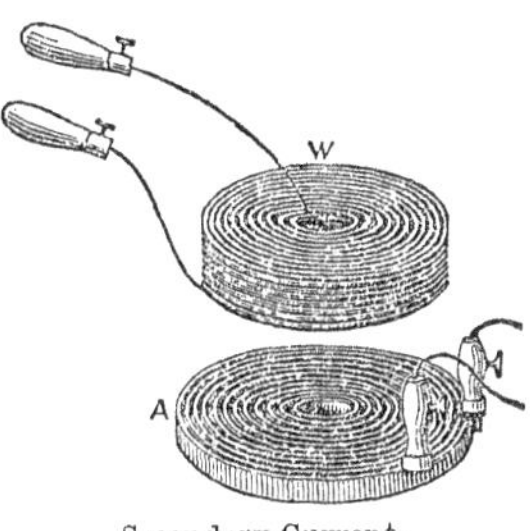

Secondary Current.

The former is termed the *initial* and the latter the *terminal* secondary current.

The effect of the shocks will be increased by rapidly breaking and closing the battery circuit, the feeble intensity of the shocks being compensated by their frequency.

Of the two coils above used, it is to be observed, that the one connecting the poles of the battery is made of large wire, or ribbon of metal, and is of moderate length, while the other, in which the secondary current is to be induced, is made of smaller wire, and is of much greater length. The secondary current in this case is one of considerable intensity (100), as is shown by the fact that it can be made to pass through the system; but if the coil had been made of a large wire, and much shorter than it is, the current induced would have had only a feeble intensity, but would have been much greater in quantity. And, in general, it is found that long coils of small wire give secondary currents of

---

QUESTIONS.—Describe the mode of obtaining a secondary current, as illustrated in the next figure. When only is this secondary current induced? What are the terms used to indicate these currents? What is their direction as compared with the primary current? What is said of the two coils used in this experiment?

great intensity, while, on the other hand, to obtain a current of quantity short coils of large wire or ribbon are required.

By arrangements altogether similar to the above, with additional coils, *tertiary* currents have been produced, and others of still higher orders, even up to the seventh or ninth.

The following mode of developing the *tertiary* current will show the method to be pursued to obtain others of still higher orders. A is a ribbon coil through which the battery current revolves as before, and B another ribbon coil in which a secondary current is induced, as already explained. A third ribbon coil, C, is placed at a little distance, but has its two ends connected by means of wires with the two ends of the ribbon of the coil B, so that the same current which is induced in B flows also in C. Above C is placed an intensity coil of small wire, W, with metallic handles connected with its extremities. Now when the battery current is established in the coil, A, a secondary current is induced in B and revolves in the coil, C, by which a *tertiary* current is induced in the coil W. As the secondary current flows in the opposite

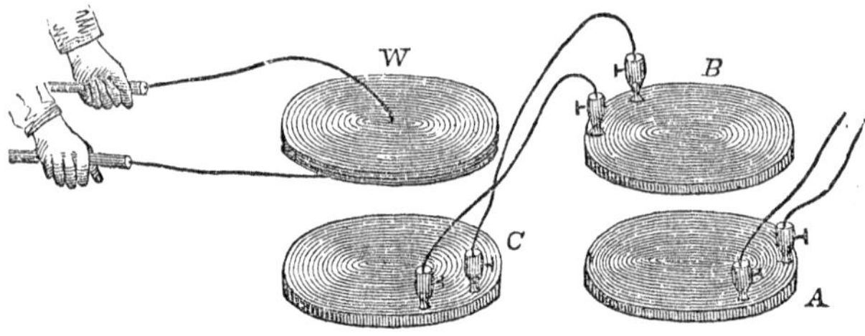

Tertiary Current.

direction from that of the battery current, so the tertiary takes the opposite direction from that of the secondary; and the same law holds good through all the series of currents. As the distance from the battery increases, the strength of the current gradually diminishes.

QUESTIONS.—Describe the mode of obtaining a tertiary current. What is said of the direction of the tertiary current as compared with that of the secondary current?

143. Induction of Electricity by Magnetism—Magneto-Electricity.—Magneto-electricity, as the term implies, is the reverse of electro-magnetism. The current of galvanic electricity circulating around a bar of soft iron, converts it, as we have seen, into a temporary magnet, and renders a bar of steel permanently magnetic. Now, *à priori*, it would seem very probable that a magnet placed in the centre of a helix or spiral of metallic wire, would develope in it a current of electricity. This is found to be actually the case; but the current can be observed only at the moment of inserting the magnet or removing it.

The development of electricity by magnetism is shown in a very simple manner, by winding the middle of the keeper or armature A B of a powerful horse-shoe magnet, with copper wire, properly bound, and bringing the two extremities of the wire into contact, one of which should be flattened a little, and amalgamated by dipping it into a solution of nitrate of mercury, and the other filed to a point. If, now, the armature be suddenly placed upon the magnet or removed from it, a spark of electricity will manifest itself every time, at the point of contact, C, of the two extremities of the wire.

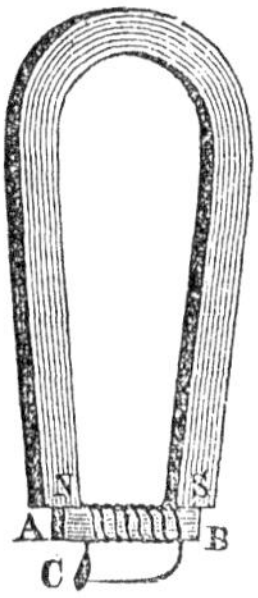

Magneto-Electricity.

The electric current flows in one direction at the moment magnetism is induced in the soft iron enclosed in the coil of wire, and in the other direction, when its magnetism is destroyed.

The same thing is accomplished, but in a manner a little different, by the apparatus figured on the next page. Externally a large coil of fine copper wire is seen, terminating in binding screws, to which wires with handles, H, are attached, to be grasped by the hands. Within this is a smaller coil of larger wire, the top of which is seen at B; the extremities of this wire are soldered to the binding screws A and D, the latter also connecting with the horizontal bar E C, which has an irregular surface, resembling that of a coarse file. Inside of the last coil is a bundle of small iron wires, W. Now when the wires from the battery are connected with the binding screws A and D, the current circulating in the

---

QUESTIONS.—143. What is the term *magneto*-electricity used to signify? Describe the method of developing electricity by means of a permanent magnet and armature. Describe the apparatus represented by the next figure, and the mode of using it.

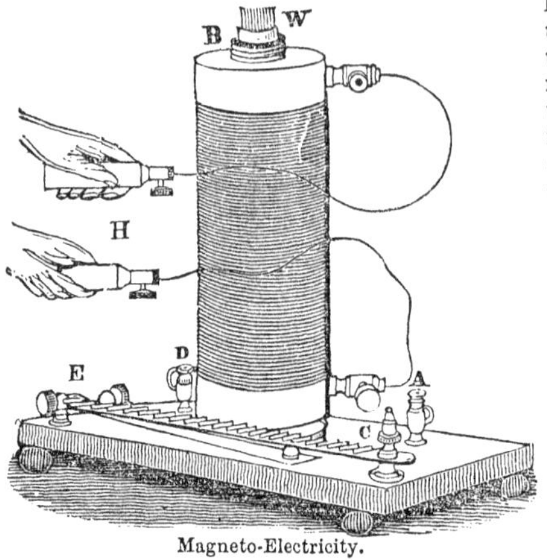

Magneto-Electricity.

helix B induces magnetism in the wires W, and this in turn at the moment when the magnetism is induced, or when it is destroyed, induces a current of electricity in the outer helix, which will be felt in the hands when grasping the handles.

To produce the greatest effect, the current from the battery should be formed and broken rapidly, which is accomplished by connecting one wire with the binding screw A, and drawing the end of the other over the rough surface E C.

144. *Magneto-electric machines* are now made, which act with great energy, producing all the effects both of quantity and intensity.

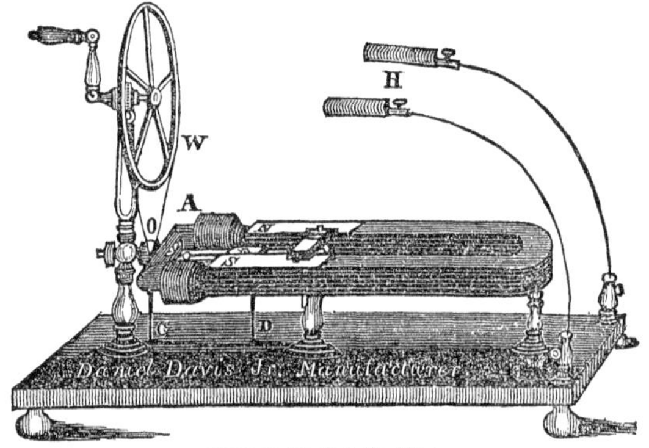

Magneto-Electric Machine.

The above figure* represents a machine of this kind. Several steel magnets, N S, are firmly fixed together upon pillars on a base board; and in front of the poles, an armature, A, in the

* From Davis's Manual of Magnetism.

QUESTION.—144. Describe the *magneto-electric machine.*

form of the letter U, having its two arms wound with 2000 or 3000 feet of fine insulated wire, is made to revolve on an axis by means of the multiplying wheel W. As the armature is made to revolve in front of the strong poles of the fixed magnets, magnetism is induced in it (121), currents of electricity being at the same time made to flow through the coils of wire wound upon its arms, in accordance with the principles just explained, which are communicated by wires, concealed under the base board, to the metallic handles, H, by the wires C and D. These currents being in one direction while the magnetism of the armature is increasing—that is, during one quarter of each revolution—and in the other direction while the magnetism is diminishing, would be scarcely appreciable but for a contrivance by which it is constantly interrupted and renewed. This is accomplished by a toothed wheel placed upon the axis at O, against which the wire C constantly presses, slipping from tooth to tooth as the wheel is turned. At every interruption of the current a powerful shock is felt by any one grasping the handles.

145. The Electro-Magnetic Telegraph.—This is an instrument for conveying intelligence instantaneously, by means of the galvanic current, to any distance, where metallic wires can be extended and properly insulated.

We have seen above (136), that the current, when made to pass around pieces of soft iron, renders them magnetic while the current is passing, but that they lose their magnetism instantly when the current is interrupted. Usually, the experiment is performed with the battery and the piece of iron in the same room, and even upon the same table, but this is not necessary. If the battery be in one room, and the piece of iron in another room, or in another building at a distance, the result will be the same, only a little increase in the power of the battery will be necessary. All that is required is, that good conducting wires, well insulated, should extend from the battery and form a helix around the iron, and on closing the circuit, whatever may be the distance

QUESTIONS.—What is the length of the wire upon the revolving armature? What is the use of the toothed wheel? 145. What is the *electro-magnetic telegraph*? May the current from a battery be made to magnetize a piece of iron at a distance? What is required for this purpose?

of the iron from the battery, it instantly becomes magnetic, and loses its magnetism again when the circuit is broken. The closing and breaking of the circuit can be performed at any point in the line, but we suppose it to be done at the battery.

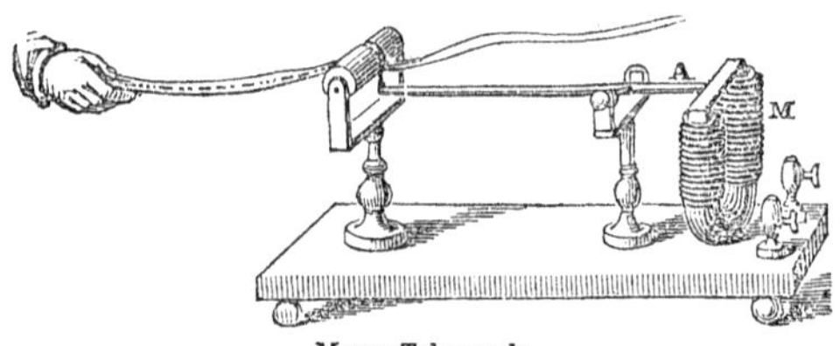

Morse Telegraph.

The above figure, which represents the recording part of Morse's telegraph, will perhaps serve better to illustrate the operation of the instrument than a full figure of it. A piece of soft iron, M, in the form of the letter U, with each arm surrounded with a helix of covered wire, is firmly fixed upon a base-board, with the ends upward, and the extremities of the wires of the helix connected with the binding screws, (seen at the right,) with which the wires from the battery are connected. A is an armature of soft iron, fixed to the short arm of a lever, which at the other end carries a blunt point of steel, capable of indenting characters upon paper when pressed against it. Above the point is a metallic roller, with a groove cut around it, into which the point plays, and a narrow slip of paper is supposed to be drawn along between them by the hand.

Now let us suppose this instrument, which we will call the register, to be in New York, and the battery in Washington, with metallic wires, well insulated, extended between them; when the operator in Washington closes the circuit, the iron M instantly becomes magnetic, drawing down the armature A, and pressing the point at the other extremity of the lever against the paper, producing a dot if the paper is at rest, but a straight line if it is in motion. When the operator in Washington breaks the

---

QUESTIONS.—Describe the figure given to illustrate the recording part of Morse's telegraph. How is a mark made upon the paper? Describe the mode in which an operator in Washington may write in New York. What are the only marks the instrument is capable of making?

circuit, the iron M loses its magnetism, and the lever drops by its own weight, leaving the paper to move along unmarked. When the circuit is again closed, the point is raised again in like manner against the paper, and immediately drops when the circuit is broken; so that the steel point in New York is perfectly under the control of the operator in Washington, for making dots and straight lines, with which an alphabet is easily constructed. Thus a dot and line (. —) is A, a line and three dots (— . . .) B, a single dot (.) E, and so on through the alphabet.

In the next figure the full instrument of Morse is represented. The paper, P, is carried along by machinery, between two rollers, in the direction indicated by the arrows, the machinery being propelled by a weight. A spiral spring, S, attached to a piece projecting downward from the pen-lever, removes the steel point from the paper when the circuit is broken. W W, the wires of

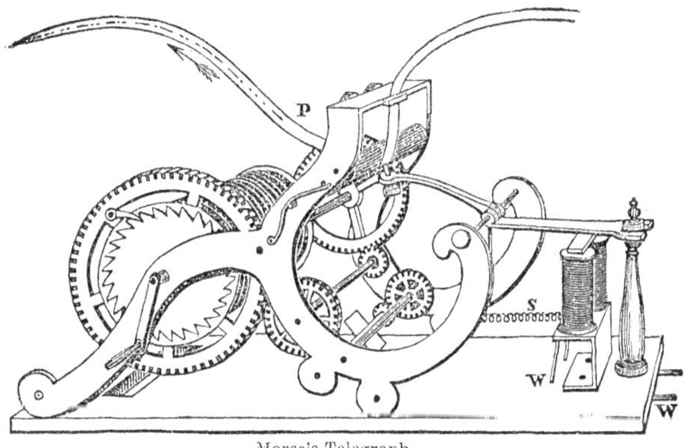

Morse's Telegraph.

the magnet. The motion of the machinery is regulated by a fan, which is partly seen in the figure.

146. In order to send communications at the same time from New York to Washington, another arrangement similar to the above is needed, with the register in Washington; but the same wires and the same battery may be used to return an answer, after

Questions.—How is the alphabet used in this telegraph constructed? 146. May the same battery and the same wires be made use of to send communications between two places in both directions?

the Washington operator has finished, merely by having a second register in the same circuit in Washington, and the operator in New York breaking and closing the circuit in that place, as before explained. Nor is it necessary, if the register is in New York, the battery should be in Washington; it may be at any point in the circuit.

In the above explanation, two wires are supposed to be extended the whole distance between the places connected by the telegraph; but experiment has shown that only one is really needed, as the earth may be made to form a part of the circuit.

Besides the above, two other electric telegraphs are in use in this country, and one or two of still different construction in Europe. House's telegraph (from the name of the inventor) is a most ingenious instrument, which performs the wonderful operation of printing in distinct capitals the messages transmitted. It is too complex to be understood without a close examination of the instrument itself.

Bain's telegraph is worked precisely like Morse's, but the marks are made on chemically prepared paper, by transmitting the current of electricity through it. Like Morse's, also, it marks only dots or straight lines.

QUESTIONS.—Are more than one wire needed to send communications between two places in both directions? Are other telegraphs besides Morse's in use?

PART II.

GENERAL CHEMISTRY.

147. THERE are two very distinct sub-divisions of the general science of Chemistry, usually denominated Inorganic Chemistry, and Organic Chemistry; the former of which treats only of the chemical history of the elements and their inorganic compounds, while the latter treats of the chemistry of animal and vegetable bodies. There are, however, certain principles equally applicable to both branches of the science, which we propose first to consider, under the above head.

THE ELEMENTS.—CHEMICAL AFFINITY.

148. The Elements.—We consider all substances as simple, or elementary, that have not been decomposed. Of the exact number of these we cannot speak with certainty, as the existence of two or three, the discovery of which has been announced, has not yet been sufficiently verified. On a following page will be found a list of sixty-two, the existence of which, it is believed, has been established. Some of these, as we shall see hereafter, are found only in very small quantities, while others are abundant, the great mass of the globe we inhabit being made up of the various compounds of some fourteen or sixteen.

The elementary substances are usually divided into the two classes of non-metallic and metallic, fifteen being included in the former, and the remainder in the latter class; but several of each class possess characters so peculiar, that it is difficult to assign them their proper place.

QUESTIONS.—147. What two sub-divisions of the general science of Chemistry are there? What is the distinction between them? Of what is it proposed to treat under the head of General Chemistry? 148. What are simple substances? How many elements are there? What is said of the quantities of these afforded by nature? What two divisions of them do we recognize?

149. Affinity.—Affinity, or chemical attraction, has already been mentioned as one of the forces with which the particles of every kind of matter seem to be naturally endowed. It is to the action of this force that all chemical changes, and the accompanying phenomena, are to be attributed. If two elements unite to form a compound, it is this force which occasions it; and if a com pound of two or more substances, already formed, is destroyed by the action of another element, it is to this force we are to look for the cause of the change.

Affinity is exerted only between the *minutest particles* of different kinds of matter, causing them to combine so as to form new bodies, endowed with new properties. It acts only at insensible distances; in other words, apparent contact, or the closest proximity, is necessary to its action. Every thing which prevents such contiguity is an obstacle to combination; and any force which increases the distance between particles already combined, tends to separate them permanently from each other. It follows, therefore, that, though affinity is regarded as a specific power, distinct from the other forces which act on matter, its action may be promoted, modified, or counteracted by them; and, consequently, in studying the phenomena produced by affinity, it is necessary to inquire into the conditions that influence its operation.

150. *The most simple instance* of the exercise of chemical attraction is afforded by the admixture of two substances. Water and sulphuric acid, or water and alcohol, combine readily. On the contrary, water shows little disposition to unite with sulphuric ether, and still less with oil; for, however intimately their particles may be mixed together, they are no sooner left at rest than the ether separates almost entirely from the water, and a total separation takes place between that fluid and the oil. Sugar dissolves very sparingly in alcohol, but to any extent in water; while camphor is dissolved in a very small degree by water, and abundantly by alcohol. It appears, from these examples, that chemical attraction is exerted between different bodies with different degrees

QUESTIONS.—149. What is affinity? Are all chemical changes to be referred to this force? Between what only is it exerted? Is it influenced in its action by other forces of nature? 150. Where do we observe the most simple instance of the exertion of this force? Is it exerted with equal intensity between all bodies?

of force. There is sometimes no proof of its existence; between some substances it acts very feebly, and between others with great energy.

Affinity is therefore said to be elective, as, when several substances, capable of combining, are mixed together, a particular compound is always formed, in preference to others. Thus, if sulphuric acid, soda, and lime are at any time mixed together, the acid will always combine with the lime, in preference to the soda.

151. *Decomposition* of a compound is often effected by presenting to it a third substance, having a stronger affinity for one of the ingredients of the compound than those ingredients have for each other. Thus, oil has an affinity for the volatile alkali, ammonia, and will unite with it, forming a soapy substance called liniment. But the ammonia has a still greater attraction for sulphuric acid; and hence, if this acid be added to the liniment, the alkali will quit the oil, and unite by preference with the acid. If water be poured into a solution of camphor in alcohol, the camphor will be set free, because the alcohol combines with the water. Sulphuric acid, in like manner, separates baryta from nitric acid. Combination and decomposition occur in each of these cases;—combination of sulphuric acid with ammonia, of water with alcohol, and of baryta with sulphuric acid;—decomposition of the compounds formed of oil and ammonia, of alcohol and camphor, and of nitric acid and baryta.

The action becomes more complex when two compounds are presented to each other, of such a nature that a transfer of elements takes place between them. Thus, in mixing a solution of carbonate of ammonia with another of hydrochlorate of lime, the carbonic acid of the first compound unites with the lime of the second, while the hydrochloric acid of the second combines with the ammonia of the first.

152. Action of Affinity Modified by Circumstances.—The action of affinity is influenced greatly by various circumstances, only a few of which can be here noticed.

QUESTIONS.—Why is affinity said to be elective? 151. How is decomposition effected by the exertion of this force? Give an illustration. 152. Is the action of affinity modified by the circumstances in which it is exerted?

The particular state of a substance, whether solid, liquid, or gaseous, is a very important circumstance, intimately affecting all its chemical relations. It is very rarely the case that two solids are capable of combining;—one of them at least must first be made liquid, either by solution in some solvent, or by melting. Thus, tartaric acid and carbonate of soda, in a dry state, are kept together for any length of time, without any action taking place between them; but if a little water be added to dissolve them, chemical action at once ensues, attended with violent effervescence. Phosphorus and iodine, though both are solids, and phosphorus and sulphur, will indeed combine when brought together in their ordinary state; but even in these cases, one or both of the substances are made liquid during the action.

153. *Cohesion* (for it is this force which unites the particles of a solid) is therefore always opposed to the action of affinity; and whatever tends to diminish it in bodies capable of acting upon each other, facilitates their union. Heated water is therefore usually a more powerful solvent than when cold; and salt in fine powder will be dissolved more rapidly in water, than when in large lumps.

154. *The gaseous state* is also unfavorable to the action of affinity. Gases do indeed, in some cases, combine spontaneously, but, usually, the introduction of an ignited body, the electric spark, or, in some cases, the influence of the sun's rays, is needed to cause the action to commence. Thus, a mixture of oxygen and hydrogen may be kept for any length of time; but the introduction of flame, or the passage of the smallest spark of electricity, will cause them to combine instantly, with a powerful explosion.

155. *Contact with a third body* sometimes will cause a union of two elements, that would otherwise remain together without combining. Thus, the introduction of a piece of spongy platinum into a mixture of oxygen and hydrogen, produces an explosion in a few seconds, without any change being produced in the metal,

---

QUESTIONS.—What is said, in this connection, of the peculiar state of a substance, whether it be solid, liquid, or gaseous? Give an example. 153. Is cohesion always opposed to affinity? 154. What is said of the gaseous state? How may gases that remain in mixture without uniting afterwards be made to combine? 155. What is *catalysis?*

except that it becomes intensely heated. This power of some bodies to produce chemical changes, merely by their presence, is termed *catalysis* (from the Greek, *kata*, by, and *luo*, to unloose).

156. *Mechanical action* also favors the action of affinity. Thus, if a piece of phosphorus be wrapped, with a few crystals of chlorate of potash, in a piece of tin foil or strong paper, a smart blow will cause a violent explosion. Common friction matches are ignited by friction against sand-paper, or other hard substance.

157. *Relative quantity of matter* is the last circumstance we mention, as affecting the action of this force. What is meant by this may be illustrated by the solution of common salt in water. If equal quantities of the salt are added in succession to the water, the first portion will disappear in less time than the second, the second in less time than the third, and so on. As the relative quantity of salt contained in solution increases, the action of the water becomes enfeebled, until full saturation takes place. If a large quantity of salt had been added at first, the full saturation of the water would have taken place much more speedily.

158. Action of Affinity always accompanied by a Change of Properties.—A change of properties, to a greater or less extent, always attends chemical action; but no means are yet known by which we can predict what any of these changes will be in any case, previous to making the trial.

Often, when two bodies unite, the characteristic properties of both will disappear, and the bodies are said to *neutralize* each other. Thus, the acids are usually sour to the taste, and possess the power of changing the blue color of some vegetables to red; while the alkalies are exceedingly caustic to the taste, and change vegetable blues to green. Now, when an acid and an alkali unite, a new substance is formed, called a salt, which is mild to the taste, and produces no effect whatever upon vegetable colors;

QUESTIONS.—156. What is said of mechanical action? Give an illustration. 157. What is said of the relative quantities of matter that are brought to act upon each other? Give an illustration. 158. Does a change of properties usually accompany chemical action? Can we from a knowledge of the substances used, predict before trial what the changes will be? When are substances said to *neutralize* each other? Give an illustration.

and, indeed, while it possesses many new properties of its own, it exhibits none of those of either of its ingredients.

159. This change may extend to any or all of the properties of bodies. (1.) Bodies, after combining, do not usually occupy the same space as they did before combination. Generally, a contraction of volume takes place, but this is not universal. A pint of water, added to a pint of sulphuric acid, will not produce a quart of the mixture; and the same will be found true of water and alcohol. When two gases combine, a very great contraction often takes place; but the result with different gases is very different. (2.) The changes of form that attend chemical action are exceedingly various. The combination of gases may give rise to a liquid, as in the union of oxygen and hydrogen to form water; or to a solid, as in the union of carbonic acid gas and ammonia to form solid carbonate of ammonia; or hydrochloric acid and ammonia, to form hydrochlorate of ammonia. Two solids may, in combining, form a liquid, as is the case when crystals of sulphate of soda and nitrate of ammonia are rubbed together in a mortar, or acetate of lead and alum. Solids may also, in combining, form gases, as is the case when gunpowder detonates. Two liquids, by uniting, may form a solid, as may be shown by pouring sulphuric acid into a solution of hydrochlorate of lime. (3.) Chemical action is frequently attended by change of color. No uniform relation has been traced between the color of a compound and that of its elements. Iodine, whose vapor is of a violet hue, forms a beautiful red compound with mercury, and a yellow one with lead. The black oxide of copper generally gives rise to green and blue-colored salts; while the salts of the oxide of lead, which is itself yellow, are for the most part colorless.

A beautiful instance of the change of color produced by chemical action is seen in mixing solutions of chloride of mercury and iodide of potassium. The solutions may be made as perfectly limpid as water, but, upon being mixed, a beautiful vermilion red is produced, by the formation of iodide of mercury. The color shortly disappears, if either solution was in excess, by the redissolving of the precipitate. About 27 parts of the chloride should be used with 33 parts of the iodide.

Questions.—159. May the change extend to all the properties of bodies? Give some illustrations.

LAWS OF COMBINATION.—ATOMIC THEORY.

160. Laws of Combination.—The relative proportions in which substances unite to form the different compounds, is governed by fixed laws. There are, however, some apparent exceptions to this rule, in which bodies seem to unite in all proportions, without reference to any law. Thus, water and alcohol seem to unite in all proportions; and the same may be said of water and the liquid acids. There are still other substances which seem to combine in any proportion within certain limits. Thus, water dissolves common salt very readily, but the quantity it is capable of holding in solution cannot exceed about four-tenths of its own weight. Below this limit, the water and salt appear to unite in every proportion.

161. The following are the laws which regulate the composition of such compounds :—

I. *The composition and properties of compound bodies are unchangeable.*

By this it is meant (1.) that any compound, while it retains its characteristic properties, must contain the same elements, united in the same proportions; and (2.) that while a compound contains the same elements, united in the same proportion, it must also possess the same characteristic properties. Thus, water, a compound of 1 part* of hydrogen and 8 parts of oxygen, possesses certain well-known properties; now, whenever and wherever a substance is found, possessing the various properties of water, we know, from this law that it must be a compound of these two substances, united in the above proportion; and whenever a compound of these substances, in this proportion, is formed, it must possess the peculiar properties of water.

A change of properties always necessarily implies a change of composition; and, conversely, a change of composition necessarily implies a change of properties. This law is of universal application, except in the case of isomeric compounds, which will be hereafter noticed.

* Parts by weight are always intended, unless it is otherwise expressed.

QUESTIONS.—160. Are the relative proportions in which bodies combine governed by any law? What example is mentioned? 161. State the first law of combination. What is water composed of? Does a change of properties always imply a change of composition?

II. *When any substances (as B, C, D, &c.) combine with a given quantity of another substance (A), then the numbers which represent the proportions in which B, C, D, &c., combine with A, will also represent the proportions in which they will combine with each other, if such combination be possible.*

This law is also of universal application, and examples to illustrate it are abundant. Thus,

8·0	parts of	oxygen	combine with 1 part of hydrogen.
35·4	"	chlorine	
16·0	"	sulphur	
127·0	"	iodine	
80·0	"	bromine	

It follows, therefore, that if oxygen and chlorine combine, it will be in the ratio of 8 parts of the former to 35·4 parts of the latter; and if chlorine and iodine combine, the compound will contain 35·4 parts of the former, and 127 parts of the latter; and so of the other substances mentioned.

But it is more common, as oxygen unites with nearly all other substances, to determine the quantities severally of these which combine with 8 parts of this element. Thus,

16·0	parts of	sulphur	combine with 8 parts of oxygen.
35·4	"	chlorine	
1·0	"	hydrogen	
6·0	"	carbon	
14·0	"	nitrogen	
39·1	"	potassium	
31·0	"	phosphorus	

Any other substance, having an extensive range of affinity, might also be selected as the basis of our table, and the results would be the same. The numbers thus obtained for the various substances are called their *combining numbers, equivalents,* or *atomic weights.* It will be noticed that the numbers used merely express the *relative quantities* of the substances they represent, that combine together; it is therefore in itself immaterial what figures are employed to express them. The only essential point is, that the relation should be strictly observed. Thus, the equivalent of hydrogen may be assumed as 10; but then the number for oxygen will be 80; that for chlorine 354, &c. Hydrogen

QUESTIONS.—State the second law of combination. Give an example in illustration. What are the combining numbers, equivalents, or atomic weights of substances?

combines with other bodies in a lower proportion than any other known substance, and is therefore, with propriety, made the unit by most writers in the English language; but on the continent of Europe, oxygen is usually considered as 100, and the tables constructed accordingly.

162. We give below a table of all the known elementary substances, with their combining numbers, or equivalents, the combining weight of hydrogen being considered as the unit. To find the corresponding numbers in a table in which the combining weight of oxygen is made 100, it is only necessary to multiply the numbers in this table by 12·5. We insert also, in the table, the symbols by which the elements are represented in chemical formulæ, a subject to which the attention of the student will be called in a future paragraph.

TABLE OF ELEMENTS, WITH THEIR EQUIVALENTS AND SYMBOLS.

Elements.	Symbols.	Equivalents.	Elements.	Symbols.	Equivalents.
Aluminum	Al	13·7	Molybdenum	Mo	46·0
Antimony (Stibium)	Sb	129·0	Nickel	Ni	29·5
Arsenic	As	75·0	Nitrogen	N	14·0
Barium	Ba	68·5	Norium	No	?
Bismuth	Bi	208·0	Osmium	Os	99·5
Boron	B	10·9	Oxygen	O	8·0
Bromine	Br	80·0	Palladium	Pd	53·2
Cadmium	Cd	56·0	Pelopium	Pe	?
Calcium	Ca	20·0	Phosphorus	P	31·0
Carbon	C	6·0	Platinum	Pt	99·0
Cerium	Ce	47·0	Potassium (Kalium)	K	39·1
Chlorine	Cl	35·4	Rhodium	R	52·2
Chromium	Cr	26·7	Ruthenium	Ru	52·2
Cobalt	Co	29·5	Selenium	Se	39·5
Columbium	Cb	?	Silicon	Si	21·3
Copper (Cuprum)	Cu	31·7	Silver (Argentum)	Ag	108·0
Didymium	Di	48·0	Sodium (Natrium)	Na	23·0
Erbium	E	?	Strontium	Sr	44·0
Fluorine	F	19·0	Sulphur	S	16·0
Glucinum	G	4·7	Tantalum	Ta	184·0
Gold (Aurum)	Au	198·0	Tellurium	Te	64·1
Hydrogen	H	1·0	Terbium	Tb	?
Iodine	I	127·0	Thorium	Th	59·4
Iridium	Ir	99·0	Tin (Stannum)	Sn	58·0
Iron (Ferrum)	Fe	28·0	Titanium	Ti	25·0
Lanthanum	La	47·0	Tungsten (Wolfram)	W	92·0
Lead (Plumbum)	Pb	103·5	Uranium	U	60·0
Lithium	L	6·5	Vanadium	V	68·5
Magnesium	Mg	12·0	Yttrium	Y	32·2
Manganese	Mn	28·0	Zinc	Zn	32·5
Mercury (Hydrargyrum)	Hg	100·0	Zirconium	Zr	34·0

QUESTIONS.—162. What are contained in the table on this page? What is taken as the unit? How may the numbers given in the table be converted into others, in which the combining weight of oxygen is 100?

Besides the preceding sixty-two elements, the existence of which seems to be well established, the discovery of four others has been announced, which are to be considered as doubtful. The names Aridium, Donarium, Ilmenium, and Thalium have been given to them.

The equivalents given in the table have been taken chiefly from Dana's table, found in the last edition of his Mineralogy, but are reduced to the hydrogen standard. In some few instances preference has been given to Regnault's numbers.

III. *When two substances combine in more proportions than one, then these different proportions will always sustain some simple ratio to each other.*

To illustrate this law, let A and B be two substances capable of combining with each other in several proportions;—the lowest proportion in which they combine will be in the ratio of one equivalent of each, or 1 A + 1 B. The other compounds they form will be of the character, 1 A + 2 B, 1 A + 3 B, 1 A + 4 B, &c.; that is, 1 equivalent of one substance will be united with 2 or 3, or some exact number of equivalents of the other substance.

The compounds of nitrogen and oxygen afford an excellent illustration of the above law. Thus, the

1st	compound	contains—Nitrogen,	14,	and	Oxygen,	8.
2d	"	" Do.	14,	"	Do.	16.
3d	"	" Do.	14,	"	Do.	24.
4th	"	" Do.	14,	"	Do.	32.
5th	"	" Do.	14,	"	Do.	40.

The law, however, admits of a more general application than these examples alone would authorize us to expect. Substances may combine—and instances of the kind are not unfrequent—in the ratio of 2 equivalents of the first to 3 or 5, &c., equivalents of the second; and even more complex ratios than these are known, as 3 equivalents of one substance to 4, 5 or 7 of the other, but they are not common. They may be expressed thus, 2 A + 3 B, 2 A + 5 B, &c., and 3 A + 4 B, 3 A + 5 B, &c.

Manganese and oxygen combine in five different proportions, as shown below. The second compound, it will be observed, contains 2 equivalents of the metal to 3 equivalents of oxygen, while the fifth contains 2 equivalents of the metal to 7 equivalents of oxygen. Thus, the

1st	compound	contains—Manganese,	28,	and	Oxygen,	8.
2d	"	" Do.	56,	"	Do.	24.
3d	"	" Do.	28,	"	Do.	16.
4th	"	" Do.	28,	"	Do.	24.
5th	"	" Do.	56,	"	Do.	56.

QUESTIONS.—State the third law of combination. Give an illustration.

These complex ratios may be expressed fractionally; thus, the expressions 2 A + 3 B, 2 A + 7 B, and 3 A + 4 B, are equal respectively to 1 A + $1\frac{1}{2}$ B, 1 A + $3\frac{1}{2}$ B, 1 A + $\frac{4}{3}$ B.

In the manganese series above given one place, it will be seen, is wanting, between the fourth and fifth, which may hereafter be filled by further research.

IV. *The equivalent of a compound substance will always b equal to the sum of the equivalents of the substances which compose it.*

Thus, water is composed of 1 equivalent of oxygen 8, and 1 equivalent of hydrogen 1; its combining number, or equivalent, will therefore be (8 + 1 =) 9. So, also, sulphuric acid, which contains 1 equivalent of sulphur (16) and 3 equivalents of oxygen (8 × 3 = 24), has for its equivalent 40.

The same is true of all compound bodies, as hydrochloric acid, which contains 1 equivalent of chlorine, 35·4, and 1 equivalent of hydrogen, 1, giving for its equivalent 36·4. The equivalent of potassium is 39·1, and as this combines with 8 of oxygen to form potassa, the equivalent of the latter is 47·1 = 39·1 + 8. Now, when any compound substances unite with each other, one equivalent of one substance combines with one, two, or three equivalents of the other substance, precisely as in the case of the elements. Thus, the hydrate of potassa, composed of potassa and water, contains exactly 1 equivalent, 47·1 parts of potassa and 1 equivalent, or 9 parts of water; and its equivalent is 56·1 = 47·1 + 9. So the sulphate of potassa is composed of 40 parts of sulphuric acid and 47·1 parts of potassa, having for its equivalent 87·1; and the nitrate of potassa, composed of nitric acid and potassa, contains 54 parts of the acid and 47·1 parts of potassa, its equivalent being 101·1.

V. *The quantities of gaseous substances, estimated by measure, which enter into combination, bear some simple ratio to each other, and also the quantities of the resulting compounds, considered as gaseous, bear some simple ratio to the sum of the volume of the ingredients.*

Water, we have seen, is composed of 1 equivalent (estimated by weight) of each of its elements, hydrogen and oxygen, which

QUESTIONS.—State the fourth law. What illustration is given?

are gases; but if we measure them before causing them to combine, we shall find that 1 pint of oxygen will unite with exactly 2 pints of hydrogen. If we use more oxygen than this in proportion, a part of it will remain after combination; and if we use less, a part of the hydrogen will be left.

So, a pint of chlorine will combine with exactly a pint of hydrogen, and a pint of nitrogen with exactly 3 pints of hydrogen form ammonia; and so of other gaseous substances, or any capable of taking this form, as sulphur or mercury.

If we represent the combining volume of oxygen by 1, as is usual, then the combining volume of each of the following substances, viz., hydrogen, mercury, nitrogen, chlorine, iodine, and bromine, will be 2. Phosphorus and arsenic have the same combining volume as oxygen, 1; while that of sulphur is $\frac{1}{3}$.

These numbers are determined by experiment precisely as the numbers representing the proportion in which substances combine by weight.

It follows from the above that there must be a close relation between the atomic weights, the combining volumes, and the densities of gaseous substances. The atomic weights of oxygen and hydrogen, we have seen, are as 8 to 1, while their combining volumes are 1 to 2; it follows, therefore, that the density of oxygen must be to that of hydrogen as 16 to 1. So the atomic weights of chlorine and hydrogen are as 35·4 to 1, but they combine in equal volumes; their comparative densities must therefore be as their atomic weights, that is, as 35·4 to 1. In fact, in all cases in which the gases combine in equal volumes, their densities respectively must be as their atomic weights. And in the cases of those which do not combine in equal volume, the simple ratio of the combining volume being observed, it is easy to calculate their relative densities from their atomic weights. Thus the combining volumes of hydrogen and sulphur vapor are, respectively, 2 and $\frac{1}{3}$, and their atomic weights 1 and 16; the

QUESTIONS.—State the fifth law of combination. What illustration is given? Is there any relation between the atomic weights, combining volumes, and densities of gaseous bodies? How is this illustrated by reference to oxygen and hydrogen? When the combining volumes and atomic weights of gases are given, may their relative densities be calculated? How is this illustrated by reference to sulphur and hydrogen?

density of sulphur vapor, compared with that of hydrogen, must therefore be 96, which is very nearly the same as it is found to be by direct experiment.

It is customary to refer the densities of gaseous bodies to that of atmospheric air (at a given temperature and pressure) taken as unity; and we have then, as the result of very many experiments, the numbers in the table below. These numbers will be of use hereafter.

One volume of Air weighs	1·000	One vol. of Iodine vapor weighs	8·716
" " Hydrogen weighs	·069	" " Bromine " "	5·400
" " Oxygen "	1·106	" " Mercury " "	6·976
" " Nitrogen "	·971	" " Sulphur " "	6·654
" " Chlorine "	2·440	" " Phosphorus vapor weighs	4·326
		" " Arsenic " "	10·370

By the second part of the fifth law, as enunciated above, the combining volumes of compound gases will bear a simple ratio to the sums of the volumes of their respective ingredients. In some cases, no change of volume takes place when gases combine, as in the case of chlorine and hydrogen, where equal volumes combine and produce two volumes of hydrochloric acid; but more frequently a change of volume is observed, as in the case of vapor of water, which is composed of 1 vol. of oxygen and 2 vols. of hydrogen condensed to 2 vols. of steam. So ammonia is a compound of 1 vol. of nitrogen and 3 vols. of hydrogen condensed into 2 vols. of ammonia.

The density of a compound gas may therefore often be calculated from the density of its ingredients and the change of volume known to take place in combination. Thus, in the case of water, which we have seen is composed of 1 vol. of oxygen and 2 vols. of hydrogen condensed into 2 vols. of steam,

1 vol. of	Oxygen =		1·106
2 "	Hydrogen =	2 × ·069 =	0·138
2 "	Steam =		1·244

One-half of this sum, 1·244, or 0·622, is therefore the weight of 1 vol., or, in other words, the density of steam. This is the same as found by direct experiment. So in the case of ammonia, mentioned above,

3 vols. of	Hydrogen =	3 × ·069 =	0·207
1 "	Nitrogen =		0·971
2 "	Ammonia =		1·178

As this forms 2 vols. of ammonia, one-half of the sum, 1·178, or, ·589 is the weight of 1 vol., or the density of this compound gas.

We add one more instance, by way of illustration. Hydrosulphuric

QUESTIONS.—Does a change of volume sometimes take place when gases combine? How is this illustrated in the case of water? May the density of a compound gas be calculated from the densities of its ingredients and the change of volume that takes place when they combine?

acid is a compound of a single equivalent of hydrogen and sulphur, the combining volume of the former being 1, and that of sulphur $\frac{1}{6}$. Now,

1 vol. of Sulphur vapor =		6·654
6 " Hydrogen =	6 × 0.069 =	·414
6 " Hydrosulphuric acid	=	7·068

As the 7 vols. in combining form but 6 vols. of the gas, the sum 7 068 is to be divided by 6, giving us for the density of hydrosulphuri acid, 1·178.

Some substances, as carbon, cannot by any means yet known be obtained in a state of vapor; still we may with some certainty calculate what the density of their vapor would be if vaporization were possible. Thus, in the case of carbon, assuming that carbonic acid is composed of 1 volume of carbon vapor and 2 volumes of oxygen, the whole condensed in 2 volumes, we have—

2 vols. of Carbonic Acid	= 2 × 1·524 =	3·048
2 " Oxygen (subtract)	= 2 × 1·106 =	2·212
1 " Carbon vapor	=	·836

One volume of carbonic acid we know contains 1 volume of oxygen, but whether the carbon vapor should be taken as 1 volume or only $\frac{1}{2}$ of a volume, we have no means of determining, except the analogy of other compound gases, and therefore our assumption above may not be correct. It is possible that carbonic acid may have a more complex organization and carbon a very different combining volume.

The combining volume of oxygen being taken as 1, those of all the elementary bodies capable of assuming the gaseous form are 1, 2, or 4, except that of sulphur, which is $\frac{1}{3}$. Of the very many compound gases whose combining volumes have been determined, that of nearly every one is either 2 or 4.

163. Atomic Theory.—It will be observed that nothing theoretical pertains to the above laws, which are simply the enunciation of well-determined facts; but such striking results, obtained by experiment, naturally incline us to inquire for their cause; and, in the absence of positive proof, the *atomic theory* has been proposed for this purpose.

This theory assumes that every simple substance is an aggregation of atoms (9), by which is meant the particles in their smallest state of division; and that the atoms of the same substance are all precisely of the same weight. It assumes, also, that when simple substances combine to form compounds, they unite by atoms; that is, one atom of one substance combines with one, two, three, &c., atoms of the second; or two atoms of the

---

QUESTION.—163. Explain the atomic theory.

first combine with three, five, or seven, &c., atoms of the second, &c. As these atoms are supposed to be absolutely indivisible, there can, of course, be no such thing as half an atom; and all compounds must be of the form 1 A+1 B, 1 A+2 B, 2 A+3 B, &c. The absolute weight of an atom of any substance has never been determined, but it is assumed that the equivalents of the various bodies do actually express their relative weights; the term *atomic weight* is therefore often used as synonymous with equivalent. Thus, the atomic weight of oxygen is said to be 8, and that of hydrogen, 1, &c.; and water is said to be a compound of one atom of oxygen and one atom of hydrogen; while sulphuric acid is a compound of one atom of sulphur and three atoms of oxygen—and so of other compounds.

Whether the assumptions of this theory are strictly true, we may never be able to determine experimentally, but, it is readily seen, that if they are a correct expression of what really takes place in nature, they afford an elegant explanation of the reason of the above demonstrated laws.

164. **Specific Heat of Atoms.**—The quantities of heat required to raise the temperature of equal weights of different bodies by a given number of degrees (59), varies greatly, according to their nature; and the numbers expressing these quantities, having reference usually to water as the standard, express also the capacities for heat or the specific heats of these bodies. Between these numbers no special relation can be traced; but if we make the calculation in reference, not to *equal* weights, but to *atomic* weights, then in some nine or ten of the elementary bodies the specific heats are found to be the same, while several others have a specific heat twice or four times as great.

It has been supposed that the specific heat of the atoms of all bodies may be the same—a view which is favored by the facts above mentioned—but most bodies, both elementary and compound, depart so widely from this supposed law that its existence is altogether improbable.

NOMENCLATURE OF CHEMISTRY—SYMBOLS.

165. **Nomenclature.**—Chemistry possesses a more systematic nomenclature than any other branch of natural science; and a thorough knowledge of it, at the very beginning of his studies, is

QUESTIONS.—Does this theory account satisfactorily for the preceding laws of combination? 164. Is the specific heat of the atoms of all bodies the same? 165. What is said of the nomenclature of chemistry?

very important to the student. This nomenclature is framed in reference to the composition of compounds, and is so contrived that the names of all compounds shall indicate the substances of which they are composed.

166. *Elementary substances* being composed of only one kind of particles, of course the above remark does not apply to them: their names are mere names; that is, mere sounds connected by usage with the things signified. Yet, in the case of newly discovered elements, names have in some instances been given that indicate some important property of the substance. Thus, oxygen (from the Greek *oxus*, acid, and *gennao*, to produce) was so named, because it was supposed to form a necessary part of all acids; and hydrogen (from *hudor*, water, and *gennao*), because it was known to enter into the composition of water. So chlorine, being of a greenish color, received its name, in consequence, from the Greek *chloros*, green; and bromine was so called from its offensive odor, from *bromos*, fetid. Potassium and sodium are so named, because they form the basis, respectively, of potassa and soda; and glucinum (from *glukus*, sweet), because of the sweet taste of some of its compounds. Other elementary substances of recent discovery have been named in like manner; but all simple substances which have been long known retain their ancient names. Thus, gold, silver, lead, copper, sulphur, carbon, are names of well-known substances, and they are retained in chemistry; but they contain, it is evident, no descriptive meaning.

167. But it is to *compound bodies* that the nomenclature especially applies; and, as above intimated, its design is to indicate their composition by their names. For this purpose, when two substances only are combined in a compound, a part of the name of one, with the termination *ide*, is made use of, while the other is expressed in full. Thus, oxygen forms oxides; chlorine, chlorides; bromine, bromides; sulphur, sulphurides, or sulphides; carbon, carbonides, or carbides, &c., of the other substance, the name of which is fully expressed; as oxide of iron, oxide

---

QUESTION.—In reference to what is this nomenclature framed? 166. What is said of elementary substances? 167. How is the termination *ide* used? Give an example.

of sulphur, chloride of hydrogen, &c. Formerly, usage required us to employ the termination *ide* for the compounds of all the elementary substances except those of carbon, sulphur and phosphorus, with which, without any apparent reason, another termination, *uret*, was used. Thus, we have been accustomed to say sulphuret of carbon, and not sulphide; phosphuret of calcium, and not phosphide; but a change in this respect has taken place, and the termination *ide* is alone made use of, as sulphide of carbon, &c.

When two elements unite in more than one proportion, numeral prefixes from the Greek or Latin are used to designate them; as protoxide of copper (from *protos*, first), and deutoxide (*deuteros*, second) or binoxide (*bis*, twice) of copper. The first of these compounds contains one equivalent of oxygen, united with one eq. of copper, while the second contains two eq. of oxygen, combined with one of copper. So the teroxide (*tertio*, third) or tritoxide (*tritos*, third) of nitrogen, is a compound of three eq. of oxygen and one eq. of nitrogen. The same rule is observed with regard to the compounds of other substances, as protosulphide and bisulphide of mercury, bicarbonide of sulphur, terchloride of gold, &c. The prefix *per* is often used to indicate the highest compound known, as peroxide of lead; and the prefix *sesqui* implies that two eq. of one of the substances is combined with three eq. of the other substance; as sesquioxide of iron, which contains two eq. of iron and three eq. of oxygen. Generally, the electro-negative element is expressed first, as chloride of sulphur, and not sulphide of chlorine; but this rule is not universally followed.

168. Most of the compounds above described, which may properly be called binary, or biclomentary, as being composed of two elements, are also capable of combining together, and forming other more complex compounds, usually called salts. In considering the relations they sustain to each other, they are usually divided into the two classes of acids and bases. The acids are

---

QUESTIONS.—How was the termination *uret* formerly used? When two elements combine in more than one proportion, how are the different compounds indicated? For what is the prefix *per* used? Which element is usually expressed first? 168. May compounds combine to form other more complex compounds?

generally more or less sour to the taste, change vegetable blues to red, and are electro-negative in relation to the other class; while the bases are electro-positive, and restore the blue colors which have been changed to red by acids. Some of the bases are soluble in water, and are exceedingly acrid and caustic.

A large proportion of all the acids are oxides, and are therefore called oxygen acids, or oxyacids; but many contain hydrogen, and are called hydracids. So, when a sulphide possesses acid properties, it is called a sulphur acid.

As most of the acids belong to the first class, or are oxyacids, special reference is had to them in the nomenclature; and they are named by using the termination *ic* or *ous* in connection with the name of the substance with which the oxygen is combined to form the acid, the termination *ic* being used for the acid containing most oxygen, when there are more than one formed from the same substance, and the termination *ous* for the one containing least. Thus we have sulphuric and sulphurous acids, the former of which contains more oxygen, and is a more powerful acid than the latter. When there are more than two acids formed from the same substance, the prefix *hypo* (*hupo*, sub, or under) is used in connection with the name of one or the other of the two already described, as the case may require. Thus, we have *hypo*sulphuric acid, which contains less oxygen than the sulphuric, and the *hypo*sulphurous, which contains less oxygen than the sulphurous.

When a compound, not containing oxygen, possesses acid properties, a part of the name of one of the substances is used as a prefix to the name of the other substance, to form a name for the acid. Thus, hydrochloric and hydrosulphuric acids are compounds of hydrogen with chlorine and sulphur respectively. So chloriodic acid is a compound of chloride and iodine. Some writers are particular to take the prefix from the name of the negative element, and say chlorohydric, sulphydric, &c.; but there is no good reason for such a distinction If no prefix is used, the acid is understood to be an oxyacid, as nitric acid, which is composed of nitrogen and oxygen.

QUESTIONS.—How are acids generally characterized? How are bases? What are oxyacids? What hydracids? How are the oxyacids named? How is the prefix *hypo* used? Give examples. When an acid compound does not contain oxygen how is it named? Give examples.

169. The *salts* are compounds of the acids with bases, as Glauber's salt (sulphate of soda), which is composed of sulphuric acid and soda. Names of the salts are formed by changing the termination of the name of the acid from *ic* into *ate*, and from *ous* into *ite*, and expressing in full the name of the base. Thus, sulphur*ic* acid, combined with bases, forms sulph*ates*; carbon*ic* acid, carbon*ates*, &c., of the bases with which they may be severally united; as sulph*ate* of lime, phosph*ate* of alumina, hyposulph*ate* of soda, &c. So sulphur*ous* acid forms sulph*ites*; nitr*ous* acid, nitr*ites*; hyposulphur*ous* acid, hyposulph*ites*, &c., of the various bases.

Many of the metallic oxides serve as bases of salts, but in expressing them (the salts), the word oxide is often omitted; thus, sulphate of iron is the same as sulphate of the oxide of iron. If a higher oxide than the protoxide forms the base of a salt, it is usually expressed in full; thus, we have the sulphate of the *sesqui*-oxide of iron.

170. *Acid* or *super* salts are such as contain an excess of acid, while *basic* or *sub* salts contain an excess of base; salts that contain no excess of either acid or base being called *neutral salts*. A *bisulphate* contains twice, and a *tersulphate* three times as much acid as a sulphate. Prefixes derived from the Greek numerals are often used to express the excess of base in the subsalts; as *di*nitrate of lead, a salt which contains 1 equivalent of nitric acid and 2 equivalents of oxide of lead. The same thing would be expressed by calling it bibasic nitrate of lead.

The above explanations will serve to illustrate the fundamental principles of the present nomenclature; but it is admitted that it applies but partially to the more complex chemical compounds, which, however, are not of frequent occurrence.

171. **Chemical Symbols.**—Instead of writing the full name of substances, it is often convenient to substitute abbreviations, which are called the *symbols* of these substances. For a simple substance, the first letter of the Latin name is generally used; but when there are two or more having the same initials, some

---

QUESTIONS.—169. What are *salts?* How are the salts named? What is said of the names of salts which have metallic oxides for their bases? 170. What are acid or super salts? What basic or subsalts? 171. What are chemical symbols? How are they formed?

other letter of the name is connected with the initial, in the symbols of all except one. Thus, O stands for oxygen, and Os for osmium; B for boron, Ba for barium, and Bi for bismuth; P for phosphorus, Pd for palladium, and Pt for platinum, &c. In the table on page 145, the symbols in general use for all the simple substances are given.

These symbols indicate single equivalents of the substances they respectively represent; and to indicate two, three, or more equivalents, a figure is placed before the symbol, as the coefficient in Algebra, or a little below it at the right. For instance, S signifies a single equivalent of sulphur, 2 S, 3 S, or S_2, S_3, &c., two, three, &c., eq.; and 4 O, 5 C, or O_4, C_5, four eq. of oxygen, five eq. of carbon, &c. To indicate that several substances are combined, their symbols are simply written side by side, as HO, or with a comma between them, as H, O, or with the plus sign (+), as H + O; all of which expressions represent a single equivalent of protoxide of hydrogen, or water. The comma and the plus sign are generally made use of only when the expression is somewhat complex; thus, NO_5, is the symbol for nitric acid, KO that for potassa, and KO, NO_5, or KO + NO_5 that of nitrate of potassa. Sometimes the plus sign is used when the substances between which it is placed are not combined, but only mixed. Generally, in writing the symbols of compounds, we express the electro-positive element first, as HO, KO, and not OH, OK, hydrogen and potassium being the positive elements of these compounds. So also we write KO, SO_3 for sulphate of potash, and not SO_3, KO.

It is to be particularly observed that small figures, placed at the right of letters, apply only to the ones to which they are attached; but large figures, placed at the left, like algebraic coefficients, affect all that follow them to the next comma or plus sign. Thus, PO_5 represents phosphoric acid; NaO, soda; NaO,PO_5, phosphate of soda; 2(NaO,PO_5), two equivalents of the same phosphate of soda; but 2NaO,PO_5 indicates a single eq. of bibasic phosphate of soda, which contains two eq. of soda, united to one of acid.

In consequence of the frequent occurrence of the double equivalent, it is often expressed by drawing a line under the symbol

QUESTIONS.—How are compounds expressed by the use of these symbols? How is a double atom often expressed?

of the single equivalent, or by a black letter. Thus, Al signifies an eq. of aluminum, and $\underline{Al}$ or **Al**, two equivalents. $\underline{Al}O_3$ or **Al**O_3 is the symbol for the sesquioxide of aluminum or alumina, and means the same as Al_2O_3. As oxygen forms an extensive list of compounds, simple dots are often used to indicate its presence in them; the above symbol for alumina would then become $\dddot{\mathbf{Al}} = Al_2O_3$. Other examples follow the same rule.

Combinations of these symbols, according to the principles above explained, are called *Chemical Formulæ;* and the great advantages of their use, in expressing forcibly complicated chemical changes, will be fully seen as we proceed.

172. **Isomerism—Polymerism.**—Isomeric compounds (161) are such as have the same ultimate composition, but differ from each other in some or all of their sensible properties. The term is derived from the Greek *isos*, equal, and *meros*, part.

Thus phosphoric acid, PO_5, is known in three different conditions, as ordinary, pyro, and metaphosphoric acids, the first of which is capable of saturating 3 eq. of a base, as soda; the second, 2 eq. of base, and the third only 1 eq. In other properties, also, this acid in its different states exhibits a diversity, but its composition in the different states is the same. There are other examples of the same kind

Polymeric compounds have the same ultimate composition, but differ in their properties; and their equivalents are multiples or sub-multiples of each other. Olefiant gas, C_4H_4, oil gas, C_8H_8, and cetine, $C_{32}H_{32}$, form a series of this kind, having the same elements in the same atomic proportion, but the equivalent of the second being twice that of the first, and one-fourth of that of the third. So, oil of turpentine, $C_{10}H_8$, and oil of lemons, $C_{20}H_{16}$, sustain to each other a similar relation. Many other examples are known.

173. **Allotropism.**—This term is used to designate the different conditions in which a substance is sometimes found, as it regards the chemical action of other bodies. Thus, iron, in its ordinary state, is readily dissolved by nitric acid; but if, before immersing a piece of iron wire in this acid, one end of it be heated to redness, or if it is connected with the positive electrode of a galvanic battery, or if it be immersed in the acid in contact with a piece of platinum,—in either of these cases, the acid fails to act upon it. So, if an aqueous solution of chlorine be prepared in the dark, it may be kept in a dark place without change for a long time; but if the sun is permitted to shine upon it a few seconds, decomposition will commence, hydrochloric acid will be formed in the water, and bubbles of oxygen rise to the surface.

Many other substances exhibit similar peculiarities, and are said to exist in different allotropic states.

QUESTIONS.—172. What are isomeric compounds? What examples are given? What are polymeric compounds? Give an example. 173. What does the term allotropism designate?

CRYSTALOGRAPHY.

174. The particles of liquid and gaseous bodies, as they unite to form solids, sometimes cohere together in an indiscriminate manner, and give rise to irregular, shapeless masses; but more frequently they attach themselves to each other in a certain order, so as to constitute solids possessed of a regularly limited form. The process by which such a body is produced is called *crystalization;* the solid itself is termed a *crystal;* and the science, the object of which is to determine and classify the forms of crystals, is *crystalography.*

175. Mode of producing Crystals.—Nature presents us with an abundance of crystals in the mineral kingdom, but they may also be produced artificially by several processes. The essential condition is, that the particles of the substance to be crystalized should be free to move among each other, which is accomplished by bringing it into the liquid state by solution, or by melting it. Alum forms beautiful octahedral crystals, by making a saturated solution in warm water, and allowing it to cool slowly. If a small tree be made of copper wire, and its branches immersed in such a solution while cooling, on removing it, the part immersed will be covered with a multitude of small shining octahedrons, like fruit. If the solution be allowed to stand after it has become cold, the crystals will gradually increase in size as the water evaporates. Common salt, blue and green vitriol, and many other substances may be crystalized in a similar manner.

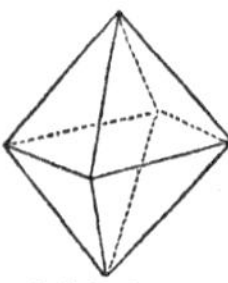
Octahedron.

Crystalization by fusion is also very common. If a quantity of sulphur be melted and allowed to cool slowly, upon breaking

---

QUESTIONS.—174. Do the particles of bodies sometimes unite so as to form solids of a regular form? What are such solids called? 175. What is the essential condition for the formation of crystals? What is the form of the crystals of alum?

the crust and pouring out all that remains liquid, a mass of crystals will be found within, shooting in every direction, as represented in the figure.

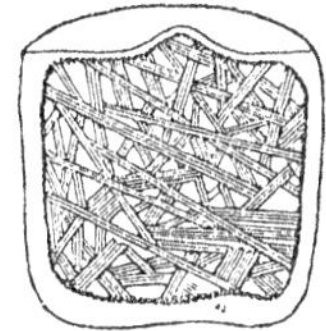

Crystalization of Sulphur.

The crystalization of many substances, as sulphur, corrosive sublimate, iodine, &c., may also be produced by sublimation.

176. Crystalization in Solids.—Even in solids, crystalization sometimes takes place. Copper wire which has been long kept is said often to lose its tenacity, in consequence of cubic crystals of the metal gradually forming in it. When sugar is melted and allowed to cool, it forms a hard, transparent mass; but by keeping some time, it gradually becomes opaque, and exhibits the ordinary white color and crystaline structure of refined sugar. Common "lemon candy," which is usually sold in small flat pieces, an inch wide and four inches long, is beautifully transparent when first formed; but after a few hours, crystalization commences in numerous points, and gradually extends through the mass, which now becomes opaque; and at the same time its flavor is improved.

177. Water of Crystalization.—Many substances, in crystalizing, absorb a large quantity of water, called their *water of crystalization*, which is essential to the existence of the crystals. It sometimes amounts to half their weight. When exposed to the air, the water often evaporates, and the crystals fall to powder. They are then said to *effloresce.* Glauber's salt affords a noted instance of this.

All substances are limited in the number of their crystaline forms. Thus, calcareous spar crystalizes in rhombohedrons, fluor spar in cubes, and quartz in six-sided pyramids; and these forms are so far peculiar to those substances, that fluor spar never crystalizes in rhombohedrons or six-sided pyramids, nor calcareous spar or quartz in cubes. Crystaline form may, therefore, serve as a ground of distinction between different substances. But the composition of substances having the same form is not necessarily

QUESTIONS.—How may sulphur be crystalized? 176. May crystalization take place in solids? Give an example. 177. What is water of crystalization? When does a substance *effloresce?* Are substances limited in the number of their crystaline forms?

the same, nor are the crystaline forms of the same substances always identical.

178. Primary Forms.—The total number of crystaline forms is unlimited; but it is found that all the more complex may be derived from thirteen of the more simple forms, which are therefore called *primary* or *fundamental forms.* These readily arrange themselves (*see Dana's excellent work on Mineralogy*) in six classes or *systems,* which are characterized by the comparative length and position of certain imaginary lines, supposed to be drawn through them, called *axes.*

1. *Monometric System.*—This system includes the *cube,* A, the *regular octahedron,* B, and the *rhombic dodecahedron,* C; all of which have three equal axes at right angles with each other. Hence the name, from *monos,* one, and *metron,* measure.

The *cube,* A, is a solid with six equal square faces; its axes, *a, b, c,*

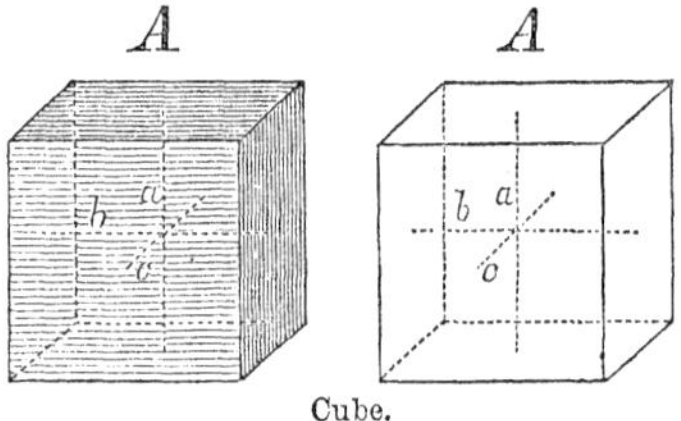

Cube.

are three lines, supposed to connect the centres of opposite faces. All its solid angles are similar, as also its edges.

The *regular octahedron,* B, is contained under eight equilateral triangles,

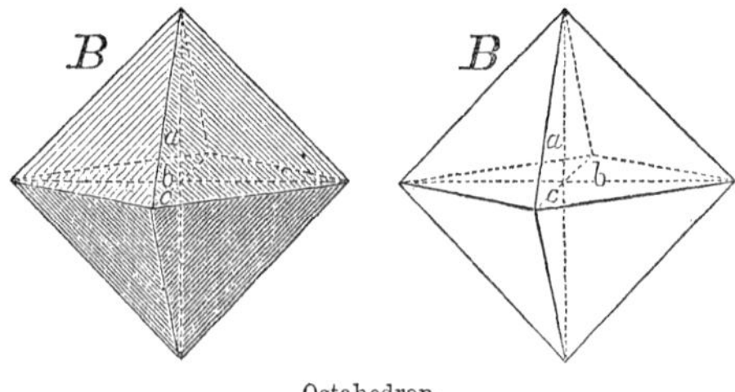

Octahedron.

Questions.—178. What are primary forms? How many of these are there? In how many classes do they arrange themselves? What are the forms of the monometric system? What is said of the axes of the forms of this system?

and its axes, *a*, *b*, *c*, connect opposite solid angles. All its angles and edges are similar.

The *rhombic dodecahedron*, C, is bounded by twelve equal rhombs; it has two kinds of solid angles, one kind, as *m*, *m*, which is composed of four acute plain angles, and another kind, as *n*, *n*, which is composed of

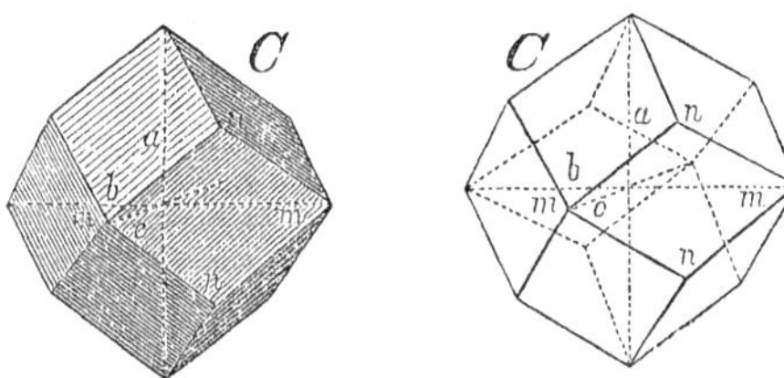

Dodehecadron.

three obtuse plain angles. The axes, *a*, *b*, *c*, connect the opposite acute solid angles.

Comparing these three forms with each other, it will be seen that the axes are the same in all, being equal in length, and making right angles with each other. If any one of them is supposed to be placed within another, the axes of the two will entirely correspond throughout.

2. *Dimetric System.*—This system includes only two forms, the *square prism*, and the *square octahedron.* These solids have three axes, at right angles to each other, but they are of two kinds, as indicated by the name, from *dis*, twice, and *metron*, measure.

The *square prism*, D, is bounded by six faces, two of which, *O*, *O*,

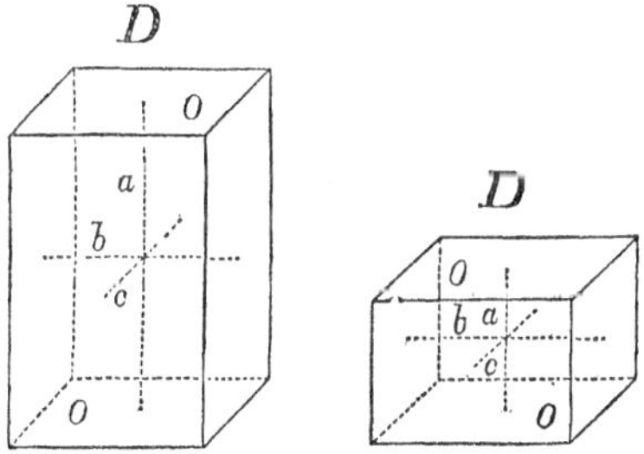

Square Prism.

called the bases, are squares, but the four other faces are rectangles, the height of which may be either greater or less than the base. The

QUESTIONS.—What are the forms of the dimetric system? How are the axes of the forms of this system characterized?

axes, *a*, *b*, *c*, connect the centres of opposite faces, as in the cube;—the two last named, *b* and *c*, are always equal, but the other, *a*, sometimes called the prismatic axis, is variable, and may be either longer or shorter than the others. Its solid angles are similar to each other, but the edges are of two kinds, the basal and the lateral.

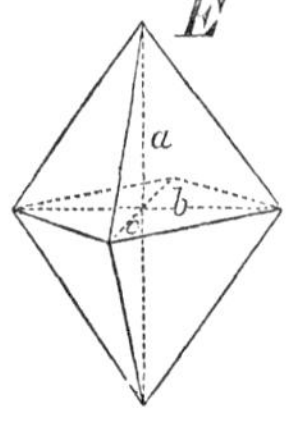

Square Octahedron.

The *square octahedron*, E, has for its faces eight equal isosceles triangles. The axes connect opposite solid angles, and correspond in every respect with the axes of the square prism. The two therefore properly constitute one system. In describing this solid the variable axis, *a*, is always supposed to be vertical, the other two being horizontal. It has two kinds of edges, and two kinds of solid angles.

3. *Trimetric System.*—The three solids of this system are the *rectangular prism*, F, the *right rhombic prism*, G, and the *rhombic octahedron*, H. The name is from *tris*, trice, and *metron*, measure. These forms have three axes intersecting at right angles, but all are variable, no two being equal.

The *rectangular prism*, F, has six faces, all of which are rectangles:—its three axes, *a*, *b* and *c*, connect the centres of opposite faces, and intersect each other at right angles, and are variable, as to their length. Its solid angles are all similar; but its edges are of three kinds.

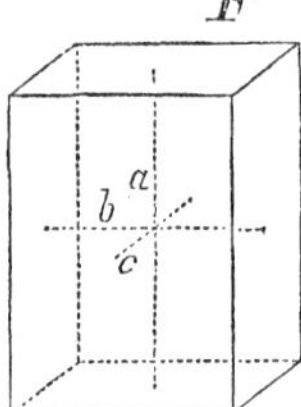

Rectangular Prism.

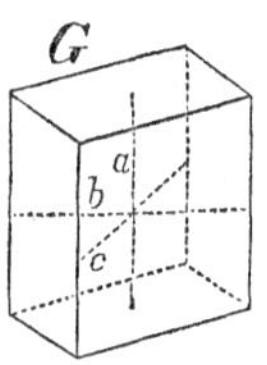

Right Rhombic Prism.

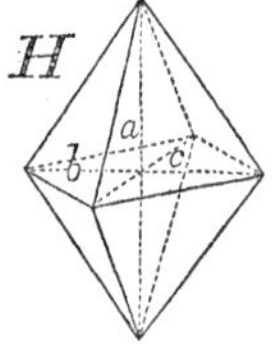

Rhombic Octahedron.

The *right rhombic prism*, G, has six faces, two of which are equal rhombs, and constitute the bases, while the four lateral faces are rectangular. Its three axes, *a*, *b* and *c*, intersect at right angles, and are variable; the first, *a*, connecting the centres of the bases, but the other two, *b* and *c*, being drawn between the centres of opposite lateral edges. Its solid angles are of two kinds and its edges of three.

The *rhombic octahedron*, H, has eight faces, which are scalene triangles; its axes, three in number, *a*, *b* and *c*, unite the apices of oppo-

QUESTIONS.—What are the forms of the trimetric system? What is said of their axes?

site solid angles, and in every respect correspond to the axes of the two solids last described. This solid may be considered as composed of two pyramids with rhombic bases, united base to base.

Comparing the three forms of this system, we see that their axes are identical; comparing the systems now described with each other, we find that they each have three axes intersecting at right angles in the centre of the solids; but in the first, or submonometric system, all are equal; in the second, or dimetric, two are equal, and the third variable; while in the third, or trimetric, all are variable.

4. *Monoclinic System.*—This system includes two forms, the *right rhomboidal prism*, I, and the *oblique rhombic prism*, J, which have three axes, *a*, *b* and *c*, two of them making their intersections at right angles, but the third is inclined. The name is from *monos*, one, and *clino*, to incline.

The *right rhomboidal prism*, I, has two rhomboidal bases, upon one of which in this figure it is supposed to stand, the other four being rectangles. The axes connect the centres of opposite faces, of which *a* and *b*, *a* and *c*, intersect at right angles, but *b* and *c* are inclined to each other.

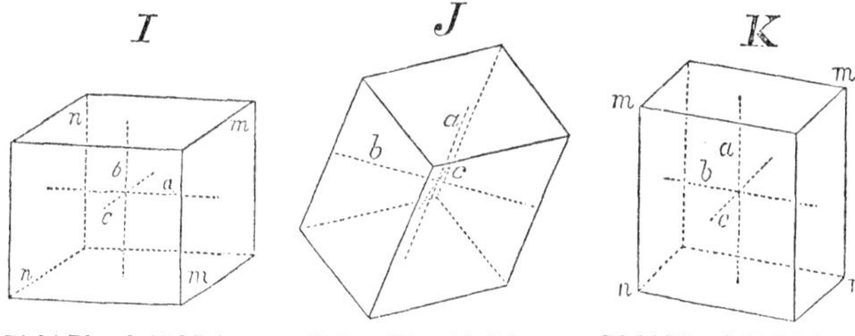

Right Rhomboidal Prism. Oblique Rhombic Prism. Right Rhomboidal Prism.

The *oblique rhombic prism*, J, has its faces rhombs, and its four lateral faces parallelograms. Its vertical axis, *a*, connects the centres of the bases, but the two lateral axes, *b* and *c*, are drawn between the centres of the lateral edges. The last two intersect at right angles, and the vertical axis, *a*, makes a right angle with *b*, but is inclined to *c*.

To compare these two forms, the right rhomboidal prism, I, must be placed upon one of its rectangular faces, as in K, making the axis, *a*, vertical; the faces *m m* and *n n*, in I, will become respectively *m m* and *n n* in K, and all the axes of the two will correspond. If now, when in

---

QUESTION.—Comparing now the axes of the forms belonging to the three preceding systems, in what do they differ? What forms are included in the monoclinic system? What is said of their axes?

this position, all the lateral edges are removed, as in K′, we shall have the oblique rhombic prism, represented within the rhomboidal, as in

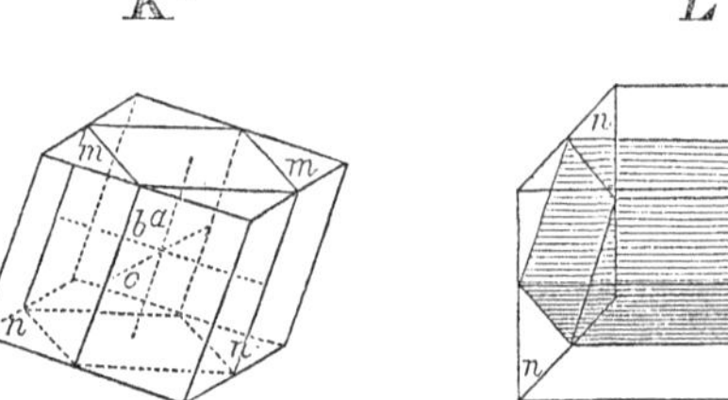

Right Rhomboidal Prism. Oblique Rhomb'l and Right Rhomb'l Prisms

figures K′ and L; the crystal in the last figure being supposed in the same position as in I.

5. *Triclinic System.*—A single form only, the *oblique rhomboidal prism*, M, belongs to this system. It is bounded by six faces, but only those directly opposite each other are equal. Its basal faces are rhomboids, and its lateral edges are inclined to the plane of its base. Its three axes are unequal, and no two of them intersect at right angles. The name is from *tris*, thrice, and *clino*, to incline.

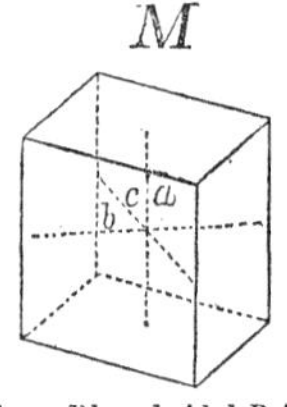

Oblique Rhomboidal Prism.

6. *Hexagonal System.*—This system includes the *rhombohedron*, N and O, and the *hexagonal prism*, T. These solids have

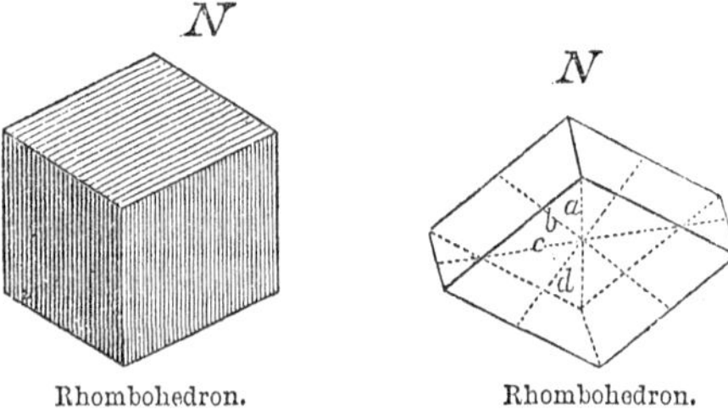

Rhombohedron. Rhombohedron.

four axes, one of which, called the vertical axis, is perpendicular

---

QUESTIONS.—What solid only belongs to the triclinic system? What is said of the axes of this form? What are the forms of the hexagonal system? How many axes have they, and how are they drawn?

to the other three, called the lateral axes. These latter intersect each other at angles of 60°, and are equal.

The *rhombohedron* is of two kinds, the obtuse, N, and the acute, O; in each all the six faces are equal rhombs, and in each there are two solid angles, diagonally opposite, which are formed by three equal plain angles. Between these, one axis, *a*, is drawn, called the vertical axis; and in describing the solid this is always supposed to be in a vertical

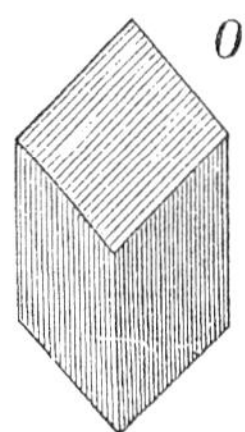

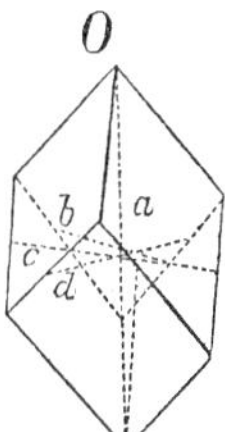

Rhombohedron.

position. The other three axes, *b*, *c* and *d*, unite the centres of opposite lateral edges.

Of the twelve edges, six (three above and three below) are connected with the extremities of the vertical axis around which they are symmetrically placed. The other six, called the lateral edges, are also situated around this axis symmetrically, and at equal distances from it. This is readily seen by looking downward upon a crystal of this form when in position, a section through the centre perpendicular to the vertical axis being plainly a regular hexagon, as in the figure P.

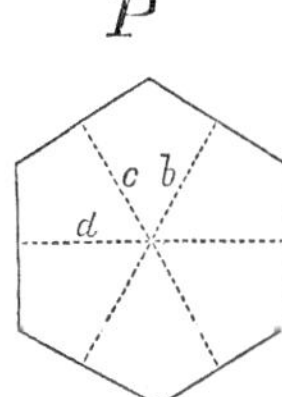

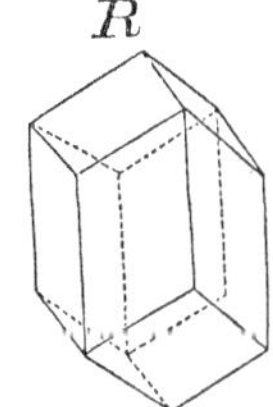

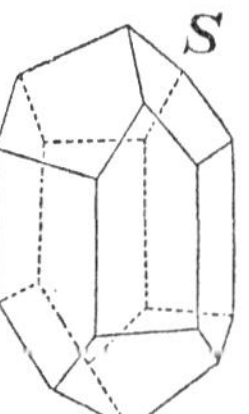

Section Rhombohedron

Secondaries.

By replacing the six lateral edges by planes parallel to the vertical axis, *a*, a hexagonal prism, terminated by three-sided pyramids, is produced, represented in R. Removing in a similar manner the lateral solid angles, we have the form, S, which is essentially the same as the

---

QUESTIONS.—How is the hexagonal prism terminated with three-sided pyramids produced from the rhombohedron?

last except the terminal pyramids; and by a removal of the terminal pyramids a regular hexagonal prism, T, will be found.

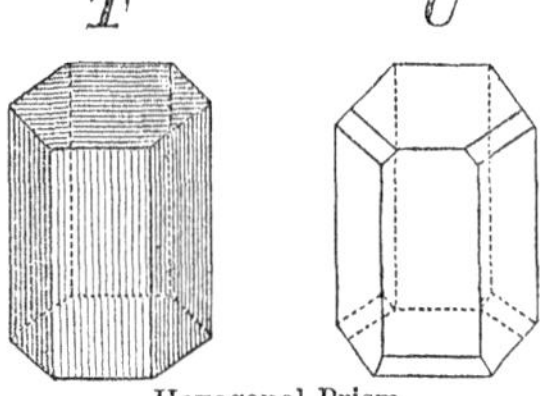

Hexagonal Prism.

In a similar manner, the rhombohedron may be derived from the hexagonal prism, by a replacement of the alternate basal edges of the prism; first in the form represented in U; and the replacement being continued until the faces of the original prism are obliterated, the resulting solid will be the rhombohedron.

179. Secondary Forms.—Secondary forms are derived from the primary, by regular, systematic changes or modifications upon the angles or edges. Their number is unlimited.

A general discussion of the laws of these modifications cannot be here introduced, and only a few further illustrations. The connection between the different forms of each of the six systems, it is believed, has been made plain; and also the modifications by which one form is evolved from another, as the oblique rhombic prism, J, from the right rhomboidal, I, in the monoclinic system.

By the same process the secondary forms are obtained from the primaries; indeed, in each system containing more than one form, any one in that system may be taken as the primary, of which the others are then to be considered as secondaries. Thus, in the dimetric system, either the square prism, or the square octahedron may be taken as the primary form, from which the other may be derived in the manner pointed out. Therefore, although it is customary to consider as primary all the thirteen forms mentioned above as belonging to the several systems, yet the whole number necessary for the derivation of all known forms is only six, one for each of the six systems.

180. Laws of Modification of Forms.—The modification of the forms of crystals always takes place in accordance with certain laws, which may be stated as follows, viz: *All the similar parts of the crystal will be similarly and simultaneously modified, or half of the similar parts will be similarly modified, independently of the other half.*

In the forms of the monometric system all the parts are similar, and all will therefore be modified in the same manner; but this is not the

---

QUESTIONS.—How may the rhombohedron be produced from the hexagonal prism? 179. What are secondary forms? In each of the above systems may all the forms be derived from one? How many fundamental forms only are then necessary? 180. Do the modifications in the forms of crystals always take place according to fixed laws? What is the general law stated? Are all the parts similar in the forms of the monometric system?

case with the forms of the other systems. In the dimetric system, for instance, the square prism, D, (page 161,) has two kinds of faces, the basal and the lateral, and also two kinds of edges, but only one kind of solid angles.

The lateral edges, being formed by the meeting of similar faces, may be truncated, that is, replaced by a plane equally inclined to the two faces; but the basal edges, being formed by the meeting of dissimilar faces, do not admit of truncation. The new plane replacing one of these edges will always be more inclined to one of the old faces than to the other.

The solid angles are all alike, but each being formed by two plane angles that are similar to each other, but unlike the third, they cannot be truncated. The figure V represents a square prism with its lateral

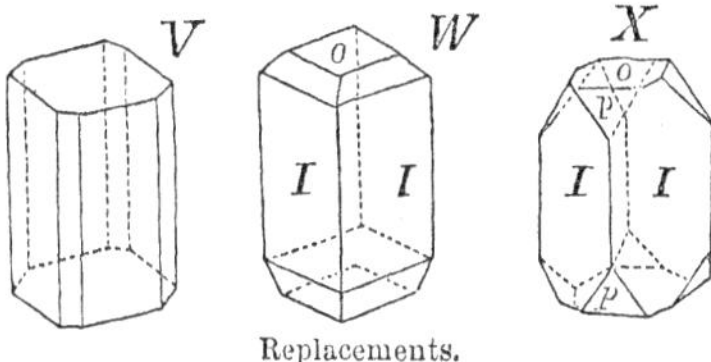

Replacements.

edges truncated, and in W, we see one of the same with its basal edges replaced, but each of the new planes is unequally inclined to the adjacent basal faces O, and lateral faces I. In the figure X, the square prism has its angles replaced, the new plane, *p*, making equal angles with the faces I, but not with O.

By continuing the replacements seen in W, or in X, until the old faces are entirely obliterated, it is evident that we shall have the square octa-

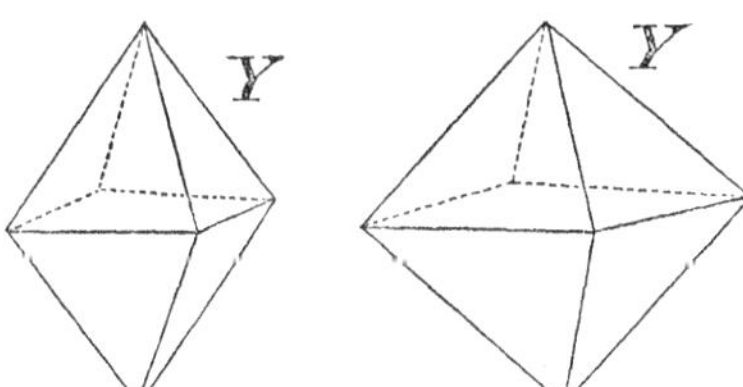

Square Octahedron.

hedron, Y, which sustains the same relation to the square prism as the regular octahedron does to the cube.

QUESTIONS.—What is said of the faces of the square prism? What is said of the faces forming the basal edges? Can these edges then be truncated? May the lateral edges be truncated? If we replace the basal edges on the angles of the square prism, continuing the replacement until the original faces are all obliterated, what form will result?

By examining the forms belonging to other systems, and applying the same general law, it becomes apparent that the variety of forms capable of being produced is really unlimited, but the subject cannot be pursued further in this place.

181. **Formation of Crystals.**—When a substance crystalizes in circum stances to admit of observation, a very small crystal is first seen, which gradually increases in size by the deposition of particles on the different faces. If this addition of matter be equal on all the faces the form is continued the same, but if there is more added on some of the faces than on others the crystal becomes more or less irregular. Changes of form, in accordance with the laws above mentioned, we may suppose to be produced in the following manner. Let us suppose a substance crystalizing in the monometric system, as galena, or common salt. A cube is formed and is gradually increasing in size by equal additions upon every face, but at length, from some cause not understood, a change takes place, and regular spaces are left at the angles or edges of the cube. We will suppose that at each edge of the cube there is left a deficiency, or *decrement*, of one row of particles for every addition upon the surface; this will occasion the formation of planes upon the edges, and, if continued, will result in the production of a rhombic dodecahedron.

Or let us suppose, as the additions are made, a decrement occurs at the angles only:—this will result in the replacement of the angles. Let A be a small cube on which deposits are taking place, as suggested above, leaving decrements at the angles. After a time it will have the form B, and by a continuation of the process will become of the form C, and then of D: but unlike the forms represented in the figures, it will be constantly increasing in size. But the form, D, is evidently that of a

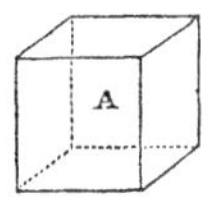

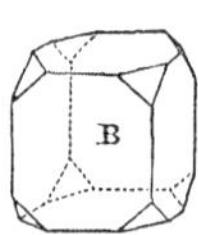

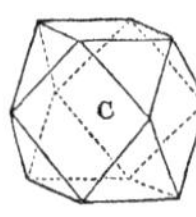

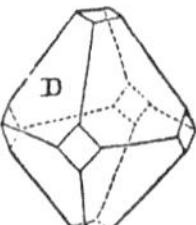

Monometric Forms.

regular octahedron with all its solid angles replaced by planes; and we suppose the process continued until the perfect octahedron would be formed. By reversing this process, commencing with the regular octahedron, and supposing additions to be made on the faces, with decrements at the angles, and continuing the process sufficiently, the cube will eventually be produced.

Whether this is the real mode by which the variously modified secondary forms are produced, we may never be able to determine fully; but, in the absence of positive proof, this has been presented as theoretically possible.

QUESTIONS.—181. How do crystals increase in size? Illustrate the mode in which a secondary form may be produced in the crystalization of common salt by a supposed decrement upon the edges?

182. Cleavage.—Crystals are generally capable of separation in certain directions by natural joints;—a property which is denominated *cleavage*. In the primary forms this usually takes place in planes parallel to the faces, but may also occur in planes parallel to any of the faces of secondary forms belonging to the species. It will usually be found in crystals that cleavage is much more easy in some directions than in others; and the general rule is that it will take place with equal facility parallel to similar faces, and only in this direction.

183. Isomorphism.—Isomorphous substances (from *isos*, equal, and *morphe*, form), strictly are substances which crystalize in the same form; but the term is generally used to indicate those bodies which, in the compounds they form, are capable of replacing each other, without changing the forms of the crystals of these compounds. Thus magnesia and lime, and the protoxides of iron and manganese, form a group, any one of which may replace another in whole or in part in the compounds they form. So also alumina (sesquioxide of aluminum), and sesquioxide of iron, form another group, and phosphoric and arsenic acids another.

In general, isomorphous substances will be found to possess other analogous properties; and the compounds they form will often closely resemble each other in all their leading properties. We have an excellent illustration of this in the *alums*, of which there are several, but all possessing the same crystaline form, and the same astringent, sweetish taste, the same solubility in water, &c. Below are the formula representing their composition

Common alum ....$KO,SO_3 + Al_2O_3, 3 SO_3 + 24 HO$.
Iron alum..............$KO,SO_3 + Fe_2O_3, 3 SO_3 + 24 HO$.
Chromium alum........$KO,SO_3 + Cr_2O_3, 3 SO_3 + 24 HO$.
Soda alum ....$NaO,SO_3 + Al_2O_3, 3 SO_3 + 24 HO$.
Ammonia alum ....$NH_4O,SO_3 + Al_2O_3, 3 SO_3 + 24 HO$.

QUESTION.—182. What is meant by cleavage? How does this usually take place in primary forms? Is cleavage made with equal facility in all directions? 183. What are isomorphous substances? Will isomorphous substances usually be found to possess other analogous properties besides those of form? What is said of the alums in this connection?

Comparing these formulæ, one after another, with the first, that of common alum, we perceive that in iron and chromium alums the sesquioxide of aluminum is replaced by the sesquioxides of iron and chromium respectively; and in the soda and ammonia alums, the potash is replaced in one by soda and in the other by ammonia (oxide of ammonium). These several sesquioxides are therefore isomorphous, as are also the potash, soda, and oxide of ammonium which replace each other in the other formula. This list of alums might be still further extended; and all constitute a single family of compounds formed on the same type and possessing very similar properties.

Many other similar groups of compounds might be given in which isomorphous substances replace each other, but the above will suffice.

The following group of simple substances that are isomorphous, is from Gmelin.

1. Carbon, phosphorus, potassium, titanium, bismuth, cadmium, lead, iron, copper, silver, and gold.
2. Potassium, sodium, lithium, calcium, zinc, lead, and silver.
3. Oxygen, sulphur, and chlorine.
4. Arsenic, antimony, and tellurium.
5. Platinum, iridium, and osmium.

184. Dimorphism.—Dimorphous substances (*dis*, double, and *morphe*, form,) are such as are capable of crystalizing in two forms belonging to different systems of crystalization. A few substances are known which, under different circumstances, crystalize in three forms, and are therefore called *trimorphous*.

Among the simple substances, carbon, sulphur, and perhaps copper, are dimorphous. Carbon, as *diamond*, takes a form belonging to the monometric system, but, as *graphite*, it is found in the hexagonal system. It is scarcely necessary to remark that in these different forms many of its properties are essentially different. As diamond, it is the hardest substance known, and is transparent, but, as graphite, it is comparatively soft, perfectly black, and opake.

Sulphur when crystalized by melting and slow cooling (175)

QUESTIONS.—How do the formula for the different alums compare with each other? 184. What are dimorphous substances? What simple substances are mentioned as being dimorphous?

the crystals take forms belonging to the monoclinic system, but the native crystals, and those obtained by dissolving sulphur in the bisulphide of carbon, belong to the trimetric system.

Many compound bodies are found to crystalize in two or more forms; and sometimes it has been ascertained that the particular form produced will depend in some degree upon the temperature of the crystalizing solution. Carbonate of lime usually crystalizes in forms of the hexagonal system, but occasionally it forms crystals of the trimetric system, and is then called arragonite. Such researches as have been made on the subject indicate that it takes the form last mentioned when crystalizing at about 212°, but forms crystals of the hexagonal system only at very low or very high temperatures.

In some cases a change of the crystaline arrangement of the particles of a body, attended by a change of color, is produced by a mere variation of temperature. Thus protiodide of mercury sublimed at a very gentle heat forms small dimetric crystals of a scarlet color, but if sublimed at high temperatures the crystals are monoclinic and of a sulphur yellow color. The scarlet crystals become yellow by being heated, but turn red again when cooled.

The yellow crystals obtained by sublimation retain this color when cooled, but if scratched or disturbed by some pointed instrument, they at once turn red at the point touched, and the change of color soon extends to the whole mass of crystals in contact.

QUESTIONS.—Upon what has the form of dimorphous substances sometimes been found to depend?

PART III.

SPECIAL CHEMISTRY—INORGANIC.

185. Classification of Elements.—Having discussed under the head of GENERAL CHEMISTRY the important principles pertaining to chemical changes generally, we are now prepared for the examination of the various simple substances, or elements, found in nature, and their inorganic compounds so far as may come within the objects of the present work. And, in accordance with general usage, we shall divide them into the two great classes of *non-metallic* and *metallic* elements, called also *metalloids* and *metals*.

Though this division is universally recognized by writers upon this science, yet it is founded upon properties which are not absolute, and are much more distinctly seen in some of the elements than in others; a degree of vagueness therefore attaches to it, and several of the simple substances are not readily determined, being referred to either of the divisions with nearly equal propriety. It is however retained because of its convenience; and, following the example of Regnault, we shall consider as non-metallic the following fifteen elements, which are further arranged in natural groups. Of these, silicon, tellurium, and arsenic have not unfrequently been classed with the metals which they—and especially the last two—strikingly resemble in some of their properties. Iodine and bromine also, though with less reason, have been sometimes regarded as metals.

All the elements may enter into the composition of inorganic compounds, which are produced in the mineral world, and may be formed artificially; but only a few of them—more especially carbon, hydrogen, oxygen, and nitrogen—form organic compounds, which originate almost exclusively in plants and animals.

QUESTIONS.—185. Into what two classes are the elementary substances divided? How many non-metallic elements are there? May all the elements enter into the composition of inorganic or mineral compounds? What four are mentioned as forming most organic compounds?

A further discussion of the relations of organic and inorganic bodies will be introduced hereafter.

186. For the sake of convenience we shall arrange the non-metallic elements in the five following groups, and introduce them for description in the order in which their names appear. Among the individuals of some of the groups a striking resemblance is manifest, but in other groups this is less apparent.

Each element will first be described, and afterwards will be introduced the more important compounds it forms with the other elements previously described.

METALLOIDS, OR NON-METALLIC ELEMENTS.

Group. I. Oxygen. Hydrogen. Nitrogen.

These substances are brought together in the same group merely for the sake of convenience; each one properly constitutes a class by itself.

The first, oxygen, is more abundant than any other element, constituting from 30 to 50 per cent. of the whole globe, or at least that part of it accessible to man. It forms compounds with nearly all other elements.

Hydrogen is a highly electro-positive substance, and in some of its properties resembles the metals.

Nitrogen is noted for its very feeble affinity for the other elements, whether electro-positive or electro-negative. It has sometimes been classed with phosphorus and arsenic, to which it has some resemblance.

Group II. Chlorine. Iodine. Bromine. Fluorine.

These electro-negative elements constitute a very distinct natural family; each forms a single acid compound with hydrogen, and all except the last form analogous acid compounds with oxygen.

Group III. Sulphur. Selenium. Tellurium.

These elements, especially the first two, closely resemble each other; they also form acid compounds both with oxygen and hydrogen, which are analogous in their constitution.

Questions.—186. In how many groups are the metalloids arranged? Name the elements of the first group. Why are these arranged together? Name those of the second group. Name the elements of the third group

GROUP IV. PHOSPHORUS, ARSENIC.

Two elements very similar in many of their properties, and forming similar compounds with oxygen and hydrogen. Many of their compounds are isomorphous. Antimony also properly belongs to the group.

GROUP V. CARBON. SILICON. BORON.

Combustible bodies incapable of being volatilized even at the highest temperatures known, and forming feeble acids with oxygen.

GROUP I.

OXYGEN HYDROGEN NITROGEN	Gaseous at all temperatures; but having few other points of resemblance. A compound of the first two constitutes *water*,—a mixture of the first and last, *atmospheric air*.

OXYGEN.

Symbol, O; *Equivalent*, 8; *Density*, 1·106.

187. History.—Oxygen (from *oxus*, acid, and *gennao*, to produce) was discovered by Priestly and Scheele, independently of each other, in 1774. It has been called *empyreal air*, because it supports combustion, and *vital air*, because necessary to respiration. It is the most abundant of the elementary substances, and constitutes from 30 to 50 per cent of the mass of the globe, or at least that part of it accessible to man; it forms also an ingredient of nearly all animal and vegetable compounds.

188. Preparation.—Oxygen is a gaseous substance, and may be obtained from several sources; but the best method to procure it, when only a small quantity is required, is to heat an ounce or less of chlorate of potash in a green glass flask or retort. This salt is composed of chloric acid, ClO_5, and potash, KO; and, when heated nearly to redness, gives up the whole of its oxygen, as shown by the following formula. Thus,

$$KO,ClO_5 = KCl + 6\,O;$$

each atom of the salt yielding one atom of chloride of potassium, and six atoms of oxygen. The process succeeds better if, before

QUESTIONS.—Name the elements of the fourth and fifth groups. 187. What important facts are mentioned in the history of oxygen? 188. Describe the mode of preparing oxygen from chlorate of potash? How is it to be collected?

heating, the salt is intimately mixed with an equal weight of powdered peroxide of manganese, or oxide of copper.

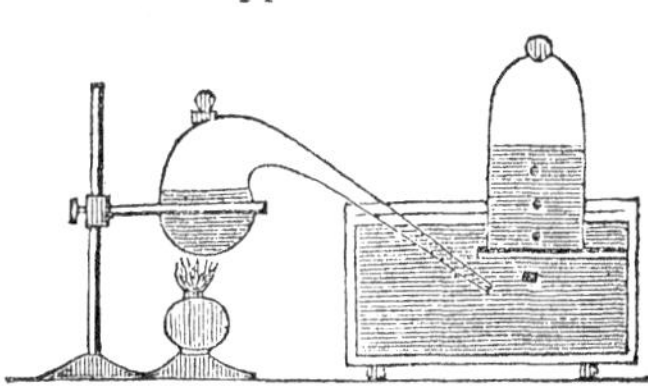
Preparation of Oxygen.

The accompanying figure will serve to illustrate the arrangement of the apparatus required for the experiment. The salt, contained in a retort, is heated by a spirit-lamp, which produces no smoke, and the gas, as it forms, passes under a receiver filled with water, and placed on a shelf in a pneumatic cistern, a section of which is shown in the figure. The receiver is open at the bottom, and it is first filled with water by plunging it in the cistern, and then bringing it to its upright position, and raising it carefully to its place upon the shelf, which is just beneath the surface of the water. The water rises in the receiver, in consequence of the atmospheric pressure (see the author's *Natural Philosophy*, p. 135); and when the gas is forced underneath it, it rises in bubbles, displacing the water, and occupying the highest part of the receiver,—the shelf being supposed to have an aperture in it, to allow the gas to pass upward, and the water also to escape.

189. We have in the above equation the first instance that occurs in this work of the use made of symbols to illustrate chemical changes. The expression constitutes a proper equation, but it has peculiar properties, in which it differs from algebraic equations. The separate symbols in the equation, of course, indicate equivalents or atoms of the elements symbolized, and the manner in which they are combined in the first member is also designed to indicate the mode of combination of the several substances expressed, before any change is produced; while the second member of the equation shows the state or mode of combination of the same elements after the change has been produced. There must therefore be found represented in each member of the equation the same elements (and no others), and the same

QUESTIONS.—189. What is the design of the equation in the preceding paragraph? What is expressed in the first member of the equation? What in the second? What is required in the two members in equations of this kind?

member of atoms of each element. Thus, in the above equation we have represented in each number 1 eq. of potassium, 1 eq. of chlorine, and 6 eq. of oxygen. Of course, if the sum of these were represented in numbers, it would be the same for each member. These remarks apply in all similar cases.

When a large quantity of oxygen is to be obtained, it is more economical to make use of saltpetre (nitrate of potash); but a much greater heat is required to decompose it, and an iron retort must therefore be substituted, instead of glass. The iron retort containing the saltpetre is placed in a furnace, the heat of which can be easily regulated, and a lead tube is connected with it, leading to the pneumatic cistern. As soon as the bottle has attained a full red-heat, the gas begins to come over; and if care is taken to prevent the heat becoming too great, very pure oxygen will be obtained. The changes that take place are illustrated by the following formula:

$$KO,NO_5 = KO,NO_3 + 2\ O.$$

Thus, each atom of nitrate of potash yields one atom of hyponitrite of potash, and two atoms of oxygen.

This gas may also be procured from the peroxides of manganese, lead, and mercury, and from other substances. A ready method to procure it from the peroxide of manganese is as follows: Let a flask of green glass be partly filled with the peroxide, and then enough strong oil of vitriol poured upon it to moisten it thoroughly, and then apply the heat of an alcohol lamp. By this process sulphate of protoxide of manganese is formed, and oxygen gas, as shown by the following equation:

$$Mn\ O_2 + SO_3 = MnO,\ SO_3 + O$$

It is thus seen that every atom of the oxide of manganese will yield one atom of oxygen. To insure purity it is well to pass the gas as it is formed through a weak solution of potassa by means of a three-necked bottle, as shown in the figure on next page.

QUESTIONS.—May oxygen be prepared from nitrate of potash? Describe the mode of preparing it by the use of peroxide of manganese and sulphuric acid.

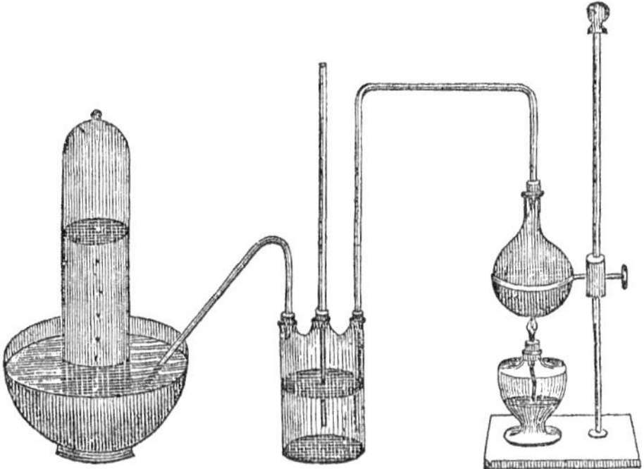

Preparation of Oxygen.

Whatever mode of preparing it may be adopted, the first por-)ns of gas that come over should always be allowed to escape, it will be mixed with atmospheric air, contained in the appa-tus at the beginning of the operation.

190. It is always easy to calculate the quantity of oxygen iich any given quantity of materials we may desire to use will 'ord.

We will suppose that 1 oz. of chlorate of potash is to be employed,—w many cubic inches of oxygen will it afford?

There are in each equivalent of the chlorate of potash 1 eq. of potash d 1 eq. of chloric acid, composed as follows, viz:

$$\left.\begin{array}{l} KO = 47{\cdot}1 \\ ClO_5 = 75{\cdot}4 \end{array}\right\} = 122{\cdot}5$$

which 48 parts, or 6 eq., are oxygen. We have then the following ɔportion:

$$122{\cdot}5 : 48 :: 480 : x,$$

ıence $x = 188$ grs. for the weight of the oxygen in 1 oz. of the chlorate. ·w 100 cubic inches of oxygen (192) weigh 34·28 grains; and we have ɜrefore this proportion:

$$34{\cdot}28 : 100 :: 188 : x,$$

$x = 548$ cubic inches, or about $2\frac{1}{3}$ gallons.

QUESTION.—190. Supposing the chlorate of potash used to prepare the ɜ, how may we calculate the quantity that may be produced from an ɑce of the salt?

Oxygen from Plants.

191. Oxygen gas is also given off by plants under the influence of light. Let a sprig of mint be placed in a white glass globe, which is then to be filled quite full of spring-water, and the mouth inverted in a tumbler of water, as shown in the figure. It is then to be placed in the direct rays of the sun; and in a short time, bubbles of gas will be seen collecting in the upper part of the glass, which is nearly pure oxygen.

192. Properties.—Pure oxygen is a colorless gas, without odor or taste, and has never yet been reduced to the liquid state by any degree of cold or pressure. It is very slightly absorbed by water, 100 cubic inches of that liquid taking up 3 or 4 of the gas. It is heavier than air, 100 cubic inches weighing 34·28 grains, while the same volume of air weighs only 31 grains, the temperature being at 60° F., and the barometric pressure at 30 inches. Its density is therefore 1·106. It has a very extensive range of affinity, entering into combination with all the other elements except fluorine.

Oxygen is a powerful supporter of combustion; and all substances that are capable of burning in the open air, burn in it with far greater brilliancy. A piece of wood, on which the least spark of light is visible, bursts into flame the moment it is put into a jar of oxygen; a recently extinguished candle, with the least spark of fire upon the wick, is relighted; lighted charcoal emits beautiful scintillations; and phosphorus burns with so powerful and dazzling a light that the eye cannot bear its impression. Even iron and steel, which are not commonly ranked among the inflammables, undergo rapid combustion in oxygen gas.

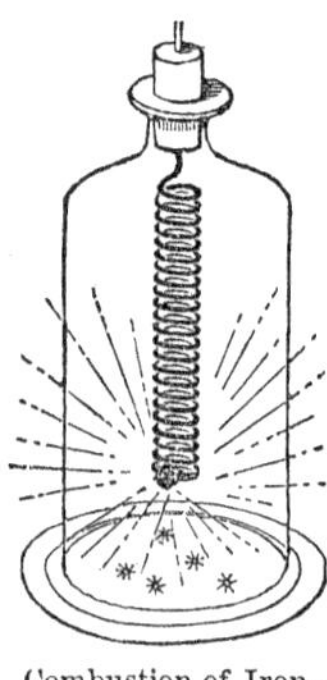
Combustion of Iron.

The combustion of iron and steel is effected by introducing it in the form of wire or thin slips,—as pieces of watch-spring,—into a vessel of the gas, as shown in the figure. The

QUESTIONS.—191. How may it be shown that plants evolve oxygen under the influence of light? 192. What are some of the properties of oxygen? Why is it called a supporter of combustion? Why is a candle recently extinguished relighted when plunged in the gas?

combustion is commenced by attaching to the lower extremity a piece of spunk or other combustible, which is ignited the moment it is to be introduced into the gas. As the combustion progresses, if the cork is tight, the water contained around the bottom of the receiver in the shallow dish is seen to rise, and more must be poured in, to prevent the entrance of air from without. This is occasioned by the absorption of the oxygen by the iron, to form oxide of iron, which falls in melted globules into the dish. When a lighted candle is let down by a wire into a jar of this gas, as in the figure, the case is different; the candle burns for a time with increased splendor, but soon the flame begins to diminish, and at length entirely disappears, without any diminution of the volume of gas contained within. If the candle be relighted and returned to the receiver, it is now instantly extinguished, the gas having lost entirely its power of supporting combustion. The reason is, that the oxygen has disappeared, and a new gas, carbonic acid, CO_2, taken its place, which has exactly the same volume as the oxygen from which it was formed. If a piece of burning charcoal had been used, the result would have been the same.

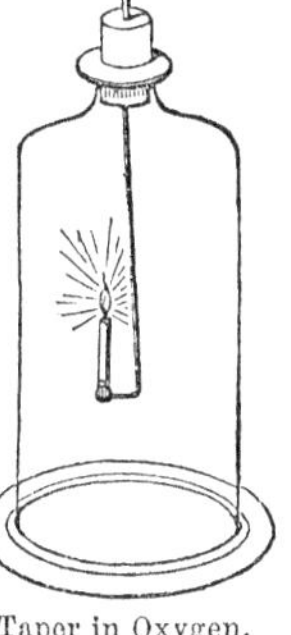
Taper in Oxygen.

193. *Ordinary combustion* consists in the union of combustible matter with oxygen, and is usually attended by the evolution of heat and light. A new substance is also formed, which may be solid, liquid, or gaseous. When iron is burned, a solid product (oxide of iron) results, which just equals the weight of the iron and oxygen together, that have disappeared during the operation. It is evident that in every case the product of the combustion must be equal in weight to that of the oxygen and other substance which have combined.

Combustion may, however, be produced without oxygen; a

QUESTIONS.—How may iron and steel be made to burn in the gas? What is produced when iron is burned in this way? Why does the water rise in the receiver as the combustion progresses? Why is not the same effect produced when a candle is made to burn in the gas? 193. What is ordinary combustion? May combustion be produced without oxygen?

piece of phosphorus or powdered antimony, let down into a receiver filled with chlorine, will take fire spontaneously, and burn with the evolution of light and heat; so that ordinary combustion can only be considered as a particular case of chemical action.

Combustion is the great source of artificial heat and light (13, 65). To produce an intense heat means are contrived to force large quantities of air (one-fifth of which is oxygen) in contact with a mass of ignited coal, the carbonic acid formed being allowed to escape freely. To produce light, we burn oil, tallow, or other substances which contain a large proportion of the same material as coal, as we shall see hereafter. By supplying the burning body with pure oxygen, the intensity of both the heat and light is greatly increased. A very considerable heat is also produced, simply by blowing with a proper blowpipe through the flame of a lamp or candle, which is sufficient for numerous small operations.

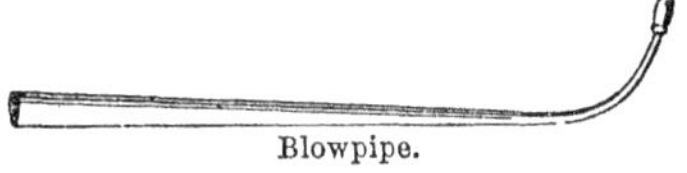
Blowpipe.

A better form of the blowpipe is represented in the next figure.

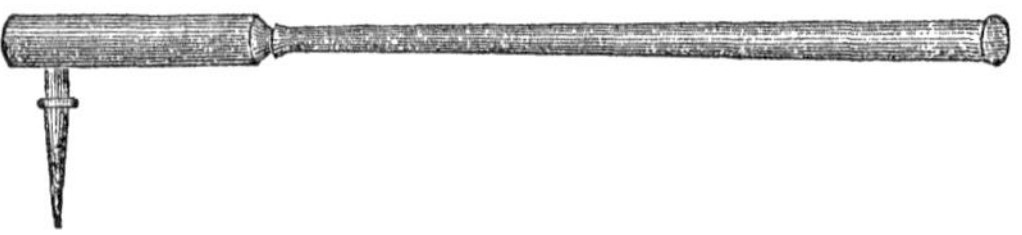
Mouth Blowpipe.

The enlargement at the angle is designed to retain any moisture that may enter the tube from the mouth.

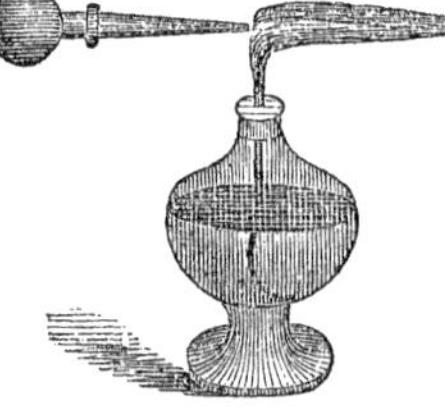
Blowpipe Flame.

In using the blowpipe the flame is driven violently to one side, as shown in the figure; and the greatest heat is found just beyond the point of the inner flame. If pure oxygen is used with the

QUESTIONS.—What is the great source of artificial light and heat? What substances do we use for fuel when light is to be produced? What is the use of the blowpipe? At what point in the flame is the heat greatest?

blowpipe, instead of atmospheric air, as may easily be done, the heat produced becomes much more intense.

194. *Respiration* is also supported by oxygen gas, which is absolutely essential to the process; no animal can live in an atmosphere that does not contain it. A small animal, as a bird, confined in a close box, feels no inconvenience for a time; but the oxygen gradually is absorbed, carbonic acid gas taking its place, and respiration becomes laborious, until at length the animal dies for the want of oxygen. Pure oxygen, however, does not answer the purposes of respiration, as it excites the vital action too much, producing various inflammatory symptoms, and at length death, as the result of the over-action.

195. *Manipulation of Gases.*—The general method of collecting oxygen, described above, answers for all the gases that are not absorbed by water. In the same mode, also, a gas may be transferred from one receiver to another. When a gas is to be collected that is largely absorbed by water, some other liquid must be used, as mercury, or a saturated solution of salt; or the air may be removed by the air-pump, and then the gas admitted.

HYDROGEN.

Symbol, H; *Equivalent*, 1; *Density*, 0·069.

196. History.—This gas was first described by Cavendish in 1766, and received from him the name of *inflammable air*, because of its combustibility. It has received its present name because it forms a part of water (from *hudor*, water, and *gennao*, to produce). It is not found free in nature, but constitutes one-ninth part of water, and enters into nearly all animal and vegetable substances.

197. Preparation.—Hydrogen gas is always procured by the decomposition of water, either directly or indirectly. The direct

---

QUESTIONS.—194. Is oxygen necessary for the support of respiration? Will pure oxygen answer the purpose? 195. What is meant by the manipulation of the gases? When a gas is to be collected that is largely absorbed by water what is the course to be pursued? 196. What facts are mentioned in the history of hydrogen? Why has it been called inflammable air? Why is it now called *hydrogen?* 197. From what is this gas always prepared?

method consists in passing the vapor of water over metallic iron, heated to redness. This is done by putting iron wire or turnings into a gun-barrel open at both ends, to one of which is attached a retort containing pure water, and to the other a bent tube. The gun-barrel is placed in a furnace, and when it has acquired a full red-heat, the water in the retort is made to boil briskly. The gas, which is copiously disengaged as soon as the steam comes in contact with the glowing iron, passes along the bent tube, and may be collected in convenient vessels, by dipping the free extremity of the tube into the water of a pneumatic trough.

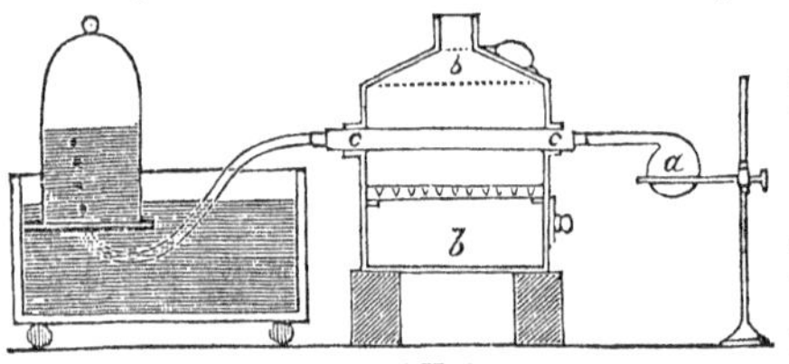

Preparation of Hydrogen.

The arrangement of the apparatus will be seen by the accompanying figure. A retort, *a*, contains the water, *b b* is the furnace, with the gun-barrel, *c c*, in which are the iron turnings to be heated, and at the left is the receiver to collect the gas as it is formed.

The second, or indirect method, which is the one usually adopted in practice, consists simply in dropping pieces of zinc (or iron) into sulphuric acid, diluted with five or six times its weight of water, and contained in a convenient retort. Action immediately commences, without the aid of heat, and the gas may be collected over water. A jar prepared as in the figure is very convenient for the purpose. The water and zinc are first introduced, and after the cork with the tubes is carefully inserted in its place, the acid is poured in through the long-necked funnel. The gas is collected by means of a tube leading from the cover to a receiver, as before.

Preparation of Hydrogen.

QUESTIONS.—Describe the mode first mentioned of preparing hydrogen gas. Describe the mode by the use of dilute sulphuric acid and zinc

Hydrogen gas is also very readily procured by the action of metallic potassium or sodium upon water. A small receiver is first filled with water, and then a piece of the metal, wrapped in bibulous paper, is quickly placed under it; as soon as the paper becomes moistened, violent action takes place, and the hydrogen that is liberated rises to the upper part of the receiver. This method, on account of the high price of potassium and sodium, is very expensive.

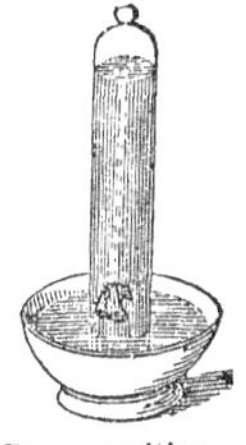
Decomposition of Water by Potassium.

198. Properties.—Pure hydrogen gas is without color, odor, or taste, and refracts light powerfully. It has never yet been reduced to the liquid state. It does not support respiration, but may be breathed, when mixed with air, without injury.

It is the lightest substance known, being sixteen times lighter than oxygen, and more than fourteen times lighter than atmospheric air, 100 cubic inches weighing only 2·14 grains. Soap-bubbles filled with it rise readily through the air. They may be formed very easily by attaching a common tobacco-pipe by its stem to a gas-bottle filled with the gas, and forcing out the gas slowly, immediately after dipping the mouth of the pipe in a strong solution of soap in warm water.

Soap Bubbles with Hydrogen.

Hydrogen gas is eminently combustible, and burns with a feeble yellowish flame. This may be shown by pouring some dilute sulphuric acid upon some pieces of zinc in a vial, and inserting a cork with a small glass tube or pipe-stem, as shown in the accompanying figure. In a short time a jet of hydrogen will issue from the tube, and may be inflamed. This is often described as the *philosophical candle.*

Hydrogen Combustible.

QUESTIONS.—Describe the mode of preparing hydrogen by the use of metallic potassium. 198. Mention some of the properties of hydrogen. What is the weight of 100 cubic inches of the gas? What is said of soap-bubbles formed with it? How may it be shown that the gas is combustible?

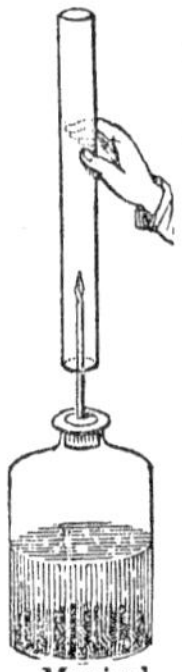

If, now, as the jet continues to burn, a glass tube, half an inch or more in diameter, and one or two feet long, be held over it, and properly managed, a clear musical note will be produced, its pitch depending upon the length and diameter of the tube. It is occasioned by successive explosions within the tube, which follow each other so rapidly as to cause the air in the tube to vibrate, as in a musical instrument.

Musical Sounds.

The following is an interesting and instructive experiment with this gas. Let a bell-glass receiver be filled with it in the pneumatic cistern, and then, carefully lifting it with the left hand, with the right hand pass up into the interior a lighted candle, as represented in the annexed figure. As the flame enters the hydrogen, it will take fire with a slight explosion, and continue to burn where it is in contact with the air; but the candle being carried farther upward into the pure hydrogen, will be extinguished. The hydrogen, being combustible, is inflamed by the burning candle, but not being a supporter of combustion, the candle is extinguished as soon as it is surrounded by it. The gas is retained in the receiver when lifted from its place, because of its being so much lighter than air.

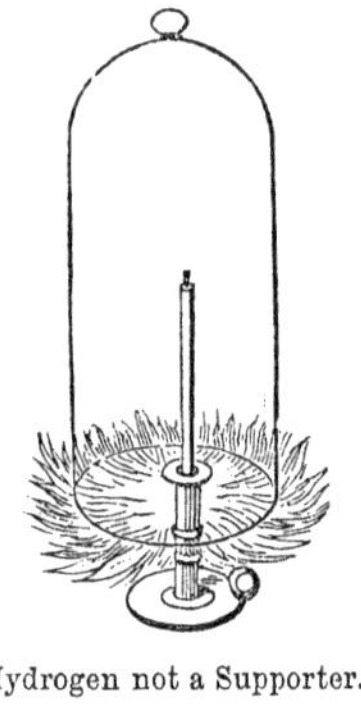

Hydrogen not a Supporter.

199. Hydrogen, being the lightest substance known, is made use of by æronauts for filling their balloons. The buoyancy of a balloon filled with it is easily calculated. Thus, 100 cubic inches of air weighing 31 grains, the weight of a cubic foot will be 535·68 grains, while the weight of an equal volume of hydrogen will be but 36·98 grains. Now 535·68 — 36·98 = 498·70 grains. This number multiplied by the number of cubic feet in a balloon, will give in Troy grains the weight required to be added to make it of equal weight with the air displaced;

QUESTIONS.—Describe the mode of producing musical sounds. How may it be shown that the gas does not support combustion? Will the hydrogen itself be inflamed in the experiment? 199. Why is hydrogen selected for filling balloons? Describe the method of calculating the weight a balloon will sustain in the air.

and if it is loaded with anything less than this it will ascend. The weight of the balloon itself is of course always to be taken into the account.

Coal gas being always on hand where there are gas-works, though much heavier than hydrogen, is now much used in balloons in consequence of its being cheaper, the balloon being of course made proportionally larger.

Compounds of Hydrogen and Oxygen.

There are only two compounds of these substances known, the protoxide, or water, and the peroxide; and the latter is altogether an artificial product, of difficult formation.

200. Protoxide of Hydrogen, or Water.—HO, or Aq.; eq., $(1 + 8 =)$ 9.—This compound, considered in all its important relations, and absolutely universal diffusion, is probably the most important substance known to man. It is the sole product of the combustion of hydrogen, whether in the open air or mixed with oxygen gas. In the experiment for producing musical sounds, the water that is formed will be seen to condense in considerable quantity on the inside of the glass tube, at the beginning of the process, but it will be evaporated when the tube becomes hot.

The affinity of hydrogen for oxygen is very great, but the two gases do not combine spontaneously, even if kept together for any length of time. We have seen above (162) that two measures of hydrogen combine with exactly one measure of oxygen; and the mixture may be exploded by the approach of flame, by the electric spark, by intensely heated metal, or by the mere presence of spongy platinum, a substance that will be described hereafter. To explode the mixed gases by the electric spark, the spark must be made to pass through them. This is accomplished in the following manner. A small metallic vessel, as a miniature cannon, has a metallic wire, W, inserted in one side through a piece of wood or ivory, so as to extend nearly through to the other side, as shown in the figure.

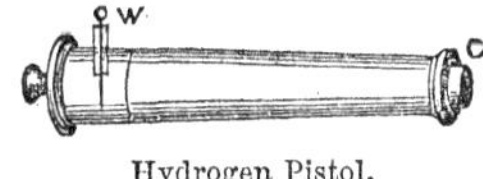

Hydrogen Pistol.

QUESTIONS.—Why is coal gas now often used as a substitute for hydrogen in filling balloons? How many compounds of hydrogen and oxygen are known? 200. What is protoxide of hydrogen? What is said of the affinity of hydrogen for oxygen? In what proportion do they combine by measure? By weight? How may the mixed gases be exploded?

The piece is then to be filled with the proper mixture of the gases, and a cork, C, inserted in the muzzle. If, now, a spark of electricity be communicated to the ball of the wire W, in escaping from the other end it ignites the gases, and the cork is forced out with a loud report. If a mixture of equal parts of hydrogen and atmospheric air is used, the effect will be nearly the same.

The explosion which accompanies the union of these gases is occasioned by the great expansion of the gases at the momen of combination, in consequence of the heat developed. This is the only rational explanation that can be given, and yet it is a fact that this combination is attended by a contraction of one-third of the volume of the mixed gases in forming the vapor of water, (162) to say nothing of the still greater contraction which takes place by the condensation of the steam that is found.

This expansion takes place apparently with great violence; but by using a strong receiver for the mixed gases it may be restrained, and then the combination takes place without any report. The figure in the margin represents a piece of apparatus for this purpose. A strong glass globe three or four inches in diameter has a metallic cap upon the neck, upon which a cover is firmly screwed after the mixed gases are introduced. From each side of the globe is a projection or neck, with a metallic cap, through which wires are inserted so as nearly to meet in the centre. When the globe is charged with the mixed gases the electric spark is passed between these wires, and the gases at once combine, but without explosion, a slight flame only being seen at the instant. The inside of the glass becomes covered with dew from the moisture produced, and when the cover is unscrewed the rushing in of the air attests the vacuum that has been formed.

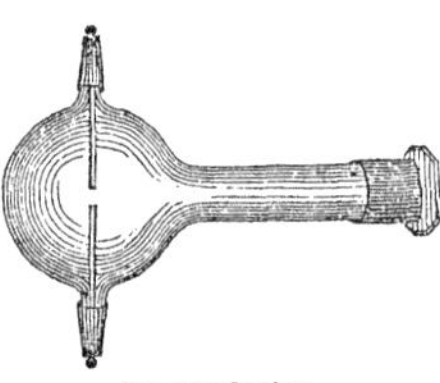
No Explosion.

The ignition of hydrogen by spongy platinum is well shown by holding a small piece of this substance in a jet of the gas, by means of a small wire twisted round it. The platinum must be

QUESTIONS.—Describe the pistol for exploding the gases by the electric spark. What occasions the great expansion that takes place? May the gases be made to unite without an explosion?

perfectly dry. It gradually becomes heated to redness, and soon the jet is inflamed. The common *hydrogen fire-apparatus* acts upon this principle. Its construction is shown in the figure in the margin. A cylindrical glass vessel, A, is partly filled with dilute sulphuric acid, and in it is a small glass receiver, B, firmly cemented at the top into a cap connected with the brass cover. By the action of the acid upon a piece of zinc, Z, suspended inside of this receiver near the bottom, it is soon filled with hydrogen gas, which, on turning the faucet F, is forced out by the rise of the liquid upon a piece of platinum sponge, contained in the cup C. Soon the platinum is heated so as to inflame the jet of hydrogen, from which a candle can be at once lighted. The platinum sponge often loses its property of inflaming hydrogen, but recovers it again by being heated.

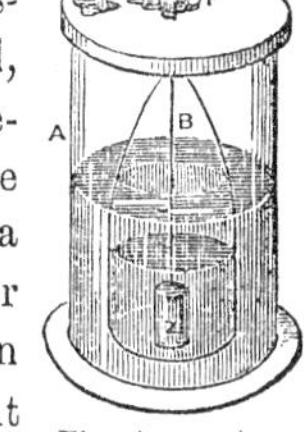

Fire Apparatus.

A metallic cup of half a pint capacity filled with the mixed gases and inverted over a small piece of platinum sponge, supported a little above the table by a wire, will be thrown upward to the ceiling by the explosion produced.

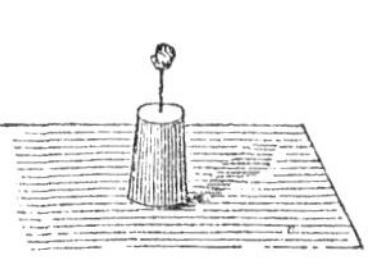
Explosion.

201. *The compound blowpipe*, which is an invention of Dr. Hare of Philadelphia, is a contrivance by which two jets, one of hydrogen and another of oxygen, are made to issue together, and are inflamed as they escape. The gases are brought by tubes from separate gas-holders, so as to discharge, as near as may be, two measures of hydrogen to one of oxygen. The heat produced by this instrument is very great, probably exceeding that produced by any other means.

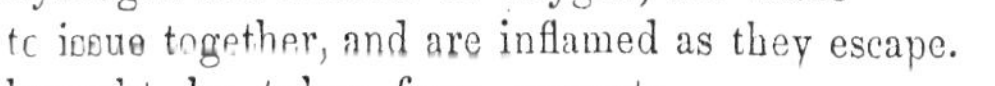

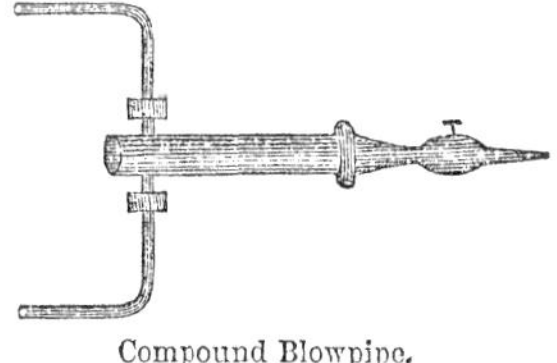
Compound Blowpipe.

QUESTIONS.—Describe the hydrogen *fire apparatus*. Describe the experiment with the spongy platinum and the metallic cup. 201. Describe the *compound blowpipe*. What is said of the heat which is produced?

Platinum and other substances incapable of fusion in the hottest furnaces, are melted, and often even volatilized by it. A small piece of lime, held in the flame, becomes intensely heated, and glows with a brilliant light, exceeding any other that can be produced artificially. It is known under the name of the *Drummond light*, and is much used for practical purposes.

202. *Water*, at ordinary temperatures, is a transparent, colorless liquid, without taste or smell. In the open air, it boils at a temperature of 212°, and freezes at 32°. The vapor produced by boiling, familiarly called steam, is perfectly colorless and transparent, and has a density of 0·622, air being 1; 100 cubic inches weighing 19·28 grs. In the form of ice, its density is 0·92. A cubic inch of pure water weighs 252·458 grs., being 814 times as much as an equal volume of air would weigh. A cubic foot weighs 1000 oz., or 62½ lbs. avoirdupois. Water is probably the most powerful solvent known. It is capable of combining, in definite proportions, with many substances, forming compounds which yet remain perfectly dry. They are usually called *hydrates*. Substances from which all water has been separated are said to be *anhydrous*.

Water is never found naturally perfectly pure; that of wells, springs, and rivers always contains carbonic acid and saline matter in solution, obtained while percolating through the soil; and that from rain or snow is impregnated with air and oxygen, and sometimes with other gases. It is obtained pure only by distillation; and even then, by standing for a time, it takes up more or less atmospheric air, which, however, does not unfit it for the ordinary operations of the laboratory in which it is required.

203. Water when it solidifies often forms beautiful crystals (25) which are best seen in snow and frost;—in ordinary ice they cannot usually be seen, though it is believed there is always

---

QUESTIONS.—What is said of the light produced when a piece of lime is held in the flame? By what name is this light known? 202. What is said of water at ordinary temperatures? What are its freezing and boiling points? What is steam? What is its density? What is the density of ice? What is the weight of a cubic inch of water? How many times is it heavier than air? What are the chemical compounds of water called? Is water ever found perfectly pure in nature? How only is pure water obtained? 203. When water solidifies does it take the form of crystals?

a proper crystaline arrangement. The primary crystals belong to the hexagonal system (178–6), but they are often grouped together in various modes, as shown in the figures. Nos. 3, 4, 5, 6, 7,

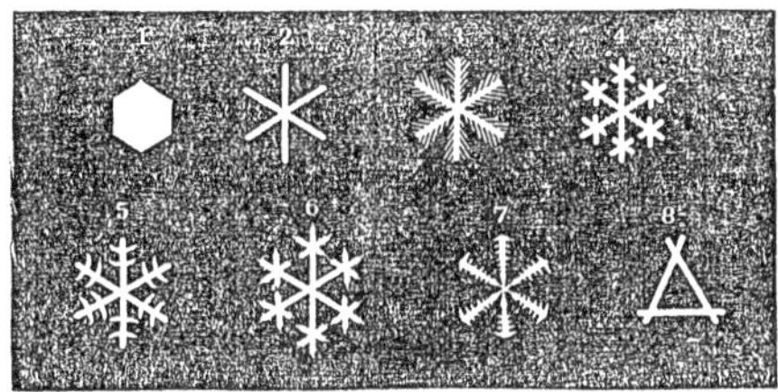

Crystals of Ice.

and 8 are more common than the simple forms Nos. 1 and 2. These latter are of more frequent occurrence in hoar-frost than in snow.

204. The different processes for procuring hydrogen, given above, require some further explanation, before we dismiss this subject. The first method is founded on the fact that iron, at high temperatures, decomposes water when presented to it in the form of steam, the oxygen combining with the iron to form protoxide of iron, and the hydrogen being set free. The changes are thus represented:

$$HO + Fe = FeO + H.$$

The changes which take place in the second and more common process for procuring this gas, are more complicated. They are represented in the following formula. Thus,

$$HO,SO_3 + Zn = ZnO,SO_3 + H.$$

In this case it will still be seen that it is the water which supplies the hydrogen. The oxygen of the water is transferred to the zinc, forming protoxide of zinc, with which the sulphuric acid immediately combines. When a piece of clean zinc is immersed

---

QUESTIONS.—What is said of frost and snow? 204. Explain the several modes of procuring hydrogen, and write the equations illustrating the reactions that take place.

in water, little action takes place, because the outside becomes coated with a thin film of oxide of zinc, which is insoluble in water; but if sulphuric acid is present, this oxide is instantly dissolved, and thus a clean surface constantly exposed to the water.

In the third process, the metal itself at once decomposes a portion of the water, forming a soluble oxide of the metal, and liberating the hydrogen. Thus,

$$HO + K = KO + H.$$

Here the affinity of metallic potassium for oxygen is sufficient to separate this element from its union with hydrogen in the water, forming oxide of potassium or potassa. Great heat is produced by the action, and the hydrogen liberated is inflamed if the experiment is made in the open air. To show this all that is necessary is to throw a small piece of potassium upon the surface of some water, as shown in the figure.

Potassium inflamed by water.

The mode of decomposing water by the galvanic current has already (111) been explained.

205. Knowing that water is formed by the union of 2 volumes of hydrogen and 1 volume of oxygen condensed into 2 volumes of steam, it is easy to calculate the density of steam, and also the quantity of each of these elements in a given quantity of water. Thus,

One volume	of oxygen	weighs		1·106
Two	" hydrogen	"	$2 \times 069 =$	0·138
Two	" steam	"		1·244

One-half of this sum, or ·622, is then the density of the vapor of water which is formed.

To calculate the quantity of each of the elements in 100 parts of water we have

$$1{\cdot}244 : 1{\cdot}106 :: 100 : x,$$

$$\text{whence } x = 88{\cdot}9 = \text{quantity of oxygen.}$$
$$\text{And } 100 - 89{\cdot}9 = 11{\cdot}1 = \text{quantity of hydrogen.}$$

QUESTIONS.—Is heat produced when water is decomposed by potassium? 205. What is the density of steam? Describe the mode of calculating it.

206. Peroxide of Hydrogen—HO_2; eq., (1 + 16 =) 17.—This substance has been obtained only in the liquid form, and when most concentrated possesses a specific gravity of 1·452. It is colorless, transparent, and without odor. It is sometimes called oxygenized water.

Though it differs from water only in containing an additional equivalent of oxygen, it acts powerfully upon the skin, producing a prickling sensation, whitening the surface, and destroying the texture, if the application is long continued.

If the temperature is raised as high as 59°, it is decomposed and converted into water and oxygen gas; a sudden elevation of temperature even causes an explosion. It is also decomposed by the action of nearly all the metals, and many of the metallic oxides. Diluted with water or mixed with some acid, it is less liable to decomposition than when pure.

NITROGEN.

Symbol, N; *Equivalent*, 14; *Density*, 0·972.

207. History.—The existence of this element has been known since 1772; and it was recognized as a constituent of the atmosphere in 1775. It was first called *azote* (from *a*, privitive, and *zoe*, life), because it does not support respiration; it receives its present name from the circumstance that it forms an ingredient of nitre.

208. Preparation.—Nitrogen gas is readily prepared, nearly pure, by burning a piece of phosphorus in a receiver over water. A small cup, C, containing a piece of phosphorus, is placed upon the surface of the water in the pneumatic cistern, the phosphorus ignited, and the receiver then placed over it. The phosphorus continues to burn, absorbing all the oxygen, and the water rises to supply its place, as shown in the figure. The glass will at first be filled with phosphoric acid fumes, which however will soon be absorbed by the water. Some vapor of phosphorus, carbonic acid, and perhaps a trace of other gases, may be contained in the

Preparation of Nitrogen.

QUESTIONS.—206. What is peroxide of hydrogen? What are some of its properties? 207. How long has nitrogen been known? What was it first called? Why has it received its present name? 208. Describe the mode of preparing nitrogen.

nitrogen thus prepared, but it will be found sufficiently pure for nearly all purposes. Other processes might be described for procuring it, as by passing a current of chlorine gas through strong aqua ammonia, but the above is the most speedy and convenient.

209. Properties.—Pure nitrogen is a colorless gas, wholly devoid of smell and taste, and is distinguished from other gases more by negative characters than by any striking quality. It is not a supporter of combustion, but, on the contrary, extinguishes all burning bodies that are immersed in it. A taper immersed in it is extinguished at once. No animal can live in it; but yet it exerts no injurious action either on the lungs or on the system at large, the privation of oxygen gas being the sole cause of death. It is not inflammable, like hydrogen; though, under favorable circumstances, it may be made to unite with oxygen. It is slightly dissolved by water, and is sometimes found in the water of mineral springs, as at Lebanon, in the State of New York. 100 cubic inches of the gas weigh 30·16 grs., giving a specific gravity of 0·97.

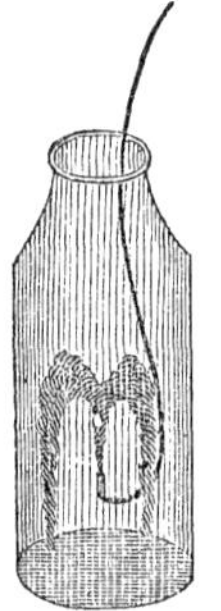

Does not support Combustion.

Atmospheric Air.

210. The earth is everywhere surrounded by a mass of gaseous matter, called the *atmosphere*, which is preserved at its surface by the force of gravity, and revolves with it around the sun. It is colorless and invisible, excites neither taste nor smell when pure, and is not sensible to the touch, unless when it is in motion. It possesses the physical properties of elastic fluids in a high degree. Its specific gravity is unity (1), being the standard with which the density of all gaseous substances is compared. It is 814 times lighter than water, and nearly 11,065 times lighter than mercury, 100 cubic inches weighing 31 grs.

QUESTIONS.—Is the nitrogen thus obtained pure? 209. What are some of the properties of nitrogen? 210. What is the *atmosphere?* What are some of the properties of atmospheric air? What is the weight of 100 cubic inches?

Atmospheric air is composed of nitrogen and oxygen, with a variable proportion of carbonic acid and watery vapor, and usually a trace of ammonia. Besides these, there may occasionally be other substances present, depending upon local causes, as the odoriferous principle of plants, and the miasmata of marshes, which is supposed to be the chief cause of disease in many unhealthy situations; but they cannot be detected by chemical tests.

Instruments for determining the relative proportion of the gases composing the atmosphere are called *eudiometers*. The following contrivance answers the purpose very well. Let a glass tube, closed at one end, and graduated to 100 parts, with a small piece of phosphorus supported in it on a wire near the top, be placed as in the figure, with the open end immersed in a vessel of water. The phosphorus gradually absorbs the oxygen of the air in the tube, and the water rises to supply its place. In one or two days, depending upon the temperature, the absorption will be complete; and the number of the division of the tube now filled with water will indicate the proportion of oxygen.

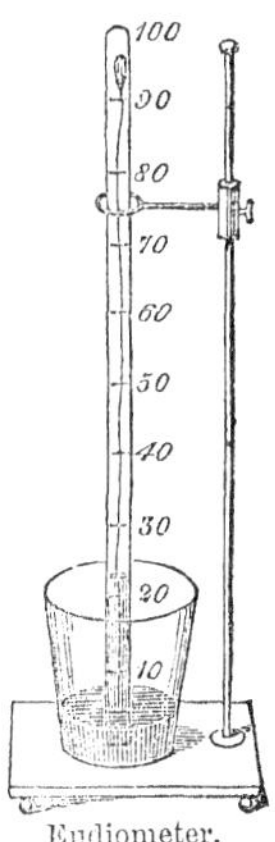

Eudiometer.

By the above and other similar modes of analysis, it is found that the atmosphere in 100 parts is composed of—

	By Weight.	By Measure.
Nitrogen	76·9	79·3
Oxygen	23·1	20·7
	100	100

The proportion of carbonic acid varies from 2 to 6 parts in 10,000 of air. A trace of ammonia is not unfrequently found, and also of carburetted hydrogen.

The atmosphere is believed to extend to the height of about forty-five miles, becoming continually less and less dense from the

QUESTIONS.—What is atmospheric air composed of? What other gases are also found in it? What is the design of *eudiometers?* Describe the mode of determining the proportion of oxygen by means of phosphorus. What is thus found to be the quantity of oxygen in 100 parts of air? What is the usual quantity of carbonic acid in 10,000 parts of air? What is the height of the atmosphere?

surface upward; and presses by its gravity upon the surface with a force equal, ordinarily, to about 15 lbs. to the square inch. It is capable of supporting a column of water about 34 feet, and a column of mercury about 30 inches in perpendicular height.

211. Atmospheric air is highly compressible and elastic, so that its particles admit of being approximated to a great extent by compression, and expand to an extreme degree of rarity when the tendency of its particles to separate is not restrained by external force. The volume of air and all other gaseous fluids, so long as they retain the elastic state, is inversely as the pressure to which they are exposed. Thus a portion of air which occupies 100 measures when compressed by the ordinary atmospheric pressure, will be diminished to 50 measures when the pressure is doubled, and will expand to 200 measures when the compression is reduced to one-half. But those gases which are susceptible of condensation by pressure into the liquid form, as the pressure approaches this point, vary from the law, the volume diminishing more rapidly than this would indicate.

The chief chemical properties of the atmosphere are owing to the presence of oxygen gas. Air from which this principle has been withdrawn, is nearly inert. It can no longer support respiration and combustion, and metals are not oxydized by being heated in it. The uses of the nitrogen are in a great measure unknown. It has been supposed to act as a mere diluent to the oxygen; but it most probably serves some useful purpose in the economy of animals and plants, the exact nature of which has not been discovered.

212. The question has often been discussed, whether the oxygen and nitrogen of the atmosphere are to be considered as chemically combined, or only in a state of mixture; but the latter opinion now generally prevails. It has been supposed that if they are merely in a state of mixture, oxygen, being the most dense, ought

---

QUESTIONS.—To what height will the atmosphere support a column of water? Of mercury? 211. Is atmospheric air compressible and elastic? What is said of the volume of air and other elastic fluids in relation to the pressure to which they may be subjected? To what are the chief chemical properties of atmospheric air owing? 212. Are the oxygen and nitrogen of the air to be considered chemically combined, or only in mixture?

to settle towards the surface of the earth; but it is found by experiment that gases, whatever may be their relative density, when brought in contact, mix (39) uniformly with each other. The mixture of the gases will even take place through thin membranes, (see author's *Nat. Phil.*, p. 22,) whether animal or vegetable; the least dense of the gases passing the most rapidly.

213. There is still one circumstance for consideration respecting the atmosphere. Since oxygen is necessary to combustion, to the respiration of animals, and to various other natural operations, by all of which that gas is withdrawn from the air, it is obvious that its quantity would gradually diminish, unless the tendency of these causes were counteracted by some compensating process. This, to some considerable extent, is accomplished by vegetation, as it is found that healthy plants, under the influence of the sun's light, are constantly absorbing carbonic acid from the air, the carbon of which is retained, while the oxygen is returned to the air, as we have before seen (191). Still, it has been calculated that the loss of oxygen employed in respiration, over the whole surface of the globe, in 100 years, would not exceed $\frac{1}{7200}$ part of the whole quantity contained in the atmosphere.

Compounds of Nitrogen and Oxygen.

214. Oxygen combines with nitrogen in five different proportions, forming the compounds NO, NO_2, NO_3, NO_4, and NO_5; the last three of which are acids.

215. *Protoxide of Nitrogen, or Nitrous Oxide*—NO; eq., $(14 + 8 =)$ 22. —This is a permanent, colorless gas, of a sweetish taste and smell. 100 cubic inches of it weigh 47·22 grs., and

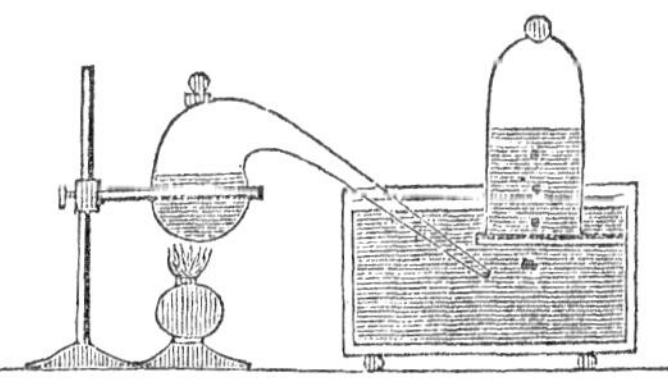

Preparation of Nitrous Oxide.

QUESTIONS.—213. Is oxygen constantly absorbed from the atmosphere by combustion and by the respiration of animals? By what reverse process is carbonic acid absorbed and oxygen restored to the atmosphere? 214. How many compounds of nitrogen and oxygen are there? 215. Describe the protoxide of nitrogen.

its density therefore is 1·527. It is best prepared as shown in the figure on the preceding page, by heating nitrate of ammonia, by means of a spirit-lamp, in a glass retort to a temperature of 450° or 480°. The sole products of the operation, when carefully conducted, so as not to raise the temperature too high, are water and the gas in question. The composition of nitrate of ammonia is NH_3,HO,NO_5—which by heat is converted into 2 atoms of the nitrous oxide in question and 4 atoms of water. Thus,

$$NH_3, HO, NO_5 = 4\ HO + 2\ NO.$$

The water of course is condensed and remains with that in the cistern, while the gas is collected as in other cases. It is slightly absorbed by water and should not therefore be allowed to remain in contact with it.

Some substances burn in this gas with great brilliancy, as a lighted candle and phosphorus; and sometimes the combustion of iron wire in it may be effected, but not without difficulty. By a pressure of about 30 atmospheres, at a temperature of 32°, it is compressed into a liquid, which freezes or becomes solid at about 150° below zero; and by the evaporation of this solid, a temperature considerably lower than this has been attained.

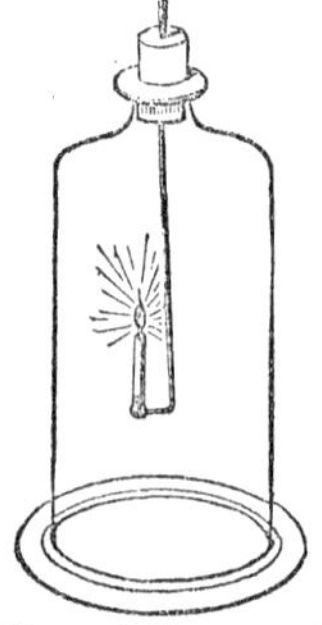

Nitrous Oxide supports Combustion.

Its action on the system, when breathed, is very remarkable, producing a species of intoxication, which has acquired for it the name of *laughing gas*. In a few cases, when it has been inspired, injurious effects have resulted; and it should never be breathed but with caution.

216. The analysis of nitrous oxide is made as follows; in a given quantity of the gas accurately measured introduce a piece of potassium, the gas being contained in a glass tube closed at one end and a little bent,

---

QUESTIONS.—How is the protoxide of nitrogen prepared? Describe the chemical changes that take place, and write the equation illustrating them. Is this gas absorbed by water? Will it support combustion? What is said of its effects upon the system when breathed? 216. Describe the mode of analyzing this gas.

(263) so as to retain the potassium while the open end is immersed in mercury. Now apply the heat of a lamp and the potassium takes fire immediately, absorbing all the oxygen and leaving the nitrogen. This being returned to the graduated tube in which the gas was first measured, will be found to have the same volume as at first. We learn therefore that the protoxide of nitrogen occupies precisely the same volume as the nitrogen contained in it would alone occupy. Now,

One volume of the	protoxide weighs	1·527
One " "	nitrogen (subtract)	·972
Half " "	oxygen, very nearly,	·555

Nitrous oxide is therefore a compound of 1 volume of nitrogen and $\frac{1}{2}$ volume of oxygen, the whole condensed into 1 volume.

The same result is obtained by passing a quantity of the gas through a heated porcelain tube by which it is decomposed, and 1 volume of nitrogen and $\frac{1}{2}$ volume of oxygen obtained.

217. *Binoxide of Nitrogen, or Nitric Oxide*—NO_2; eq., $(14 + 16 =)$ 30.—This is also a gaseous substance, and is easily obtained by pouring nitric acid upon pieces of copper contained in a glass retort. The arrangement shown in the figure is very convenient for the purpose. Into the glass vessel *a* put some clean pieces of metallic copper, and then introduce the cover, through which passes the glass tube *b*, with a funnel at top, and extending nearly to the bottom of the vessel, and a lead tube, *c*, bent at right angles, to convey away the gas as it is formed. The cover must fit very accurately, in order to prevent the escape of the gas, which is rapidly formed as soon as a little nitric acid is introduced by the funnel and tube *b*. It may be collected over water, but a small proportion is absorbed.

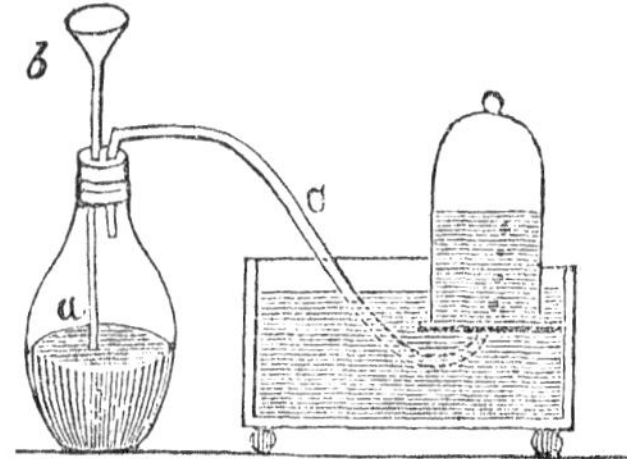

Preparation of Nitric Acid

The changes which take place between the copper and the acid are indicated as follows:—

$$4\ NO_5 + 3\ Cu = 3\ (CuO,NO_5) + NO_2.$$

QUESTIONS.—217. How is the binoxide of nitrogen prepared? What are the chemical changes that take place?

17*

Thus, from 4 atoms of nitric acid and 3 of copper, there are formed 3 atoms of nitrate of protoxide of copper, and one of the binoxide of nitrogen.

218. Binoxide of nitrogen is a colorless gas, of a density of 1·039, 100 cubic inches weighing a little more than 32 grs. Its most striking property is its strong affinity for oxygen, although most substances introduced into it in a state of combustion are extinguished. Charcoal and phosphorus, however, if well ignited, burn in it with great splendor, though the latter by careful heating may be melted in it without ignition. When allowed to escape into the air, it forms, with the oxygen of the air, dense orange fumes of nitrous acid. Thus, if a bell-glass receiver, filled with it over water, be suddenly inverted in the air, the dense orange fumes of nitrous acid formed by its union with the oxygen of the atmosphere will rise for a few moments from its interior, like smoke from a chimney.

Experiment with Nitric Oxide.

Let a receiver filled with this gas be placed in a basin of pure water, as shown in the figure; and then let a few bubbles of oxygen gas be forced in by a pipe leading from a gasometer, or by means of a gas-bag. As the oxygen mixes with the gas within, nitrous acid fumes are produced, which are immediately absorbed by the water, giving it a slight acidity, as may be determined by the usual tests. By continuing to force in the oxygen gradually, and supplying the basin with water, all the gas will at length disappear, being entirely absorbed by the water as nitrous acid.

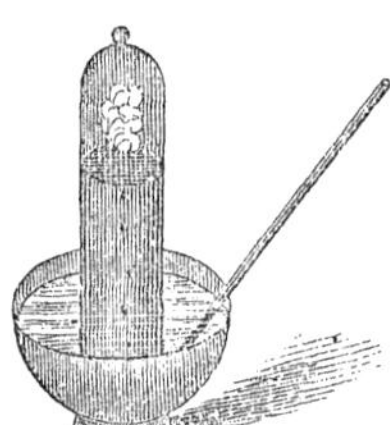

Nitric Oxide and Oxygen.

QUESTIONS.—218. Describe binoxide of nitrogen. What is produced when it comes in contact with atmospheric air or oxygen? Describe the experiment with a receiver filled with the gas over water and oxygen gas.

Binoxide of nitrogen may be analyzed in the same manner as the protoxide (216), by heating in a small quantity of it, accurately measured, a small piece of potassium or sodium, which absorbs the oxygen, leaving the nitrogen free, the volume of which will be found just one-half of that of the compound gas. The binoxide therefore is composed of

1 vol.	of nitrogen,	which weighs	0·972
1 "	oxygen,	"	1·106
2 "	binoxide,	"	2·078

The weight of 1 vol. or the density is therefore, as above stated, $2{\cdot}078 \div 2 = 1{\cdot}039$.

219. Hyponitrous Acid — NO_3; eq., (14 + 24 =) 38.—This acid is formed by mixing 4 measures of binoxide of nitrogen with 1 of oxygen, both perfectly dry, and subjecting the mixture to a cold at least as low as zero. It is a colorless liquid, which is at once decomposed by water into nitric acid and the binoxide. This acid does not combine directly with bases, but its salts may be formed by heating carefully the corresponding nitrates. Thus by heating nitrate of potassa, a part of the oxygen of the nitric acid is expelled, and the nitrate is converted chiefly into hyponitrite of potassa, as indicated in the following formula:

$$KO,NO_5 = KO,NO_3 + 2\,O.$$

The heat should be carefully applied, and the process arrested, as soon as the oxygen which comes over begins to be mixed with nitrogen By treating the mass after cooling with strong alcohol the hyponitrite is dissolved out, and may be obtained very pure.

220. Nitrous Acid—NO_4, eq., (14 + 32 =) 46.—This substance is formed, as we have seen, whenever binoxide of nitrogen comes in contact with oxygen. It may also be prepared by heating dry nitrate of lead in a retort, and passing the vapor formed through a tube surrounded with a freezing mixture to condense it. A piece of apparatus like that represented in the

QUESTIONS.—What is the effect when a piece of potassium is heated in a tube containing nitric oxide? What is the volume of nitrogen that is left? 219. What is hyponitrous acid? Does it combine directly with bases? How may the hyponitrites be formed? 220. How may nitrous acid be formed?

figure answers the purpose well. It is to be placed in a proper vessel, containing the freezing mixture, and connected with the mouth of the retort.

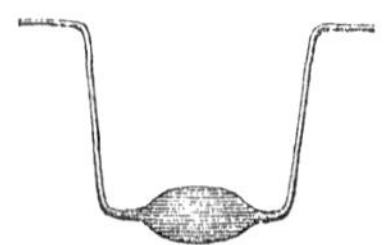

As thus obtained, this acid has a density of 1·42, and at 60° F. is of an orange color, but becomes yellow when cooled to 32°, and is nearly colorless at zero. Though it requires a temperature nearly as low as zero to condense it to the liquid form, the liquid boils at 82°. It does not combine with bases, but forms compounds with some of the stronger acids. It is decomposed by contact with water.

221. Nitric Acid—NO_5; eq., (14 + 40 =) 54.—Nitric acid, or *aquafortis*, is usually seen as a liquid, and is best obtained by decomposing nitrate of potash or nitrate of soda by strong sulphuric acid, by the aid of heat. The salt, previously well dried, is placed in a retort of hard glass, A, with an equal weight of strong sulphuric acid, in a furnace, and surrounded at the bottom with sand. A moderate heat is applied, and the nitric acid, as it is separated, distils over into the receiver, B, where it is condensed. To render the condensation more complete, the receiver may be surrounded with a net-work, and cold water from a pipe, *i*, made to fall constantly upon it. The water escapes by the troughs C C and *cd*. A furnace is not absolutely necessary; the heat of a lamp is sufficient in all laboratory operations.

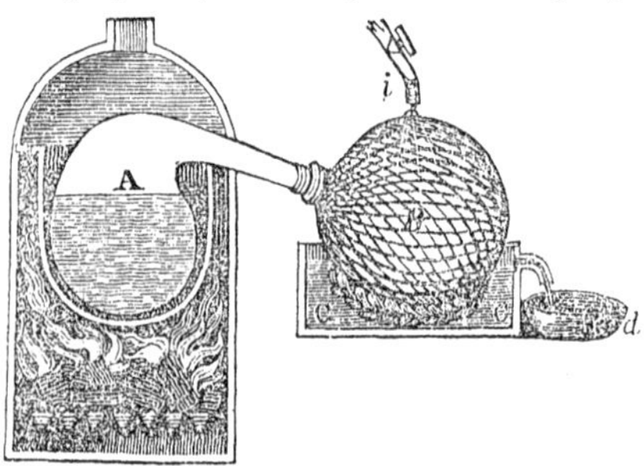

Preparation of Nitric Acid.

Nitric acid, as thus formed, is a dense liquid, of a yellow or orange color, and always contains more or less nitrous acid mixed

QUESTIONS.—May nitrous acid be obtained in the liquid form? Does it combine with bases? 221. How is nitric acid obtained? What is the common or commercial name of this acid? Describe nitric acid.

with it. In its most condensed state it has a density of 1·52, and boils at 188°; its composition being then NO_5,HO. There is another definite compound of the acid and water, $NO_5,4HO$, which has a specific gravity of 1·42, and boils at 253°. The first contains 14 per cent. of water, and the latter 40 per cent.

By decomposing dry nitrate of silver by dry chlorine, nitric acid may be obtained as a white crystaline solid, which is readily soluble in water, forming the common *aquafortis.*

Strong nitric acid, when exposed to the atmosphere, emits dense, white fumes, which are exceedingly suffocating. Mixed suddenly with water, it occasions a considerable rise of temperature; but mixed with snow, it produces cold by the rapid liquefaction which is occasioned.

Nitric acid may be frozen by cold. The temperature at which congelation takes place varies with the strength of the acid. The strongest acid freezes at about 58° below zero. When diluted with half its weight of water, it becomes solid at —1½°. By the addition of a little more water, its freezing point is lowered to — 45°.

It is one of the most energetic acids known, and oxydizes many substances powerfully. Powdered charcoal and oil of turpentine are ignited by it, and most animal and vegetable bodies disorganized. Phosphorus is oxydized so rapidly as to produce an explosion. A small drop on the skin will, in a few seconds, destroy its vitality, and produce a permanent yellow spot. There are two varieties of it in commerce, called *single* and *double aquafortis,* the latter of which is much the strongest, and usually is of a deep orange color.

222. This acid is sometimes produced by the union of the oxygen and nitrogen of the atmosphere, as the effect of lightnings, and is found as nitrate of ammonia in rain-water after a shower in summer. The same effect may be produced by the electric spark, but the presence of water and some base to combine immediately with the acid is necessary. Let a tube be bent, as in the figure on next page, and the two ends covered with metallic cups, one of them opening by a screw, so that the tube may be half filled with a weak solution of potash. Now insert the screw

---

QUESTIONS.—Does nitric acid always contain water when in the liquid form? How may it be obtained in the solid form? May aquafortis be frozen? What is said of its oxydizing power? 222. How may nitric acid be formed by the direct union of its elements?

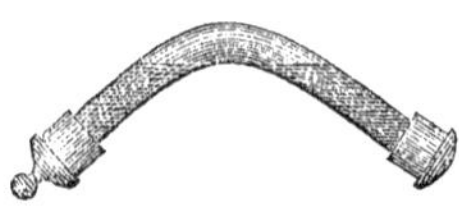

Nitric Acid Produced.

and divide the solution into two equal parts, as nearly as may be, and connect it with the prime conductor of an electrical machine, and cause a succession of sparks to pass through the portion of air contained in the central part of the tube. Upon examination, the solution will now be found to contain nitrate of potassa; the nitric acid of which has been formed by the union of the constituent gases of the atmosphere, as above explained.

223. Gold-leaf answers well, in most cases, as a test for nitric acid. The substance supposed to contain it is mixed with a little hydrochloric acid, and the mixture, if nitric acid is really present, should then be capable of dissolving the leaf. If it is a salt that is to be tested, it should first be dissolved in water, and the solution treated in the same manner with hydrochloric acid and gold-leaf.

Nitric acid is much used in the arts for etching on copper, as a solvent for the metals, &c., and as a tonic in medicine. In the laboratory of the chemist, it is in constant and most important use in a great variety of operations.

Compounds of Nitrogen and Hydrogen.

224. Ammonia—NH_3; eq., $(14 + 3 =) 17$.—This gaseous substance is the only compound of nitrogen and hydrogen that is known. It has been called by a variety of names, as *hartshorn*, *spirits of hartshorn*, *volatile alkali*, &c. It has sometimes been detected in rain-water, either alone or in combination with nitric acid (222); but it is readily

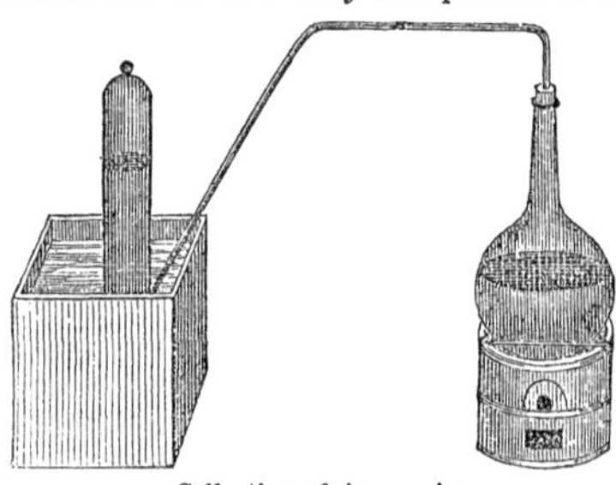

Collection of Ammonia.

QUESTIONS.—223. What test of free nitric acid is mentioned? 224. What is the composition of ammonia? By what names is it known?

obtained by heating the common solution of ammonia, called *aqua ammoniæ*, or a mixture of sal ammonia and recently slaked lime. When the latter substances are mixed the odor of ammonia is at once perceived, but it is evolved freely when the mixture is moderately heated. It is rapidly absorbed by water, and should therefore be collected over mercury. The reaction is as follows:

$$NH_3,HCl + CaO = Ca, Cl + HO + NH_3.$$

The two gases do not combine directly, but their union often takes place when they are presented to each other in their nascent state (226), in various chemical operations. Thus, during the rusting of iron, when moisture is present a portion of ammonia is frequently produced, and when zinc is dissolved in dilute nitric acid the solution will usually be found to contain a little nitrate of ammonia.

Ammonia is also produced during the destructive distillation of animal substances; and the name *hartshorn* came into use from the circumstance that deer's horns were made use of for the purpose.

225. Ammonia is a colorless gas, and possesses a very pungent odor, by which it may always be distinguished. By a pressure of six or seven atmospheres it is compressed into a liquid; it also takes the liquid form by the application of intense cold, under the ordinary atmospheric pressure. A lighted candle plunged into it is extinguished, but a small jet of it burns in oxygen gas. Its density is about 0·59, 100 cubic inches weighing 18·29 grs.

Ammonia is composed of one volume of nitrogen and three volumes of hydrogen, condensed into two volumes

1 vol. of	nitrogen	weighs	0·972
3 "	hydrogen	"	0·207
2 "	ammonia	"	1·179

One-half of this, or 0·589, is the weight of one volume, or the theoretical density of the gas.

QUESTIONS.—How may ammonia be obtained from aqua ammonia? What is said of its absorption by water? What is said of the formation of ammonia during the rusting of iron? 225. What are some of the properties of ammonia?

This gas is largely absorbed by water, which at 32° is capable of dissolving 500 or 600 times its own volume of it, and then forms the *liquid ammonia*, or *aqua ammoniæ* of commerce. As the gas is absorbed, the water increases considerably in volume, so that, when saturated, its density is only 0·87, and it contains 32 per cent. of the gas.

Liquid ammonia is best prepared by the use of a Woulf's apparatus, which is constructed as follows. First a vessel of

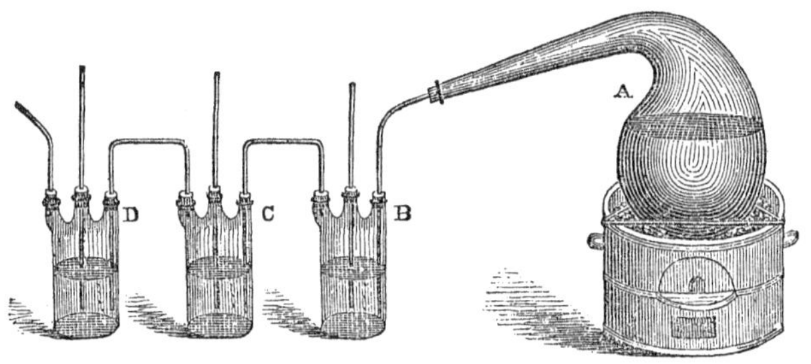

Preparation of Aqua Ammonia.

glass or metal, A, containing the substances from which the gas is to be evolved, is placed upon a furnace or other source of heat, and connected by a tube with a series of three-necked bottles, B, C, D, &c., each partly filled with water. The first tube from the vessel containing the materials, it will be observed, terminates below the surface of the water, in the bottle B, so that the gas coming through it is made to rise through the water; but the tube connecting this with the next bottle C terminates in B near the top, but extends downward beneath the surface of the water in C. The same thing will also be observed of the tubes connecting together the other bottles. Now when the gaseous ammonia is evolved it passes over, and rising through the water in C, is absorbed, but any other gas passes on in like manner to the next bottle, and so on until it escapes in the air. If any of the ammonia escapes from the first bottle, as will be the case after the water has become in part saturated. it will be likely to be con-

QUESTIONS.—What is said of the absorption of ammonia by water? Describe the mode of preparing the liquid ammonia.

densed in the next, or farther on in the series. In the centre of each bottle is a tube, open at both ends, but extending downward so as to terminate below the surface of the liquid. It is designed merely as a safety tube; it is plain that if, in using the apparatus, any obstruction occurs in one of the tubes, the effect will be to force out the liquid through these central tubes without danger to the apparatus; so also any external pressure will at once be equalized by the entrance of the air through these tubes. This apparatus is used for many similar purposes.

In order to ensure the full saturation of the water in the bottles they must be kept cold by means of ice.

Ammonia has all the properties of an alkali in a very marked manner. Thus it has an acrid taste, and gives a brown stain to turmeric paper; though the yellow color soon reappears on exposure to the air, owing to the volatility of the alkali. It combines also with acids, and neutralizes their properties completely.

This substance is used extensively in the laboratory of the chemist, and in medicine. Solution of ammonia, taken internally in considerable quantity, has been known to produce death; and the gas, if inspired too long, is apt to occasion inflammation in the throat and lungs.

226. **Nascent State.**—This phrase, which occurs above, is used to indicate the state of a body as it is evolved from a compound containing it. In the case alluded to, the rusting of iron in the presence of moisture, the oxygen which unites with the iron is mostly supplied by the water; and the hydrogen, as it is thus set free, is said to be in its *nascent state*, and just at the moment is capable of combining with the nitrogen of the air to form ammonia, which it will not do under other circumstances. Many other substances, when thus separated from compounds of which they form a part, manifest a disposition to form compounds with bodies with which they will not combine directly.

QUESTIONS.—What is the design of the central tube in each bottle? What is said of the alkaline properties of ammonia? 226. What is meant by the *nascent state* of a body? What is said of the properties possessed by substances in this state?

GROUP II.

CHLORINE, IODINE, BROMINE, FLUORINE } Electro-negative elements, which constitute a very distinct natural family. Each forms a single acid compound with hydrogen; and all combine with the metals, forming compounds that are similar.

CHLORINE.

Symbol, Cl; *Equivalent*, 35·4; *Density*, 2.44.

227. History.—Chlorine was discovered by Scheele, in 1774, and for many years was regarded as a compound, but its true character, as a simple element, is now universally admitted. It is not found uncombined in nature, but is abundant in its compounds, especially common salt, which is so generally distributed over the surface of the earth. It receives its name from the Greek *chloros*, green, because of its yellowish-green color.

228. Preparation.—Chlorine gas is easily prepared by pouring strong hydrochloric acid upon half its weight of peroxide of manganese, in a glass retort, and applying a gentle heat. It may be collected over hot water, but not without some being absorbed by it. It should always be collected in bottles with good stoppers, in which it may be kept for some time. The following formula indicates the changes that take place:—

$$MnO_2 + 2HCl = MnCl + 2HO + Cl.$$

A cheaper process, and nearly as convenient, is to mix, intimately, three parts of common salt with one of the peroxide, and pour over it two parts of sulphuric acid, previously diluted with an equal weight of water. In this case,

$$MnO_2 + NaCl + 2SO_3 = MnO,SO_3 + NaO,SO_3 + Cl.$$

QUESTIONS.—What are the substances of the second group? What is said of them as a group? 227. What facts are mentioned in the history of chlorine? From what is the name derived? 228. Describe the method first mentioned for preparing chlorine. Why should hot water be used in collecting it? Describe the mode of preparing it by the use of common salt. What are the chemical changes that take place?

The chlorine by this process is evidently supplied by the salt (chloride of sodium) and there are formed sulphate of the protoxide of manganese and sulphate of soda.

It is afforded also by various other preparations.

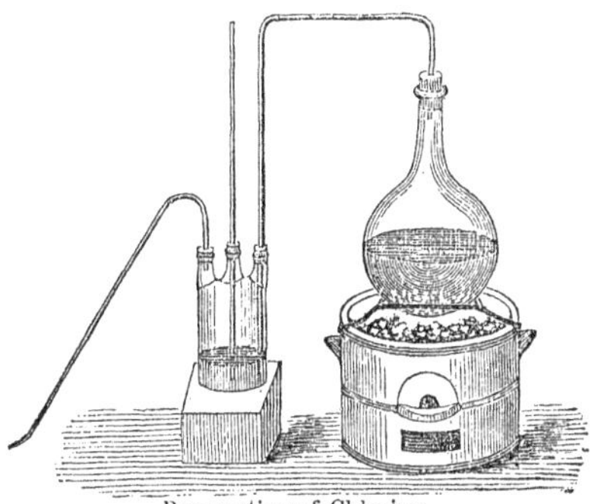

Preparation of Chlorine

229. Properties.—Chlorine is a dense gas, of a yellowish green color, as above stated, and has an astringent taste, and rather disagreeable odor. The weight of 100 cubic inches is 76 grains, which gives a density of 2·44. By moderate pressure it is converted into a liquid, which has a density of about 1·33. As cold water absorbs the gas readily it cannot be collected over it without great loss; but, in consequence of its great specific gravity, it may be collected in an open vessel, by direct displacement of the air. Let *a* be the flask containing the materials, and *c* a receiver in which the gas is to be collected, the tube from the flask extending nearly to the bottom. The chlorine gas being so much heavier than air, fills up the receiver just as water would if conveyed into it in the same manner; and if the process is expeditiously conducted, the chlorine may be collected nearly pure.

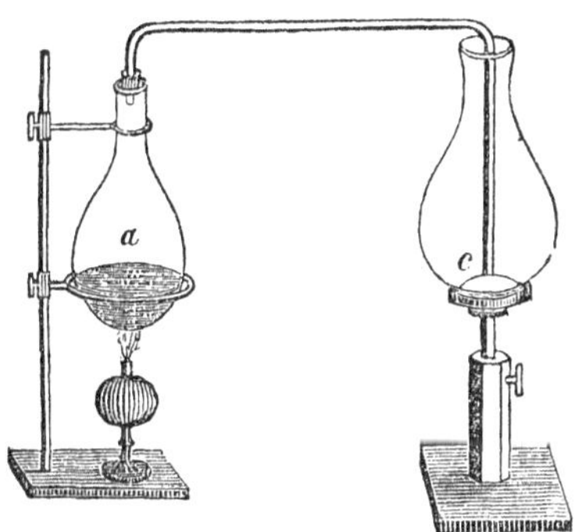

Collection of Chlorine by Displacement of the Air.

QUESTIONS.—229. Describe the properties of chlorine How may it be collected by displacement of the air?

Prepared by either of the modes described above the gas is always mixed with watery vapor, from which it is readily separated by passing it through a tube containing lumps of dry chloride of calcium. After being evolved from the materials used it should be made to pass through a bottle containing a small quantity of water, to separate any hydrochloric acid that may

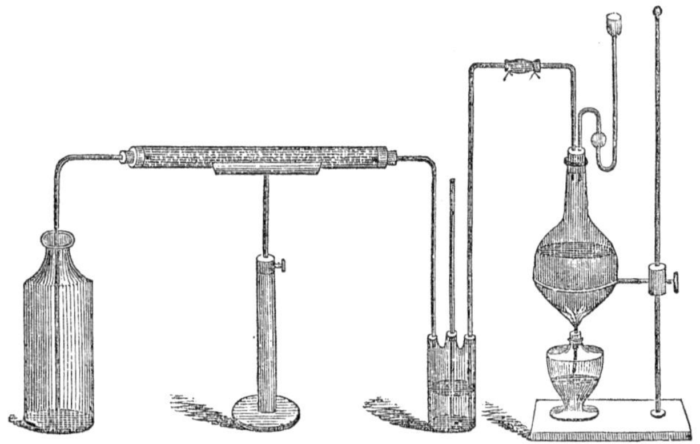

Preparation of Dry Chlorine.

have come over, and then through the chloride of calcium tube, which will absorb all the watery vapor; it may then be collected in a bottle by displacement of the air, as just described. The bottle should be provided with a well-ground glass stopper.

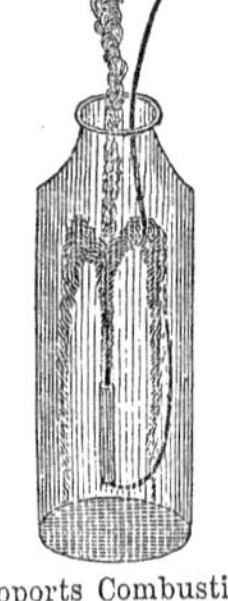

Supports Combustion.

Chlorine is allied to oxygen in many of its properties; a lighted candle continues to burn in it for a time with a dull red flame, giving off a dense black smoke of uncombined carbon; and phosphorus takes fire in it spontaneously, burning with great brilliancy. Some of the metals also, if in thin leaf or fine powder, take fire in it spontaneously. For this purpose, a

QUESTIONS.—How may the gas be obtained free from moisture? Will chlorine support combustion? How is phosphorus affected by it?

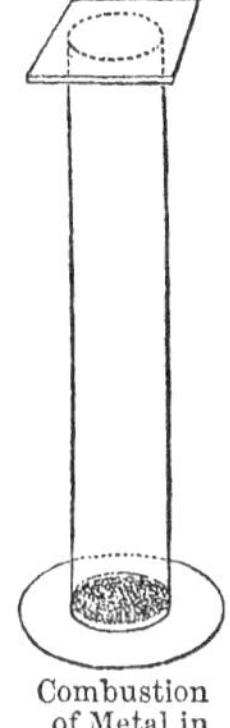
Combustion of Metal in Chlorine.

tall jar should be used, as represented in the figure on the left; and the bottom should be covered with sand, to prevent the breaking of the glass by the heated chloride falling upon it. It has a strong affinity for hydrogen, and a mixture of the gases explodes by the approach of flame, and by the electric spark, precisely as a mixture of oxygen and hydrogen. If a piece of paper, moistened with oil of turpentine, be suspended in a bottle of chlorine, it takes fire spontaneously, by the chlorine combining with the hydrogen of the turpentine; at the same time the carbon of the turpentine is liberated, in a state of minute division, as a dense black smoke. The bottle containing the chlorine should have a large mouth.

Turpentine Inflamed.

One of the most important properties of chlorine is its bleaching power, all vegetable and animal coloring matters being speedily destroyed by it. Powdered indigo, slightly moistened and dropped into a bottle containing it, even if it is considerably diluted with air, soon loses its color entirely; and pieces of calico, of different colors, suspended in it, are affected in the same manner, without, in the least, injuring the texture of the cloth. It is therefore much used in preparing rags for the manufacture of writing-paper, and also in bleaching cotton and linen goods. Writing done with common ink is easily removed by it; but printers' ink, being an oily preparation, is not attacked by it.

Chlorine is also a powerful disinfecting agent, removing at once all offensive effluvia from sewers, vaults, and other places where they may have collected. For this purpose, bleaching-powder (to be described hereafter) is moistened with water, and placed in shallow dishes in the apartments to be fumigated.

The aqueous solution of chlorine is best prepared by the use

QUESTIONS.—What is said of the affinity of chlorine for hydrogen? What is its effect upon oil of turpentine? What occasions the dense black smoke? What is said of the bleaching properties of chlorine? What is said of it as a disinfecting agent?

of a Woulfe's apparatus (225). The solution readily dissolves gold-leaf, forming trichloride of gold.

If one of the bottles is kept cold by means of ice, a crystaline solid will be deposited, which is a compound of one atom of chlorine and ten atoms of water.

230. By dexterously putting some of these crystals in a glass tube and sealing it hermetically, liquid chlorine may be obtained. By slight elevation of temperature the solid melts and the water and chlorine separate, the latter of which, in consequence of the pressure, takes the liquid form. While sealing the tube the end containing the crystals must be kept cold by means of ice.

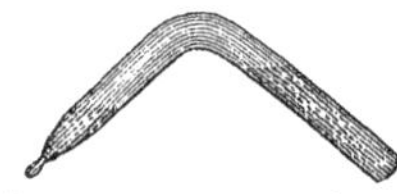

Preparation of Liquid Chlorine.

The aqueous solution, if prepared in the dark and kept constantly from the action of light, undergoes no change (173), but if once exposed for a few moments only to the sun's rays, decomposition of the water commences, hydrochloric acid is formed, and oxygen set free.

The proper test of chlorine, either free or in combination, is solution of nitrate of silver; silver and chlorine always forming a dense white precipitate, quite insoluble in water, but readily soluble in aqua ammoniæ.

Compounds of Chlorine and Oxygen

231. The compounds of chlorine and oxygen are five in number, as follows, viz.:—ClO, ClO_3, ClO_4, ClO_5, and ClO_7, all of which are acids; but as the affinities of these substances for each other are very feeble, their compounds are all decomposed by slight causes.

232. Hypochlorous Acid—ClO; eq., (35·4 + 8 =) 43·4.—This is a gaseous substance, of a yellowish-green color, like that of chlorine, but a shade deeper. It exists in combination with lime in the common *bleaching-powder*.

233. Chlorous Acid—ClO_3; eq., (35·4 + 24 =) 59·4—Is prepared by the action of oil of vitriol upon chlorate of potash.

234. Hypochloric Acid—ClO_4; eq., (35·4 + 32 =) 67·4.—This compound is also gaseous, and of a deep yellowish color. It is formed by the action of sulphuric acid upon chlorate of potash. If a few grains

---

QUESTIONS.—How is the aqueous solution of chlorine prepared? May the compound of chlorine and water be crystalized? 230. How may these crystals be used to form liquid chlorine? Can the aqueous solution be long preserved without decomposition? What test for chlorine is mentioned? 231. What compounds of chlorine and oxygen are there? What is the number of equivalents of oxygen in each of these?

of chlorate of potash are placed in a wine-glass, and a little sulphuric acid poured in, the glass will be soon filled with the gas, which will be recognised by its color. If now a rag, wet with oil of turpentine, be presented to it, on the end of a wire or a stick, it will be inflamed, and the gas at the same time exploded.

235. Chloric Acid—ClO_5; eq., (35·4 + 40 =) 75·4.—Chloric acid is obtained by passing a current of chlorine through a strong solution of potash, with which it combines as it is formed, producing chlorate of potash. At the same time with the chlorate of potash there is also produced much chloride of potassium; the reaction takes place between 6 equivalents of chlorine and 6 equivalents of potash, according to the following equation :—

$$6KO + 6Cl = 5KCl + KO,ClO_5.$$

The chlorate being only slightly soluble in water, some separates in crystals by a little evaporation of the water, while the chloride remains in solution.

To obtain the acid in a separate state, chlorate of baryta is first produced and then decomposed carefully by sulphuric acid. The BaO,SO_3 which forms is separated by filtration, and the chloric acid obtained as a syrupy liquid. It is decomposed at a temperature of 104°.

Perchloric Acid possesses no characters that render it of any special interest.

Compounds of Chlorine and Hydrogen.

236. Hydrochloric Acid—HCl; eq., (35·4 + 1 =) 36·4.—This is the only known compound of chlorine and hydrogen; and, in solution in water, has long been used in the arts, under the names of *muriatic acid*, and *spirit of salt*. It is formed by the action of diluted sulphuric acid upon common salt. The sulphuric acid should be diluted with about an equal weight of

QUESTIONS.—235. How is chlorate of potassa formed? How is chloric acid procured from this salt? 236. What is the only compound of chlorine and hydrogen that is known? How is hydrochloric acid prepared?

water, and be allowed to cool before being used. The changes which take place are as follows:—

$$NaCl + SO_3,HO = NaO,SO_3 + HCl.$$

From this it appears that the water of the oil of vitriol is essential to the process; it is decomposed, yielding its oxygen to the sodium to form soda, which combines with the acid, and its hydrogen to the chlorine to form the hydrochloric acid.

Abundant fumes of the gaseous acid will be given off, which should in no case be allowed to diffuse themselves in a room where there are articles of delicate apparatus made of metal, as they will be sure to be corroded after a little time, if not immediately.

It is also formed by the direct union of its elements. When equal measures of chlorine and hydrogen are mixed together, and an electric spark is passed through the mixture, instantaneous combination takes place, heat and light are emitted, and hydrochloric acid is generated. A similar effect is produced by flame, by a red-hot body, and by spongy platinum. Light also causes them to unite. A mixture of the two gases may be preserved, without change, in a dark place; but if exposed to the diffused light of day, gradual combination ensues, which is completed in the course of twenty-four hours. The direct solar ray, like flame and the electric spark, produces an explosion by a sudden inflammation of the whole mixture; but to insure the success of the experiment, the gases should be very pure, and the chlorine recently prepared over warm water. The glass vial containing the mixed gases, after being filled, should be instantly covered with a black cloth, which can be suddenly removed by a stick, or wire, after it is placed in the sun's rays.

Hydrochloric acid is, of course, a chloride of hydrogen. When pure it is a colorless gas, of which 100 cubic inches weigh 39·38 grains, giving it a density of 1·25. By strong pressure it may

---

QUESTIONS.—Explain the equation. Is the presence of water essential to the formation of hydrochloric acid? May it be formed by the direct union of its elements? What several means are mentioned by which the gases may be made to combine? What are some of the properties of hydrochloric acid?

be compressed into a liquid. It is quite irrespirable, and incapable of supporting combustion. Water absorbs it with avidity, taking up, under favorable circumstances, no less than 480 times its own volume. During the absorption it increases considerably in volume, and the saturated solution has a density of 1·21, and contains about forty-two per cent. of the acid. In preparing the liquid hydrochloric acid, a Woulfe's apparatus (225) is used, only a very little water being contained in the first bottle, in which most of the impurities mixed with the gaseous acid will be deposited. In the open air, copious fumes of the gas constantly arise from the liquid, which produce a cloud of smoke if any ammonia be present in the air. Thus, let a little aqua ammoniæ be poured into a glass vessel, A, with a large mouth, and then invert over it a tumbler, the inside of which has been thoroughly moistened with common hydrochloric acid. The two gases, coming in contact, unite, and fill both glasses with a dense, white smoke, which is solid hydrochlorate of ammonia (159), in a finely divided state. The same thing is shown when a glass rod moistened with hydrochloric acid is brought near an open vessel containing aqua ammoniæ.

Hydrochloric Acid and Ammonia.

Gaseous hydrochloric acid is composed of equal volumes of chlorine and hydrogen united without condensation.

One volume of chlorine weighs	2·440
One " hydrogen "	069
Forming two vols. hydrochloric acid	2.509

The weight of 1 vol., or the theoretical density of the gas, is therefore $\frac{2·509}{2} = 1·254$.

237. *Aqua regia*, so called because of its ability to dissolve gold and platinum, is a mixture of two parts of hydrochloric to one of nitric acid. Let a single leaf of gold be placed in a wine-

QUESTIONS.—What is said of the absorption of hydrochloric acid by water? What is said of the fumes produced by the gas with ammonia? What is said of the volumes of chlorine and hydrogen which unite to form this acid? 237. What is *aqua regia?* Why is it so called?

glass containing a little hydrochloric acid, and another leaf in a separate glass with some nitric acid; the gold will remain undissolved in both glasses for any length of time; but, on mixing the contents of the glasses, the whole of the gold will be speedily dissolved. The real solvent in this case is the chlorine, which is liberated by the action of the acids upon each other. The reaction is shown by the following equation. Thus,

$$NO_5 + 2HCl = NO_3 + 2HO + Cl.$$

This decomposition, however, proceeds only so far as to saturate the liquid with chlorine, but if heat is applied to expel the chlorine, or a metal placed in the liquid with which it will unite, new quantities of the acid are decomposed. Nitrohydrochloric acid, therefore, is a source of chlorine in a very concentrated state, and is capable of dissolving several substances which are not attached by any single acid.

Hydrochloric acid may readily be distinguished by its odor and volatility, and by its giving, with a solution of nitrate of silver (230), a precipitate of the white chloride of silver, which is blackened by exposure to the light.

Liquid hydrochloric acid is extensively used for various purposes in the arts, and is one of the most important chemical agents of the laboratory.

Compounds of Chlorine and Nitrogen.

238. Terchloride of Nitrogen.—NCl_3; eq., $(14 + 106{\cdot}2 =)$ $120{\cdot}2$.—There is known only a single compound of these elements, the terchloride of nitrogen. It is prepared by passing a current of chlorine through a solution of the hydrochlorate, or other salt of ammonia, and takes the form of a dense yellow liquid, which is seen in small globules at the bottom of the solution. The reaction is expressed in the following equation:—

$$NH_3,HCl + 6Cl = 4HCl + NCl_3.$$

QUESTIONS.—What compound of gold is formed? How may hydrochloric acid be known? 238. What compound of chlorine and nitrogen is there? What is said of it? Explain the equation.

The liquid has a density of 1·65, and may be distilled under a pressure less than one atmosphere, but at 212° it explodes with the utmost violence. So also it detonates violently by the mere contact with phosphorus, many of the oils, &c., and never should be handled but with the greatest care.

IODINE.

Symbol, I; *Equivalent*, 127; *Density*, 4·95

239. History.—Iodine from (*iodes*, violet color) was discovered in the ashes of sea-plants, from which it is still prepared, by M. Courtois of Paris, in the year 1812. It is found also in certain ores of silver and zinc, in sea-water, and in the water of certain mineral springs.

240. Preparation.—As stated above, the iodine of commerce is obtained from the ashes of sea-plants, especially the *fucus*

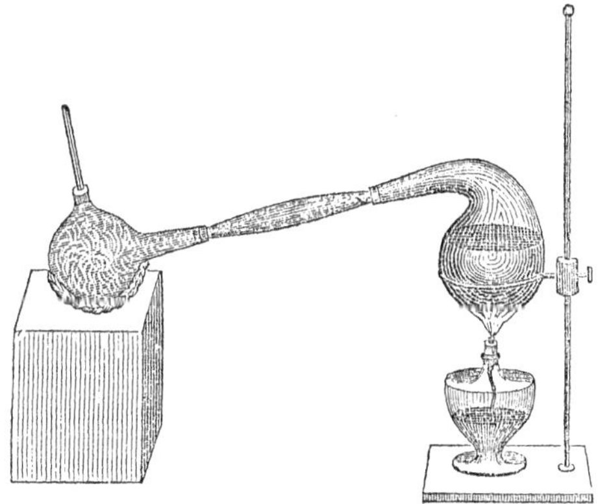

Preparation of Iodine.

palmatus. The ley obtained by lixiviating the ashes is first evaporated, to separate a portion of the carbonate of soda and

QUESTIONS.—239. When was iodine discovered? In what is it found? 240. What is the mode of preparing iodine from the ashes of sea-plants?

other saline compounds it contains, which are less soluble than the iodine compounds—chiefly the iodides of sodium and magnesium—and are therefore first to crystalize out from the solution; the residue is then mixed with peroxide of manganese and sulphuric acid, and a gentle heat applied, when the iodine distils over, as a beautiful violet-colored vapor, into a receiver prepared for the purpose, where it is condensed. A simple apparatus like that represented in the figure on the preceding page answers well for the distillation.

A good method to show the evolution of iodine, is to heat in a glass globe, over a lamp or ignited charcoal, a small quantity of sulphuric acid, and throw suddenly into it 25 or 30 grains of iodide of potassium. A large quantity of iodine will instantly be set free, and its vapor fill the globe.

241. Properties.—Iodine, at ordinary temperatures, is a soft, friable, nearly black solid. Usually it is in small shining crystals, which have a metallic lustre, and a density of 4·95. Heated a little above 212°, it melts, and is then converted into a beautiful violet-colored vapor, which has a density of 8·70 (air being 1), and 100 cubic inches weigh 270·1 grains. Iodine is a non-conductor of electricity and heat, and is allied to oxygen and chlorine in many of its properties. Its odor resembles that of chlorine, but is less offensive. It is sparingly soluble in water, requiring about 7000 times its own weight of this liquid for complete solution; but alcohol and ether dissolve it freely, forming a deep brown solution. A few of the crystals pressed upon the skin produce a deep stain, which however soon disappears.

Starch affords a delicate test of iodine, forming with it a beautiful blue. The starch should be prepared by dissolving it in hot water, and allowing it to cool before using. Let a little hot water be poured upon ashes obtained by burning a piece of sponge, and, after filtering, add a drop or two of solution of starch; then pour in a few drops of sulphuric or nitric acid, and stir it gently, and almost always the blue color will be observed, indicating the presence of iodine.

QUESTIONS.—241. Describe the properties of iodine. What is said of its solubility in water? In alcohol and ether? What test of iodine is mentioned? How may iodine be detected in sponge?

Iodine has not been much used in the arts, but is largely employed in medicine. In the Daguerreotype process (76), it is essential; and recently it is said to have been employed in dyeing.

Compounds of Iodine and Oxygen, Hydrogen, &c.

242. Iodine and oxygen combine in three proportions, producing iodous IO_4, iodic IO_5, and periodic IO_7, acids; neither of which however possesses any special interest.

243. Hydriodic Acid, HI, (iodide of hydrogen,) is a gaseous substance, of a specific gravity 4·39, and in many of its properties strongly resembles the corresponding chloride of hydrogen (hydrochloric acid). It is absorbed by water, and then forms the liquid hydriodic acid.

It is best prepared by placing in a glass tube alternate layers of iodine, powdered glass (to prevent too rapid action) and posphorus, slightly moistened with water. Iodide of phosphorus is first formed, which is decomposed by the water, producing phosphorous acid and hydriodic acid, the last of which being gaseous makes its escape, and may be collected in a dry bottle by displacement of the air, or it may be condensed in water.

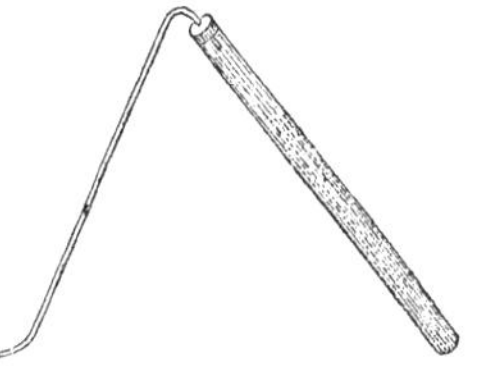

Preparation of Hydriodic Acid.

244. Teriodide of Nitrogen—NI_3.—From the weak affinity that exists between iodine and nitrogen, these substances cannot be made to unite directly. But when iodine is put into a solution of ammonia, the alkali is decomposed; its elements unite with different portions of iodine, and thus cause the formation of hydriodic acid and teriodide of nitrogen. The latter subsides in the form of a dark powder, which is characterized, like chloride of nitrogen, by its explosive property. It often detonates spontaneously as soon as it is dry, and even when moist, by the slightest causes. Heat and light are emitted during the explosion, and iodine and nitrogen are set free, the former of which may be seen at the instant in the form of vapor.

QUESTIONS.—What use is made of iodine? 242. What compounds of iodine and oxygen are mentioned? 243. Describe hydriodic acid How is it formed? 244. Describe teriodide of nitrogen.

Chlorine combines with iodine when made to pass over it in a dry glass tube, or when passed through water in which crystals of iodine are diffused. The two substances form several different compounds, but they are not at present well understood. One of these, probably the proto-chloride, ICl, has been used in the Daguerreotype process.

BROMINE.

Symbol, Br; *Equivalent*, 80; *Density*, 2·97.

245. History.—Bromine was discovered in 1826, in sea-water; and received its name, *bromine* (from *bromos*, offensive odor), in consequence of its exceedingly disagreeable smell. Recently it has been obtained in large quantities from the waters of some of the salt-springs in Pennsylvania and Virginia.

246. Preparation.—The usual mode of preparing bromine is a little complex. First, the brine from the spring is evaporated, and the common salt removed by crystalization, then the mother-liquor, or *bittern*, as the uncrystalizable residue is called, is treated with a current of chlorine to decompose the bromides of magnesium, sodium, &c., and sulphuric ether afterwards added, by which the bromine that has been separated from its compounds by the chlorine is taken up, and rises to the surface as a solution of bromine in ether.

Another method is to mix with the solution sulphuric acid and peroxide of manganese, as in the preparation of iodine; and then distilling with a gentle heat. The same apparatus (240) may be used as in the preparation of iodine, but the receiver must be kept cool by a current of cold water.

247. Properties.—Bromine is a liquid of a blackish-red color, and specific gravity 2·97. At a temperature a little below zero it is frozen, and boils at about 117°, forming a vapor of a beautiful blood-red color, and specific gravity 5·39, air being 1. It

---

QUESTIONS.—What is said of the compounds of chloride and nitrogen? 245. When was bromine discovered? From what is the name derived? 246. From what is it obtained? What is the mode of preparing it? 247. Describe its properties.

stains the skin yellow, like iodine, but less intensely. Vapor of bromine ignites phosphorus spontaneously, and a lighted candle burns in it a short time.

Bromine, in many of its properties is closely allied to chlorine and iodine. Taken into the system it is highly poisonous, and its vapor possesses considerable bleaching power. With starch it forms a yellow color.

Like chlorine, it forms a compound with water, which crystalizes when exposed to the cold of a freezing mixture of salt and snow. The compound is Br,10HO.

Bromine is sometimes used in medicine, and much more extensively in photography, especially in the Daguerreotype process.

Compounds of Bromine, with Oxygen, Hydrogen, &c.

248. Bromic Acid, BrO_5, is the only well determined compound of bromine and oxygen. It is a liquid, and may be procured of the consistency of syrup. It is very corrosive, and sour to the taste, and by a temperature of 212° is decomposed.

249 Hydrobromic Acid, HBr, is a colorless gas of a density 2·73. To prepare it, a tube of the form of the letter W is provided, and the part at the left filled with pieces of phosphorus mixed with pounded glass, and the whole moistened with water. Into the end at the right some bromine is then poured, and a cork firmly inserted; and into the other end a smaller tube is fixed, by means of a perforated cork, to convey away the hydrobromic acid as it is formed. The whole being ready, a gentle heat is applied to the part containing the bromine; and the vapor as it is formed, attacking the phosphorus, first produces bromide

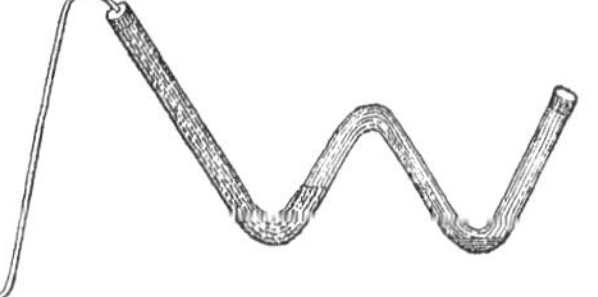

Preparation of Hydrobromic Acid.

QUESTIONS.—What other elements does bromine resemble? Does it combine with water? What use is made of it? 248. Describe bromic acid. 249. Describe hydrobromic acid, and the mode of preparing it

of phosphorus, which is at once decomposed by the water, forming phosphorous and hydrobromic acids, the latter of which, being gaseous, passes off, and may be collected over mercury.

The gas is rapidly absorbed by water, like hydrochloric acid, and the concentrated solution gives off fumes in the air. In most of its properties it closely resembles hydrochloric acid.

By intense cold and pressure the gas may be condensed to the liquid form.

250. Hydrobromic acid is composed of equal volumes of vapor of bromine and hydrogen combined without condensation. Thus,

1 vol. vapor of bromine weighs	5·390
1 " " hydrogen "	069
2 vols. hydrobromic acid,	5·459

One volume of the acid therefore weighs 2·729.

FLUORINE.

Symbol, F; *Equivalent*, 19; *Density*, 1·29.

251. History and Properties.—Fluorine has long been known to exist, but it has not, until recently, been obtained in a separate state. It is found in nature in considerable abundance in the mineral called *fluor spar*, which is a compound of this substance with calcium, the metallic base of lime. It is a brownish-colored gas, of specific gravity about 1·29 (probably), and bleaches like chlorine.

Such is its affinity for other substances that it attacks them with violence, even gold and platinum; and can be prepared and kept only in vessels made of fluor spar, which, being already saturated with the substance, is not acted on by it.

QUESTIONS.—What is said of the absorption of hydrobromic acid by water? 250. Considered as gaseous, what is its composition? 251. Give the history and properties of fluorine. What is said of its affinity for other substances? Of what only can vessels be made to contain it? Why is not this substance acted on by it?

Compounds of Fluorine.

Fluorine seems to be incapable of uniting with oxygen, but combines with hydrogen, forming the acid compound HF.

252. Hydrofluoric Acid—HF; eq., (19 + 1 =) 20.—This acid is formed by subjecting powdered fluor spar, moistened with strong sulphuric acid, to a very gentle heat in a leaden vessel. The acid distils over as a pungent, corrosive, vapor, but may be condensed in a leaden receiver, that is kept surrounded with ice. The reactions are as follows :—

$$CaF + SO_3,HO = CaO,SO_3 + HF.$$

As thus formed, the acid has a density of 1·07, and manifests a strong affinity for water, with which it combines with great energy. It attacks glass powerfully, combining with its silica, and may therefore be used to etch it. This is done by spreading a thin coat of bees'-wax or varnish upon the glass, and tracing the design upon it, taking care to cut quite through the wax. The liquid acid is now poured over the coated surface, or it is exposed a few minutes to the acid vapor, and the wax afterwards removed ; the design will then be found beautifully traced upon the glass.

This acid attacks animal substances powerfully, and, therefore, should always be handled with great care.

Fluorine unites also with chlorine, iodine, bromine, and some others of the elements, but the compounds are not important.

QUESTIONS.—Does fluorine combine with oxygen ? 252. Describe the mode of preparing hydrofluoric acid. Describe its properties. How may it be used for etching upon glass? What is said of its action upon animal substances ? Does fluorine combine with chlorine, iodine, &c. ?

Group III.

Sulphur Selenium Tellurium	Elements in many of their properties closely resembling each other, and forming similar acid compounds with oxygen and hydrogen.

SULPHUR.

Symbol, S; *Equivalent*, 16; *Density*, 1·99.

253. History and Preparation.—Sulphur, called also *brimstone*, has been known from the remotest antiquity. It occurs, as a mineral production, in many parts of the world, particularly in volcanic regions, as in the neighborhood of Naples, in some of the Sandwich Islands, and in the island of Sicily. In combination with several of the metals, as iron, lead, copper, &c., it is still more abundant, and is found in almost every place. From one of its compounds with iron, called *iron pyrites*, it is procured in large quantities, for the purposes of commerce. It is found, also, in many organic bodies, as in eggs, in the hair, horns, and hoofs of animals, and in the seeds of black mustard.

The island of Sicily furnishes a large part of all sulphur of commerce; the native sulphur here occurs in immense beds mixed more or less with gypsum, lime, and other earthy matter. This sulphurous earth is first heated in pots so as to melt the sulphur, which is dipped out with ladles, the earthy matter settling to the bottom. The sulphurous earth remaining as sediment is then heated in earthen pots, which are arranged in double rows, and entirely inclosed in mason-work, except at the top, where is an opening by which they are charged and emptied. At the sides, in the open air, pots are arranged to receive the sublimed sulphur, which, taking the liquid form, passes finally into buckets situated as shown in the figure on next page.

Questions.—What elements constitute group 3d of the metalloids? 253. Give the history of sulphur. With what is it found combined? In what organic bodies is it contained? Describe the mode of separating it from the earthy matter with which it is mixed.

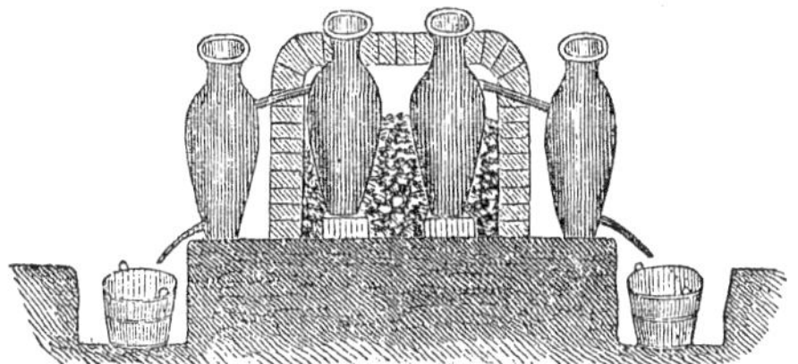

Separation of Sulphur.

The figure represents a section of the mason-work with two of the included pots, and also the external receivers with which they are connected, and the buckets.

254. Properties.—Sulphur is a brittle solid, of a greenish-yellow color, emits a peculiar odor when rubbed, and has little taste. It is a non-conductor of electricity, and is excited negatively by friction. It fuses at 226°, and becomes nearly as liquid as water; but if the heat be raised as high as 430°, it becomes so tenacious that the vessel containing it may be inverted without spilling it, and is then of a dark molasses color. When heated to at least 428°, and then poured into water, it becomes a ductile mass, which may be used for taking the impressions of seals. After some time, it changes into its ordinary state.

Fused sulphur has a tendency to crystalize in cooling. A crystaline arrangement is perceptible in the centre of common roll sulphur; and, by good management, regular crystals may be obtained. For this purpose, several pounds of sulphur should be melted in an earthen crucible; and, when partially cooled, the outer solid crust should be pierced, and the crucible quickly inverted, so that the inner and as yet fluid parts may gradually flow out. On breaking the solid mass, when quite cold, a confused arrangement of prismatic crystals will be found in the interior.

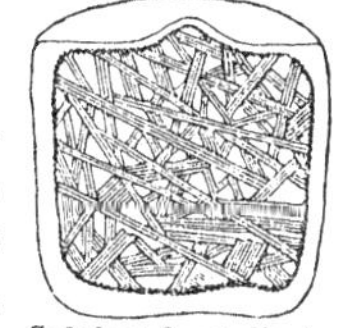

Sulphur Crystalized.

Sulphur is dimorphous (184); that is, it is capable of crystalizing in two distinct primary forms, the oblique rhombic

QUESTIONS.—254. Describe the properties of sulphur. How may it be crystalized? Why is it said to be dimorphous?

prism, and the rhombic octahedron; the first belonging to the monoclinic, and the second to the trimetric system.

Sulphur is very volatile, and begins to rise in vapor even before it is completely fused. At about 750°, it boils, and the vapor, if in a close vessel, will be condensed on any cold surface, forming the *flowers of sulphur*. The density of its vapor is about 6·65. When vapor of sulphur is brought in contact with vapor of alcohol, they unite; but solid sulphur is quite insoluble in alcohol, or water, but dissolves in boiling oil of turpentine, and in sulphide of carbon.

By melting the flowers of sulphur and pouring it into moulds, the *roll-sulphur* of commerce is formed. In this form it is very brittle, and will sometimes break by the heat of the hand. It is the only substance known that always becomes negatively excited by friction, whatever may be the nature of the substance used as the rubber.

The vapor of sulphur combines readily with iron and other metals, attended with all the phenomena of combustion. Let the breech of a gun-barrel be heated to redness, and a lump of sulphur dropped into it, and then let the muzzle be instantly closed by a cork; a jet of vapor of sulphur will issue violently from the touch-hole, which will be inflamed as it enters the air; and a bunch of small iron wire, held in it, will burn freely, forming sulphide of iron, which will fall in drops.

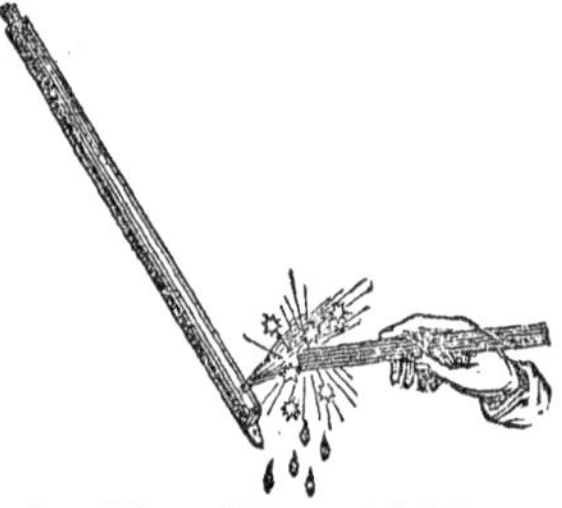

Iron Wire and Vapor of Sulphur.

The sulphur of commerce generally has an acid reaction, probably in consequence of a slight oxidation that is gradually taking place. Organic substances in contact with sulphur are always more or less acted upon by the acid or acids thus produced.

QUESTIONS.—What is the boiling point of sulphur? What are the flowers of sulphur? Is it soluble in water or alcohol? How is roll-sulphur formed? What is said of its electrical state when rubbed? How may iron wire be made to burn in vapor of sulphur? Why does the sulphur of commerce usually have an acid reaction?

Uses of Sulphur.—Sulphur is used extensively in the arts, and in medicine. It is employed in the manufacture of gunpowder, sulphuric acid, the different kinds of matches, vermillion, &c., and for taking impressions of seals. In medicine, it is used in cutaneous diseases, and as a cathartic and alterative

Compounds of Sulphur and Oxygen.

255. Sulphur and oxygen form as many as seven different compounds, viz. :—

1. Hyposulphurous acid.............................. S_2O_2.
2. Trisulpho-hyposulphuric acid.................... S_5O_5.
3. Bisulpho-hyposulphuric acid..................... S_4O_5.
4. Monosulpho-hyposulphuric acid................... S_3O_5.
5. Sulphurous acid................................. SO_2.
6. Hyposulphuric acid.............................. S_2O_5.
7. Sulphuric acid.................................. SO_3.

Of these, the fifth and last are by far the most important; and only these will be here described.

256. Sulphurous Acid—SO_2; eq., (16 + 16 =) 32.—This substance is gaseous at ordinary temperatures, and is the sole product of the combustion of sulphur in the open air, or in dry oxygen gas. It is more conveniently prepared, however, by heating strong sulphuric acid in contact with mercury or pieces of copper. One equivalent of the sulphuric acid gives up one equivalent of its oxygen to unite with the metal, and the oxide thus formed is immediately dissolved by a second atom of the sulphuric acid, while the sulphurous acid passes off as a gas. Thus,

$$Hg + 2SO_3 = HgO,SO_3 + SO_2.$$

Another very easy method is to heat in a retort a mixture of 6 parts of powdered peroxide of manganese and 1 part of sulphur.

QUESTIONS.—What use is made of sulphur in the arts? 255. How many compounds of sulphur and oxygen are mentioned? 256. Describe sulphurous acid, and the mode of preparing it.

By this mode, the peroxide gives up one-half of its oxygen, which unites with sulphur to form the sulphurous acid, and protoxide of manganese remains in the retort.

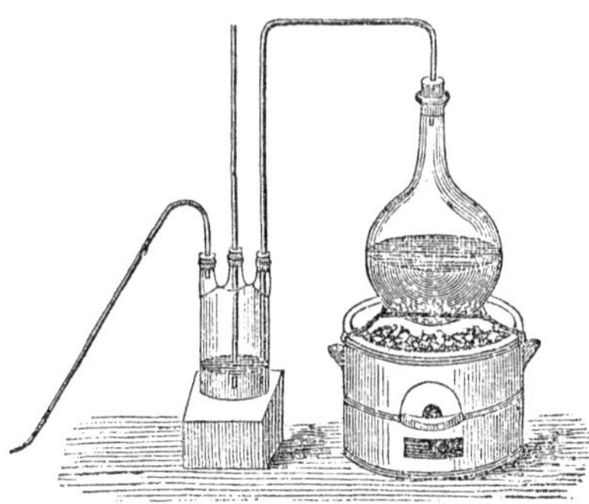
Preparation of Sulphurous Acid.

An arrangement like that figured in the margin, answers well for its preparation, by either mode. As the gas forms it is made to pass through a little water, to condense any vapor of sulphuric acid that may have come over, and it may then be collected over mercury.

Sulphurous acid is a dense, colorless gas, 100 cubic inches of which weigh 68·55 grains, giving it a specific gravity of 2·24. It is distinguished from all other gases by its suffocating odor, which every one has recognized in burning sulphur. It is absorbed largely by water, and may be condensed into the liquid form by moderate pressure, or by a cold of zero. A little of the liquid may be obtained very easily, by putting a small quantity of mercury and sulphuric acid in a bent tube, as represented in the figure, sealing it hermetically, and supplying heat to the extremity, *a*, which contains the materials, while the other, *b*, is kept cool by means of ice, or the evaporation of ether. The liquid will be soon found to collect in the cool part of the tube. Care should be taken not to heat the tube too much, lest it should burst.

Preparation of Sulphurous Acid in Liquid Form.

A better method to procure the liquid is to prepare a tube, (as in the figure on next page,) by closing one end, and drawing out a part in the middle to a capillary bore, and then inserting it in a freezing mixture of snow and salt. If, now, a current of the gas, first dried by passing through a chloride of calcium tube, be made

QUESTIONS.—What is said of the odor of sulphurous acid? How may it be obtained in the liquid form?

to enter the open end of the tube, it will be condensed and collected in the lower part. When this part is nearly filled, still keeping in the freezing mixture, the upper part may be separated and the lower part hermetically sealed, and the liquid preserved for any length of time. Or a tube like that represented in the figure on the right margin may be used. The bulb at the centre part is to be surrounded with a freezing mixture in a suitable vessel, as a glass tumbler; and when a sufficient quantity of the fluid has accumulated, the two ends of the tube may be hermetically sealed before it is removed from its place in the freezing mixture. Under the ordinary atmospheric pressure it becomes liquid at about 14°, but at a temperature of 59° it requires a pressure of about 2 atmospheres. The liquid has a density of 1·42. When allowed to escape in the open air it evaporates rapidly, producing a cold of —60° to —76°, according to the temperature of the air and other circumstances. By severe cold it may be frozen, and with water it forms a compound ($SO_2 9HO$) which may be solidified.

Preparation of Liquid Sulphurous Acid.

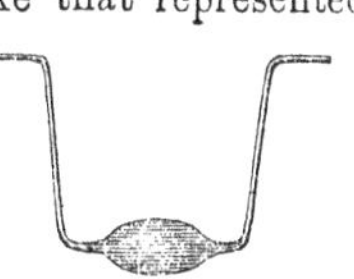

Preparation of Liquid Sulphurous Acid.

257. Sulphurous acid is composed of 1 volume of oxygen and $\frac{1}{6}$ of a volume of vapor of sulphur, condensed into 1 volume. If we burn a small quantity of sulphur in a glass globe over mercury, the sulphur is converted into sulphurous acid, but after cooling the volume of gases is found to be unchanged. The acid therefore occupies the same space as the oxygen entering into its composition Therefore if we subtract the weight of the volume of oxygen from that of a volume of sulphurous acid, there should remain $\frac{1}{6}$ of a volume of vapor of sulphur. This we find to be the case very nearly. Thus,

1 vol. sulphurous acid weighs	2·247
1 " oxygen (subtract) "	1·106
$\frac{1}{6}$ " vapor of sulphur,	1·141

QUESTIONS.—Describe the two modes mentioned for collecting sulphurous acid in glass tubes. At what temperature does it become liquid? What is the density of liquid sulphurous acid? What is said of the cold produced by its evaporation? 257. What is the composition of one volume of sulphurous acid?

We have heretofore (254) taken the volume of sulphur vapor to be 6·654, one-sixth of which is 1·109. The discrepancy in the results is occasioned by the great difficulty in obtaining accurately the real weight of the volume of sulphur vapor.

258. Sulphurous acid is much used for bleaching, especially articles of straw; which, in a moist state, are suspended in an atmosphere charged with the gas. For this purpose, the gas is formed by burning sulphur in the air, in some enclosure, as a box or empty cask, in which the articles to be bleached are suspended.

259. Sulphuric Acid—SO_3; eq., (16 + 24 =) 40.—This acid is always seen as a dense liquid, not unlike oil in appearance; and, having been formerly obtained altogether by the distillation of green vitriol (sulphate of iron), it received the name, *oil of vitriol*, by which it is now often known. It is prepared at the present time, at Nordhausen, Germany, by the same mode. Green vitriol is thoroughly dried by heat, and then distilled, at a high temperature, by which it is decomposed, and the acid passes over and condenses as a brown oil-like liquid, which still contains one eq. of water for every two eq. of the acid. Its composition, therefore is $2SO_3,HO$. Its density is 1·9, or nearly twice that of water. When this liquid is again distilled, at a moderate heat, a dry, silky solid is obtained, which is the pure compound, SO_3; but it possesses no acid properties until water is added, which changes it to common sulphuric acid. This solid has a strong affinity for water, and hisses like a hot iron when thrown into it.

260. The common method of preparing the oil of vitriol of commerce is, to burn a mixture of sulphur and nitrate of potash, or soda, in a furnace so contrived that the current of air which supports the combustion conducts the gaseous products into a large leaden chamber, the bottom of which is covered to the depth of several inches with water. Numerous complicated changes take place in the leaden chamber, during the combustion of the sulphur, by which the oxygen from the air is transferred

QUESTIONS.—258. What use is made of sulphurous acid? 259. Describe sulphuric acid. How is the Nordhausen acid prepared? How may the solid acid be obtained from it? What is the composition and density of the Nordhausen acid? 260. What is the mode of preparing the common oil of vitriol of commerce?

to the sulphurous acid formed by the combustion of the sulphur, converting it into sulphuric acid.

In the first place, as the mixture burns, sulphurous acid is formed (256), and binoxide of nitrogen;—the latter by the decomposition of the nitric acid of the salt;—and the two gases with a current of air are carried together into the leaden chamber, the binoxide of nitrogen, NO_2, at the same time absorbing oxygen, and being thus converted (218) into nitrous acid (NO_4).

Secondly, these two gases, in the absence of water, are capable of combining to form a crystaline solid, the composition of which is $NO_4,2SO_2$, and which, on coming in contact with water, is at once decomposed into binoxide of nitrogen and sulphuric acid. Thus,

$$NO_4,2SO_2 = 2SO_3 + NO_2.$$

Or, more properly,

$$NO_4,2SO_2,\ 2HO = 2(SO_3,HO) + NO_2.$$

Thirdly, if at the same time as the mixed gases from the combustion of the sulphur enter the chamber, steam be also forced in, the crystals just alluded to are not formed, but all the reactions described take place in the order mentioned, the hydrated sulphuric acid, as it forms, falling like rain to the bottom of the chamber.

The binoxide of nitrogen, being set free, combines again with atmospheric oxygen present, forming nitrous acid, NO_4, as before, and this, with the sulphurous acid, by the reaction of water, again producing sulphuric acid, and so on indefinitely.

These are the essential reactions that take place in the process, but they are often, perhaps always, accompanied by others, which however tend to the same result, viz., the conversion of the sulphurous acid, formed from the burning sulphur, into sulphuric acid.

QUESTIONS.—Describe the reactions which take place. Firstly? Secondly? Thirdly? Do other reactions tending to the same result usually accompany the above?

Firstly, a portion of the nitrous acid, NO_4, in the leaden chamber, when but little moisture is present, is converted into monohydrated nitric acid, and hyponitrous acid, NO_3. Thus,

$$2NO_4 + HO = NO_5,HO + NO_3.$$

If a large quantity of moisture is present, the nitrous acid is converted into hydrated nitric acid and binoxide of nitrogen. Thus,

$$6NO_4 + nHO = 4NO_5,nHO + 2NO_2.$$

Any hyponitrous acid, NO_3, that may have been formed, as indicated in the second equation above, when the supply of moisture is increased, undergoes a similar change, producing nitric acid and binoxide of nitrogen.

Secondly, sulphurous acid, SO_2, by the reaction of hydrated nitric acid, is converted into hydrated sulphuric acid, while the nitric acid, NO_5, by the loss of oxygen is changed to nitrous acid, NO_4. Thus,

$$SO_2 + NO_5 + nHO = SO_3,nHO + NO_4.$$

We have, then, as the result, sulphuric and nitrous acids, the former of which mixes with the water at the bottom of the chamber, while the latter, remaining in the gaseous state, is ready again to go through the same reaction as before, being thus made a vehicle for conveying oxygen from the atmosphere to the sulphurous acid.

In some manufactories, no nitrate of potash is used, but the sulphur is burned alone, and nitric acid, in proper vessels, is placed in the leaden chambers in such a situation that it shall be evaporated by the heated sulphurous acid and other gases entering from the furnace; reactions similar to those above take place with the same results.

The figure following will serve to illustrate the process. C C is a section of the chamber lined inside with sheet lead, and supported at the ends by mason-work. At A is the furnace for

---

QUESTIONS.—Describe the other reactions tending to the same result which accompany those already mentioned.

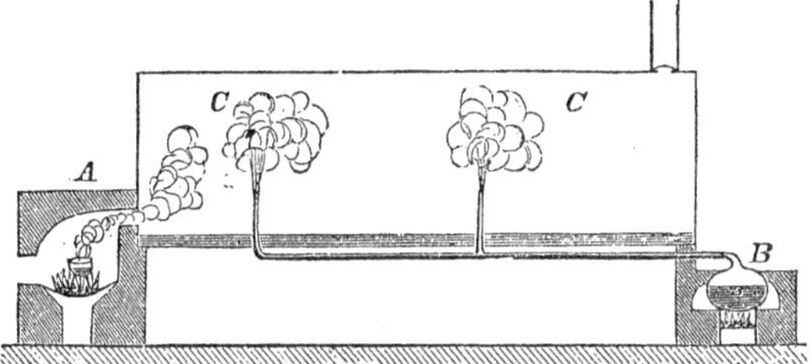

Manufacture of Sulphuric Acid.

burning the mixture of sulphur and nitric salt, the fumes of which are carried directly into the chamber, now filled only with air, and at B a steam boiler from which steam in one or more jets is constantly entering the chamber, for the purposes described above. A valve at the the top allows the escape of the spent gases.

261. For a class experiment, the apparatus figured below answers well to illustrate the mode of manufacturing this acid. A large

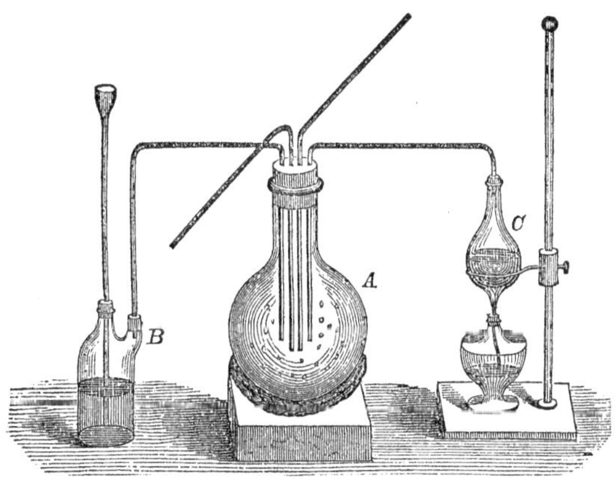

Illustrates the Preparation of Oil of Vitriol.

balloon glass, A, is provided, containing a small quantity of water, a two-necked flask, B, partly filled with small pieces of

QUESTIONS.—Describe the first figure on this page. 261. Describe the apparatus figured in connection with this paragraph. What is illustrated by it?

copper, and a second flask, C, containing mercury and strong oil of vitriol. These two flasks, B and C, are connected with the balloon, A, by means of tubes inserted in perforated corks, as shown in the figure; and a lamp then applied to C, from which sulphurous acid fumes will soon be made to pass over to the balloon, A. Into the flask, B, some nitric acid, a little diluted with water is now poured by the long-necked funnel, which, acting upon the copper (217), will soon supply binoxide of nitrogen to mix with the other gases contained in A. The red fumes of nitrous acid will at once appear in A, and all the changes take place described above, resulting in the production of a small quantity of oil of vitriol. Through the cork in the mouth of the balloon glass, A, two glass tubes bent at right angles are inserted, by which the air within may be changed, by blowing into one of them by the mouth.

In the manufacture of this acid on a large scale, the acid, as it comes from the leaden chambers, always has an excess of water, its density varying from 1·35 to 1·50. It is then heated in leaden pans until its density becomes about 1·75, and, finally, in platinum retorts, by which all excess of water is expelled, and its density is brought to 1·84. Its boiling point is then 617°, and its freezing point —30°.

Oil of vitriol, we thus see, always contains water; the most concentrated, that of Nordhausen, as stated above, containing one atom to every two atoms of the acid; or, expressed by symbols, its composition is $2SO_3,HO$. As prepared by the ordinary mode, it contains one atom of acid to one of water, or its composition is SO_3,HO. If to about 49 parts of common oil of vitriol we add 9 parts of water, we have an acid the specific gravity of which will be about 1·78, and its composition $SO_3,2HO$. It will then freeze at about the same temperature as water, but on the application of heat the solid does not melt until the temperature rises to 45°.

A fourth compound of water and sulphuric acid is $SO_3,3HO$, which has a specific gravity of 1·63. It boils at about 338°.

QUESTIONS.—What is the density of the acid as it is drawn from the leaden chambers? How is a portion of the water expelled? What is the proportion of water in the Nordhausen acid? In common oil of vitriol? Are there other definite compounds of this acid and water?

Common oil of vitriol is the monohydrated acid, SO_3,HO. Exposed to the open air it rapidly absorbs moisture, and increases in volume, so that a small vessel partly filled with it, if left open, is often found running over after a few days.

Place a few drops of it in a watch glass in the pan of a small balance, and exactly counterpoise it by weights placed in the other pan; in a very few moments the effect of the absorption of moisture will be seen. It may therefore be used to separate moisture from gases which are not acted on by it, by passing the gas through tubes containing pieces of pumice stone moistened with the acid.

Sulphuric acid is, perhaps, the most important of all the acids, as by its aid nearly all the others are produced. Its acid properties are very decided; aided by heat, it decomposes animal and vegetable substances, causing a deposition of charcoal, and formation of water, which it absorbs. Its affinity for water is very great, and the combination of the two substances is attended with the production of considerable heat. If a mixture of four parts of the acid and one of water is stirred with a test-tube containing sulphuric ether, the heat generated will be sufficient to cause the ether to boil.

Mixture of SO_3HO and HO produces Heat.

Free sulphuric acid is occasionally found in the water of springs, as at Byron, Genesee County, New York; but such cases are rare.

Uses.—Sulphuric acid is applied in the arts, and in the laboratory, to very many important uses; as, in the preparation of the other acids, the extraction of soda from common salt, the manufacture of alum, sulphate of iron, chlorine, &c. It is also used as a solvent for indigo, and in the various manufactures of the metals.

Test.—Chemists possess an unerring test of the presence of sulphuric acid. If a solution of chloride of barium is added to a

---

QUESTIONS.—How may the absorption of water from the atmosphere by oil of vitriol be shown? What is said of the affinity of this acid for water? What is the effect when it is mixed with water? What use is made of it? What test of its soluble salts is mentioned?

liquid containing free sulphuric acid, or any sulphate in solution, it causes a white precipitate, sulphate of baryta, which is characterized by its insolubility in acids and alkalies.

Compounds of Sulphur and Hydrogen.

262. There is only one well-determined compound of sulphur and hydrogen, the protosulphide, or hydrosulphuric acid, HS; though by a particular process a second compound is obtained, as a heavy, yellowish liquid, which is supposed to be a bisulphide, HS_2. The former only will be here described.

263. Hydrosulphuric Acid—HS; eq., (16+1=) 17.—This substance, often called *sulphuretted hydrogen*, is gaseous, and may easily be prepared by the action of diluted sulphuric acid upon powdered protosulphide of iron, formed by intensely heating a bar of iron, and then rubbing it with a roll of sulphur, or by heating intensely common iron pyrites (native bisulphide of iron) for some time in a covered crucible.

One mode of preparing hydrogen, it will be recollected (197, 204), is by the action of dilute oil of vitriol upon metallic iron; but if instead of iron we use sulphide of iron, a particle of sulphur being liberated at the same time as the particle of hydrogen, the two combine, and gaseous sulphide of hydrogen is evolved. The reactions are as follows:—

$$FeS + SO_3,HO = FeO,SO_3 + HS.$$

An apparatus like that represented in the figure on the next page is convenient for the purpose. The sulphide of iron is first introduced, and after the cork with the tubes is inserted, the acid is added by means of the long-necked funnel.

The gas may also be prepared by the action of hydrochloric acid upon native sulphide of antimony finely pulverized, aided

---

QUESTIONS.—262. What only well-determined compound of sulphur and hydrogen is mentioned? 263. How is it prepared? How is sulphide of iron prepared for the purpose? Describe the reactions that take place.

by a gentle heat. The reactions then are as expressed in the following equation :—

$$SbS + HCl = SbCl + HS.$$

Hydrosulphuric acid is a colorless gas, of a most offensive odor, similar to that of putrefying eggs; 100 cubic inches of it weigh 36·93 grains, giving it a density of 1·191. At the ordinary summer temperature, it takes the liquid form under a pressure of about 15 atmospheres, and by a cold of —122° is frozen. A jet of it burns readily in the open air, forming water and sulphurous acid.

Hydrosulphuric Acid.

Cold water absorbs 2 or 3 times its own volume of the gas, and acquires its peculiar odor and taste; but the gas is all given off again when the water is boiled. Water impregnated with the gas, if kept for a time, becomes milky from the decomposition of the gas, and the separation of the sulphur contained in it.

Sulphur-springs, which occur in many places in New York, Virginia, and other States, are springs, the waters of which are naturally impregnated with hydrosulphuric acid. They may always be recognised by the offensive odor, which extends to a distance around them, and by their blackening pieces of silver coin, by the formation of sulphide of silver. Water, possessing all the properties of that of the most noted sulphur-springs, may be prepared artificially, by passing a current of this gas, for a few minutes, through cold water. Let a little diluted sulphuric acid be poured upon some powdered sulphide of iron, in a small bottle, and then insert a cork with a bent tube, as shown in the figure, the other

Preparation of Sulphur Water.

QUESTIONS.—What are some of the properties of hydrosulphuric acid? What is said of its absorption by water? What are *sulphur springs?* How may sulphur water be prepared artificially?

end of which shall dip in a vial of cold water. After the gas has bubbled through it a few minutes, it will be found fully impregnated.

This gas blackens many colorless metallic salts, by the formation of metallic sulphides. An amusing experiment may be performed in the following manner: Let a picture be traced on white paper with a solution of sugar of lead, which is perfectly colorless, and the picture, at a little distance, will be invisible. Let the back of the paper be now moistened by means of a wet sponge; and, after tacking it to the wall, let a current of this gas be directed against it, and all the parts traced by the lead solution will instantly become dark brown, or black, by the formation of sulphide of lead on the paper.

Picture.

264. The composition of this gas is easily determined by heating some tin-foil in a measured quantity of the gas. Let the tube be of the form represented in the figure, and let the tin-foil be placed in the upper part by means of a wire after the gas, carefully measured, has been introduced; and then having the open end of the tube immersed in mercury, heat the tin-foil by means of a spirit-lamp. All the sulphur will be immediately absorbed by the metal, and there will remain only the hydrogen, which will however occupy the same space as before. Now,

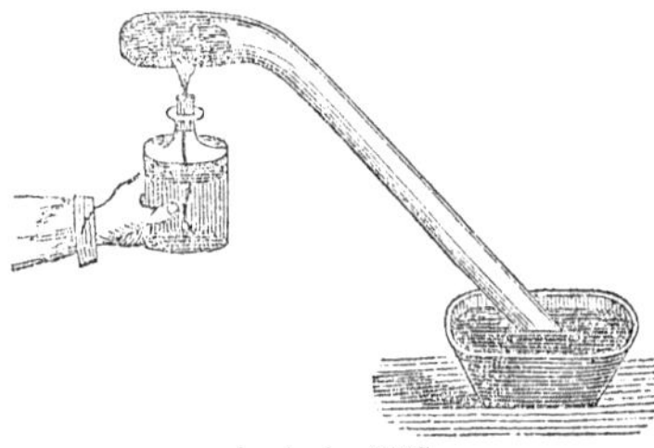
Analysis of HS.

One volume of hydrosulphuric acid gas weighs	1·191
One " hydrogen (subtract)	·069
There remains for the sulphur,	1·122

This is very nearly (162) the weight of one-sixth of a volume of vapor of

QUESTIONS.—What is said of the action of sulphur water upon many colorless metallic salts? 264. How may the composition of hydrosulphuric acid be determined?

sulphur; so the composition of hydrosulphuric acid must be one volume of hydrogen and one-sixth of a volume of sulphur vapor, condensed to one volume.

Compounds of Sulphur and Chlorine.

265. There are, it is believed, several compounds of these two elements, but two only (or perhaps, three) have been obtained in a separate state, the dichloride, S_2Cl, and the chloride, SCl.

266. Dichloride of Sulphur—S_2Cl; eq., $(2 \times 16 + 35{\cdot}4 =)$ $67{\cdot}4$.—The formation of this compound requires an apparatus which is somewhat complex. To prepare the chlorine, a flask, A, containing some peroxide of manganese, is provided, a tubulated retort, D, containing a quantity of sulphur, and a three-necked bottle, B, for the purpose of washing the chlorine. These are connected together by tubes, as shown in the figure, and some hydrochloric acid poured into A by means of the crooked tube,

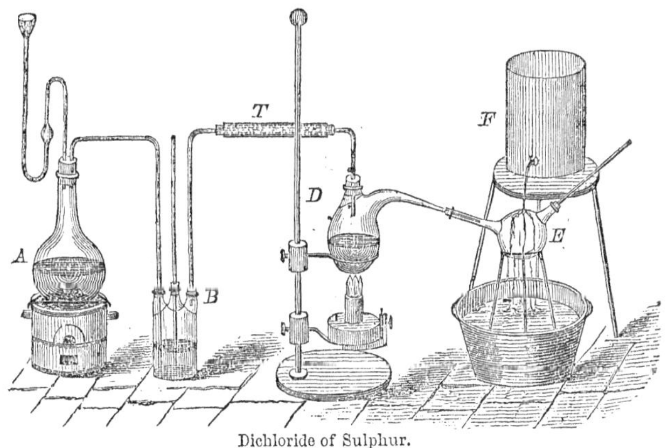

Dichloride of Sulphur.

the design of which is to prevent any escape of the gas, as would be the case if the liquid was poured directly in. Everything

QUESTIONS.—265. What is said of the compounds of sulphur and chlorine? 266. How is dichloride of sulphur formed?

being ready, heat is applied to the sulphur in D, so as to melt it, and also a gentle heat to A, to cause a slow evolution of chlorine. This gas, after being washed in B, is dried by passing through a chloride of calcium tube T, and finally comes in contact with the vapor of sulphur in D, where the compound in question (S_2Cl) is formed, and passes as a vapor into the receiver, E. This being kept cool by a stream of cold water from the vessel F, the dichloride is condensed, and any atmospheric air or other gaseous matter passes off by the waste-tube inserted in E. The liquid thus obtained must be separated from a little sulphur it contains by a second distillation.

Dichloride of sulphur is a reddish-yellow liquid, having a disagreeable, and very peculiar odor, which boils at about 280°. Its specific gravity is 1·69, and that of its vapor 4·668, It is immediately decomposed by contact with water.

Dichloride of sulphur, considered in the gaseous state, is composed of one volume of chlorine, and one-third of a volume of sulphur vapor, condensed into one volume. Thus,

One volume of chlorine weighs	2·440
One-third vol. sulphur vapor weighs $\frac{6 \cdot 654}{3}$	2·218
One vol. dichloride (or its density),	4·658

The density thus obtained, called the theoretical density, it will be seen, differs but slightly from that given above, obtained by direct experiment.

267. Chloride of Sulphur—SCl; eq., (16 + 35·4 =) 51·4.—This compound is prepared by passing a current of chlorine through a quantity of the preceding, until it is entirely saturated, and then distilling at a temperature of 147°. It is a deep red fluid, having a density of 1·62.

The density of its vapor is 3·549. Considered in the gaseous state, it is composed of one volume of chlorine and one-sixth of a volume of sulphur vapor, condensed to one volume.

One volume of chlorine weighs	2·440
One-sixth of a vol. sulphur vapor weighs	1·109
One volume of chloride,	3·549

The compounds of sulphur with nitrogen, iodine and bromine, being of little interest, are not here described.

QUESTIONS.—Describe the properties of dichloride of sulphur. 267. How is chloride of sulphur formed? Describe its properties.

SELENIUM.

Symbol, Se; *Equivalent*, 39·5; *Density*, 4·32.

268. History, etc.—Selenium was discovered, in 1817, by Berzelius, and received its name from *selene*, the moon. It is usually found associated with sulphur, in some of its compounds with other substances, especially sulphide of iron (iron pyrites). It is found also in combination with copper, lead, mercury, silver, and other metals. From any of these it is prepared by several different processes.

Selenium, at ordinary temperatures, is a solid of a deep brown color, the shade varying a little, according as it is seen in a powder or in a solid mass. When melted and suddenly cooled, it has a vitreous conchoidal fracture, and becomes negatively electrical by friction in very dry air. When heated to 212°, it becomes partially fluid, and perfectly so at a temperature a little higher than this. Heated to 700°, it sublimes like sulphur, which it closely resembles in many of its properties. At a temperature of 212°, or a little higher, especially if it has been heated considerably above this and again cooled down, it is viscid, and may be worked like softened sealing-wax, and drawn out into small threads.

When heated in the open air, it readily takes fire and burns, exhaling a strong odor not unlike that of decaying horse-radish,—a character by which it may always be distinguished.

Compounds of Selenium and Oxygen.

269. Selenium forms with oxygen three compounds, SeO, SeO_2, and SeO_3. The latter two are acids, and in many of their properties quite similar to sulphurous and sulphuric acids, to which they correspond in composition.

QUESTIONS.—268. In what is selenium found? Describe its properties. 269. What compounds does it form with oxygen?

270. Selenous Acid—SeO_2; eq., (39·5 + 2 × 8 =) 55·5.—To prepare this acid, a retort containing a mixture of chlorate of potash and peroxide of manganese is connected, as represented in the figure, with a tube bent downward so as to receive a quantity of selenium at the lowest part. Heat is then applied to the chlorate by which oxygen gas is evolved, and also to the selenium in the tube. As the selenium becomes heated, it burns slowly, with a blue flame, and the selenous acid is collected in the upper part of the tube in the form of white acicular crystals, which are very soluble in water.

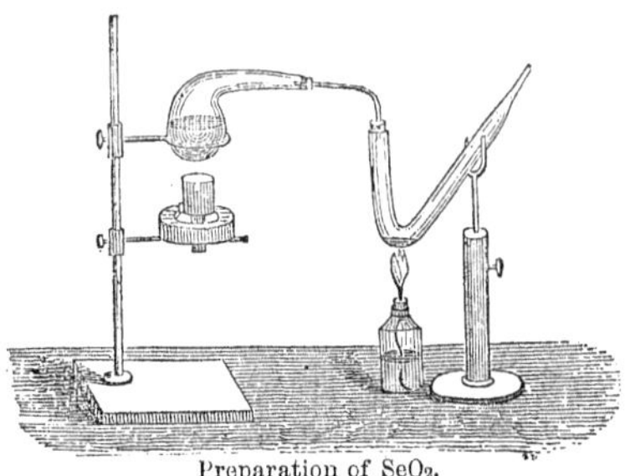

Preparation of SeO_2.

271. Selenic Acid—SeO_3; eq., (39·5 + 3 × 8 =) 63·5.—Selenic acid is prepared by burning selenium with nitrate of potash, when seleniate of potash is formed, from which the acid may be obtained in the liquid form. Its chief interest is found in its close resemblance to the corresponding sulphur acid, SO_3, and in the fact that it is capable of dissolving gold.

With hydrogen selenium forms a compound, hydroselenic acid, HSe, which is gaseous, and irritating to the eyes, nose and lungs. It is absorbed by water, like hydrosulphuric acid, and the solution, like that of hydrosulphuric acid, is decomposed by contact with the air.

The compounds of selenium with sulphur, chlorine and bromine are not of sufficient interest to require attention in this work.

TELLURIUM.

Symbol, Te; *Equivalent*, 64·5; *Density*, 6·2.

272. History, etc.—Tellurium is a rare substance, which has sometimes been found native, but is usually combined with the metals, as gold, silver, bismuth and lead. It is generally prepared from the telluride of bismuth, which is found in Schemintz in Hungary.

QUESTIONS.—270. Describe the mode of preparing selenous acid. 271. Describe selenic acid. What compound does selenium form with hydrogen? 272. Describe tellurium.

Tellurium, in some of its physical properties, closely resembles the metals with which it is often associated; but in its chemical properties it is more nearly allied to the non-metallic elements, especially sulphur and selenium.

When pure, it has a clear white color, and bright metallic lustre; and in its general appearance is not unlike antimony. It melts at a dull red heat, and, by slow and careful cooling, may be obtained in crystals, the primary form of which is the rhombohedron. At a very high temperature it becomes gaseous.

Compounds of Tellurium with Oxygen, Etc.

273. Tellurium forms with oxygen two acid compounds, viz., tellurous acid, TeO_2, and telluric acid, TeO_3, which, as will at once be seen, are similar in composition to the corresponding compounds of sulphur and selenium.

With hydrogen, also, like the two elements just named, it forms a single gaseous compound, hydrotelluric acid, HTe, the smell of which is even more offensive than that of hydrosulphuric acid.

Tellurium combines with chlorine, iodine, bromine, sulphur, selenium, etc.

GROUP IV.

PHOSPHORUS ARSENIC	Two elements which are solid at ordinary temperatures, and similar in many other properties. Many of their compounds are isomorphous.

PHOSPHORUS.

Symbol, P; *Equivalent*, 32; *Density*, 1·8 to 2.

274. History.—Phosphorus was discovered by an alchemist of Hamburg, in 1669; and received its present name (from *phos*, light, and *pherein*, to carry) from the circumstance that, at ordinary temperatures, it always appears luminous in the dark.

QUESTIONS.—273. What is said of the compounds of tellurium with oxygen? 274. Give the history of phosphorus.

It is not found in nature in a separate state; but in combination with oxygen and lime, it is very generally diffused, being contained in all fertile soils, without exception, and in many vegetable and animal substances.

275. Preparation.—Phosphorus, at the present time, is prepared entirely from bones, which are first heated in the open air until they become white, so as to destroy all the animal matter they contain. More than half their weight remains, which is chiefly phosphate of lime. This is then ground to a fine powder, and digested, for one or two days, with dilute sulphuric acid, in the ratio of 3 parts of the bone ashes to 2 parts of acid and 15 or 20 parts of water.

The action of the sulphuric acid upon the phosphate of lime, is to take away a part of its lime, forming with it sulphate of lime; and as the whole of the phosphoric acid of the original phosphate is then in combination with only a part of its lime, it is plain that this must now be a super-phosphate of lime. This latter is soluble in water, while the sulphate of lime which has been formed is insoluble; more water is therefore added, and a clear liquid obtained, which is evidently solution of super-phosphate of lime. This liquid is now evaporated until it begins to be quite thick, when it is mixed intimately with charcoal, in fine powder, and thoroughly dried. It is next introduced into an earthern retort, *a*, which is placed in a proper furnace, as represented in the figure; and to the neck of the retort, a wide copper tube, *b*, is attached, which connects with a vessel of water. The heat is then gradually raised, when the phosphorus distils over, and is condensed in the water. Much combustible gaseous matter, also,

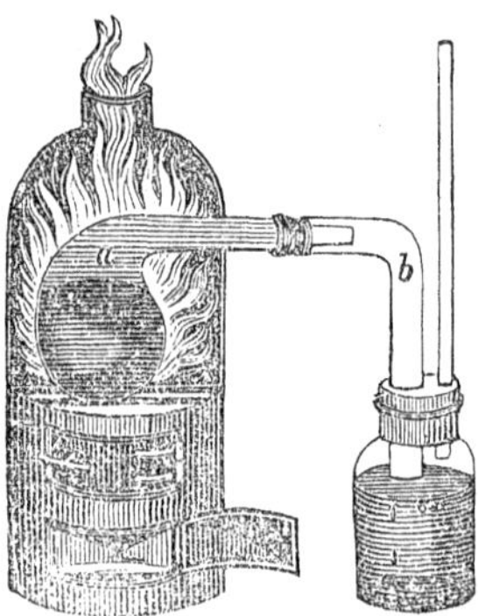

Preparation of Phosphorus.

QUESTION.—275. Describe the mode of preparing phosphorus.

comes over and escapes by the second tube, inserted in the water-vessel.

The affinity of charcoal for oxygen at low temperatures is not very considerable, but when highly heated it is intense, and sufficient to abstract the oxygen from the phosphoric acid of the acid phosphate of lime, causing the liberation of the phosphorus. This, taking the gaseous state, distils over and is condensed to the liquid form, and finally solidified.

The phosphorus thus procured is still impure, and is to be melted in hot water, and pressed through porous leather.

276. Properties.—Pure phosphorus is of a light flesh-color, and nearly transparent. At common temperatures, it is a soft solid, of specific gravity about 2, and may easily be cut with a knife. At 108° it fuses, and at 554° is converted into vapor, which has a density of 4·326. It is soluble, by the aid of heat, in naphtha, in fixed and volatile oils, and in some other liquids. By the fusion and slow cooling of a considerable quantity, it may be crystalized, and also from its solution in bisulphide of carbon. The crystals belong to the monometric system.

It is usually seen in long, slender sticks, of a waxy lustre, which are made by melting the phosphorus under water and pouring it into glass tubes.

Phosphorus may be distilled without difficulty. For a small operation, fit a green glass retort to a tube bent, as represented in the figure, and having its extremity drawn out to a fine point, but not closed. Separate the retort and tube, and put in the first a small quantity of phosphorus, and in the latter a little water; and again connect them firmly together. If now the retort is carefully heated, the phosphorus will gradually distil over and collect in the water in the lowest part of the tube.

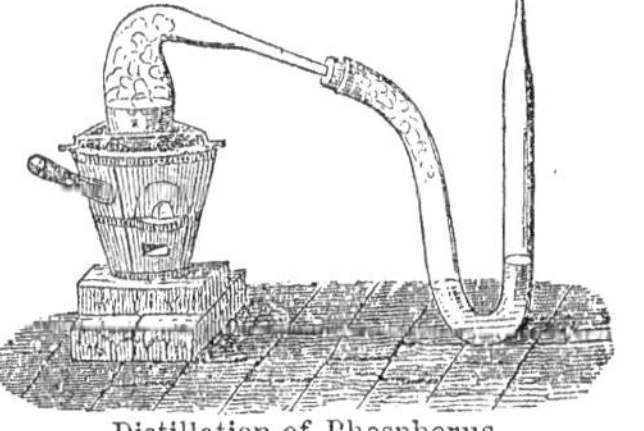

Distillation of Phosphorus.

QUESTIONS.—What is the effect produced by the charcoal? 276. Describe phosphorus. In what is it soluble? In what form is it usually seen?

Phosphorus is exceedingly inflammable. Exposed to the air, at common temperatures, it undergoes slow combustion, emits a white vapor of a peculiar alliaceous odor, appears distinctly luminous in the dark, and is gradually consumed. On this account, phosphorus should always be kept under water. In the open air, even the heat of the hand, aided by the slightest friction, is sufficient to inflame it; and it should therefore always be handled with the greatest caution. It burns in the air with a brilliant, yellowish-white light and intense heat; but in oxygen gas, its combustion is particularly splendid. A good method for performing the experiment is to place a piece of phosphorus in a small cup on a stand, a few inches high, in a basin of water, and, having ignited the phosphorus by touching it with a piece of heated wire, dexterously to place over it a large bell-glass, previously filled with oxygen. By careful management, but little of the oxygen will be lost. It may also be made to burn under warm water, by forcing a current of oxygen upon it by means of a gas-bottle, or a flexible tube, leading from a gasometer. A red suboxide is formed which readily takes fire in the open air.

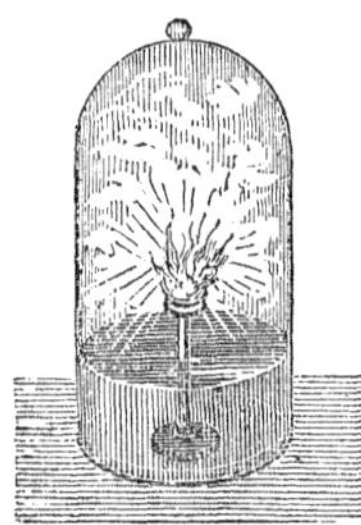
Combustion of Phosphorus in Air or Oxygen.

Combustion of Phosphorus by Oxygen under Water.

The red crust which forms upon the surface of pieces of phosphorus exposed to the action of light, is believed to be a peculiar isomeric, or allotropic (173) condition of this substance. It may be obtained by keeping a quantity of phosphorus for several hours, at a temperature of 450° to 480°, in a gas which does not act upon it, as nitrogen or hydrogen.

This red or amorphous phosphorus differs essentially from ordinary phosphorus. Its melting point is about 482°, and it is non-luminous in the air at ordinary temperatures. Heated to 500°, it changes to ordinary phosphorus.

QUESTIONS.—How is phosphorus affected in the open air? How is it usually preserved? May it be distilled? What is said of its combustion in oxygen gas? How may it be made to burn under water?

277. **Uses.**—Phosphorus is now used in large quantities in the manufacture of friction-matches, which ignite by slight friction. For this purpose it is made into a paste, with gum or glue, by which it is made to adhere to small pieces of wood or paper previously dipped in melted sulphur, and is also protected from the action of the air.

The paste is sometimes mixed with nitrate or chlorate of potash, or oxide or nitrate of lead, but this is not necessary. Occasionally, fine emory or powdered glass is mixed in the paste, to increase the friction.

Phosphorus is of great service in the laboratory, and has been sometimes used in medicine.

Compounds of Phosphorus and Oxygen.

278. There are four compounds of phosphorus and oxygen, the atomic constitution of which appears to be P_2O,PO,PO_3 and PO_5. The last three are acids; but only one of these, the last, will be specially described.

279. **Dinoxide of Phosphorus**—P_2O.—This compound is prepared by the combustion of phosphorus under hot water, as in paragraph 276. Atmospheric air may be substituted for the oxygen. The oxide appears as flocculi of a brick-red color. It absorbs oxygen from the air, and mixed with phosphorus it renders it more combustible.

280. **Hypophosphorous Acid**—PO.—May be obtained as a syrupy liquid, but cannot be crystalized. It cannot be obtained separate from water.

281. **Phosphorous Acid**—PO_3.—Phosphorous acid may be prepared by the action of the air upon sticks of phosphorus, at ordinary temperatures. For this purpose, place a few sticks of phosphorus in a funnel under a bell-glass, as represented in the figure. The glass is supported a little above the table, to allow the air to enter. Regnault advises to put the sticks of phosphorus in small glass tubes, having capillary apertures for the acid to pass through as it is

Preparation of PO_3.

QUESTIONS.—277. What use is made of phosphorus? How are friction-matches made? 278. What compounds of phosphorus and oxygen are there? 279. How may dinoxide of phosphorus be prepared? 281. How may phosphorous acid be prepared?

formed, but they are not essential. Too large quantities of phosphorus should not be used at once.

282. Phosphoric Acid—PO_5; eq., (32 + 40 =) 72.—This acid is formed by burning phosphorus in air or in oxygen gas, as in the experiment given above (276). To prepare it perfectly anhydrous, the receiver should be placed over mercury, and the oxygen or air supplied should be perfectly dry. The acid appears as a dense white vapor, which is gradually precipitated, and may be collected. In the open air it at once absorbs moisture. If the white flakes are collected and ignited, the mass, after cooling, is semi-transparent, and is called *glacial phosphoric acid.* This acid may also be formed from calcined bones.

To prepare the common acid, a very good mode is to digest phosphorus in nitric acid, with the aid of heat. One part of phosphorus, with 13 parts of the acid, of specific gravity 1·20, is placed in a glass retort which connects with a receiver that is kept cold by a stream of cold water, as shown in the figure, and heat applied. Red fumes of nitrous acid are given off, and the phosphorus rapidly consumed. The liquid collected in the receiver is now to be poured back into the retort, and heated until the water and any remaining nitric acid are expelled. On cooling it will become solid, and present the same appearance as glacial phosphoric acid just described. The acid thus obtained contains 1 eq. of water for each eq. of the acid, and is called monohydrated acid.

Preparation of PO_5.

The anhydrous phosphoric acid has a very strong affinity for water, and when thrown into it, unites with it with great energy, often producing slight explosions, in consequence of the heat pro-

QUESTIONS.—282. What is the composition of phosphoric acid? How is it prepared? How by the use of nitric acid? What is said of its compounds with water?

duced. With water it forms three different compounds, as follows, the first two of which have been called, respectively, metaphosphoric, and pyrophosphoric acids.

Monohydrated, or metaphosphoric acid,	PO_5,HO.
Bihydrated, or pyrophosphoric acid,	$PO_5,2HO$.
Trihydrated, or common phosphoric acid,	$PO_5,3HO$.

These three acids, or rather compounds of acid and water, cannot be distinguished from each other by external appearance, but dissolved in water they manifest chemical characteristics which render them quite distinct; they also form salts, which, though much alike in their general properties, are nevertheless easily distinguished from each other by the proper tests.

Compounds of Phosphorus and Hydrogen.

There are several compounds of phosphorus and hydrogen, but one only will claim attention from us, the common phosphuretted hydrogen, PH_3. The others are P_2H, and PH_2.

283. Phosphide of Hydrogen—PH_3; eq., $(32 + 3 =)$ 35.—This gaseous substance, called also *phosphuretted hydrogen*, is best prepared by heating some sticks of phosphorus in a strong solution of caustic potash, in a small glass retort, which, at the beginning of the operation, should be quite filled with the materials. If, then, the mouth of the retort is made to dip slightly in a basin of water, each bubble of the gas, as it breaks into the air, will burst into a flame,

PH_3 spontaneously combustible.

QUESTIONS.—283. What phosphide of hydrogen is mentioned? How is it prepared?

with the formation of beautiful wreaths of smoke of phosphoric acid, as shown in the figure.

In the presence of potash phosphorus has the property of decomposing water, especially if the temperature be raised; the oxygen and the hydrogen both combining with separate portions of the phosphorus, producing the teroxide of phosphorus (phosphorous acid) and the terphosphide of hydrogen. Thus,

$$2P + 3HO + KO = KO,PO_3 + PH_3.$$

Instead of potash, milk of lime may be used with the same result.

Still another method of procuring it is to decompose phosphide of calcium by water, or dilute hydrochloric acid; but when this acid is used, the gas which is given off is not spontaneously inflammable. It seems to be very well determined that the spontaneous combustion of this gas, in certain cases, is owing to the presence of the vapor of another phosphide of hydrogen containing proportionably more phosphorus, which may be separated by passing the gas through a tube surrounded by a freezing mixture.

PH^3 spontaneously combustible.

Other combustible gases, as hydrogen, are made to inflame spontaneously by receiving a portion of the vapor of this inflammable phosphide.

To prepare phosphide of calcium, select a tube of green glass, half an inch in diameter and a foot long; seal one end hermetically, and bend it a little, as represented in the figure. In the sealed end put some pieces of phosphorus, and then, holding the straight part in a horizontal position, introduce some lumps of well burned lime. The part containing the lime is next to be heated to redness by means of burning charcoal, when heat is also to be applied to the phosphorus sufficient to volatilize it; and the vapor coming in contact with the heated lime, at once unites with it to form the phosphide of calcium. This should be preserved in bottles with close stoppers; but even then it gradually undergoes decomposition.

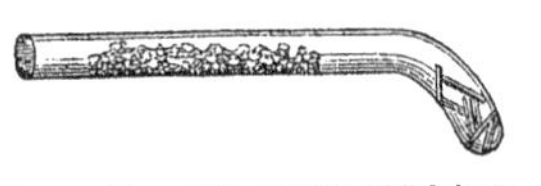

Preparation of Phosphide of Calcium.

Terphosphide of hydrogen is a colorless gas, with a very disagreeable

QUESTIONS.—Describe some of the properties of phosphide of hydrogen. To what is its spontaneous combustion owing?

odor,* and is irrespirable; 100 cubic inches of it weigh 36·75 grains, giving it a specific gravity of 1·18.

1 vol. of phosphorus vapor weighs	4·326
6 vols. of hydrogen weigh (·069 × 6)	·414
Forming 4 vols. of the terphosphide,	4·740

One volume therefore weighs, or the density of the gas is, 1·185

Other Compounds of Phosphorus.

Phosphide of Nitrogen, N_2P, is a white solid, which is infusible even at a red heat.

284. Chlorides of Phosphorus.—There are two chlorides of phosphorus, PCl_3, and PCl_5, corresponding in composition to phosphorous and phosphoric acids.

The Terchloride of Phosphorus is formed in the same manner, and by using the same apparatus, as chloride of sulphur (266). The chlorine

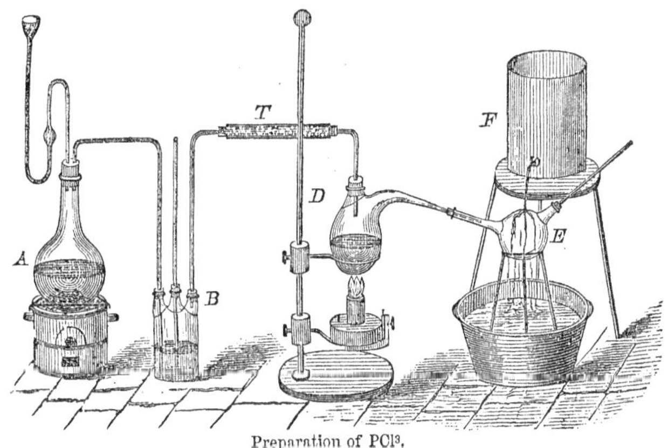

Preparation of PCl^3.

is formed in the flask A, and after being washed in water and dried in the chlorine of calcium tube, is brought in contact with the vapor of phos-

* Those who have observed the odor of this gas, and that of the liquid emitted by the American skunk (*Mephitis Americana*) when disturbed, cannot but have noticed the resemblance between them; which seems to render it probable that the fluid emitted by the skunk contains, in solution, a portion of the gas, or some other nearly-related compounds of the same substances. This is rendered still more probable from the fact, that the fluid, when emitted by the animal in the dark, is distinctly *phosphorescent.—See Godman's Natural History, vol.* 1, 289: *Philadelphia Edition*, 1829.

QUESTION.—284. What chlorides of phosphorus are mentioned?

phorus in the retort, D, where the chloride is formed, and afterwards condensed in the receiver E, which is kept cool by a stream of cold water It is a colorless liquid, of a density of 1·45, which boils at about 172°.

285. Perchloride of Phosphorus is formed by saturating the preceding with chlorine. It is a solid, having its point of fusion, and also its boiling point, at about 298°.

Prep. of Iodide of Phosphorus.

Bromine and sulphur combine readily with phosphorus, but the compounds are not important.

Iodide of Phosphorus is formed by bringing the two substances together in a vessel where as little air may have admission as possible. It forms a dark-colored mass. These two substances afford one of the few instances in which reaction takes place between two solids. Let a few crystals of iodine be dropped into a wine-glass, upon a small piece of phosphorus, and immediately place over it a bell-glass. By the heat produced, the phosphorus will be inflamed and a portion of the iodine sublimed; and the white cloud of phosphoric acid (152), mingling with the dense iodine vapor, presents to the eye a very pleasing appearance.

ARSENIC.

Symbol, As; *Equivalent*, 75; *Density*, 5·88.

286. History and Preparation.—Arsenic has very generally been classed with the metals, chiefly on account of its metallic lustre, and comparatively high specific gravity; but, in its chemical properties, it is much more closely allied to the metalloids, with which it is here classed.

Arsenic sometimes occurs native, but usually it is found in combination with the metals, and especially with iron and cobalt. The substance itself, or some of its compounds, seems to have been known from the earliest times.

It may readily be prepared by heating the mineral called *mispickel*, which is a natural compound of arsenic, sulphur and iron, in close vessels, by which the arsenic is expelled and the sulphide

---

QUESTIONS.—285. How is iodide of phosphorus prepared? What is said of the action of iodine and phosphorus upon each other? 286. Why has arsenic often been classed with the metals? Why is it here classed with the metalloids? With what is it usually found combined?

of iron remains. By again heating it with black-flux, the arsenic is obtained nearly pure. If the pulverized mineral is well mixed with black-flux at the beginning, and no air admitted into the apparatus, a single operation will afford it in great purity.

Let a common Hessian crucible be half filled with the mixture, and then place another crucible, a size smaller, in an inverted position above it, as shown in the figure, carefully luting them at their junction. A moderate heat should then be applied to the lower crucible and very gradually raised. The arsenic will be sublimed from the mixture and condensed in small crystals in the inverted crucible, which should have a very small aperture in the bottom, to allow the air to escape as the heat is raised.

Preparation of Arsenic.

287. Properties.—Arsenic is a brittle substance, of a dark color, and feeble metallic lustre. Heated to about 356°, it is sublimed, without first melting, as is the case with most solids. Its vapor has a strong garlic odor, by which its presence may be recognised, and a density of 10·37. Heated in the open air, it readily takes fire and burns with a livid flame.

Arsenic is often sold under the very improper names of *cobalt* and *fly-powder*.

Compounds of Arsenic and Oxygen.

288. Two compounds only of arsenic and oxygen are known, both of which are acids, and in composition correspond to phosphorous and phosphoric acids.

289. Arsenious Acid—AsO_3; eq., $(75 + 3 \times 8 =)$ 99.—This compound is the *arsenic*, or *rats' bane*, of commerce, well known as a destructive poison. It is always produced when arsenic or its ores are heated in the open air. It is usually sold in a state

QUESTIONS.—How may arsenic be separated from its compounds? 287. Describe arsenic. What is said of its odor? 288. What compounds of arsenic and oxygen are known? 289. By what names is arsenious acid often known.

of fine white powder; but when first sublimed, it is in the form of brittle masses, more or less transparent, colorless, of a vitreous lustre, and conchoidal fracture. This glass, which may also be obtained by fusion, gradually becomes opaque without undergoing any apparent change of constitution, but becomes more soluble in water than before. Its specific gravity is 3·7. At 380° it is volatilized, yielding vapors which do not possess the odor of garlic, and which condense unchanged on cold surfaces. If thrown on burning charcoal, the garlic odor is perceived, because of the reduction of the oxide by the carbon.

Destructive as this substance is to the animal system, in minute doses it is sometimes used in medical practice; and in some countries, as in parts of Austria and Hungary, it is habitually used much in the same manner as narcotics, and even administered to horses. The effect is said to be to give a roundness and fulness of form, and clearness and freshness to the complexion. Horses accustomed to receive it in their food, have a fat and plump appearance, and bright and glossy skins. It also improves their breathing.

This substance is so frequently used to destroy life, that its detection in suspicious cases becomes an important object; but we reserve our remarks on this point until some others of the many compounds of arsenic have been described

290. Arsenic Acid—AsO_5; eq., (75 + 8 × 5 =) 115.—This acid may be formed by dissolving arsenious acid, just described, in nitric acid mixed with a little of the hydrochloric, and evaporating to dryness. It is a powerful acid, much resembling phosphoric acid (247); with which it is isomorphous. Its salts are also isomorphous with the salts of phosphoric acid.

Compounds of Arsenic and Hydrogen.

291. Two compounds of these elements are known, one of which is solid, and the other gaseous. Of the former, little is known, with certainty, and we therefore do not further allude to it.

292. Arsenide of Hydrogen, Arseniuretted Hydrogen—AsH_3; eq., (75 + 3 =) 78.—This gas is evolved when arsenide of tin or zinc is treated with strong hydrochloric acid, or when sulphuric or hydrochloric acid is made to act upon zinc or iron in the presence of any soluble compound of arsenic.

To prepare it, pour upon some pieces of zinc diluted sulphuric acid with a few drops of solution of arsenious acid; the gas, which burns with a feeble blue flame, will be at once rapidly evolved.

QUESTIONS.—Describe arsenious acid. For what purpose is it often used? 290. To what other acid is arsenic acid analogous? 292. How is arsenide of hydrogen prepared?

When thus prepared, the reactions are as indicated in the following formula, the zinc being oxydized at the expense of the oxygen both of the water and the arsenious acid. Thus,

$$6Zn + 3HO + AsO_3 + 6SO_3 = 6\ (ZnO,SO_3) + AsH_3.$$

The gas has a peculiar nauseating odor, and is exceedingly poisonous Its density is 2·69, and by a cold of —22° it is converted into a liquid under the ordinary atmospheric pressure. By chlorine it is instantly decomposed, chloride of arsenic and hydrochloric acid being formed. By solution of blue vitriol it is rapidly absorbed, and arsenide of copper precipitated.

The equivalent of this gas, AsH_3, answers to 4 vols., which is thus constituted:

1 vol. of arsenic vapor weighs	10·370
6 " hydrogen "	·414
4 vols. arsenide of hydrogen,	10·784

Weight of one vol., or the calculated density of the gas, 2·696.

We shall have occasion to speak of this compound again in connection with the detection of arsenic.

Compounds of Arsenic with Sulphur and Other Elements.

293. Sulphides of Arsenic.—The *bisulphide of arsenic*, AS_2, is found native, and called *realgar* by mineralogists. It may also be formed by art. It is of a dull red color. The tersulphide, AsS_3, is the *orpiment*, or *king's yellow* of commerce. It is formed artificially by passing a current of hydrosulphuric acid through an arsenic solution containing a little free acid. It is also found as a natural production, and is of a bright yellow color, and is sometimes called *sulpharsenious acid.* A third compound of these elements, the *pentasulphide of arsenic*, AsS_5, called also *sulpharsenic acid*, is formed by mixing solutions of hydrosulphuric and arsenic acids. It forms slowly, and some days are often required before the whole is precipitated.

The compounds of arsenic with chlorine, iodine, phosphorus, &c., will be found described in larger works.

QUESTIONS.—Describe the reactions which take place in the preparation of arsenide of hydrogen. What are some of the properties of this gas? 293. What is *realgar?* What is *orpiment?*

Detection of Arsenic.

294. Poisoning by arsenious acid is at the present day, unfortunately, very common; and it therefore becomes a matter of special importance to be able with certainty to detect the instrument of death.

295. There are as many as ten or twelve different tests for arsenic, but we shall confine our remarks to some of the most important. A single test should never be relied on, but several different ones should always be applied to separate portions of the suspected substance.

I. Marsh's Test.—Put into a two or four ounce vial some pieces of clean zinc, and pour on them a small quantity of dilute oil of vitriol (oil of vitriol 1 part, and water 8 parts), and insert a cork with a small tube, as shown in the figure. Very soon hydrogen gas will begin to be evolved, as in the preparation of hydrogen. After a little time the jet of hydrogen may be inflamed; and if all the materials used were pure, a piece of glass or porcelain held in the flame will receive no stain, but only a deposition of moisture from the combustion of the hydrogen.

Marsh's Test.

The cork and tube being now removed, introduce some of the suspected substance, or water in which the suspected substance has been digested, with the aid of heat if necessary, and immediately replace the cork and tube. In a little time the jet of gas may be relighted; and if any appreciable quantity of arsenious acid is present, a piece of clean glass or porcelain held in the flame will at once receive a black stain upon its surface, caused by a deposition of arsenic.

If the quantity of arsenious acid present is large, the flame will be of a pale color, and the blackening of the glass held in the flame will be instantaneous; but if the quantity be small, the blackening effect will be produced only after a little time.

The deposition of arsenic here is from the arsenide of hydrogen, which is formed in the manner heretofore (292) explained. The cold substance held in the flame causes the deposition of the arsenic while the hydrogen is consumed.

This is perhaps the most delicate test of arsenic known, but in using it some precautions must always be observed to avoid mistake. In some cases, where organic substances are present, spots similar to those produced by arsenic may be formed, that may be mistaken for arsenic by

---

QUESTIONS.—295. What is said of the number of tests for arsenic? Describe *Marsh's test.* From what is the arsenic deposited? What is said of the delicacy of this test?

the inexperienced; and antimony will form spots very much like those of arsenic. Means must therefore be adopted to test the material forming the dark spot upon the porcelain. For this purpose, it will generally be sufficient to hold the spot a few minutes in the flame of a spirit-lamp; if the deposite be arsenic it will be volatilized, and disappear, but if produced by antimony or other substances, it will remain. So also arsenic spots, exposed a few minutes, at a moderately elevated temperature, to vapor of iodine, become yellow, and then subsequently disappear by exposure to the air.

II. Reinch's Test.—In a portion of the suspected liquid, made acid by hydrochloric acid, place a piece of metallic copper, previously filed perfectly bright, and heat the whole nearly to the boiling point. If any appreciable quantity of arsenious acid be present, the arsenic will be deposited upon it as a gray crust of a metallic lustre.

III. Test by Hydrosulphuric Acid.—Through a portion of the suspected substance, supposed to be in the liquid form, acidulated with muriatic acid, pass a current of hydrosulphuric acid gas for half an hour, and then boil it a few moments; if arsenic be present, a yellow precipitate—orpiment (293)—will be formed. The mode of passing the current of gas through the liquid will be seen by the accompanying figure. The materials for producing the gas are put into a flask, and a tube, bent twice at right angles, is inserted through a cork, so as to be air-tight; the other end is then immersed in the liquid, contained in a glass vessel, so as to reach near the bottom, and the gas, as it escapes, bubbles through the liquid. The mode is the same as before described (263). The precipitate (orpiment) thus formed, is entirely soluble in aqua ammoniæ, and in solutions of the alkalies.

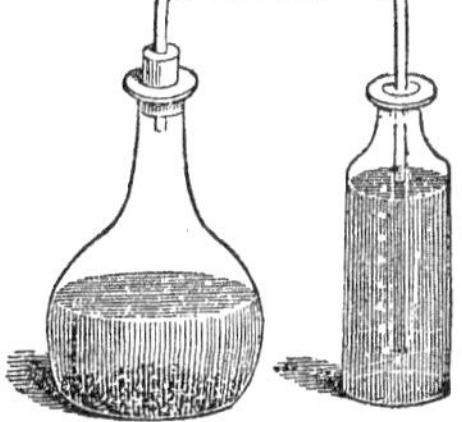

Test by Hydrosulphuric Acid.

IV. Test by Ammonia-Nitrate of Silver.—Nitrate of silver forms with arsenious acid solutions a precipitate of arsenide of silver, which is of a peculiar canary yellow color, and is soluble in nitric acid. To ensure the formation of this precipitate, the arsenious solution should be slightly alkaline, and therefore the ammonia-nitrate of silver is used in preference to the simple nitrate. This is prepared by pouring into the nitrate of silver solution aqua ammonia, until the precipitate at first thrown down is *nearly* all dissolved.

A precipitate very similar in its appearance to the above would be produced by phosphoric acid in the suspected liquid; so that the precipitate formed by use of this test should always be further examined, before any reliance is placed upon it.

V. Test by Ammonia-Sulphate of Copper.—Solution of sulphate of copper produces in neutral or alkaline solutions of arsenious compounds a beautiful green precipitate, sometimes called *Scheele's* green. In making

---

QUESTIONS.—Describe *Reinch's test.* Describe the test with *hydrosulphuric acid.* Describe the test with *ammonia-nitrate of silver.* Describe the mode of testing with *ammonia-sulphate of copper.* What will be the color of the precipitate?

the experiment, it is best to use the ammonia-sulphate of copper, which is prepared by pouring into a solution of blue vitriol, aqua ammoniæ, until the precipitate at first formed is nearly all redissolved, as in the corresponding preparation of ammonia-nitrate of silver.

But this test also may form with other substances a precipitate similar in appearance to the above, so that further examination should always be made.

VI. Flandin and Danger's Test.—Dry the suspected substance (supposed to be organic, as sugar or starch,) and treat it with one-fourth of its weight of the strongest oil of vitriol, and apply heat until it is quite dry;—the whole will now be reduced to a black, friable mass, which can easily be pulverized, and is then to be boiled with strong nitric mixed with a little hydrochloric acid. By this process the arsenic, in whatever form it may be, is converted into arsenic acid, which after the whole has been again evaporated to dryness, to expel any remaining nitric acid, and redissolved in pure water, may be examined by the appropriate tests, not for arsenious but for arsenic acid.

For this latter purpose, the ammonia-nitrate of silver may be used, which gives with arsenic acid a brick-red precipitate.

Or, the arsenic acid being obtained in solution, may be precipitated as arseniate of lime by lime-water, and from this precipitate pure arsenic with its metallic lustre may be obtained by the process next to be described.

VII. Reduction of the Arsenic.—When arsenic is present in any appreciable quantity, it may always be obtained in a separate state, so as to be recognised by its peculiar metallic lustre and garlic odor; and no chemist in any particular case will positively swear to its presence unless he is able thus to procure it!

For this purpose, provide a tube of hard glass, a quarter of an inch in diameter and three or four inches long, with one end hermetically sealed; and fill it to the depth of half an inch with a mixture of the suspected substance, charcoal, and carbonate of soda, the whole being previously well dried at a moderate heat, and ground together to a fine powder. After wiping the inside of the tube with a little cotton attached to a wire, to remove any dust or remaining moisture, a strong heat is applied to the end of the tube containing the mixture, by which the arsenic will be separated, to be again condensed upon the sides of the tube a little above the heated part. The bright metallic lustre will at once be recognised, and by breaking the tube and heating the part coated, as in the figure, the peculiar garlic odor will be perceived. Other tests may also be applied to it if desired.

Test of Arsenic.

This last test may be applied to any of the precipitates obtained by the previous tests.

QUESTIONS.—Describe *Flandin and Danger's* test. Describe the mode of testing by reduction of the arsenic. How does the arsenic show itself? May the last mode be applied to the precipitates obtained by the other modes?

296. In these directions we are supposed, as a general thing, to be operating with pure arsenical solutions, but in cases of actual poisoning it will usually be otherwise; and it often becomes an important object to be able to separate the organic matter contained in the suspected substance. Sometimes the substance will be soluble, as sugar, which will not interfere badly with the operation; and at others it will be of such a character that it can be removed by passing through it a current of chlorine, or it may be that it can be removed only by heating it with strong oil of vitriol, as heretofore (VI.) described. In any particular case, the mode of proceeding to be pursued must be adapted to its peculiar circumstances.

GROUP V.

CARBON, SILICON, BORON } Combustible bodies, and incapable of being volatilized even at the highest temperatures.

CARBON.

Symbol, C; *Equivalent*, 6; *Density (crystalized)*, 3·52.

297. **History.**—Carbon, though rarely met with in nature perfectly pure and uncombined, is one of the most important of the elements, forming, as it does, an essential ingredient of nearly all vegetable and animal bodies. It is found in a variety of forms; and when uncrystalized and uncombined, its color is always black.

298. **Preparation and Properties.**—Carbon presents itself to us in a variety of forms, as the *diamond*, *graphite or plumbago*, *mineral coal*, *charcoal*, *gas coal*, and perhaps we may add *lamp-black*, though the latter very probably differs from charcoal only in being in a state of fine division.

QUESTIONS.—296. In these directions what are we supposed to operate with? Will this usually be the case in practice? Will it often be necessary to separate organic matters from the suspected substance? What elements constitute the fifth group? How are they characterized? 297. Give the history of carbon. 298. What are some of the varieties of carbon?

The diamond is pure crystalized carbon, and is the hardest substance known in nature. The crystals are of the form of the regular octahedron, but the faces are frequently a little convex, as shown in the figure. Such crystals, properly set, are used for cutting glass, a purpose for which they are admirably adapted. Heated intensely in the flame of the compound blowpipe, the diamond is entirely consumed, forming carbonic acid, just as if the same weight of pure charcoal had been consumed. Diamonds are generally very small, the largest ever found weighing less than six ounces. A single diamond has been sold for more than half a million of dollars. It is generally found in the same situations as gold and platinum. A few crystals of little value have been discovered in the vicinity of the gold mines in some of the Southern States. It is a powerful refractor of light, and seems to have the faculty of absorbing light, and giving it out again after a time. A diamond held in the sun's rays a few seconds, and then removed at once to a dark room, phosphoresces very distinctly for a few seconds.

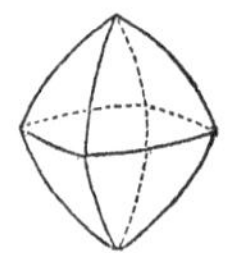
Diamond.

Graphite, or *plumbago*, called also, very improperly, *black lead*, is a variety of carbon, containing usually a little iron. It is often found crystalized in thin scales, of a hexagonal form. It is not unfrequently formed as an artificial production in iron furnaces, and is sometimes quite free from iron.

It is used for the manufacture of pencils, and in the construction of crucibles that are to be exposed to a very intense heat. For this purpose, it is ground to a fine powder, and mixed thoroughly with fire-clay. These crucibles are used chiefly for melting metals.

Mineral coal is of two kinds; the *bituminous*, and the non-bituminous, or *anthracite*.

Bituminous coal is distinguished by its softening, like wax, when heated, and giving off much gas, which burns with flame.

QUESTIONS.—What is the diamond? What is said of its hardness? What use is made of it? What is the effect of the intense heat of the compound blowpipe upon it? What is said of the size of diamonds? What is said of the absorption of light by the diamond? What is *graphite* or *plumbago?* What use is made of it? What two kinds of mineral coal are there? How is bituminous coal distinguished?

It is also much lighter than anthracite, and more easily ignited. Some of the different varieties of bituminous coal are *caking*, *splint*, *cherry*, and *cannel* coal. *Jet*, also, which is used in jewelry, is a bituminous coal; and in the same family may be included *wood* or *Bovey* coal, sometimes called *lignite*.

Anthracite, or *stone-coal*, differs from the above varieties, in containing no bituminous matter; and, therefore, it yields no inflammable gas by heat. Its sole combustible ingredient is carbon; and, consequently, it burns without flame. It is found in different countries, but nowhere in such profuse abundance as in the eastern part of the State of Pennsylvania, which supplies most of the northern and eastern parts of the United States with fuel.

All the varieties of mineral coal are believed to have been formed from vegetable substances, which, in the changes the earth's surface has undergone, have become buried beneath it.

When bituminous coal is subjected to a high temperature in close vessels, or with only a limited supply of atmospheric air, the volatile or bituminous manner is expelled, and the remaining porous carbon is called *coke*. It is used for many important purposes in the arts.

Charcoal is prepared by exposing vegetable matter, and especially wood, to a high temperature in close vessels, or in such circumstances as to avoid the presence of atmospheric air. By the heat a large quantity of water, acetic acid, tar, and other matters, is expelled, and the carbon, with any mineral matter which has been absorbed from the soil, remains. The latter constitutes the *ashes* which remain after the combustion of the coal in the open air.

The usual method of preparing charcoal for ordinary purposes, is to ignite large heaps of wood, which are covered with earth so as to admit only a limited supply of atmospheric air; and the result is *to char* or convert into coal a large part of the wood, by the heat occasioned by the combustion of the other part.

QUESTIONS.—What is said of anthracite or stone-coal? Where is it found in this country? What is *coke?* How is charcoal prepared? What constitutes the ashes?

The figure, diminished from Knapp's Technology, represents a section of a coal-pit ready to be ignited. The stake at the centre

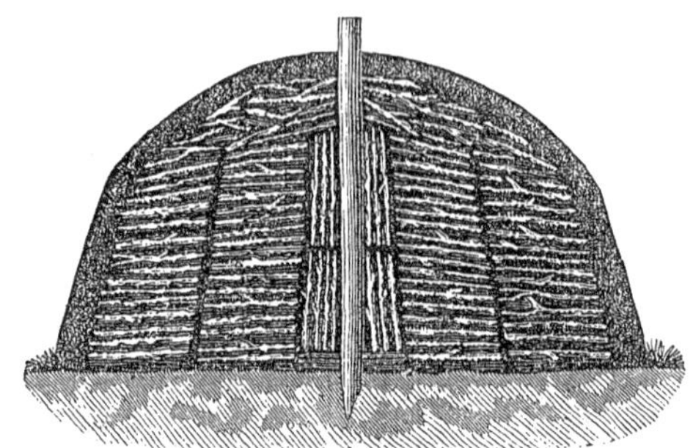

Preparation of Charcoal.

serves as a support for beginning the heap; and by one side of which space is left to kindle the fire, by dropping in pieces of very dry wood and burning coals.

Charcoal is a black, hard, brittle substance, perfectly insoluble in every liquid, but attacked and oxidized by strong nitric acid. It is a good conductor of electricity, but a non-conductor of heat; is little acted upon by air and moisture, and is perfectly infusible in the most intense heat that can be applied to it. Heated in the open air, it takes fire and burns freely, especially if in large masses, leaving only a small residue of *ashes.*

299. Charcoal possesses the property of absorbing a large quantity of air, or other gases, at common temperatures, and of yielding the greater part of them again when it is heated. Recently-burned charcoal absorbs air and moisture so rapidly, for a few days, as materially to increase its weight. Both are absorbed and retained with such force, that a red heat is required to expel them. This absorption of air may be readily shown in the following manner:—Let a piece of charcoal, of moderate size, be heated to redness for a few minutes, and then quenched under

---

QUESTIONS.—Describe charcoal. 299. What is said of the absorption of gases by charcoal? How may the absorption of air be shown?

mercury, and placed under a receiver, over the mercurial cistern. The mercury will shortly begin to rise, in consequence of the absorption of the air within; and the process will continue for several hours.

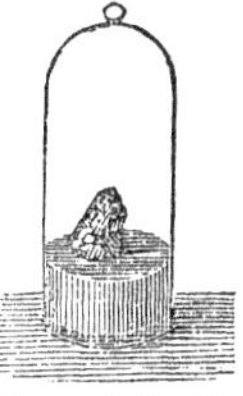
Charcoal absorbs Gases.

Charcoal, likewise, absorbs the odoriferous and coloring particles of most animal and vegetable substances. When colored infusions of this kind are digested with a proper quantity of charcoal, a solution is obtained which is nearly, if not quite, colorless. Tainted flesh may be deprived of its odor by this means, and foul water be purified by filtration through charcoal. The substance commonly employed to decolorize fluids is animal charcoal reduced to a fine powder. It loses the property of absorbing coloring matters by use, but partially recovers it by being heated to redness.

At very high temperatures charcoal has a higher affinity for oxygen than any other substance, and is therefore often heated with oxides of the metals to deoxidize them, or deprive them of their oxygen.

Lampblack is minutely divided carbon, prepared by burning rosin or tar in a confined portion of air, so that the hydrogen only of the material is consumed, and the carbon remains as an exceedingly fine powder. It is used as a pigment, and for other purposes.

Gas coal is a deposite of nearly pure carbon upon the inside of the large retorts used in the manufacture of illuminating gas. It is very hard and black, and a good conductor of electricity.

Uses.—Carbon is used as fuel; in forming gunpowder; as a pigment; in the formation of steel; as a polishing-powder; and in medicine as an antiseptic, &c., &c.

QUESTIONS.—What is said of the affinity of charcoal for oxygen at high temperatures? What is lampblack? What use is made of it? What is gas coal?

Compounds of Carbon and Oxygen.

300. Carbon combines with oxygen in two proportions, forming carbonic oxide, CO, and carbonic acid, CO_2.

301. Carbonic Oxide, Protoxide of Carbon—CO; eq., $(6 + 8 =)$ 14.—This is a gaseous substance, and is best prepared by heating a mixture of equal parts of dry powdered chalk and iron-filings in a gun-barrel. The chalk, which is carbonate of lime, when heated, gives off its carbonic acid (the compound next to be described) in contact with the heated iron, by which one-half of its oxygen is instantly absorbed, and the carbonic oxide thus produced passes on, and may be collected over water. Thus,

$$CaO,CO_2 + Fe = CaO + FeO + CO.$$

Another method of preparing it, is to heat gently a mixture of oxalic acid and five or six times its weight of oil of vitriol, by which both carbonic oxide and carbonic acid are produced in equal volumes; but the

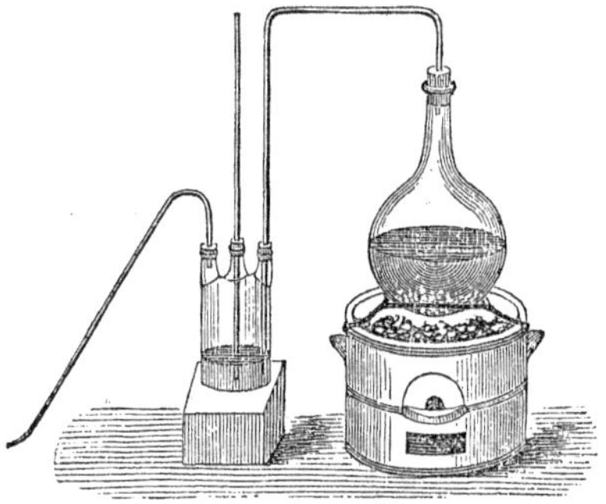

Preparation of CO.

latter may readily be separated by passing it through a solution of caustic potash or milk of lime, in the three-necked bottle, and collected over water. The changes which take place are as follows, viz:—

$$C_2O_3,3HO + SO_3,HO = SO_3,4HO + CO_2 + CO.$$

The density of the gas is about 0·97; 100 cubic inches weighing 30·20 grains. It is highly combustible, and burns with a beautiful blue flame.

QUESTIONS.—300.—What compounds of carbon and oxygen are there? 301. Describe the mode first mentioned for preparing carbonic oxide. The second mode. Is carbonic oxide combustible? What is the color of the flame?

It will not support respiration or combustion; and a lighted candle being immersed in it, as heretofore described in connection with hydrogen (198), is instantly extinguished.

It is this gas which gives the blue flame seen about the fire of the blacksmith, and in anthracite stoves, when the door is suddenly opened soon after the fire has been kindled, and in furnaces for the reduction of the metals from their ores.

Carbonic oxide is believed to be composed of 1 vol. of carbon vapor and 1 vol. of oxygen combined without condensation. Thus,

1 vol. of carbon vapor weighs	·836
1 vol. of oxygen "	1·106
2 vols. of carbonic oxide "	1·942
1 vol. of the oxide therefore "	·971

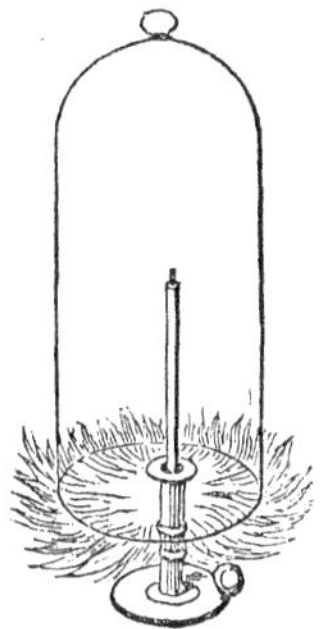

CO is Combustible.

302. Carbonic Acid—CO_2; eq., (6 + 16 =) 22.—Carbonic acid is remarkable as being the first gaseous substance recognised, after atmospheric air, which must always have been known. It was first described by Dr. Black, in 1757, and called, by him, *fixed air*, because he found it *fixed* in common limestone and magnesia; from which it may be expelled by heat, or by the action of any strong acid. It may be collected over water, but a portion will be absorbed. A gas-bottle, of the form shown in the figure, is convenient for preparing it. Some fragments of marble, and water, are placed in the bottle, and the cover put on, and then strong hydrochloric acid is poured into the long-necked funnel.

Preparation of CO_2.

As thus prepared, carbonic acid is a colorless, inodorous gas, of specific gravity 1·52; 100 cubic inches weighing 47·14 grains.

It is considered a compound of 1 vol. of carbon vapor and 2 vols. of oxygen condensed to 2 vols. Thus,

Questions.—302. When was carbonic acid discovered? What was it first called? How may it be prepared? What is its specific gravity? What is it composed of?

1 vol. of carbon vapor weighs		·836
2 vols. of oxygen	" (1·106 × 2)	2·212
2 vols. of carbonic acid "		3·048
The weight of 1 vol. therefore is		1·524

Or we may consider it a compound of 2 vols. of carbonic oxide and 1 vol. of oxygen, condensed to 2 vols., as follows:

2 vols. of carbonic oxide weigh	(·971 × 2)	1·942
1 vol. of oxygen	"	1·106
Giving for the weight of 2 vols. of CO_2,		3·048
And for the weight of 1 vol., as before,		1·524

Carbonic acid is so much heavier than atmospheric air, that it may be poured from one vessel to another without difficulty. Let a bottle, with a wide mouth, be filled with the gas, and then plunge into it a piece of lighted paper, or other substance, so that some smoke may be mixed with it and render its motions visible. Then hold the bottle in the hand, as if pouring a liquid from it (as represented in the figure), and the motion of the gas, as it is emptied from it, will be made apparent to the eye.

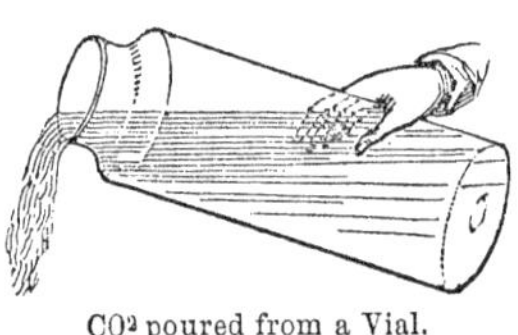

CO^2 poured from a Vial.

Another instructive experiment of the same character may be performed as follows: Provide a glass jar with a large mouth, and place at the bottom a lighted taper, as shown in the figure. Then having filled another jar of about equal capacity with carbonic acid gas, carefully remove the cover and gradually pour the contents into the first-mentioned jar;—the flame of the taper will first be considerably disturbed by the motion occasioned by the descending gas, and will finally be extinguished.

Candle Extinguished.

QUESTIONS.—How may the high specific gravity of carbonic acid be shown by pouring it from a vessel? How may the flame of a candle be extinguished by it?

By a pressure of 36 atmospheres, at 32°, it is converted into a beautiful transparent liquid, which may be frozen by intense cold, in the manner already explained.

303. It is capable of supporting neither combustion nor respiration;—a burning candle plunged into it is instantly extinguished; and a living animal, thrown into a vessel containing it, even though considerably diluted with air, soon dies. Carbonic acid is always produced by ordinary combustion; and lives have often been lost by persons placing an open dish of burning charcoal in their bed-rooms before retiring to rest. The oxygen of the air in the room is taken up by the carbon, and the gas in question takes its place, producing the effects described. It is produced, also, by the decay of animal and vegetable substances, and sometimes is found collected in caves and wells, and is called *choke-damp*.

Though this gas does not support combustion, as the experiment is ordinarily made, yet potassium, sodium, and some other of the metals may be made to burn in it. For this purpose, fill a flask with dry carbonic acid gas, and drop into it a small piece of potassium, and apply the heat of a lamp at the point where the metal lies, by means of a blowpipe. When it has become very hot the metal takes fire and burns brilliantly, the carbon of the carbonic acid decomposed being deposited upon it as a black powder.

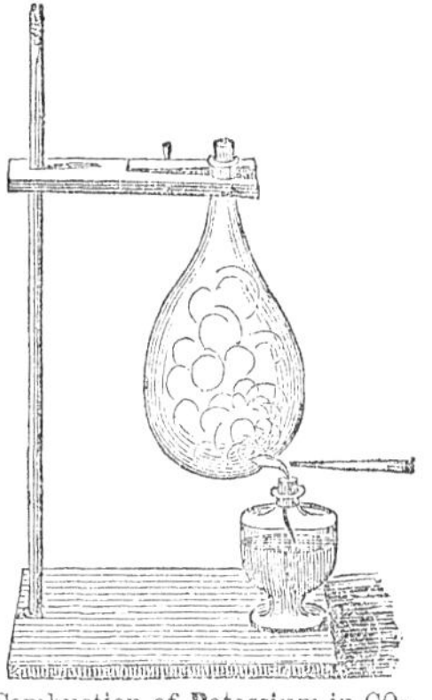

Combustion of Potassium in CO_2.

304. *Soda-fountains* are formed by compressing a large quantity of this gas in water, contained in a strong vessel adapted to the purpose. When the tube leading from the fountain is opened, the water is forced out by the pressure, and effervesces violently by the escape of the gas. Soda-powders, &c., often used to produce an agreeable drink, in the absence of a soda-fountain, consist of bicarbonate of soda and tartaric acid, which, when mingled

QUESTIONS.—May carbonic acid be compressed to the liquid form? 303. Is it always produced in ordinary combustion? How may potassium be made to burn in it? What becomes of the carbon of the carbonic acid? 304. How are soda-fountains formed?

together in solution, produce, by chemical action, tartrate of soda, the carbonic acid passing off into the air with effervescence. So, also, the effervescence which takes place on opening a bottle of beer, cider, or champagne wine, is owing to the escape of this gas, which has been produced by the fermentation of the liquid. All kinds of spring and well-water contain it in small quantity, and become insipid to the taste by boiling, in consequence of the gas having been expelled. It is also always present in the atmosphere, and in some cases accumulates in considerable quantities, as at the *Grotto del Cane*, in Italy, through which a man may pass without danger, but a dog on entering it is instantly suffocated. The carbonic acid here constantly issues from the earth, and accumulates at the bottom, while the air above remains comparatively pure.

Blowing through Lime-water.

305. *Lime-water* is an excellent test for carbonic acid; and a vessel of it being allowed to stand a few hours, becomes coated with a pellicle of carbonate of lime, by absorbing this gas from the air. So lime-water becomes milky by blowing into it with a tube from the lungs, for the same reason. A portion of the lime is changed into carbonate of lime, which is insoluble, and gives the water its milkiness.

Compounds of Carbon and Hydrogen.

306. Carbon and hydrogen combine in a number of different proportions, producing compounds, several of which are of special interest, because of their isomeric character; but we shall here describe only two, both of which are gaseous, viz., *light carburetted hydrogen*, C_2H_4, and *olefiant gas*, C_4H_4.

QUESTIONS.—What is said of the *Grotto del Cane* in Italy? 305. How is lime-water affected by blowing through it with the mouth? Give the reason for the milkiness produced? 306. What is said of the compounds of carbon and hydrogen?

307. Light Carburetted Hydrogen—C_2H_4; eq., (12 + 4 =) 16.—This gas, called also *fire-damp, marsh gas, hydrocarburet*, and *dicarburet of hydrogen*, is formed by the slow decomposition of wood, and woody substances, under water, especially in warm weather; and may be obtained by stirring the mud and other matters at the bottom of stagnant pools (see figure), and collecting the bubbles of gas in a receiver, as they rise. It sometimes accumulates in large quantities in coal mines, where it is formed by the action of water upon the coal.

Collecting Marsh Gas.

It is best prepared by heating in a flask, made of hard glass, a mixture of 2 parts of acetate of soda, 3 parts of caustic potash, and 3 of quick-lime. The composition of the acetic acid is $C_4H_4O_4$, which, it will be perceived, is precisely equal to 2 eq. of carbonic acid, and 1 of the hydrocarburet in question. Thus,

$$C_4H_4O_4 = 2CO_2 + C_2H_4.$$

The use of the lime is to protect the gas from the action of the potash.

Light carburetted hydrogen is a colorless, transparent gas, 100 cubic inches of which weigh 17·37 grains, giving it a specific gravity of 0·56.

One volume contains 2 vols. of hydrogen, ½ of a vol. of carbon vapor. Thus,

2 vols. of hydrogen weigh (·069 × 2)	·138
½ vol. vapor of carbon $\left(\frac{\cdot 836}{2}\right)$	·418
1 vol. of the hydrocarburet,	·556

A burning candle is extinguished by it, but it is, itself, highly combustible, and burns with a feeble, yellow flame. Mixed with twice its own volume of oxygen, or seven or eight times its volume

QUESTIONS.—307. In what situations is light carburetted hydrogen naturally formed? How may it be collected? How may it be prepared artificially?

of air, it explodes violently by the electric spark, or on the approach of flame.

308. Olefiant Gas, or Heavy Carburetted Hydrogen—C_4H_4; eq., (24 + 4 =) 28.—This gas was first described in 1796, by some Dutch chemists, who gave it the name, *olefiant gas*, because of its forming with chlorine a peculiar oil-like liquid. It is colorless and tasteless, and but slightly absorbed by water; 100 cubic inches weigh 30·41 grains, so that its density is 0·98.

Its volume contains 2 vols. of hydrogen, and 1 vol. of vapor of carbon, as follows:

2 vols. of hydrogen weigh (·069 × 2)	·138
1 vol. of vapor of carbon,	·836
Giving for 1 vol. of olefiant gas,	·974

Olefiant gas is prepared by mixing, in a capacious retort, one part of alcohol with four of concentrated sulphuric acid, and heating the mixture, as soon as it is made, by means of a lamp or ignited charcoal. The acid soon acts upon the alcohol, effervescence ensues, and olefiant gas passes over, mixed with other substances, chiefly sulphurous acid, from which it may be purified

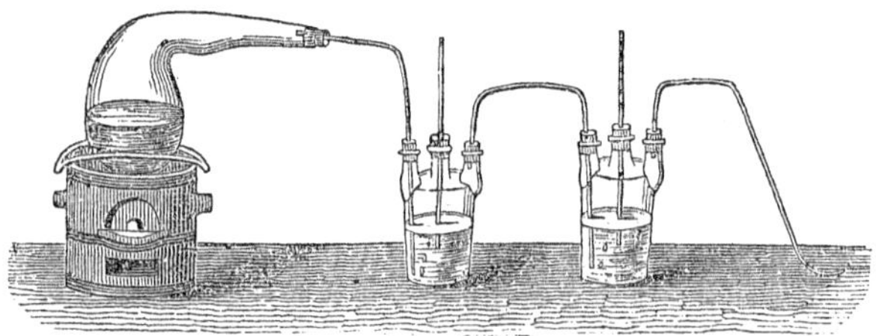

Preparation of Olefiant Gas.

by washing it with solution of lime or caustic potassa in several of Woulf's bottles, as shown in the figure.

As might be expected, olefiant gas does not support combustion; but a jet of it burns in the air, or in oxygen gas, with a

QUESTIONS.—Does light carburetted hydrogen form explosive mixtures with oxygen or atmospheric air? 308. By whom was olefiant gas discovered? Why was it so called? Describe it. How is it prepared? What is said of the light produced by its combustion?

brilliant white light. Mixed with oxygen, or air, in proper proportions, it explodes violently, like the preceding compound.

309. Illuminating Gas is usually a mixture of olefiant and light carburetted hydrogen gases, and is formed by distilling, in large cast-iron retorts, rosin, tar, or other resinous or oily substances, or bituminous coal. Besides the gases mentioned, there are also formed other hydro-carbons, but in less quantity. Illuminating gas is used in immense quantities in large cities, for lighting the streets, and for fixed lights in stores and other buildings.

Illuminating gas is now chiefly prepared from bituminous coal, and as it passes from the retorts is mixed with tar, carbonic acid, sulphuretted hydrogen, salts of ammonia, and other matters, from which it must be freed before being admitted to the burners. For this purpose, it is washed by passing it through water, and then further purified by passing through vats containing recently-slaked lime.

If this gas is made to pass through a heated tube it is decomposed, and a part of its carbon is deposited as a coating upon the inside of the tube. In this way large deposits of pure carbon are often formed in the retorts of gas-works. It has been described above (299) as *gas carbon.*

310. *The History* of gas manufacture, for illuminating purposes, possesses much interest, as showing the great benefit conferred by science on the arts, and domestic and public economy. In 1680, Mr. Clayton, of Yorkshire, England, observed that a brilliant light was produced by igniting the gas which issued from a close vessel containing bituminous coal when heated, but it was a century after this before any direct experiments were made with it, with reference to its use in the arts. In 1785, the preparation of gas for illumination, from the destructive distillation of wood, was suggested; but, in 1792, some buildings were actually illuminated with gas, in Cornwall, England; and the same thing was repeated in 1798, at a foundry in Birmingham. In 1805, some of the cotton-mills in Manchester were first lighted with gas, by means of permanent fixtures, prepared for the purpose; and this date may be assumed as the beginning of the use of gas-lights for practical purposes. In a half century, therefore, has this manufacture attained its present importance; and the time is not distant when the quantity annually consumed in every civilized country will be greatly increased.

QUESTIONS.—309. What is illuminating gas? How is it usually prepared? What is the substance now generally used for producing it? 310. What is said of the history of the use of illuminating gas for practical purposes?

Nature of Flame.—The Safety Lamp.

311. What we term the combustion of a substance is occasioned by its entering into combination with some other substance, usually the oxygen of the atmosphere, and then taking another form as a compound (193). Of the two substances thus required to produce combustion, one is called the *combustible* substance, and the other the *supporter* of the combustion; but the action is really mutual between them, and neither can burn without the other. The supporter of ordinary combustion, oxygen, is always gaseous, but the combustible may be either solid, liquid, or gaseous.

Flame is gaseous matter in a state of combustion, and is made incandescent by the intense heat of the combustion. Two gases are needed to produce it, one of which must be combustible, and the other, of course, a supporter of combustion. The action is mutual between them;—neither will burn alone;—and a jet of either will burn in the other.

In the common lamp or candle, the combustible gases are supplied from the oil, or tallow, which is gradually raised, by the capillary action of the wick, into the flame, where it is decomposed by the heat. As these gases, thus produced, escape from the wick, and come in contact with the oxygen of the atmosphere, the two combine, producing the phenomena of light and heat, with which all are familiar. A careful inspection of the flame of a lamp or candle, as it burns quietly, will show, that it is composed of three parts, viz:—1st, a central part, *a*, surrounding the wick, and extending a little above it, of gaseous matter that has emerged from the wick, and is making its way outward to the atmosphere, which it has not yet reached, and therefore has not yet become ignited; 2d, the bright part of the flame, *b b*, which, in the form of a conical shell, incloses the part *a*, and consists of gaseous matter in a state of rapid combustion, the combustible particles, as they reach the air, uniting

Flame of Candle.

QUESTIONS.—311. What is ordinarily termed combustion? Must there be two substances to produce combustion? What are they called? What is flame? In the common lamp or candle, how are the combustible gases supplied? Of what several parts is the flame of a candle composed? What is the dark interior part?

with its oxygen, with the evolution of much light and heat; and, 3d, the part, *c c*, outside of the part last mentioned, composed chiefly of heated air, and mixed with a small portion of combustible matter in a state of ignition.

That the dark, interior portion, *a*, is composed of combustible gas, may be shown by inserting, in the centre of the flame, one end of a small glass tube, as shown in the figure, and conveying away a portion, and igniting it as it escapes at the other end. So, when the flame of a candle is suddenly extinguished, the heat in the wick continues, for a short time, sufficient to decompose the tallow, and the combustible gases continue to rise in the form of smoke; and may often be again relighted by applying the flame of another candle to the ascending smoke, several inches above the wick.

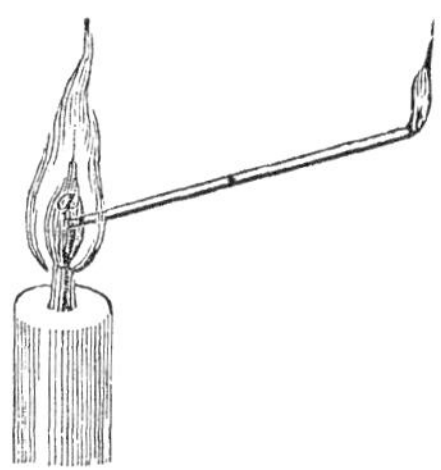

Gas from centre of Flame.

In the flame of a jet of gas, precisely the same phenomena, in every particular, will be observed;—the dark central cone of unconsumed gas, surrounded by the brilliant hollow cone of flame, and this enveloped in still another less brilliant cone. In the latter case, the gas is previously formed and consumed as it issues into the air, but in the case of the candle or lamp, it is formed in the wick, and instantly consumed as it escapes.

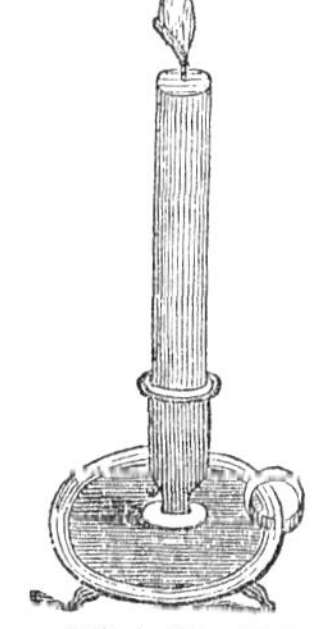

Effect of Braided Wick.

312. That there is really no combustion in the dark central part of the flame, appears from the fact that the wick remains there unconsumed, and is even protected by the gas existing there from being attacked by the oxygen of the air. In burning ordinary tallow candles, the wick occasionally becomes too long, and requires to be snuffed; but the wicks of the best spermaceti and wax candles, being plaited,

---

QUESTIONS.—How may the real character of this gas be shown? Will the same phenomena be shown in the flame of a jet of gas? 312. Why does not the wick of a candle burn off quite down to the tallow?

the end curls outward when heated by the flame, and coming in contact with the oxygen of the air, is gradually consumed as the candle burns away. The necessity of snuffing is therefore avoided.

The plaited or braided wick, while the candle is burning, will always bend toward that side in which the direction of the separate strands is upward and inward, and of course from the other side in which the direction is upward and outward. Generally the simple braiding of the wick is sufficient, but sometimes a cord or bobbin is braided in with one of the strands, as represented in one of the figures in the margin; and sometimes, also, a cord is bound to the braided wick, by a thread wound spirally around it.

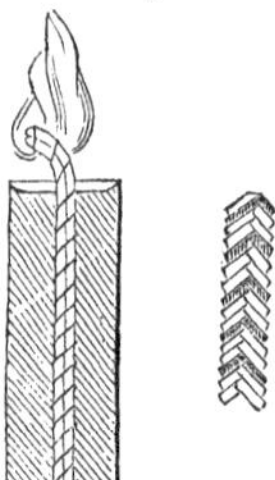

Braided Wicks.

The intensity of the light from any flame, other things being equal, will depend upon the intensity of the combustion, and this will depend in turn upon the regular and abundant supply of the combustible and the supporter. When the wick of a candle becomes too long, or that of a lamp is too high, only the hydrogen of the gases formed from the decomposed oil is consumed, the carbon escapes in a finely divided state, as a dense, black smoke. This is because of the too rapid supply of the combustible, and the usual remedy is to diminish the length of the wick by snuffing or otherwise, but the same thing would be accomplished by increasing the supply of the supporter, oxygen. This last purpose is effected by the use of a glass chimney, by which a current of air is supplied more rapidly to the flame.

By the Argand burner (so named from the inventor), a current of air is also supplied to the centre of the flame, the wick being in the form of a hollow cylinder, as shown in the figure on next page. Both through the centre of the flame and around

---

QUESTIONS.—How are the wicks sometimes made to bend outward in the air, so that the end is consumed? Upon what will the intensity of the light from a flame depend? What occasions the disagreeable smoke sometimes produced by a lamp? How is the defect usually remedied? What is the benefit of a chimney to a lamp? Why are hollow wicks often used?

the outside strong currents of air are established, as shown by the arrows. The effect is to produce a perfect combustion of all the oil or other combustible material supplied to the flame.

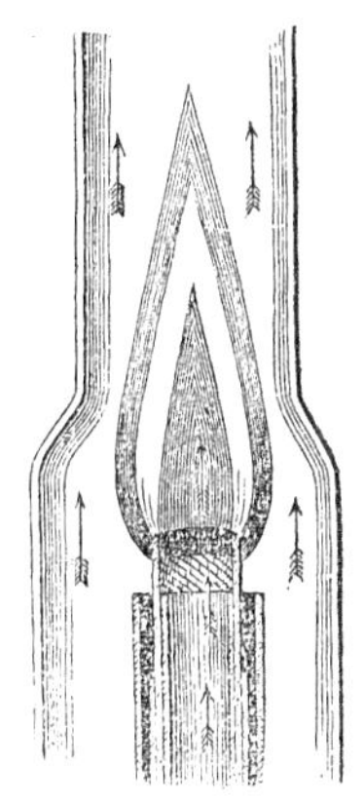

Lamp with Hollow Wick and Glass Chimney.

It is found that the intensity of light produced by a flame depends very much upon the amount of carbon consumed. The flame of pure hydrogen is very feeble, as is also that of the vapor of alcohol and ether;—and these latter contain comparatively little carbon. But add to alcohol one-fourth of its volume of camphene, which is rich in carbon, and a brilliant flame is produced. This mixture constitutes the common *burning fluid.*

313. The *blowpipe* (193) is, as we have seen, simply a contrivance to supply air to the flame from the lungs. The instrument is usually applied to one side of the flame, and as the current of air is forced through it, it is bent towards the opposite side. By means of this instrument a very intense heat may be produced, sufficient for many important purposes. Much will depend in any particular case upon the mode of using the flame; and the inexperienced student, before commencing, will consult works that treat at length of this subject.

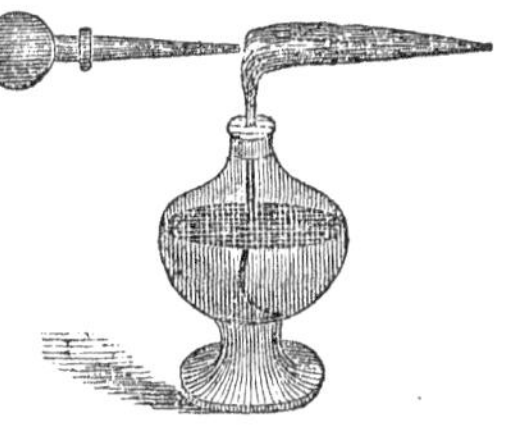

Use of Blowpipe.

314. Safety-Lamp.—The safety-lamp is the invention of Sir H. Davy, to avoid the danger of explosions from mixtures of the above gases with air, which often occur in coal mines, when unprotected lamps are made use of. It consists simply of a com-

---

QUESTIONS.—What is the composition of *burning fluid?* 313. Describe the mouth blowpipe. May a very intense heat be produced by it? 314. For what special purpose was the *safety-lamp* invented?

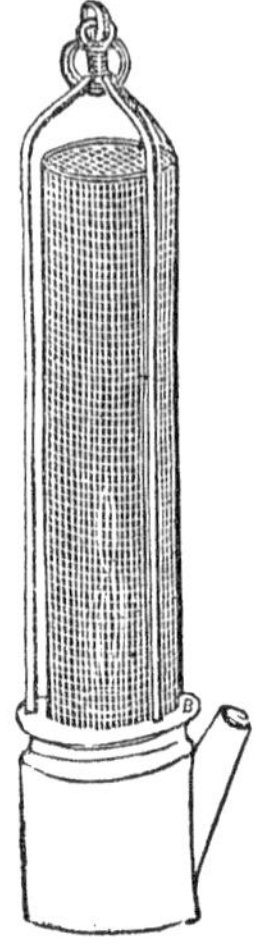
Safety-Lamp.

mon lamp, the flame of which is surrounded by wire gauze, through which, it is found, flame will not ordinarily pass. This lamp, as it is usually constructed, is represented in the figure on the left. Its action in arresting flame is easily understood when the nature of flame is considered. We have seen that this is simply gaseous matter in a state of combustion, and therefore is intensely heated;—now if by any means we can diminish the heat of this gaseous matter, so that it shall fall below the point of ignition, the combustion, and consequently the flame, must cease. And this effect is produced by the wire gauze.

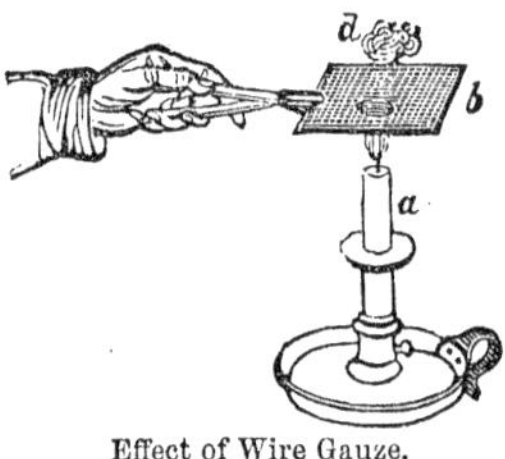

Effect of Wire Gauze.

To show the effect, let a piece of such gauze, *b*, be held in the flame of a candle, *a*; the flame appears to be cut off by the gauze, and the gases pass through unconsumed, as shown at *d*, and, by dexterous management, may be relighted.

The occurrence of combustible gases in coal mines is occasioned by the action of water upon the coal; and they often collect in large quantities in places not properly ventilated, forming mixtures with the air, ready to explode with excessive violence on the first approach of the unprotected candle of the miner. Before the invention of the safety-lamp, such accidents were of frequent occurrence; and coal miners lived and worked in perpetual fear! All this is avoided by the use of the safety-lamp;—and, in view of what has been said above, the mode in which it operates to afford the desired protection is easily understood. When this

QUESTIONS.—Describe the safety-lamp. What is flame? Will ordinary flame pass through wire gauze? What is the reason given for this fact? How may this be shown by a lighted candle and a piece of wire gauze? How are inflammable gases formed in coal mines? What is the effect when these gases, mixed with atmospheric air, come in contact with the flame of a lamp? What is the effect when the safety-lamp is used?

lamp is carried into an atmosphere containing a considerable proportion of fire-damp, this latter immediately takes fire, and burns freely within the gauze, but the flame is not communicated to that without. At first, the flame of the lamp seems to be simply enlarged, but soon it leaves the wick entirely, and the whole space immediately inside the gauze seems filled with flame. When this takes place, the miner is obliged to retire, lest, by the intense heat, the wire of the gauze should be melted or oxydized, and the flame communicated to the mixed gases, without. The same effect might also be produced by strong currents of air, which sometimes occur, forcing the concentrated flame against a particular part of the gauze, and causing it to break, or heating it so as to allow the passage of flame through it.

The mode in which the safety-lamp operates may be shown quite well, by pouring a little sulphuric ether into a common glass receiver, which should be inverted and agitated a little, so that it may be filled with a mixture of air and vapor of ether, and then letting the lighted lamp down into it. The mixture of air and vapor of ether entering through the gauze, burns brilliantly within the gauze, but the flame is not communicated to that without.

It should be remarked, that wire gauze serves as a protection against explosive mixtures of atmospheric air and the carburetted hydrogen only; a mixture of atmospheric air or oxygen with pure hydrogen may be exploded through a very narrow tube of great length.

Compounds of Carbon and Nitrogen.

315. There are several compounds of these two substances, but we shall notice only one, the *bicarbonide of nitrogen*, C_2N, or *cyanogen* (from *kuanos*, blue, and *gennao*, to produce, because it is an ingredient of Prussian blue).

316. Bicarbonide of Nitrogen, or Cyanogen—C_2N, or Cy; eq., $(12 + 14 =)$ 26.—This is a gaseous substance, and is readily formed by heating bicyanide of mercury (to be hereafter described) in a glass retort by a

QUESTIONS.—Will it answer for the miner to remain with his lamp in the explosive mixture? Describe the experiment for showing the operation of the safety-lamp. Will wire gauze in this manner prevent the explosion of mixtures of hydrogen and oxygen? 315. What is said of the compounds of carbon and nitrogen? From what does cyanogen derive its name? 316. How is bicarbonide of nitrogen, or cyanogen prepared?

spirit-lamp. It is colorless, has a very pungent odor, and is easily compressed into a liquid; and by the cold produced by a mixture of solid carbonic acid and sulphuric ether, this liquid may be frozen. Of the pure gas, 100 cubic inches weigh 56·47 grains, giving it a density of 1·82.

Cyanogen is composed of 1 vol. of vapor of carbon and 1 vol. of nitrogen, as follows, viz.:

1 vol. vapor of carbon weighs	0·836
1 " nitrogen "	0·972
Weight of 1 vol. of cyanogen,	1·808

Cyanogen, though a compound, is remarkable for combining with the elementary bodies in the same manner as an element, forming a class of compounds which are called *cyanides*. Further remarks concerning it will be deferred to Organic Chemistry.

Compound of Carbon and Sulphur.

317. These elements combine in only one proportion, forming the following compound:

318. Bisulphide of Carbon—CS_2; eq., (6 + 32 =) 38.—This compound, called also *sulpho-carbonic acid*, is formed by heating to redness pieces of charcoal in a porcelain tube or retort, and then causing vapor of sulphur to come in contact with it. The figure on next page represents a good arrangement for the purpose, even when a considerable quantity is to be prepared. The retort, R, of stone ware, is first filled with pieces of well-burned charcoal, and the tube T inserted nearly to the bottom, and luted well around the neck, to prevent any escape of gaseous matter. It is then to be placed in a suitable furnace, and connected with a Liebig's condenser, C, and this with a receiving vial, V, partly filled with water. A small tube of glass inserted into the cork allows the air to escape as the vial is filled.

QUESTIONS.—Describe bicarbonide of nitrogen or cyanogen. For what is it remarkable? 317. How many compounds of carbon and sulphur are known? 318. Describe the mode of preparing bisulphide of carbon, or sulpho-carbonic acid.

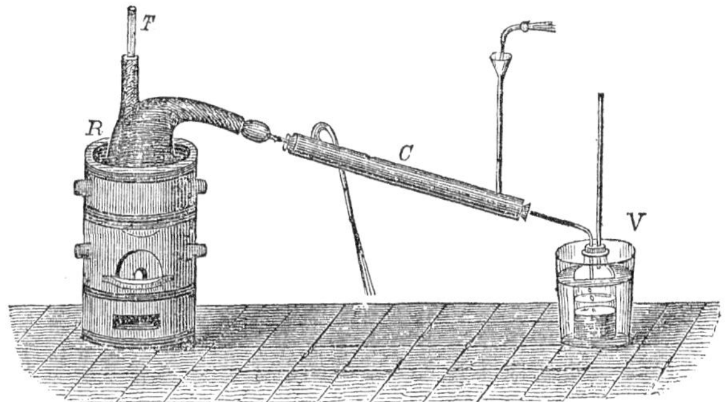

Preparation of CS_2.

When everything is ready, a fire is kindled in the furnace, and when the retort becomes well heated, pieces of sulphur are occasionally dropped into the tube T, and a cork immediately inserted. The sulphur being sublimed by the heat comes in contact with the heated charcoal, with which it combines, producing the compound in question. This now passing into the condenser, takes the liquid form, and is collected in the receiver, V.

It will be observed that this process is a case of combustion (193), the carbon being the combustible, and the vapor of sulphur the supporter; and the compound formed, CS_2, is analogous to carbonic acid, CO_2, which is formed when carbon is burned in oxygen. And as carbonic acid combines with oxybases, to form salts, the composition of which is RO,CO_2, so sulphide of carbon combines with sulphur bases, to form compounds of the analogous form, RS,CS_2. The letter R is here used indefinitely, for any element whatever.

Bisulphide of carbon is a colorless liquid, which boils at about 112°, and has a density of 1·27. It does not mix with water, but dissolves readily in alcohol or ether. Its odor is excessively fetid and nauseous. Both sulphur and phosphorus are dissolved in it, and may be obtained in crystals by the gradual

QUESTIONS.—What is said of the relation of bisulphide of carbon to carbonic acid? May this be considered as a case of combustion? Which of the substances is to be considered as the combustible body, and which the supporter? What are some of the properties of bisulphide of carbon?

evaporation of the solution. By its evaporation in the open air a very considerable cold is produced, but under the receiver of the air-pump its evaporation is so rapid as to occasion a cold of $-76°$. It burns freely in the open air, producing carbonic and sulphurous acids, and with oxygen its vapor forms an explosive mixture.

SILICON.

Symbol, Si; *Equivalent*, 21·3; *Density*, ?.

319. History.—Silicon was first obtained by Berzelius, in 1824. It was then considered a metal, and named *silicium*, but is now generally ranked with the non-metallic elements. It is never found in its separate state in nature, although it is really very abundant in every place in silicic acid (silica), and the various siliceous compounds which constitute the rocks and soils. Next to oxygen it is the most abundant element found upon the earth.

320. Preparation.—To prepare silicon a somewhat complex substance is selected, the double fluoride of silicon and potassium, which is a white powder like starch.

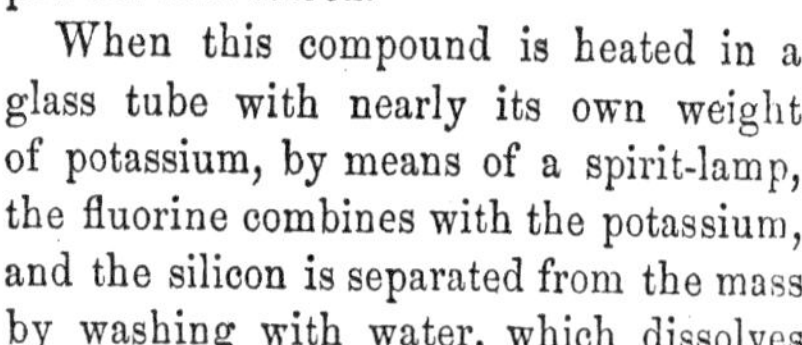

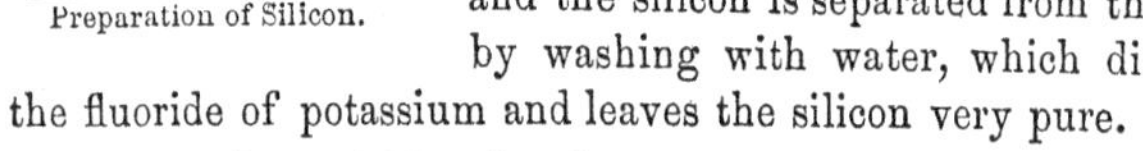

Preparation of Silicon.

When this compound is heated in a glass tube with nearly its own weight of potassium, by means of a spirit-lamp, the fluorine combines with the potassium, and the silicon is separated from the mass by washing with water, which dissolves the fluoride of potassium and leaves the silicon very pure.

The reactions which take place are represented by the following equation:

$$3KF,2SiF_3 + 6K = 9KF + 2Si.$$

QUESTIONS.—319. Has silicon sometimes been ranked with the metals? Is it abundant in nature? 320. How may it be prepared?

321. Properties.—Silicon, as thus obtained, is a powder of a dark, nut-brown color, without metallic lustre, and is a non-conductor of heat and electricity. It stains the fingers, and adheres to everything that comes in contact with it; and when heated in the atmosphere or oxygen gas, it takes fire and burns, with the formation of silicic acid.

In close vessels, silicon, like charcoal and boron, is capable of enduring a very high temperature without fusion, but is rendered harder and more compact. It is now incombustible, even when highly heated in the air or in oxygen gas, and is unaffected by the blowpipe, even in contact with chlorate of potassa.

Compound of Silicon and Oxygen.

322. There is known only a single compound of these elements, the teroxide, SiO_3.

323. Silicic Acid, or Silica—SiO_3; eq., (21·3 + 24=) 45·3. —This is one of the most abundant compounds in nature, and is found quite pure in quartz, flint, calcedony, agate, &c.; and, in combination with other substances, in the material of all soils, and nearly all rocks.

Alone, silica is nearly infusible, but may be melted in the intense heat of a compound blow-pipe. It is not acted upon by any acid except the hydrofluoric (252), but, at high temperatures, enters into combination with the alkalies and earths. It is very hard, and has a harsh feeling to the fingers, even in fine powder. In quartz crystals, which are usually six-sided prisms, terminated by six-sided pyramids, it is familiar to every one.

By heating silica in fine powder, with four times its weight of carbonate of potash, in a platinum crucible, the carbonic acid is displaced, and silicate of potash formed; which, being treated

---

QUESTIONS.—321. Describe the properties of silicon. 322. What compound only of silicon and oxygen is there? 323. What is said of the abundance of silicic acid, or silica? What is said of its fusibility? Is it acted upon by any of the acids? What is the usual form of its crystals? What is the effect when silica in fine powder is heated with carbonate of potash?

with diluted hydrochloric acid, yields a gelatinous precipitate of hydrated silica. This is soluble in water, and constitutes *soluble glass*, or *liquor of flints*,—terms sometimes used. It is sometimes found in the waters of hot springs, as the Geysers of Iceland; and, in fact, in very small quantities, in most natural waters. In this state it is taken up by the rootlets of plants, and is found in their ashes after incineration. It is especially abundant in the grasses and the straw of the cereal grains, and in the stalks of rushes and reeds.

In the gelatinous state, silica has occasionally been found in the cavities of minerals, when they have been broken. Exposed to the air, and especially if heated, the water evaporates, and dry, harsh, insoluble silica only remains.

All the different varieties of *glass* are silicates of different bases, or mixtures of these silicates. This subject will be treated of more at length hereafter.

It is in its power thus to combine with the bases, so as perfectly to neutralise them, that this compound, oxide of silicon, or silica, evinces its claims to be considered as an acid.

Other Compounds of Silicon.

324. Terchloride of Silicon—$SiCl_3$; eq., (21·3 + 106·2 =) 127·5.—This compound is formed by heating silicon in dry chlorine, or by passing a current of the latter over a mixture of silica and carbon when heated in a porcelain tube. It is a colorless, volatile liquid, which boils at about 138°, and has a density of about 1·52. By contact with water it is decomposed, producing silica and hydrochloric acid.

325. Terfluoride of Silicon—SiF_3; eq., (21·3 + 57 =) 78·3.—This fluoride, called also, *fluo-silicic acid*, is obtained by gently heating a mixture of equal parts of dry and finely-powdered glass and fluor-spar, made into a paste with strong oil of vitriol. The oxygen of the silica and the fluorine of the fluor-spar here exchange places, while the sulphuric acid combines with the lime that is formed, thus, neglecting the water of the sulphuric acid,

$$3CaF + SiO_3 + 3SO_3 = 3(CaO,SO_3) + SiF_3.$$

QUESTIONS.—Is silica found in nature? Is it taken up by the roots of plants? Of what are all the different kinds of glass composed? 324. How is the terchloride of silicon formed? Describe its properties. 325. Describe the terfluoride of silicon.

Terfluoride of silicon is a colorless gas, with a pungent, acid odor, and in the open air forms dense white fumes by combining with the moisture it contains. Its density is 3·57.

By contact with water it is at once decomposed, gelatinous silica is deposited, and the water is found to contain a peculiar compound, called *hydrofluosilicic acid*, the composition of which is $3HF, 2SiF_3$.

The experiment is best performed by putting the materials in a dry retort, A, connected with a receiver, R, partly filled with water and

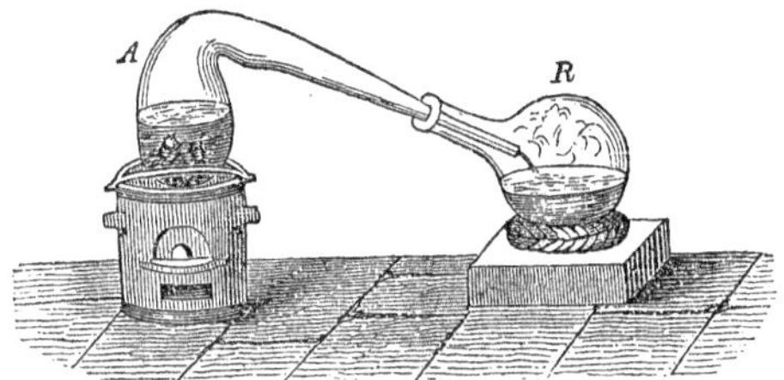

Preparation of Gelatinous SiO_3.

placed so that it can be gently shaken occasionally, in order that the film of silica forming upon the surface may be broken up, and a fresh surface of water exposed.

When hydrofluosilicic acid is saturated with a base, as potassa, the hydrogen is replaced by an equivalent quantity of the metal of the base. Thus,

$$3HF, 2SiF_3 + 3KO = 3KF,2SiF_3 + 3HO.$$

The compound $3KF,2SiF_3$ is the double fluoride of silicon and potassium (320) used in the preparation of silicon.

Silicon is capable also of forming compounds with sulphur and bromine.

BORON.

Symbol, B; *Equivalent*, 10·9; *Density*, 2.

326. History.—Boron, in a separate state, was first obtained by Davy, in 1807. It is found in nature only in combination, and in comparatively small quantities. Though it occurs in

QUESTIONS.—How is terfluoride of silicon affected by contact with water? What is the composition of hydrofluosilic acid? 326. When and by whom was boron discovered? Is it found naturally in a separate state?

several minerals, as datholite and boracite, it is obtained chiefly from the waters of certain hot springs, especially in Tuscany and the Lipari Islands, where it occurs in solution, as boracic acid.

327. Preparation.—To prepare boron, first make a saturated solution of borax in boiling water, and, while hot, pour in sulphuric or hydrochloric acid until the solution tastes distinctly sour. As it cools the boracic acid will separate in white, shining scales, which are to be thoroughly washed and dried by fusion in a platinum crucible. We thus obtain boracic acid, which is then to be mixed with potassium (or sodium), and heated in a glass tube, as in the preparation of silicon. A part of the boracic acid yields its oxygen to the potassium, and the potassa so formed enters into combination with the rest of the boracic acid to form borate of potassa. By treating the mass with cold water the latter compound is dissolved, and the boron is seen floating in the liquid as a fine powder of a brownish color, and may be collected on a filter. Before filtering, some sal ammonia should be dissolved in the mixture, which has the effect to prevent the escape of the finely-powdered boron through the filter.

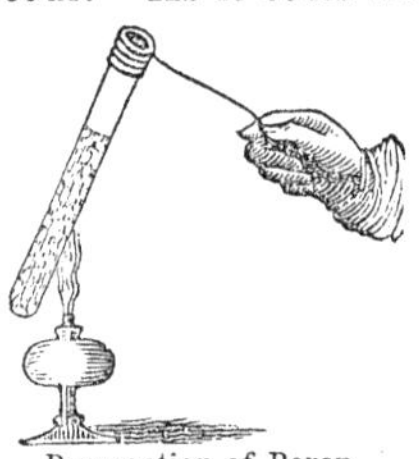
Preparation of Boron.

The mode in which the sal ammonia operates to produce this effect is not understood.

328. Properties.—Boron is a dark olive-green powder, without taste or smell, and incapable of fusion in the strongest heat. It is insoluble in water or alcohol. Heated in the open air to about 600°, it takes fire and burns brilliantly, forming boracic acid. It is a non-conductor of electricity and heat.

Questions.—327. How is boron prepared? Describe the reactions which occur. 328. Describe boron. How is it affected when heated?

Compound of Boron and Oxygen.

329. Boracic Acid—BO_3; eq., (10·9 + 24 =) 34·9.—This acid may be obtained from the common borax of commerce, as just described in the preparation of boron; it is also produced by the combustion of boron in the air.

Boracic acid is slightly soluble in water and in alcohol; and when the latter solution is inflamed it communicates to the flame a beautiful green tinge, which is characteristic of this substance. To make the experiment, fill a common dropping tube, of the form A B in the figure, with the solution; and, inserting a cork firmly in A, apply the heat of a lamp to the bulb, and inflame the jet of liquid as it issues from the capillary orifice, B. The flame will be of a beautiful green.

Green Flame produced by BO_3.

The color of the flame may also be shown very well simply by moistening one end of a pine stick in the solution, and inflaming it.

Other Compounds of Boron.

330. Terchloride of Boron—BCl_3; eq., (10·9 + 106·2=) 117·1.—This compound is formed by preparing an intimate mixture of boracic acid and carbon, and causing a current of dry chlorine to pass over it in a porcelain tube, kept at a red heat. It is a colorless gas, of a density of 4·03;—in the open air it forms dense white fumes with the moisture contained therein, and in contact with liquid water it is instantly decomposed, forming hydrochloric and boracic acids.

331. Terfluoride of Boron—BF_3; eq., (10·9 + 57 =) 67·9.—Terfluoride of boron, or *fluoboracic acid*, is a colorless gas, of specific gravity 2·37. It is prepared by heating in a gun-barrel a mixture of 2 parts of pul-

QUESTIONS.—329. How is boracic acid procured from borax? Describe this acid. What color does it give to the flame of alcohol when dissolved in it? How may it be shown? 330. How is the terchloride of boron formed? 331. How is the terfluoride of boron formed?

verized fluor spar (fluoride of calcium), and 1 part of fused boracic acid. It has a strong affinity for water, which dissolves 700 or 800 times its own volume of the gas, and acquires a sour taste and a strong acid reaction, and chars wood like oil of vitriol when brought in contact with it.

Boron combines also with sulphur, forming a tersulphide, BS_3.

THE METALS.

General Properties.

332. Characteristic Properties.—The metals are generally good conductors of electricity and heat, and possess a peculiar lustre called the metallic lustre, which can scarcely be imitated by other substances. When their compounds are decomposed by the galvanic battery, they always make their appearance at the negative electrode (112), and are therefore, without exception, to be considered as electro-positive. But these properties are much more distinct in some metals than in others. Most of the metals are also characterized by their great density, as gold and platinum; but two at least, potassium and sodium, are lighter than water. All of them except mercury are solid at ordinary temperatures.

The ancients were acquainted with only seven metals, viz.: gold, silver, iron, copper, mercury, lead, and tin; but there are now known, with certainty, forty-seven; and the discovery of several others has been announced, though perhaps not fully proved.

333. Sources of the Metals.—Though several of the metals are found in animal and vegetable substances, they seem to form no necessary part of any organic compounds. Most of them, as iron, lead, and zinc, are found in the earth in combination with other non-metallic elements, as oxygen, sulphur, or arsenic, and are said to be *mineralized*, and the compound is called an *ore*; but a few, as platinum, gold, and sometimes copper, silver, bis-

---

QUESTIONS.—332. What are some of the characteristic properties of the metals? How many metals were known to the ancients? Are all the metals solid at ordinary temperatures? 333. Do any of the metals form any part of organic compounds? What constitutes a metallic *ore?*

muth, &c., occur in the metallic state, and are said to be found *native*. Some are found in the form of salts, especially as silicates, carbonates, and sulphates.

The metals occur sometimes in *beds*, which are more or less parallel with the earthy strata containing them, but more frequently in *veins* or *lodes*, which traverse the strata in every direction. The veins are more abundant in the older than in the newer rocks; and their appearance indicates that they are fissures which have been produced in the strata subsequent to their deposition, and then filled by filtration of the metallic matter from above, or by injections from below. Besides the metallic compounds, or ores, these veins always contain other minerals, as quartz, carbonate of lime, heavy spar, and fluor spar, which constitute the *gangue*, or *vein-stone*.

334. Relations to Light.—The relations of the metals to light are in some respects peculiar and interesting. Their peculiar lustre, called the metallic lustre, has already been alluded to; this lustre entirely disappears when they are reduced to a fine powder, but reappears when the substance is rubbed with a burnisher.

All the metals except gold are perfectly opake, even when reduced to the thinnest laminæ possible. Gold in the form of leaf transmits a feeble green light. The best method to observe this light is to place an unbroken leaf upon a piece of white plate glass, and press it gently with a bunch of cotton to make it adhere. It may then be preserved for any length of time.

The *color* of most of the metals, seen in mass, is grayish-white, as platinum, iron, lead, and potassium, but seen in powder the color is a deep gray. By burnishing the particles, the original grayish-white is restored. Gold in mass has a beautiful yellow color, and in powder a deep purple, almost black. Copper and titanium are red.

335. Malleability and Ductility.—Some metals possess the property of *malleability*, that is, admit of being beaten into thin

QUESTIONS.—When is a metal said to be found *native?* 334. What is said of the lustre of the metals? Is this lustre seen in a metal when it is reduced to powder? Are most of the metals opake even when reduced to thin laminæ? What is the color of most of the metals? What is said of their color when in fine powder? What metal is yellow? What metals are red? 335. What malleable metals are mentioned?

plates or leaves by hammering. The malleable metals are gold, silver, copper, tin, platinum, palladium, cadmium, lead, zinc, iron, nickel, potassium, sodium, aluminum, and frozen mercury. The other metals are either malleable in a very small degree only, or, like antimony and bismuth, are actually brittle. Gold surpasses all the other metals in malleability; one grain of it may be extended so as to cover about 52 square inches of surface, and the film will have a thickness of only $\frac{1}{250,000}$th of an inch.

Nearly all malleable metals may be drawn out into wires, a property which is expressed by the term *ductility.* The only metals which are remarkable in this respect are gold, silver, platinum, iron and copper.

The process of wire-drawing consists in drawing the metal through round holes made in plates of steel for the purpose. The steel plate is made with a number of holes, of different diameters, through which the rod of metal is made to pass successively, its diameter at each operation being a little reduced and its length increased.

A machine for this purpose is represented by the accompanying figure. A B is the plate of steel through which the wire is drawn,—it is held

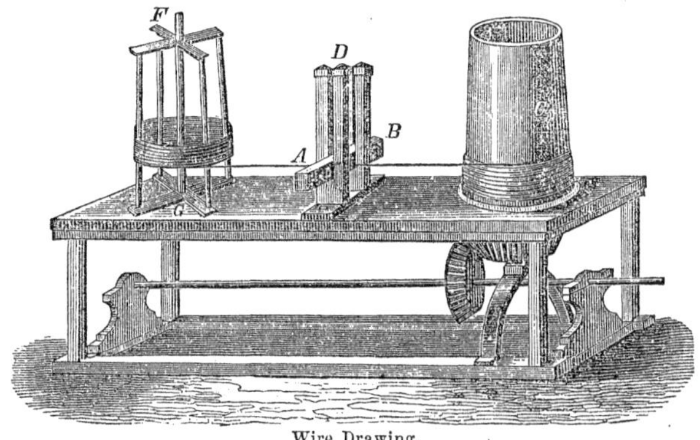

Wire Drawing.

firmly in its place by a support, D. A rod from the rolling mill, in the form of a coil, or a coil of wire, is placed upon the reel, F G, which turns

---

QUESTIONS.—What metals are mentioned as being brittle? What is said of the malleability of gold? Are the malleable metals also ductile? Describe the process of wire-drawing?

easily upon its axis; and the end of the wire, drawn out a little with the hammer, is passed through the plate and attached to the drum, C, which is turned by machinery, and by its motion winds the wire around itself, at the same time, of course, withdrawing it from the reel. To form small wire it is in this way passed many times through the plate; and, to prevent its becoming hard and brittle, it is several times annealed during the process. The passage of the wire through the plate is facilitated by dipping the coils occasionally in a moderately strong solution of sulphate of copper, by which it receives a thin coating of metallic copper.

336. The *malleability* and *ductility* of any substance would seem to depend very nearly upon the same properties, but it is found in the case of metals that they are not precisely the same. In the following table, in the first column, several of the metals are arranged in the order of their malleability, beginning with the most malleable; in the second column, the same metals are arranged in the order of their ductility. The table is from Regnault.

Malleability.	*Ductility.*
1. Gold.	1. Gold.
2. Silver.	2. Silver.
3. Copper.	3. Platinum.
4. Tin.	4. Iron.
5. Platinum.	5. Nickel.
6. Lead.	6. Copper.
7. Zinc.	7. Zinc.
8. Iron.	8. Tin.
9. Nickel.	9. Lead.

Both the malleability and ductility of several of the metals vary with the temperature. Thus iron, though partially malleable and ductile at the ordinary temperature of the atmosphere, is much more so at a red heat; and zinc is very malleable from 212° to 350°, but loses this property if cooled down to 32°, or heated to 600°. At the latter temperature it is decidedly brittle.

When a metal has been hammered or rolled, or drawn out into wire, its hardness as well as density is increased; and it becomes less malleable. This property, however, is restored by *annealing* it, which consists in heating it to redness and then cooling it

QUESTIONS.—336. Do the malleability and ductility of a substance depend upon the same properties? Are the malleability and ductility of some of the metals affected by their temperature? What is said of iron and zinc in this connection? What is meant by the annealing of a metal? Why is this often necessary in working many of the metals?

slowly. But the tenacity is often very much diminished by the process, sometimes even more than one-half.

337. The *tenacity* of a metal is determined by ascertaining the greatest weight a wire made of it, of a given diameter, will sustain. The determination is made by attaching one end of the wire to a firm support, and to the other end fixing a pan to receive the weights. Wires 0·787 of a line in diameter (= to about $\frac{7}{100}$th of an inch), made of the several metals mentioned below, were loaded gradually until they broke with the weights placed opposite to their names in the following table. These numbers therefore indicate their relative tenacities.

	Pounds.		*Pounds.*
Iron wire	549	Gold	151
Copper	302	Zinc	109½
Platinum	274	Tin	34½
Silver	187	Lead	27½

When a metallic wire is tested in this manner its length is increased with the weight, but if the load has not been too great it contracts again to the same length as at first, on the removal of the load. If the load be increased beyond a certain point, a permanent elongation is produced; and usually no great further increase is needed to break the wire.

338. The *specific gravity* of the metals, like many of their other properties, are very dissimilar. The specific gravities of some of the more important of them are contained in the following

Table of the Specific Gravity of Metals at 60°, compared with Water as Unity.

Platinum	20·98	Cobalt	8·53
Gold	19·02	Nickel	8·27
Tungsten	17·65	Manganese	8·01
Mercury	13·56	Iron	7·78
Palladium	11·05	Tin	7·29
Lead	11·35	Zinc	6·86
Silver	10·47	Antimony	6·70
Bismuth	9·82	Chromium	5·09
Uranium	9·00	Titanium	5·03
Copper	8·09	Aluminum	3·07
Cadmium	8·60	Sodium	0·97
Molybdenum	8·06	Potassium	0·86

QUESTIONS.—337. How is the tenacity of a metal determined? What metal of those mentioned has the greatest tenacity? What is the second? 338. What is said of the specific gravity of the metals? What metal is mentioned as having the greatest specific gravity? What is the second?

339. Crystaline Characters.—Many of the metals have a distinctly crystaline texture. Iron, for example, is fibrous; and zinc, bismuth, and antimony are lamellated. Metals are sometimes obtained also in crystals—and most of them, in crystalizing, assume the figure of the cube, the regular octahedron, or some form allied to it. Gold, silver, and copper occur naturally in crystals, while others crystalize when they pass gradually from the liquid to the solid condition. Crystals are most readily procured from those metals which fuse at a low temperature; and bismuth, from conducting heat less perfectly than other metals, and, therefore, cooling more slowly, is best fitted for the purpose.

Several of the metals form small but very beautiful crystals, as they are slowly separated from their solutions by the galvanic current. If in a solution of sulphate of copper we place two plates of copper, and connect them with the two electrodes of a galvanic battery in feeble action, after some time the plate on the negative side will become covered with small crystals of metallic copper, while the other plate will be gradually dissolved.

It is believed that some of the metals, in certain peculiar circumstances, may assume a crystalline structure, even when in the solid state.

340. Alloys.—Many of the metals are capable of combining with each other, forming compounds called *alloys*, which will be described in their proper places. They possess all the characteristic physical properties of the pure metals, and many of them are of great service in the arts. Generally, alloys are more fusible and more oxidable than their constituents separately. Their malleability and ductility usually are much less, and their hardness greater, than those of the metals of which they are composed.

Compounds of mercury with other metals are called *amalgams.*

341. Metallurgy.—The reduction of the metals from their ores, and the methods adopted in working them in the arts, con-

---

QUESTIONS.—339. Do any of the metals possess a crystaline texture? What metals are mentioned as being sometimes found in crystals? Describe the mode of obtaining crystals of copper by the galvanic process? 340. What are *alloys?* Do alloys possess the usual physical properties of the metals? What is said of their fusibility? What of their malleability and ductility? What are *amalgams?* 341. What is metallurgy?

stitutes a distinct branch of chemical science, called by this name. Most of the metals have an extensive range of affinity; many of them form compounds with nearly all the non-metallic elements, and all without exception combine with oxygen.

As many of the metallic ores are oxides, the reduction of this class of compounds becomes particularly important. It is effected in several different modes, as,

1. By mere heat. By this method the oxides of gold, silver, mercury, and platinum may be decomposed.

2. By the united agency of heat and combustible matter. Thus, by transmitting a current of hydrogen gas over the oxides of copper or iron heated to redness in a tube of porcelain, water is generated, and the metals are obtained in a pure form. Carbonaceous matters are likewise used for the purpose with great success. Potassa and soda, for example, may be decomposed by exposing them to a white heat, after being intimately mixed with charcoal in fine powder. A similar process is employed in metallurgy for extracting metals from their ores, the inflammable materials being wood, charcoal, coke, or coal. In the more delicate operations of the laboratory, charcoal and *black flux** are employed.

3. By the galvanic battery. This is a still more powerful agent than the preceding; since some oxides, such as baryta and strontia, which resist the united influences of heat and charcoal, are reduced by the agency of galvanism.

4. By the action of deoxdizing agents on metallic solutions. Phosphorous acid, for example, when added to a liquid containing oxide of mercury, deprives the oxide of its oxygen, metallic mercury subsides, and phosphoric acid is generated. In like manner one metal may be precipitated by another, provided the affinity of the latter for oxygen exceeds that of the former. Thus, when mercury is added to a solution of nitrate of oxide of silver, metallic silver is thrown down, and oxide of mercury is dissolved

* *Black flux* is prepared by deflagrating nitre with twice its weight of cream of tartar When equal weights are used, it constitutes *white flux.*

QUESTIONS.—What is said of the affinity of the metals? What are most of the metallic ores? What are some of the different means by which these ores may be reduced?

by the nitric acid. On placing metallic copper in the liquid, pure mercury subsides, and a nitrate of the oxide of copper is formed; and from this solution metallic copper may be precipitated by means of iron.

To reduce the other compounds of the metals, other modes are adopted, which cannot here be particularly described.

The relations of the various metalloids to the metals will be described as each of these latter elements shall come in review before us, so far as demanded by the special object of the work.

342. Classification of the Metals.—The metals have been variously classified by different writers; but the following arrangement into six groups will probably answer our purpose as well as any we can adopt.

1. Metals, the protoxides of which are alkalies.
2. Metals, the protoxides of which are alkaline earths.
3. Metals, the protoxides or sesquioxides of which are earths.
4. Metals which, at a red heat, decompose the vapor of water, but are not acted upon by liquid water.
5. Metals which are incapable of decomposing water, and whose oxides are not reduced by the mere action of heat.
6. Metals whose oxides are reduced by a red heat.

Saline Compounds, or Salts.

343. A salt is a compound of two other binary compounds, which sustain to each other the relation of acid and base; the former being electro-negative in reference to the latter, which is electro-positive. Thus, sulphate of soda, NaO,SO_3, is a compound of sulphuric acid (teroxide of sulphur) and soda, which is the protoxide of sodium; the former being electro-negative in reference to the latter, which is electro-positive.

The acids take their name from the fact that when soluble their taste is very generally sour,—but this may not always be

QUESTIONS.—342. What are the metals included in the first group? In the second group? In the third? In the fourth? In the fifth? In the sixth? 343. What is a salt? Illustrate by an example. How are the acids generally characterized?

the case; they are mostly combinations of two metalloids, as the sulphuric, nitric, and hydrochloric acids, but sometimes they are formed of a metal and metalloid, as the manganic and chromic acids, and the sulphides of tin and antimony, which are capable of acting as acids. Most of them have the property of changing the blue solution of litmus to red.

The bases, or electro-positive binary compounds, are always formed by the combination of a metal with a metalloid, as the protoxide of potassium (potassa), the protosulphide of potassium, and the protoxides of iron, copper, and silver.

344. Oxysalts.—A large majority of all known acids and bases are oxides, called *oxacids* and *oxybases;* and the salts formed by their union are therefore called *oxysalts.* They are therefore double oxides.

The metallic oxides, in regard to their disposition to combine with each other as acids and bases, and with the oxides of the metalloids, to form salts (as suggested by Regnault), may be divided into several very distinct classes, as,

1. *Basic oxides,* which combine readily with acids and form definite, crystalizable salts. Most of them are protoxides, as potassa, KO, soda, NaO, protoxide of silver, AgO, and the protoxide of iron, FeO. A few are sesquioxides or suboxides.

2. *Acid oxides,* which, as the name implies, possess acid properties;—they combine with bases to form salts, and very generally change the blue solution of litmus to red. Most of them are bi- or teroxides, as plumbic acid, PBO_2 (binoxide of lead), chromic acid, CrO_3, and manganic acid, MnO_3. Occasionally they are still higher oxides, as permanganic acid, Mn_2O_7.

3. *Neutral oxides,* which may combine with either acids or bases, as alumina, Al_2O_3, and the sesquioxide of antimony, Sb_2O_3; or with neither, as the binoxides of silver, potassium, barium, and strontium. Some of these, by the action of acids, especially if heated, are decomposed.

4. *Saline oxides,* which result from the combination of a basic metallic oxide with a higher oxide of the same metal, as the magnetic oxide of iron, Fe_3O_4, which is really a compound of the protoxide, acting as a base, with the sesquioxide, acting as an acid; and its proper formula is therefore FeO,Fe_2O_3. The oxides of manganese, Mn_3O_4, and of chromium, Cr_3O_4, furnish other examples.

QUESTIONS.—Are most of the acids formed by combinations of the metalloids? Do some of the metals form acid compounds? How do the acids generally affect the blue solution of litmus? Of what are the bases mostly composed? 344. What is said of a majority of all the known acids and bases? What four classes of metallic oxides are mentioned?

345. Sulphur Salts.—Two sulphides often combine, forming double sulphides, analogous to the double oxides. They possess, in many respects, similar characters to the double oxides or oxysalts, and are called *sulphur salts.*

Some of the principal *sulphur bases* are the protosulphides of potassium, sodium, lithium, barium, strontium, calcium, and magnesium; and some of the more important *sulphur acids* are the sulphides of arsenic, antimony, carbon, tin, gold, and hydrogen.

The sulphur salts generally are so constituted that if the sulphur they contain were replaced by an equivalent quantity of oxygen, oxysalts would be produced. Thus, the carbosulphide of potassium, KS,CS_2, when the sulphur is replaced by oxygen, becomes carbonate of potassa, KO,CO_2; and, in like manner, the double sulphide of potassium and arsenic, KS,AsS_5, by a similar substitution, becomes arseniate of potash, KO,AsO_5.

346. Chlorosalts.—The chlorosalts are double chlorides, one of the simple chlorides acting as a chlorobase, and the other as a chloroacid, as the double chloride of gold and potassium, $KCl,AuCl_3$, called also aurochlorate of chloride of potassium, and platinochlorate of chloride of sodium, $NaCl,PtCl_2$.

In the same manner two iodides (an iodobase and an iodoacid) bromides, or fluorides, may unite to form in each case a series of salts analogous to the preceding, but not much is known of them.

It is to be observed that these combinations take place only between members of the same series, as oxides with oxides, sulphides with sulphides, &c.; it is evidently possible that members of different series may unite, as an oxide with a chloride, a chloride with a sulphide, &c.; but it is a question not fully settled whether such compounds are ever really formed.

347. Haloid Salts.—Besides the compounds recognised by the above remarks as salts, the binary compounds of chlorine, iodine, bromine, and fluorine with many of the metals, have been classed with the salts by Berzelius and other eminent chemists, because of their resemblance, in many of their properties, to the salts,

---

QUESTIONS.—345. What are *sulphur salts?* What is said of the constitution of the sulphur salts? Give an example. 246. What are *chlorosalts?* May the iodides, bromides, &c., form similar series of salts? Are these combinations always between members of the same series? 347. What are *haloid salts*, so called? Are these compounds salts, according to the definition of a salt adopted above?

properly so called. To distinguish them from the salts proper, they have been called *haloid salts.* But we prefer to limit the definition of a salt, as given above, although in so doing we of course exclude common salt, chloride of sodium, from the class.

The elements, chlorine, iodine, &c., which, by combining with metals, form the so called haloid salts, combine also with hydrogen, forming acid compounds called hydracids (168), as the hydrochloric, hydriodic &c. Now when powerful hydrous acids are brought in contact with the haloid salts, important reactions take place, dependent upon the water of the acid. Thus, by the action of oil of vitriol upon chloride of potassium, we obtain sulphate of potassa and hydrochloric acid, as shown in the following equation:

$$KCl + SO_3,HO = KO,SO_3 + HCl.$$

Here the water of the acid is decomposed, yielding its oxygen to the potassium to form potassa, and its hydrogen to the chlorine to form hydrochloric acid.

So, on the other hand, the action of a hydracid on an oxybase results in the production of a haloid salt and water. Thus,

$$KO + HCl = KCl + HO.$$

348. Neutralization of Acid and Base.—When an acid and base combine to form a salt, the peculiar and characteristic properties of each, in a great measure, disappear, the new compound formed not being characterized by the properties of either of its ingredients, but possessing others entirely distinct. Thus, an acid is generally sour to the taste, and changes vegetable blues to red, but the salts it may form with bases possess neither of these properties; so, also, the alkalies are caustic to the flesh and taste, and combine with oils to form soaps, but the salts which they form with the acids are entirely destitute of these properties. Most of the other peculiar properties of both acids and bases, when the two combine, are affected in the same manner, and they are therefore said to be *neutralized.*

349. Salts are often spoken of as divided into the three classes of *neutral salts, super* or *acid salts, and sub* or *basic salts;* but the distinction

---

QUESTIONS.—What is said of the elements chlorine, iodine, &c., which form the so-called haloid salts by combining with the metals? What reactions take place when powerful hydrous acids act upon these salts? Give an illustration. 348. When an acid and base combine do the peculiar properties of each disappear? Give some further illustration 349. What three classes of salts are mentioned?

cannot always be made with accuracy. In general, a *neutral salt* is formed by the union of a single equivalent of the acid and base, while an *acid salt* contains two or more equivalents of acid to one of base; and a *sub* or *basic salt* contains two or more equivalents of base to one of acid. But this is to be understood as liable to many exceptions. Frequently salts denominated neutral contain as many equivalents of acid as there are equivalents of oxygen in the base. Thus, the neutral sulphate of soda is NaO,SO_3, and sulphate of the protoxide of iron FeO,SO_3, but the sulphate of the peroxide of iron is $Fe_2O_3,3SO_3$.

Some few acids have the property of combining with bases in such a manner that one equivalent of acid will hold in union with it one, two, three (or even more) equivalents of base, and yet these several salts are considered as neutral; such acids are called *polybasic acids*. Phosphoric acid (282) furnishes an instance of the kind.

350. Solubility and Crystalization of the Salts.—Nearly all the salts are solid at ordinary temperatures, and very many of them are susceptible of crystalization. Very generally they are more soluble in warm than in cold water, and often their solubility increases in proportion as the temperature of the water is elevated. The following table shows the quantity of nitrate of potash soluble in 100 parts of water at several different temperatures:

Temperatures.	*Parts soluble in* 100 *parts water.*
32°	13·3
64°	29·0
113°	74·6
207°	236·0

In some cases, the solubility rapidly increases as the temperature is raised, up to a certain point, and then diminishes. Sulphate of soda (Glauber's salt) furnishes an instance of this kind. Of this salt 100 parts of water, at 32°, take up 12 parts; at 77°, 99 parts; at 91°, 322 parts;—but above this temperature the solubility diminishes, so that at the boiling point only about 212 parts will be held in solution.

QUESTIONS.—What in general is a neutral salt? What is a super or acid salt? What is a sub or basic salt? Do salts denominated neutral often contain as many equivalents of acid as there are equivalents of oxygen in the base? Give an example. What are polybasic acids? 350. Are the salts usually solid at ordinary temperatures? Are they more soluble in warm or cold water? What is said of the solubility of nitrate of potash in water at different temperatures? How is the solubility of sulphate of soda in water affected, as the temperature of the water is raised?

Crystals of soluble salts may generally be obtained by making a saturated solution at an elevated temperature and allowing it to cool slowly. Thus, a saturated solution of alum at 100° or 150°, on cooling deposits beautiful octahedral crystals; and if a tree or basket made of copper wire is placed in the solution when warm, the crystals will attach themselves to it in various positions, presenting a very beautiful appearance.

351. Salts soluble in water,—and even some that are insoluble—often retain in their crystals a portion of water, which is called *water of crystalization.* Thus sulphate of soda in crystals contains 10 equivalents of water, and its proper formula is $NaO,SO_3 + 10HO$. This is the case however only when the crystalization takes place at a temperature below 92° or 93°; when it crystalizes at a higher temperature the crystals are anhydrous.

The same salt will often combine with very different quantities of water when deposited from its solution at different temperatures. Sulphate of protoxide of manganese, crystalized from an aqueous solution, at 43°, has the formula, $MnO,SO_3 + 7HO$; when crystalized between 43° and 68° its formula is $MnO,SO_3 + 6HO$; and again when crystalized between 68° and 86°, its formula is $MnO,SO_3 + 4HO$.

When a salt of the kind last mentioned is heated in the open air, it gives up its water in successive portions, as the temperature is raised. Sometimes the last equivalent of water is held by a much stronger force than the rest, and can be driven off only by a red heat, and then an entire change in the nature of the salt takes place. The water in such a case is called *constitutional water.*

352. Sometimes a salt in a dry atmosphere loses a part or all of its water of crystalization, and falls to powder; it is then said to *effloresce.* Again, some salts absorb moisture from the air, and are then said to *deliquesce.* Common pearlash (carbonate of pot-

---

QUESTIONS.—How may crystals of salts soluble in water often be obtained? 351. What is water of crystalization? Give an illustration. Will the quantity of water of crystalization often depend upon the temperature of the solutions from which the crystals are deposited? What will be the effect when a salt of this kind is heated? 352. When is a salt said to effloresce? When to deliquesce?

ash) is an instance of a deliquescent salt. Left in the open air for a time, it will absorb sufficient water to effect its solution.

353. Water holding a salt in solution invariably has a higher boiling point than pure water. The following table shows the boiling points of saturated solutions of several of the salts.

Salts.	Parts of Salt in 100 parts of water.	Boiling Point.
Chlorate of potash	61·5	220°
Common Salt	41·0	227°
Nitrate of potash	335·0	241°
Nitrate of Lime	362·0	304°

But the steam from such boiling solutions, as it escapes into the open air (omitting any regard to variations of atmospheric pressure), will always be of the same temperature of 212°.

When a hydracid, as the hydrochloric, HCl, acts upon a metal, as zinc, the acid is decomposed, and a chloride of the metal formed, the hydrogen being set free; thus,

$$Zn + HCl = ZnCl + H.$$

So when liquid hydrochloric acid is poured upon potash, the reactions are

$$KO + HCl,HO = KCl + HO + H.$$

In both these cases, therefore, as will be seen by an examination of the formulæ, the particle of hydrogen of the hydracid has simply been displaced, and a particle of metal substituted. Now all the oxyacids, or nearly all,—at least in their *active* state,—contain water, as the sulphuric acid, SO_3,HO, and nitric acid, NO_5,HO. They may indeed be obtained free from water, as SO_3, and NO_5, but they are then solid and quite inert, possessing none of the active properties of the liquid acids, which, however, they immediately assume on the addition of water.

The reaction between zinc and oil of vitriol is

$$Zn + SO_3,HO = ZnO,SO_3 + H.$$

QUESTIONS.—353. What is said of the boiling point of water holding a salt in solution? What is said of the temperature of the steam formed? When a hydracid acts upon a metal what are the reactions that take place? What when hydrochloric acid acts upon potash? In these cases is the hydrogen simply replaced by the metal? Is the result the same when the hydrated oxyacids act upon the metals? Explain by reference to the action of sulphuric acid upon zinc.

Now as hydrated sulphuric acid only, SO_3HO, is capable of acting upon this metal, and not the dry acid SO_3, may we not in this case also, in like manner, consider the metal as simply replacing the hydrogen?

If this view be adopted, as it has been by some eminent chemists, then it is required that the formulæ of the hydrated acids, the sulphuric, nitric, &c., should be changed accordingly, sulphuric acid being not SO_3,HO, but SO_4,H, or, as generally written, H,SO_4. So also nitric acid is to be considered, not NO_5,HO, but H,NO_6, and so of other acids.

We shall not discuss the subject further, but simply give the formulæ for several well known salts, according to the old and generally received view, and also according to the new view.

Salts.	Common Theory.	New Theory.
Sulphate of Soda............	NaO,SO_3	Na,SO_4.
Sulphate of Zinc............	ZnO,SO_3	Zn,SO_4.
Nitrate of potash............	KO,NO_5	K,NO_6.
Chlorate of potash..........	KO,ClO_5	K,ClO_6.

Separate Description of the Metals.

354. The classification of the metals adopted in this work has already been indicated (342).

GROUP I.

POTASSIUM SODIUM LITHIUM	*Metals, the protoxides of which are alkalies.*—The protoxides of these metallic elements are very soluble in water, and possess in an eminent degree the peculiar properties denominated *alkaline* properties.

POTASSIUM.

Symbol, K (*Kalium*); *Equivalent*, 39·2; *Density*, 0·865.

355. History.—Potassium was first obtained by Sir H. Davy in 1807, from potash, which is the protoxide of the metal. He

QUESTIONS.—If this view be adopted, how should the formula for common sulphuric acid, or oil of vitriol, be written? How that for nitric acid? 354. What metals are included in the first group? How are they characterized? 355. By whom was potassium discovered?

procured it by subjecting a piece of caustic potash, slightly moistened, to the action of a powerful galvanic current, when oxygen made its appearance at the positive, and the metal potassium at the negative, electrode. Previous to this time, potash and the other alkalies and earths had been considered as simple substances. Potassium, in combination with oxygen and other bodies, is very generally diffused in the rocks and soils of every place, but is never found in nature in a separate state. It is contained in most vegetable and many animal substances.

356. Preparation.—This metal is best prepared by heating intensely dry carbonate of potash mixed intimately with half its weight of fine charcoal-powder and iron filings. The potash, being an oxide of potassium, at a high temperature yields its oxygen to the charcoal and iron, and itself distils over into a receiver prepared for the purpose.

The following apparatus answers well for the operation. A common mercury bottle, A, covered with an infusible lute, and well dried, is three-fourths filled with cream of tartar that has been previously charred and mixed with one-fourth or one-fifth of its weight of pulverized charcoal, and placed in a proper furnace, with a piece of gun-barrel, B, about a foot in length, extending outward and connecting with a receiver, C.

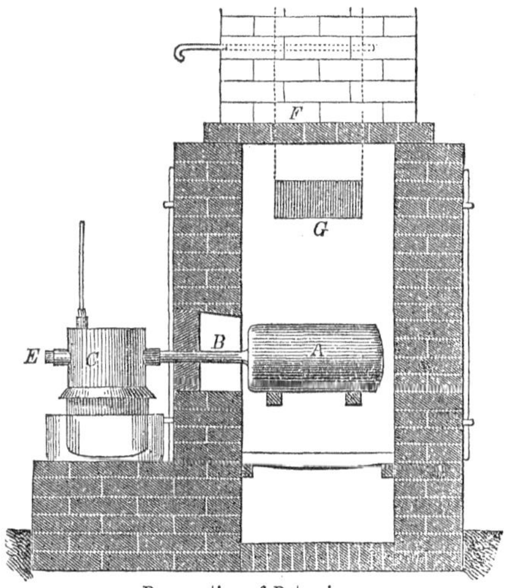

Preparation of Potassium.

QUESTIONS.—How did Sir H. Davy procure potassium? How were potash and the other alkalies considered previous to this time? Is potash very generally diffused in nature? 356. How is potassium now usually prepared? What is the effect produced by the charcoal? Describe the apparatus figured in the margin, and the mode of using it.

A fire of hard coal is then kindled in the furnace (represented in section in figure), which should have a good draft, by connecting with a chimney, G F, of sufficient height, in order to produce the greatest heat possible; and the metallic potassium, as it is liberated in the gaseous state, escapes through the iron tube, B, and is condensed and collected in the bottom of the receiver, C. In this receiver some naphtha is placed, to protect it from the atmosphere.

Charred cream of tartar is used because it furnishes an intimate mixture of carbon and carbonate of potassa, but instead of it dry carbonate of potash (pearlash) may be mixed intimately with half its weight of pulverized charcoal.

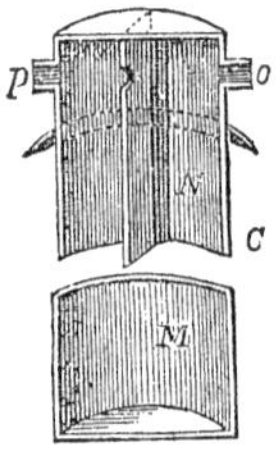

Prep. of Potassium.

The receiver, C (seen in section), is best made of sheet copper, in two parts, M and N, the upper part, N, being without a bottom, and fitting accurately into the lower part, M. The upper part, N, is also divided into two parts by a vertical partition, as shown in figure; it is also provided with two tubulures, *o* and *p*, directly opposite each other, the one, *o*, to receive the end of the gun-barrel, B (in the first of the accompanying figures), and the other, *p* (corresponding to E in the first figure), to allow the insertion of an iron wire to remove the obstructions that will occasionally collect in the gun-barrel leading from the iron bottle. A small opening is also made for this purpose in the partition of the receiver, N.

During the operation much incondensible gaseous matter passes over and escapes by a tube of glass (shown in the first figure) provided for the purpose; and as it is necessary that the receiver should be kept cold, a small stream of cold water is made to fall upon it constantly, which is prevented from entering the lower part, M, by a rim of metal passing round it, as shown in the figures. The tubulure, *p*, is to be kept closed by a cork, except as it is necessary to remove it for a moment to insert the iron wire, when obstructions occur in the gun-barrel.

The potassium, after the action has ceased, will be found in irregular masses in the naphtha at the bottom of the receiver;

QUESTION.—Where will the potassium be found?

but it will not be pure. It is now collected in a bag of cloth, through which it is squeezed by compressing the bag with pincers while held in a cup of warm naphtha. If necessary, it may be further distilled in a small iron retort.

357. Properties.—Potassium is a solid, in color and lustre much resembling lead. At 150° it melts, and at a dull red heat may be distilled in vessels void of gases capable of combining with it. It is the lightest metal known, having a density of only 0·865, and floats upon the surface of water. At ordinary temperatures it is soft like wax, but at 32° it becomes quite hard. But its most characteristic property is its affinity for oxygen;—when thrown upon the surface of water it absorbs the oxygen so rapidly as to be inflamed, and burns with a beautiful rose-colored flame. In the open air, the freshly-cut surface absorbs oxygen so rapidly as to be tarnished instantly; and heated even in carbonic acid (303), it takes fire and burns by absorbing the oxygen. In consequence of its affinity for oxygen it can be preserved only in tubes hermetically sealed, or under some liquid that does not contain oxygen, as naphtha, which is found to answer the purpose well.

Combustion of Potassium upon Water.

Binary Compounds of Potassium.

358. Protoxide of Potassium—KO; eq., $(39{\cdot}2 + 8 =) 47{\cdot}2$.—This is *potash*, or *potassa*, and is always formed when the metal is exposed in the open air, or oxygen gas, or acted on by water. In the latter case, the potash formed is immediately dissolved by the water, as will be found by applying the usual tests. Pure potash is a white, inodorous substance, with a pungent, caustic taste, and very soluble in water. It absorbs carbonic acid

QUESTIONS.—357. Describe some of the properties of potassium. What is said of its affinity for oxygen? What is the effect when a small piece of it is thrown upon water? How is it preserved? Why is naphtha selected for this purpose? 358. What is the common name for protoxide of potassium? Describe potassa? Does it usually contain water?

and water rapidly from the air, and should therefore be kept in close bottles.

When prepared by the slow oxydation of potassium in dry air or oxygen gas it is anhydrous, but when formed by the oxydation of the metal in water, or from any of its salts, it is always in the state of a hydrate.

Potash forms with water two compounds, the monohydrate, KO,HO, and the pentahydrate, $KO,5HO$; the former of which is the *caustic potash* of commerce. To prepare it, pearlash (carbonate of potash) is dissolved in ten times its own weight of water, and half its weight of recently-slaked lime mingled with it, in successive portions, and the whole boiled briskly for half an hour. It is then allowed to settle, and the clear liquid is drawn off. The lime, in this process, decomposes the carbonate of potassa, and forms insoluble carbonate of lime, which settles to the bottom; and the clear liquid contains nearly pure potassa. This may be preserved for use in well-closed bottles, or it may be evaporated to dryness in a vessel of such a form as not to allow any considerable accession of air. To insure perfect purity, it must be again dissolved in absolute alcohol, and the solution filtered and evaporated as before. The hydrate is soluble in alcohol, which is not the case with sulphate of potash, usually present in pearlash, or carbonate, portions of which may have escaped decomposition by the lime.

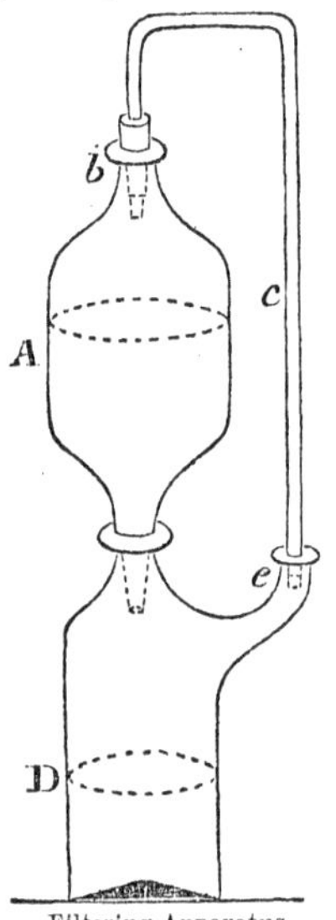

Filtering Apparatus.

359. To prevent the absorption of carbonic acid from the air while filtering, an apparatus like that figured in the margin is used. It consists of two vessels, A and D, of equal capacity, and connected with each other. The throat of the upper vessel or

QUESTIONS.—What hydrates of potash are mentioned? How is pure caustic potash prepared from the carbonate? Why should it be dissolved in alcohol and filtered? 359. Describe the mode of filtering the alcoholic solution.

funnel, A, is obstructed by a piece of coarse linen loosely rolled up, and not pressed down into the pipe through which the solution is filtered. The pipe, *c*, extending from *e* to *b*, serves for the air to pass from the lower vessel to the upper; and the operation goes quietly on, free from contact with the atmosphere, except the little contained within the apparatus at the beginning of the process.

Solution of potassa is highly caustic, and its taste intensely acrid. It possesses alkaline properties in an eminent degree, converting the vegetable blue colors to green, and neutralizing the strongest acids. It absorbs carbonic acid gas rapidly, and is consequently employed for withdrawing that substance from gaseous mixtures.

Potassa is employed as a reagent in detecting the presence of bodies, and in separating them from each other. The solid hydrate, owing to its strong affinity for water, is used for depriving gases of hygrometric moisture.

Caustic potash attacks all animal substances with avidity, and neutralizes all acids. With the fats and oils it combines readily, forming the well-known compound called *soap*, of which we shall have occasion to speak again hereafter. This property is characteristic of all the alkalies.

The *pentahydrate* possesses no special interest.

360. Peroxide of Potassium, KO_3, as is shown by the formula, is a teroxide. It is of a dull yellow color, and is formed by burning potassium in an excess of dry oxygen gas. Thrown into water, it gives up two-thirds of its oxygen, and solution of caustic potash is formed.

361. Chloride of Potassium, KCl, is readily obtained by neutralizing carbonate of potash, KO, CO_2, by hydrochloric acid. The reactions which take place have already (347) been explained. Like common salt (chloride of sodium), it crystalizes in cubes, which are anhydrous.

Iodide of Potassium, KI, is formed by heating potassium in contact with iodine; or, by digesting iodine in a hot solu-

QUESTIONS.—What is said of the action of potash upon animal substances? What does it form with the fats and oils? 360. What is peroxide of potassium? 361. What is chloride of potassium? How is iodide of potassium formed?

tion of caustic potash, and exposing the mass, when dry, to a red heat, and subsequently crystalizing from solution in water or alcohol. It is a white solid, very soluble in water, usually seen crystalized in cubes, and is often sold under the name of *hydriodate of potash.*

Solution of iodide of potassium has the property of dissolving iodine, and becomes of a brown color; it also dissolves other iodides, as the iodides of mercury.

Iodide of potassium is much used in medicine, and in certain photographic processes.

Bromide, KBr, and *fluoride*, KF, of potassium, like the chloride and iodide, crystalize in cubes.

362. **Sulphides of Potassium.**—There are at least five sulphides of potassium, KS, KS_2, KS_3, KS_4, and KS_5, which, as shown by their formulæ, contain, respectively, 1, 2, 3, 4, and 5 equivalents of sulphur in combination with 1 equivalent of the metal.

The most important of these is the *pentasulphide*, KS_5, which is easily prepared by heating gently a mixture of carbonate of potash and sulphur, or by boiling a solution of caustic potash with an excess of sulphur.

It is a yellowish-brown solid, very soluble in water, and is used in medicine, especially in cutaneous diseases, under the name of *hepar sulphuris*, or *liver of sulphur*. The *tersulphide* is also used in the same manner.

Salts of Potash.

363. Carbonate of Potash, KO,CO_2.—Carbonate of potash, or *pearlash*, is prepared for the purposes of commerce by leaching the ashes of forest trees, and evaporating the *lye* thus obtained to dryness, and then heating the dry mass to redness for a time in open vessels, to burn out the combustible matter which is contained in it. As thus procured, it is a white spongy mass, very caustic to the taste, and absorbs moisture rapidly from the air, so that it must be kept in close vessels. It is very soluble in water, but insoluble in alcohol, and is easily fused at a red heat. That found in commerce is very impure, being mixed with silica and other substances. It is manufactured in large quantities in the United States, in the Canadas, and in Russia.

QUESTIONS.—What use is made of iodide of potassium? 362. What is said of the sulphides of potassium? 363. What is *pearlash?* How is it prepared? Describe it. What countries furnish it in large quantities?

If pure carbonate of potash is required, as is often the case in the laboratory, the tartrate (cream of tartar) or oxylate is heated to redness in the open air, treating the charred mass thus formed with warm water, and filtering. The solution may then be evaporated to dryness, if the dry salt is desired.

By slowly evaporating a solution of carbonate of potash the salt may be crystalized, though not without difficulty.

The article sold as *potash* is the same as pearlash, except that it is not subjected to the last process of calcination. It is a dark-colored mass, and contains much caustic potash, as well as carbonate, and is used extensively for the manufacture of soap. Pearlash is employed in the manufacture of glass and paints, and for various other purposes.

364. Bicarbonate of Potash, $KO,2CO_2$.—This salt is prepared by subjecting pearlash in solution, for some time, to an atmosphere of carbonic acid, which is absorbed in large quantity. Though less caustic to the taste than pearlash, it is still highly alkaline. It is less soluble also than pearlash, and less deliquescent.

It may be obtained in crystals with less difficulty than the carbonate, and the crystals contain a single equivalent of water. Their formula may therefore be written $KO,CO_2 + HO,CO_2$, that is, the salt may be considered as a double carbonate of potash and water.

This salt is extensively used under the name of *salæratus*.

365. Nitrate of Potash, KO,NO_5.—This salt, the *saltpetre*, or *nitre*, of commerce, is formed, in this country, by decomposing the nitrate of lime, which abounds in the caverns of some of the Western States, by carbonate of potash, and filtering and evaporating the solution thus obtained. In some parts of Europe it is prepared in nitre-beds, which are made by heaping together old mortar, refuse animal matter, wood-ashes, &c., in which it gradually forms by the action of the atmosphere. The mass is

QUESTIONS.—How may pure carbonate of potash be prepared? What is the article known in commerce as *potash?* 364. How is bicarbonate of potash prepared? May it be obtained in crystals? By what name is it familiarly known? 365. By what names is nitrate of potash known in commerce? How is it prepared in this country? How in some parts of Europe?

lixiviated with hot water; and the solution by evaporation yields crude nitre.

Nitrate of potash is usually seen in long six-sided prisms. It is a colorless salt, of a cooling saline taste, and is very soluble in water. Its density is about 1·93. Heated to redness, it first melts and is then decomposed, giving off, at first, pure oxygen, and afterwards, if the heat is increased, nitrogen and nitric oxide. Thrown on burning charcoal it is decomposed, producing violent deflagration, by which it may always be distinguished from sulphate of soda, for which it has sometimes been mistaken. It is a powerful antiseptic, and is used with common salt in the preservation of meat and other substances.

But the chief use of nitre in the arts is in the manufacture of *gunpowder*, which is composed of nitre six parts, and charcoal and sulphur each one part, the whole being moistened and thoroughly ground together, and subsequently pressed and granulated. When fired, the nitre, by its decomposition, furnishes oxygen, which combines the carbon, forming carbonic acid, the sulphur at the same time uniting with the potassa. The action of gunpowder depends upon its generating, when decomposed, a large quantity of gaseous matter at a high temperature. The gases are chiefly nitrogen and carbonic acid, which, at the moment of explosion, occupy more than 1000 times the volume of the powder from which they are formed. The formation of the gases is not instantaneous, but occupies a certain time, and the ball is forced from the gun with a velocity due to the ultimate effect of the whole.

When made for particular purposes, the proportion of the ingredients is sometimes considerably varied. We may, in fact, consider gunpowder as of three kinds, which may be called sporting powder, war powder, and blasting powder; the former of which is required to detonate more rapidly and violently than either of the others. The composition of the three kinds is usually about as follows:—

QUESTIONS.—Describe nitrate of potash? How does heat affect it? What is the effect when it is thrown upon burning charcoal? What uses are made of it? What is the composition of *gunpowder?* What are the chemical reactions that take place when it is fired? Upon what does the action of gunpowder depend? What different kinds are mentioned?

For sporting powder	Nitre	76·9	parts.
	Sulphur	9·6	"
	Carbon	13·5	"
		100·0	"
For war powder	Nitre	75·0	parts.
	Sulphur	12·5	"
	Carbon	12·5	"
		100·0	"
For blasting powder	Nitre	62·0	parts.
	Sulphur	20·0	"
	Carbon	18·0	"
		100·0	"

Theoretically, it would seem that the best proportion would be 1 equivalent of nitre, 1 equivalent of sulphur, and 3 equivalents of carbon. Thus,

$$KO,NO_5 + S + 3C = KS + N + 3CO_2.$$

The proportion by weight of the several ingredients would then be as follows:—

		In 100 parts.
Nitre, 1 equivalent	101·2	74·85
Sulphur, 1 "	16·0	11·84
Carbon, 3 "	18·0	13·31
		100·00

This is very nearly the same proportion as that given for war powder above; but the reactions which take place in the combustion of the powder not being necessarily the same precisely as indicated in the above formula, nor, indeed, in any two cases with the same powder, considerable variation in the proportion of the ingredients may be allowed. For the best powder, the materials should be freed from all impurities before using them.

366. The comparative force of different specimens of gunpowder, or the initial velocity a given quantity of it will communicate to a ball of a given weight, is determined by several modes, only one of which, called the ballistic pendulum (see figure on next page), will be described, and that very briefly. It is composed of two parts, the *ballistic pendulum*, P, and the *pendulum-gun*, G, the former of which, P, consists of a conical box, containing a mass of lead, suspended by an iron rod, in the manner of a

---

QUESTIONS.—What is the difference in the composition of the three kinds of gunpowder? Is the composition of gunpowder exactly the same as would seem to be required by the chemical formula? 366. How is the comparative force exerted by different kinds of gunpowder, when fired, determined? Describe the ballistic pendulum.

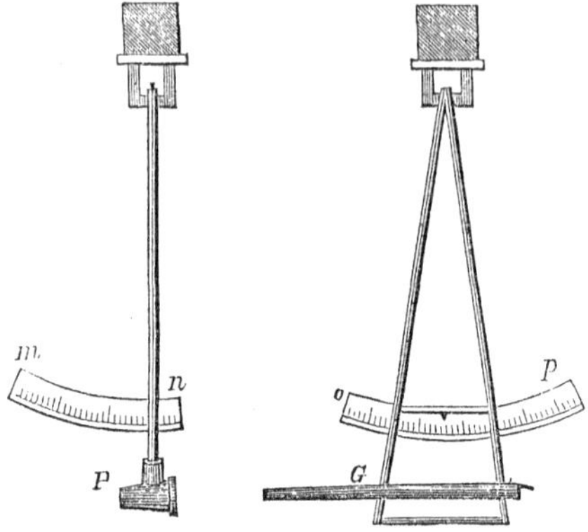

Ballistic Pendulum.

pendulum, from some very firm support. G is a common gun-barrel supported in an frame, which is also suspended in the manner of a pendulum, the gun-barrel, when hanging freely, being made to point to the mass of lead. At *m n* and *o p* are graduated arcs, each having attached to it a slider, which is carried along with the movement of the rod to the furthest point to which it oscillates, but does not return with it.

The gun-barrel is now charged with the proper quantity of the powder to be tested, and the ball placed in it, and fired; and the ball being projected into the mass of lead causes it to move to the left (as the apparatus is represented in the figure) to a certain distance, which will be exactly shown by the slide upon the graduated arc, *m n*. At the same time the gun-barrel will recoil to the right, as will be shown by the slide upon *o p*. The initial velocity of the ball may now be calculated from the distance either slide has traversed, by means of the proper mathematical formulæ.

To insure accuracy, attention must be paid to several particulars in the adjustment of the apparatus, which are not here alluded to.

367. Sulphate of Potash, KO,SO_3.—This salt is easily prepared artificially by neutralizing carbonate of potassa with sulphuric acid; and it is procured abundantly by neutralizing with carbonate of potassa the residue of the operation for preparing nitric acid (221). Its taste is saline and bitter. It generally crystalizes in six-sided prisms, bounded by pyramids with six sides, but its primary form is the right rhombic prism. The crystals contain no water of crystalization, and suffer no change by exposure to the air. They decrepitate when heated, and enter into fusion at a red heat. They require 16 times their weight of water at 60°, and 5 of boiling water for solution.

Questions.—367. How may sulphate of potash be prepared artificially? What are some of its properties?

368. Bisulphate of Potassa, $KO,2SO_3$, is easily formed by exposing the neutral sulphate with half its weight of strong sulphuric acid to a heat just below redness, in a platinum crucible, until acid fumes cease to escape. It has a strong sour taste, and reddens litmus paper, and in crystals may be considered a double sulphate of potassa and water, $KO,SO_3 + HO,SO_3$. It is much more soluble than the neutral sulphate, requiring for solution only twice its weight of water at 60°, and less than an equal weight at 212°. It is resolved by heat into sulphuric acid and the neutral sulphate.

369. Chlorate of Potash, KO,ClO_5.—This salt is formed, together with chloride of potassium, by passing a current of chlorine through a strong solution of pearlash. The reactions are as follows:—

$$6KO,CO_2 + 6Cl = 5KCl + KO,ClO_5 + 6CO_2.$$

The carbonic acid escapes during the process, and the chloride of potassium being very soluble remains in solution, while the chlorate crystalizes in shining white scales.

The chlorine, before entering the potash solution, should be made to pass through a three-necked bottle containing a little

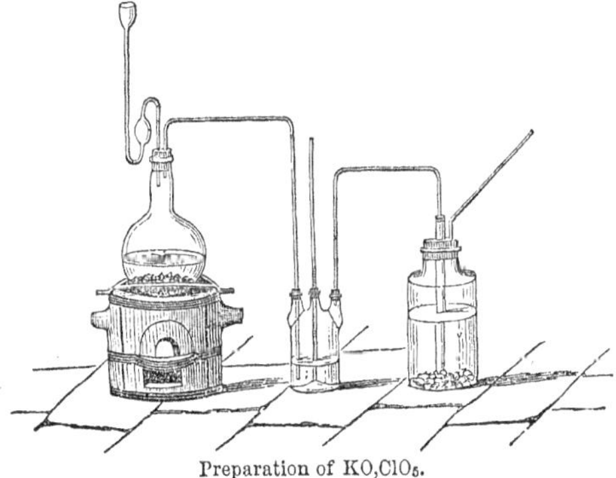

Preparation of KO,ClO_5.

water, to separate any sulphuric acid which may pass over with it. The proper arrangement is represented in the figure.

QUESTIONS.—368. How is bisulphate of potash formed? 369. How is chlorate of potash formed? Describe the reactions that take place.

Its taste is not unlike that of nitre, but it is much less soluble than that salt. Heated moderately, it melts, and at a red heat is decomposed, giving up the whole of its oxygen. Thrown on burning charcoal, it deflagrates like nitre, but more energetically. A few of the crystals, wrapped in tin-foil, with a piece of phosphorus, or a little sulphur, explode violently by a blow from the hammer. It was formerly used in the manufacture of Lucifer matches, and has been substituted for nitre in gunpowder; which, however, when thus prepared, is liable to explode from causes so slight that its manufacture is dangerous.

The *silicates* of potash will be described hereafter.

370. **Sulphur-Salts of Potassium.**—Protosulphide of potassium, as the electro-positive element, combines with many other electro-negative sulphides, forming true salts (345), but we shall here describe only the following two:—

Hydrosulphate of Potassium, or, more properly, the **hydrosulphate of sulphide of potassium**, KS,HS, is prepared by passing a current of hydrosulphuric acid (263) through a solution of potash. When the solution is concentrated the salt may be obtained in crystals.

Carbosulphate of potassium, KS,CS_2, may also be crystalized. It is prepared by pouring bisulphide of carbon into an alcoholic solution of protosulphide of potassium.

SODIUM.

Symbol, Na (*Natron*); *Equivalent*, 23; *Density*, 0·972.

371. **History and Preparation.**—Sodium was discovered in 1807, by Davy, a few days after the discovery of potassium. The first portions of it were obtained by means of galvanism; but it may be procured in much larger quantity by chemical processes, precisely similar to those just described for obtaining potassium. Its preparation is less difficult than that of potassium.

QUESTIONS.—Describe the properties of chlorate of potash. What is the effect when it is thrown on burning charcoal? What use has been made of it? Why may it not be used in the formation of gunpowder? 370. How is hydrosulphate of potassium formed? What is its composition? What is the composition of carbosulphate of potassium? 371. How is sodium prepared?

372. Properties.—Sodium has a strong metallic lustre, and in color is very analogous to silver. It is so soft at common temperatures, that it may be formed into leaves by the pressure of the fingers.

Sodium soon tarnishes on exposure to the air, though less rapidly than potassium. Like that metal it is instantly oxydized by water, hydrogen gas in temporary union with a little sodium being disengaged. When thrown on cold water, it swims on its surface and is rapidly oxydized, though in general without inflaming; but with hot water it scintillates, or even takes fire, and burns with a beautiful yellow flame, which readily distinguishes it from potassium.

By throwing two pieces, one of sodium and another of potassium, at the same time, into a vessel of water, both will usually be inflamed; and the characteristic colors of their flames will be seen together.

Sodium is preserved under naphtha in the same manner as potassium.

Binary Compounds of Sodium.

373. Protoxide of Sodium—NaO; eq., (23 + 8 =) 31.—This compound, usually called *soda*, is formed by the oxydation of sodium, as potassa is from potassium. With water it forms a solid hydrate, which is easily fusible, and very soluble both in water and alcohol. It is a powerful alkali, and very similar in all its properties to potassa. Hydrate of soda is prepared from the carbonate in the same manner as the hydrate of potash. The hydrate is known as *caustic soda*;—it contains a single equivalent of water, and its formula is therefore NaO,HO.

Peroxide of Sodium, NaO_3, is formed by burning sodium in dry air or oxygen gas.

QUESTIONS.—372. Describe the properties of sodium. What is the effect when it is thrown upon water? How is it preserved? 373. What is soda? Describe its properties. By what name is the hydrate of soda known? What is peroxide of sodium?

374. Chloride of Sodium—$NaCl$; eq., (23 + 35·4 =) 58·4.—This is the *common salt* of commerce. It may be formed by burning sodium in chlorine, but is obtained in great abundance as a solid deposit, called *rock salt*, in various parts of the world, as in England, Poland, and at Abingdon in Virginia; and in solution in the waters of brine springs, which abound in New York, Pennsylvania, Virginia, Kentucky, Ohio, and other States, as well as in other countries. Sea-water contains about 2·7 per cent. of this substance, while the water of the great salt lake in Utah Territory contains about 20 per cent. Around this latter body of water are plains, many miles in extent, which, in the dry season, are covered with incrustations of very pure salt to the depth of more than half an inch. In the rainy season it is more or less dissolved.

Rock salt is sometimes mined of sufficient purity for use, and requires only to be pulverized; but more frequently it is mixed with clay and oxide of iron, and must then be dissolved, and the pure liquid drawn off and evaporated.

In warm countries, as on the coast of Portugal, in the south of France, and the West India Islands, this substance is obtained by the spontaneous evaporation of sea-water, which is allowed, on the rise of the tide, to flow into shallow basins, being passed from one to another, as it becomes more concentrated; and finally, the evaporation is finished by means of artificial heat.

In cold countries, as on the borders of the White Sea, the process is commenced in a very different manner:—the sea-water is exposed to the cold atmosphere, by which a large part of the water is separated in the form of ice, and the remaining liquid portion is drawn off and evaporated.

Pure chloride of sodium has an agreeably saline taste. It fuses at a red heat, and becomes a transparent brittle mass on cooling. It deliquesces slightly in a moist atmosphere, but undergoes no change when the air is dry. In pure alcohol it is insoluble. It

---

QUESTIONS.—374. What is the common name for chloride of sodium? Where is it found in the solid state? What are brine springs? What is the proportion of it contained in sea-water? In the water of the great salt lake in Utah Territory? How is this substance procured in certain warm countries? How in certain cold countries? Describe some of the properties of common salt?

requires twice and a half its weight of water at 60° for solution, and its solubility is not increased by heat. It crystallizes in cubes, which are anhydrous, and have a density of about 2·13 — when its solution is slowly evaporated in the open air *hopper-shaped* crystals, as figured in the margin, are of common occurrence. Their formation may be explained as follows:—first a small cubical crystal forms at the surface of the solution, which tends to sink and depress the surface, as shown in the figure A. Soon

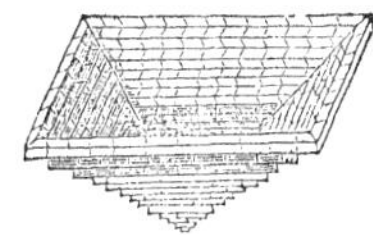

Hopper-form Crystals.

A B

Cryst. Common Salt.

other small crystals form and attach themselves to the first one at its four upper horizontal edges, by which it is a little more depressed, as represented in the figure B; a further addition in the same manner causes a further depression, as seen in C, and so on.

C

Cryst. Common Salt.

A saturated solution of common salt does not freeze even at 0°, but hydrated crystals are formed which have the formula, $NaCl + 4HO$. It fuses at a red heat, and may even be sublimed without change.

The uses of this substance are well known. Besides the ordinary purposes to which it is applied, in preserving meat from putrefaction, and in seasoning food, it is used extensively in the arts, in glazing pottery-ware, in the manufacture of bleaching-salt, carbonate of soda, hydrochloric acid, &c.

The name, *salt* (Lat., *sal*), was originally given to this substance alone, but was subsequently extended to an immense class of compounds which have been known as salts. By our present

---

QUESTIONS.—Describe the mode in which hopper-formed crystals are sometimes produced. What are the uses of this substance? Does it belong to the family of *salts*, according to the definition of the word which we have adopted?

technical arrangement (347), it is entirely excluded from the class.

Sodium forms definite compounds with *iodine*, *bromine*, *fluorine*, *sulphur*, &c., but they are not described in this work.

Salts of Soda.

375. Sulphate of Soda, NaO,SO_3.—Sulphate of soda, or *Glauber's salt*, is sometimes found native in dry situations, but more frequently in solution in the waters of mineral springs. It is also obtained in the manufacture of hydrochloric acid (236). It was first made known by Glauber, from whom it received its name, although he himself called it *sal-mirabile*. It has a cooling, saline, and somewhat bitter taste; and is very soluble in water at a temperature 91° or 92°, but less so in water that is very cold or very hot (350). The crystals of this salt usually contain more than half their weight of water of crystalization, which escapes when they are exposed to the open air, and they crumble into a white powder. Their proper formula is NaO,SO_3+10HO. The water therefore constitutes nearly 56 per cent., as will be found by making the calculation.

Sulphate of soda is used in medicine as a cathartic, and for the preparation of the carbonates of soda.

Bisulphate of Soda may be obtained in crystals which have the formula, $NaO,2SO_3+3HO$. Deprived of its water of crystalization, it may be used in preparing anhydrous SO_3 (259).

376. Hyposulphite of Soda, NaO,S_2O_2.—This salt, which is much used in the daguerreotype process, for removing the sensitive coating from the silver plate, after being taken from the mercurial vapor bath, is prepared by first passing a current of sulphurous acid gas through a solution of carbonate of soda, to form sulphite of soda, and then dissolving sulphur in a concentrated hot solution of the sulphite.

It may be obtained in crystals, which, according to some, contain 5, and according to others, 10 equivalents of water. It is also called *dithionate of soda*.

QUESTIONS.—375. What is Glauber's salt? What are some of its properties? What is said of its water of crystalization? What use is made of it? 376. What use is made of hyposulphite of soda?

377. Carbonate of Soda, NaO,CO_2.—The carbonate of soda of commerce was formerly obtained by lixiviating the ashes of sea-weeds, in the same manner as the carbonate of potassa is obtained from the ashes of land-plants. But it is now manufactured altogether from common salt, which is first converted into sulphate of soda by sulphuric acid, and then the sulphate, mixed with charcoal and carbonate of lime, is heated intensely in a wind-furnace.

The materials, which consist of about 2 parts of the anhydrous sulphate, 2 parts of chalk (carbonate of lime), and 1 of charcoal, are well ground together, and introduced upon the hearth, H H, of a reverbatory furnace, similar to that represented in section in the figure; and by continued action of the heat, carbonate of soda, NaO,CO_2, oxysulphide of calcium, CaS,CaO, and carbonic oxide, CO, are formed:—the latter compound being gaseous, of course passes off. The oxysulphide of calcium being insoluble in water, it is now only necessary to digest the black mass which comes from the furnace in warm water, and filter, and a solution of carbonate of soda is obtained. This is now evaporated to dryness. It may be obtained in crystals, which always contain much water of crystalization, and effloresce in dry air. It is the *sal soda* of commerce.

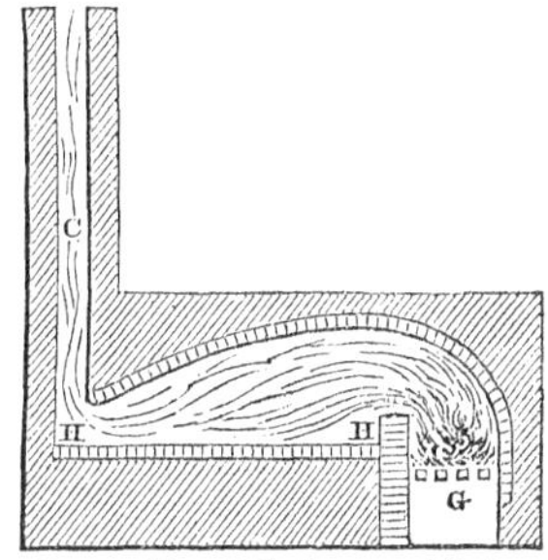

Preparation of Carbonate of Soda.

The mixture as taken from the furnace is the *soda ash*, or *British barilla* of commerce, and has been sometimes used as a manure.

Carbonate of soda is extensively used in the manufacture of glass and hard soap, and for other purposes.

Sesquicarbonate of Soda, $2NaO,3CO_2$, called *trona*, is found in the waters of certain lakes in Egypt, in Hungary, and in this country in springs among the Rocky Mountains.

QUESTIONS.—377. From what was carbonate of soda formerly obtained? How is it now manufactured? Describe the process.

378. Bicarbonate of Soda, $NaO,2CO_2$.—This salt is formed by exposing the carbonate in solution to an atmosphere of carbonic acid, in the same manner as the bicarbonate of potash. Like the corresponding salt of potash, it always contains one atom of water, and may be considered a double carbonate of soda and water, according to the formula, NaO,CO_2+HO,CO_2. It is often used by bakers as a substitute for sal-æratus.

379. Biborate of Soda, $NaO,2BO_3$.—This salt occurs in solution in the waters of certain lakes in Thibet and the East Indies, from which it is obtained by evaporation, and was formerly imported into this country and England under the name of *tincal.* When refined, by solution and recrystalization, it is sold as *borax,* a substance well known for its extensive use in the arts in various metallurgic operations. At present, most of the borax of commerce is obtained from Tuscany, where it is prepared by adding carbonate of soda to the native boracic acid (326) of the hot springs which abound in an extensive volcanic district of that country. To obtain it pure, several recrystalizations are required.

Ordinary borax crystalizes in right rhombic prisms, which contain 10 equivalents of water; but when crystalized from a hot solution the crystals are octahedrons, and contain only 5 atoms of water.

When borax is heated, it first loses its water of crystalization, which causes it to froth up very much; and at a red heat fuses into a clear transparent liquid, which on cooling has the appearance of glass. At high temperatures, it dissolves most of the metallic oxides, and becomes colored.

Borax is used as a flux in metallurgic operations, in the preparation of certain kinds of glass, and in medicine.

380. Nitrate of Soda, NaO,NO_5, resembles nitrate of potassa in many of its properties, but cannot be substituted for it in the manufacture of gunpowder, because of its tendency to absorb

---

QUESTIONS.—378. What is bicarbonate of soda? What use is made of it? 379. Where is biborate of soda obtained? What is it often called? Where is most of the borax of commerce obtained at the present time? What use is made of it? 380. For what is nitrate of soda used?

moisture from the atmosphere. It is used instead of nitrate of potash in the preparation of nitric acid, and sometimes as a manure.

381. Phosphates of Soda.—Phosphoric acid forms with soda (and the bases) three series of salts, viz., *tribasic* or *ordinary phosphates*, *bibasic* or *pyrophosphates*, and *monobasic* or *metaphosphates*, corresponding to the three states (282) of the acid.

I. *Tribasic Phosphate of Soda.*—Of this there are three varieties, viz.: 1. The salt, $3NaO,PO_5$. 2. The salt, $(2NaO + HO) PO_5$. 3. The salt, $(NaO + 2HO) PO_5$. The three varieties are tribasic, but the base of the first consists of 3 eq. of soda; the base of the second of 2 eq. of soda and 1 eq. of water; the base of the third of 1 eq. of soda and 2 eq. of water. The water serving as base in such salts is called *basic water*.

The first two varieties usually crystalize with 24 eq. of water, and the third with 2 eq. of water. If heated, they readily give up their water of crystalization, but a red heat is required to expel the basic water. All of them in solution give a yellow precipitate, $3AgO,PO_5$, with solution of nitrate of silver.

II. *Bibasic Phosphate of Soda—Pyrophosphate of Soda.*—This compound furnishes two varieties, viz., 1. The salt, $2NaO,PO_5$; 2. The salt, $(NaO + HO) PO_5$. Both varieties are bibasic, but the base of the first consists of 2 eq. of soda, and that of the second of 1 eq. of soda and 1 eq. of water. Bibasic phosphate of soda crystalizes with 10 eq. of water;—the solution of both varieties gives, with nitrate of silver, a white precipitate, $2AgO,PO_5$.

III. *Monobasic Phosphate of Soda—Metaphosphate of Soda* NaO,PO_5.—This phosphate in solution gives a white precipitate with solution of nitrate of silver, AgO,PO_5; but its composition, it will be observed, differs from that procured from the bibasic phosphate. Solution of this phosphate also has the property of

---

QUESTIONS.—381. What is said of the salts formed with soda by phosphoric acid? What varieties of tribasic phosphate of soda are there? How may they be tested when in solution? What varieties of bibasic phosphate of soda are there? What is said of the precipitate they give with nitrate of silver? What is metaphosphate of soda?

coagulating the whites of eggs, an effect not produced by the other phosphates.

For the methods of preparing these varieties and sub-varieties of phosphate of soda, the inquiring student will consult larger works on this science, especially the excellent one of Regnault; the object of the present work not permitting so much minute detail.

With other bases phosphoric acid probably forms similar series of salts, but the subject has not been fully investigaged.

382. **Characteristics of Potash and Soda Salts.**—All the salts of potash and soda that are soluble are distinguished from other metallic salts, except the salts of lithia, which are very rare, by giving no precipitate with solutions of the alkaline carbonates. It is therefore sufficient, practically, to be able to distinguish between these two classes of salts; for which the following tests will suffice.

With tartaric acid potassa forms a sparingly soluble salt, which, if the solution is moderately concentrated, appears as a white precipitate; but with soda no precipitate is formed, as the corresponding salt of soda is very soluble.

With potash a strong solution of chloride of platinum forms a yellow precipitate, the double chloride of potassium and platinum, which becomes more copious by the addition of alcohol; but in the same circumstances no precipitate is formed by the salts of soda, because of the solubility of the double chloride of platinum and sodium.

LITHIUM.

Symbol, L; *Equivalent*, 6·4; *Density*, —?

383. **History, Etc.**—Lithium is a very rare substance, and is found only in a few minerals, as *spodumeme*, and the variety of mica called *lepidolite*. It is obtained from these in combination with oxygen as the protoxide, *lithia*. From this the metal may be procured with some difficulty by means of galvanism. It is a white metal, like sodium. The protoxide, lithia, is a powerful alkali, like potash or soda, but is less soluble The name is from the Greek, *lithos*, a stone, in allusion to the source from which it is obtained.

The salts of lithia, heated before the blowpipe, give a red tinge to the flame.

QUESTIONS.—382. How are the soluble salts of potash and soda distinguished from other salts? Describe the mode of distinguishing a soluble salt of potash from one of soda by means of tartaric acid. By means of solution of chloride of platinum. 383. What is said of lithium? From what is the name derived?

Ammonium. (*Not Isolable.*)

384. History, Etc.—This name is given to a supposed compound of nitrogen and hydrogen, NH_4, which has never yet been obtained in a separate state, but is believed by many to enter into the composition of most of the ammoniacal compounds, and to possess in some respects the characters of a metal. Its symbol is written NH_4, or Am.

When strong aqua ammoniæ (225), in contact with a little mercury, which is connected with the negative electrode of the galvanic battery, is subjected to the action of a strong electrical current, oxygen is liberated at the positive electrode and the mercury increases very much in volume, and becomes less fluid, having the consistency of butter or soft lard, but still retains perfectly its metallic lustre. This has very much the character of an amalgam; and we may suppose that under the influence of the current, ammonium, NH_4 (equal to $NH_3 + H$) has been formed from the ammonia and the hydrogen of the water, and at once united with the mercury, the oxygen escaping at the other electrode.

The same compound may also be obtained simply by combining a little potassium or sodium with 50 or 100 times its weight of mercury, and pouring on it a strong solution of sal ammoniac, NH_3,HCl. In this case the reaction seems to be

$$KHg + NH_3,HCl = KCl + NH_4,Hg.$$

This compound, called *ammoniacal amalgam*, when removed from the solution in which it was formed, rapidly undergoes spontaneous decomposition, yielding ammonia and hydrogen in the proportion of 2 volumes of the former and 1 volume of the latter. After a little time the mercury alone is found entirely unchanged.

QUESTIONS.—384. What is the compound to which the name ammonium is given? Has it been obtained in a separate state? What is the mode of preparing ammoniacal amalgam by the use of the galvanic battery? Explain the reactions by which we may suppose the new substance to be produced. Describe the mode of producing it by the use of sal ammoniac in solution. What are the reactions that appear to take place? What is the effect when the amalgam is removed from the solution? What are obtained when it decomposes?

If this amalgam is subjected to a temperature of 32°, it crystalizes in cubes, and its decomposition is retarded.

Although this compound, NH_4, which has received the name of ammonium, has not been obtained in a separate state, its existence in combination with other bodies would seem to be established; and also its peculiar metallic character. It is capable of replacing potassium and sodium in combination, and is therefore isomorphous with them.

Ammonia, NH_3, has heretofore (224) been described. It has been called *volatile alkali*, because of its reactions with other substances, and especially with the acids, being the same as those of the other alkalies, the protoxides of potassium, sodium, and lithium. We introduce the subject again for the purpose of describing some of its more important compounds, which we shall do according to this *ammonium theory*, as it is called; because it affords us, in the present state of our knowledge, the most simple and lucid view of this important class of bodies that can be presented. But, at the same time, it should always be kept in mind that the existence of ammonium, NH_4, even in combination with other substances, is not to be considered as fully determined.

Binary Compounds of Ammonium.

385. Protoxide of Ammonium, NH_4O, or AmO.—As is the case with ammonium, the existence of this compound is hypothetical. But all the ammonia salts of oxygen acids, contain an atom of water in their composition, which appears to be essential to their existence. This atom of water, combined with the ammonia, NH_3, forms the compound in question, protoxide of ammonium, NH_4O, which unites with the acid to form the salt. Thus, the composition of nitrate of ammonia, as formerly supposed, is NH_3,NO_5,HO, which evidently is the same as NH_4O,NO_5, except as to the mode of the arrangement of the particles.

QUESTIONS.—How is this amalgam affected by a cold of 32°? What is said of the relation of ammonium to potassium and sodium? Why has ammonia been called volatile alkali? 385. What is said of protoxide of ammonium?

As ammonium is isomorphous with potassium and sodium, so this compound is isomorphous with potash and soda, which it is capable of replacing in many of their compounds.

386. Chloride of Ammonium, NH_4Cl.—This is the compound often called *sal-ammoniac,* and hydrochlorate of ammonia. If it be considered as a proper hydrochlorate of ammonia, its formula of course will be NH_3,HCl.

It may be obtained by neutralizing carbonate of ammonia by hydrochloric acid; but for use in the arts it is procured from the liquor obtained in the distillation of bones, in preparing animal charcoal, and also from that which condenses in the manufacture of coal-gas. The latter affords it in large quantities.

Sal-ammoniac is a white solid, very tough, and difficult to pulverize, and has a density of about 1·45. It has a pungent, saline taste, and is very soluble in water; and sublimes without fusion at a temperature below redness. Triturated with recently-slaked lime, it yields ammonia, which is easily recognised by its pungent odor. It is used for various purposes in the arts and in medicine.

Salts of Ammonia.

387. Carbonates of Ammonia.—There are several carbonates of ammonia. The one best known is the *sal-volatile* of the shops, which is a *sesquicarbonate.* It is a semi-transparent solid, which is very soluble in water, and has the pungent odor of ammonia. Its composition, on the "ammonium theory," is $2NH_4O,3CO_2$; but considered without reference to this theory, its formula is usually written $2NH_3,3CO_2+2HO$. By long exposure to the air it is converted into a bicarbonate, $NH_4O,2CO_2 + HO$.

Besides these, there is also a neutral carbonate of ammonia.

Sulphate of Ammonia, NH_4O,SO_3.—Sulphate of ammonia is a soluble salt, isomorphous with sulphate of potassa. It is some-

QUESTIONS.—Is ammonium isomorphous with potassium and sodium? 386. What is sal-ammoniac? How is it obtained? Describe its properties. 387. What carbonates of ammonia are mentioned? Describe sulphate of ammonia.

times found in the lava of volcanos, and may be formed artificially by saturating aqua ammoniæ or solution of carbonate of ammonia with sulphuric acid. It is sometimes used as a manure.

Nitrate of Ammonia, $NH_3,NO_5 + HO$, or, as it is now considered, *nitrate of oxide of ammonium*, NH_4O,NO_5, is prepared by neutralizing nitric acid with ammonia, or its carbonate. It is a white salt, very soluble in water, and destitute of any ammoniacal odor. It is used in preparing nitrous oxide (215).

388. **Phosphate of Soda and Ammonia,** $(NaO,NH_4O,HO)PO_5$.—This is the compound called *microcosmic salt*, and much used as a flux in blowpipe operations. Its crystals contain 8 eq. of water of crystallization, which is given up at a very moderate heat; and, at a high temperature, both the basic water and ammonia are expelled, and the very fusible metaphosphate of soda only remains.

It is prepared by dissolving in 2 parts of hot water 6 or 7 parts of phosphate of soda, and then adding 1 part of sal-ammoniac. On cooling, the salt in question crystalizes, while chloride of sodium remains in solution.

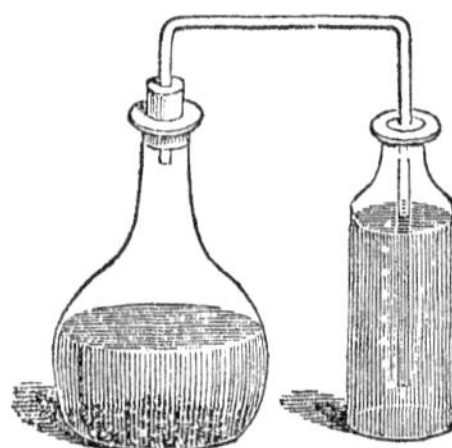

Preparation of NH^4S,HS.

389. **Hydrosulphate of Sulphide of Ammonium,** NH_4S,HS.—This compound, it will be observed, is a sulphur salt (345), being composed of two sulphides, sulphides of ammonium and hydrogen. It is prepared by passing a current of hydrosulphuric acid through aqua ammoniæ, which is to be kept cool during the operation. The apparatus represented in the figure will answer for the purpose.

It is much used in the laboratory as a test for several of the metals.

Silicates of Potash and Soda—Glass.

390. Silica, or, more properly, silicic acid, combines at high temperatures with the alkalies and earths apparently in indefinite proportion, producing compounds which at very high temperatures are more or less liquid, but at a lower heat have a pasty consistency, and when cold are hard, uncrystaline, and more or less

QUESTIONS.—388. What use is made of phosphate of soda and ammonia? How is it prepared? 389. How is hydrosulphate of sulphide of ammonium prepared? To what class of salts does it belong? 390. How is silica made to combine with the alkalies and earths? What is said of the compounds produced?

transparent. All these compounds are known under the name of *glass*.

There are several kinds of glass, all of which are double silicates of potassa and soda, or one of these with silicate of lime, lead, magnesia, baryta, alumina or iron, but the proportions are variable.

Silicic acid has no action upon the bases at ordinary temperatures, but it readily combines with them in a state of fusion, or with their carbonates; in the latter case of course expelling the carbonic acid.

391. When silica is fused with 2 or 3 times its own weight of carbonated potash or soda, a compound is formed which is soluble in hot water, and has been called *soluble glass*, or *liquor silicum*. The solution, when applied to wood and other combustible substances, soon dries and forms a transparent coating which protects them from the air and renders them less combustible when exposed to great heat.

If the proportions are reversed, and 2 parts of silica and 1 part of carbonate of potash or soda are used, a proper glass is formed, which is quite insoluble in water, and nearly all acids.

The proportions of the ingredients in the different varieties of glass are exceedingly variable, but common *window*, or *crown glass*, is always a mixture of silicate of potash or soda and lime.

Crystal, or *flint glass*, is a silicate of potash or soda and oxide of lead; it is softer than other kinds, and more fusible and dense, and therefore better adapted for optical purposes. When a thread of it is heated in the flame of a lamp, it is blackened by the reduction of the oxide of lead.

392. *Bohemian* glass, which is very infusible, contains only silicate of potash and lime. It is much used in the manufacture of chemical apparatus.

The finest kinds of glass are made only of the purest materials; but impure materials, containing alumina and the oxides of iron and manganese, answer for such glass as that of which *green bottles* are made.

QUESTIONS.—What is glass? 391. How is soluble glass formed? Of what is crystal glass composed? 392. What is said of Bohemian glass?

Though, as above stated, glass is considered insoluble in water and the acids, except such as contain fluorine (252), yet certain varieties of it are sometimes attacked by acids, solutions of the alkalies, or of their carbonates, and even by pure water, especially at a boiling temperature. Glass that has been a long time buried in the earth, is sometimes found with a pearly incrustation upon its surface, in consequence of the separation of its alkalies; and is sometimes quite soft, so as to be cut with a knife.

All these different varieties of glass have a density varying from 2·4 to 3·7, but glass may be made of a density as high as 5·4.

Enamel, used for various purposes, and especially for watch and clock faces, is made of silica and potash, or soda, and oxide of lead, and rendered opake by oxide of tin.

Colored glass is made by adding to any of the varieties metallic oxides, as those of cobalt, copper, manganese, antimony, gold, &c. A white opake glass, in imitation of porcelain, is made by adding to the glass when in fusion arsenious acid.

393. Manufacture of Glass. — The materials for glass are first to be fused together, at a high temperature, and in such a manner that no impurities from the smoke of the fire or other source shall be mixed with it. This is done by using pots made of fire-clay, and entirely enclosed within the walls of the furnace, except the projecting mouth. The figure in the margin represents the section of one, with its opening or mouth towards the left hand. Usually several are placed in a circle in the same furnace, and heated by the same fire.

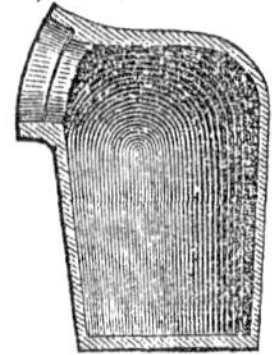
Melting Glass.

The principal instrument used is an iron tube or pipe, four or five feet in length;—one end of this being dipped into the melted glass, which has now the consistency of soft wax, a portion adheres to it, and is removed from the pot. As its shape is irregular, it is first rolled on an iron plate, called a *marver* (see figure), to give it a cylin-

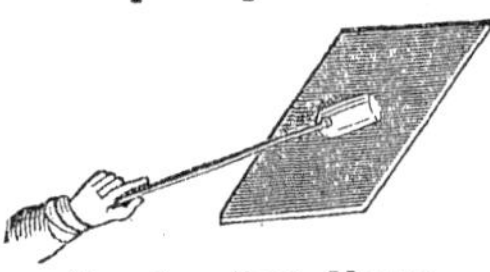
Glass Operations--Marver

QUESTIONS.—Is glass entirely insoluble in water and the acids? Of what is enamel made? How is colored glass formed? 393. Describe briefly the mode of manufacturing window glass.

drical form, and is then blown into a pear-shape (as seen in figure A), by forcing air through the pipe from the lungs. As the glass is still soft, to prevent it from inclining in one direction or another, as it would inevitably if held still for a moment, it is kept constantly whirling, by rolling the pipe in the hands. If held in the position A, the hollow mass will gradually become elongated, and if inverted and held in the position B, the upper part sinks, and it takes the form here seen; — of course, in any particular case, the workman will be guided in the mode of handling by a regard to the form which he wishes to give it. As rapid cooling is constantly taking place, the glass has to be re-heated frequently, which is done by holding it in a heated furnace provided for the purpose.

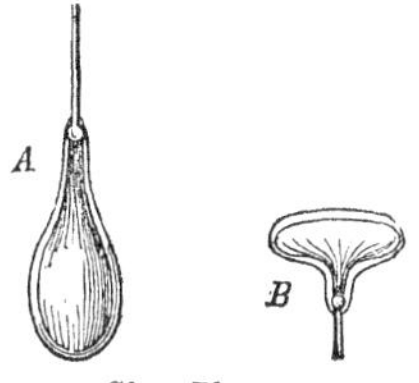

Glass Blowing.

For common window glass, the mass in the form of B, being re-heated to soften it, is held by the rod with the glass downward, and swung backward and forward in the manner of a pendulum, until it is sufficiently elongated, and has the form C;—

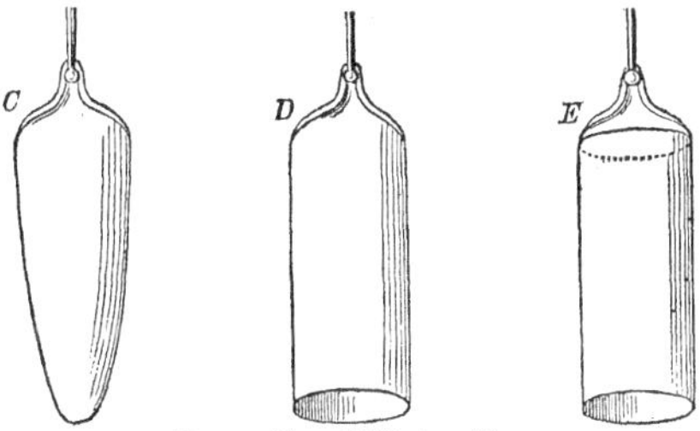

Preparation of Window Glass.

by this time it has partially cooled, and by holding the extreme point a little time in the furnace it is softened so that a blast of air from the lungs is forced through it, and an assistant with shears accurately removes the lower part, giving it the form D. The cylindrical part is now to be separated from the rod by a section around the upper part, as shown in E, and subsequently a longitudinal fracture is made in the hollow cylinder thus obtained

28

through its whole length; and it now only remains to open the cylinder thus prepared in order to reduce it to a perfect plane. This is done by softening it in a proper furnace and pressing the part gently with an iron rod, as represented in the figure.

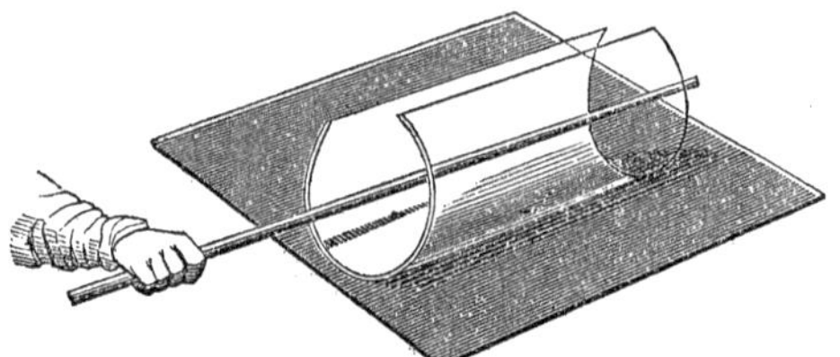

Preparation of Window Glass.

394. The variety of window glass called *crown glass* is prepared in a different mode. The melted mass taken from the melting-pot is first blown into the form of a globe, and then an iron rod attached to it on the side opposite that to which the tube adheres, and the tube separated by applying a little cold water. This, of course, leaves an opening, which becomes enlarged by softening in the heating furnace, and giving it a rapid rotary motion by means of the iron rod held in the hand. By heating it several times in this way, the rapid rotary motion being continued, the globe is at length opened out, and becomes a circular disc, which, after the proper annealing, is cut into panes by a diamond.

395. Many articles now made of glass, are blown in metallic moulds prepared for the purpose, or are pressed between two moulds, one of which shuts into the other, so as to give the proper shape. This is called *pressed glass.*

Plate glass, used for mirrors and for large windows, is poured, when in a state of fusion, upon a plane surface, and a roller passed rapidly over it, to reduce it to the proper thickness. The surfaces are then ground down to a perfect plane by means of

---

QUESTIONS.—394. How is crown glass manufactured? 395. Describe the mode of forming articles of pressed glass. Describe the mode of forming plates for mirrors.

friction with sand and fine emery, and then finely polished by friction with colcothar, or red oxide of iron.

Small articles of glass may readily be made of glass tube before a blowpipe, which is blown by a bellows worked by the foot. The best fuel to be used is burning fluid, consisting of four parts of strong alcohol and one part of camphene; but oil or tallow will answer. A very little experience will enable one to bend even quite large glass tubes, and to perform many other operations of great importance in the laboratory.

All articles made of glass, if suddenly cooled, are exceedingly brittle, and liable to fracture from the slightest causes, even trifling changes of temperature. They are therefore *annealed* by placing them in a furnace prepared for the purpose, and, after becoming quite hot, are allowed to cool very slowly. By this means such a change is produced in the molecular arrangement of the particles, that this tendency to fracture is much diminished.

Articles of glass that have not been annealed, sometimes break in a singular manner;—a tumbler half filled with liquid, and grasped by the hand, will break quite around at the surface of the water by the slight heat of the hand; or a thick glass tube several inches in length will split through its whole length simply by being wet, especially if merely touched by a hard substance, as a piece of wire. Occasionally, after being handled and laid aside, they break spontaneously.

396. The *Bologna*, or *philosopher's vial*, is made in the form of an ordinary vial, but with thicker sides and a very thick bottom, and is not annealed. A smart blow may be given to it by a piece of lead or of wood, or a leaden shot dropped into it, without producing any effect; but by dropping into it a small angular piece of flint, it almost invariably falls to pieces. Sometimes even coarse sand will produce this effect. In this, and in the following case, the result is due to the want of annealing.

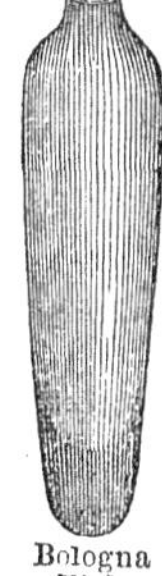
Bologna Vial.

Prince Rupert's drops are simply drops or tears of glass, which are made by allowing the glass when melted to drop from the end of a rod into water, by which they are suddenly cooled. They are perfectly solid, but when the small end is broken off, the whole mass falls to powder, with a slight explosion.

QUESTIONS.—Why are all articles made of glass annealed before using? What is likely to be the effect if the process is omitted? 396. Describe the Bologna vial. What are Prince Rupert's drops?

GROUP II.

BARIUM STRONTIUM CALCIUM MAGNESIUM	*Metals, the protoxides of which are alkaline earths.*—These latter are called *baryta*, *strontia*, *lime*, and *magnesia*. They possess the same properties which characterize the alkalies, but in less degree.

BARIUM.

Symbol, Ba; *Equivalent*, 68·5; *Density*, —?

397. History, Etc.—This metal is procured by passing vapor of potassium over baryta (oxide of barium) at a red heat, or by passing the galvanic current through hydrate of baryta in contact with mercury, the latter forming the negative electrode of the battery. The amalgam thus obtained is carefully heated in a glass tube through which a current of dry hydrogen is constantly passing, and the mercury expelled. The barium will be left in small globules. It has the color and lustre of silver, and melts at a red heat, but is not easily volatilized. In the air it is rapidly oxydized, and when heated burns with a red flame. It is also rapidly oxydized when thrown into water. Its name is from the Greek *barus*, heavy; its compounds generally possessing this characteristic property.

Binary Compounds of Barium.

398. Protoxide of Barium, BaO; eq., (68·5 + 8 =) 74·5.—This compound has been known many years as *barytes* or *baryta*. It may be obtained by decomposing nitrate of baryta by heat, which is best done by using an earthern retort in a furnace (as represented in the figure on next page), and applying the heat as long as gaseous matter is evolved.

QUESTIONS.—What metals are included in the second group? What do their protoxides form? Do these earths possess alkaline properties? 397. How is barium procured? How is the metal affected in the air? From what circumstance or property is the name derived? 398. How may the alkaline earth, baryta, be obtained?

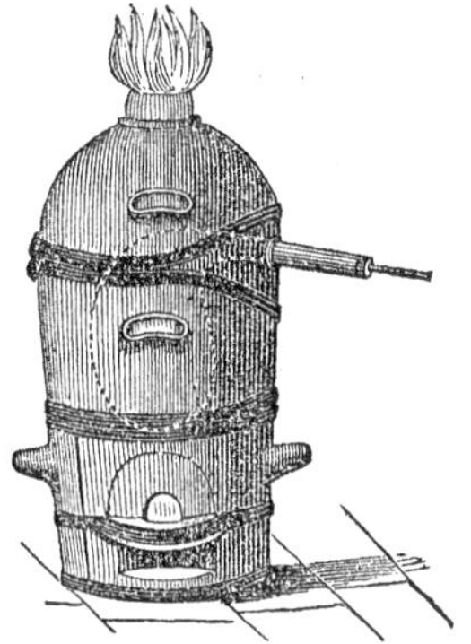
Preparation of Baryta.

Baryta is a gray powder, which slakes like lime when water is poured upon it, and becomes very hot. It dissolves readily in water, but is less soluble than potash or soda — a property by which the alkaline earths are distinguished from the alkalies. It is very caustic to the taste, and affects vegetable colors in the same manner as the alkalies.

Peroxide of Barium, BaO_2.—This oxide may be formed by passing a current of dry oxygen gas over baryta, at a low red heat; or by simply heating baryta in an atmosphere of oxygen. It is used only for the purpose of preparing peroxide of hydrogen (206).

Chloride of Barium, BaCl; eq., (68·5 + 35·4 =) 103·9.—This compound is formed by dissolving the native carbonate of baryta in diluted hydrochloric acid, and by other modes. It crystalizes in white scales, which contain two atoms of water. It is very soluble in water, and is much used as a test for sulphuric acid, with which baryta forms an insoluble sulphate

Salts of Baryta.

399. Carbonate of Baryta, BaO,CO_2, is found native, and called *witherite* by mineralogists. From it all the other salts of baryta may be prepared.

Sulphate of Baryta, BaO,SO_3.—Sulphate of baryta is found abundantly in various places, often in beautiful crystals. It has a density of about 4·4, and is insoluble in water. When pow-

QUESTIONS.—Describe the properties of baryta. Describe the mode of preparing peroxide of barium. How is chloride of barium formed? For what purpose is it used? 399. Is carbonate of baryta found native? What is said of the occurrence of native sulphate of baryta?

dered and mixed with charcoal, and heated intensely, it is converted into sulphide of barium, from which the other salts of baryta may be prepared, as from the native carbonate. By mineralogists, it is called *heavy spar*, because of its great weight. Ground to a fine powder, it is used as a substitute for white lead, either alone or mixed with white lead. All the soluble com pounds of baryta are poisonous.

Nitrate of Baryta, BaO,NO_5.—This salt of baryta is prepared by digesting the native carbonate, or the sulphide, obtained as just described, in nitric acid. Its only use is in certain chemical analyses, and in procuring the earth baryta.

STRONTIUM.

Symbol, Sr; *Equivalent*, 44; *Density*, — ?

400. History, Etc.—Strontium is obtained from its oxide, strontia, in the same manner as barium; and in its appearance it is said very much to resemble that metal. Like barium, also, it decomposes water with the evolution of hydrogen, and oxydizes rapidly in the open air. It receives its name from Strontian, a village in Scotland, near which it was first obtained.

Binary Compounds of Strontium.

401. Protoxide of Strontium, SrO; eq., $(44 + 8 =) 52$.—This compound, which is the earth *strontia*, is formed by the oxydation of strontium. It is prepared also by heating the nitrate of strontia to redness, by which the acid is expelled. It much resembles baryta, seeming to sustain much the same relation to it that soda sustains to potash.

QUESTIONS.—How is sulphate of baryta affected when heated with charcoal? What is it called by mineralogists? For what purpose is it used? How is nitrate of baryta formed? 400. Give the history, &c., of strontium? From what is the name derived? 401. How is the alkaline earth, strontia, procured? What is said of its relation to baryta?

Salts of Strontia.

402. Carbonate of Strontia, SrO,CO_2, is found native;—it is the *strontianite* of mineralogists.

Sulphate of Strontia, SrO,SO_3, is also found native, and is called *celestine.* Treated in the same manner as described for sulphate of baryta (309), the other salts of strontia may be prepared from it.

Nitrate of Strontia, SrO,NO_5, is prepared by dissolving the native carbonate in diluted nitric acid, or from the sulphide, as described under sulphate of baryta. It is much employed in fire-works, to give a beautiful red color to the flame. To show this *red fire,* mix intimately 40 parts of this salt, 13 of sulphur, 5 of chlorate of potash, and 4 of sulphide of antimony, and burn the mixture upon a dry brick, or marble slab, in a dark room. The nitrate contains water, and should be well dried before mixing with the other ingredients. All the compounds of strontia communicate a red tint to flame in which they are heated.

CALCIUM.

Symbol, Ca; *Equivalent,* 20; *Density,* — ?

403. History, Etc.—Calcium is the metallic base of lime, from which it has been obtained, but only in very small quantity. The process is precisely the same as that given above for obtaining barium from baryta. It is said to be of a brilliant white color, and rapidly oxydizes in the air. Little is known of its other properties.

Binary Compounds of Calcium.

404. Protoxide of Calcium—Lime, CaO; eq., $(20+8=)$ 28. —This compound, commonly known by the name of *lime* and *quicklime,* is obtained by exposing carbonate of lime to a strong

QUESTIONS.—402. What is the mineralogical name for carbonate of strontia? Sulphate of strontia? How is nitrate of strontia prepared? What use is made of it? 403. Give the history, &c., of calcium. 404. What is the common name for protoxide of calcium? How is it prepared from the native carbonate?

red heat, so as to expel its carbonic acid. If lime of great purity is required, it should be prepared from pure carbonate of lime, such as Iceland spar, or Carrara marble; but to obtain lime for ordinary purposes, common limestone is used.

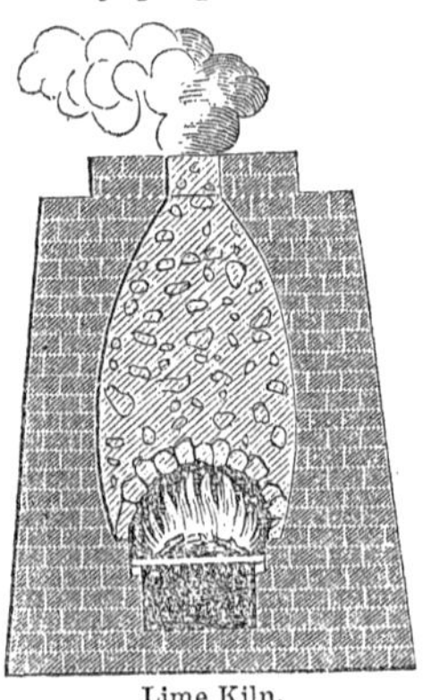
Lime Kiln.

The calcination of the carbonate, to procure lime for common purposes, is effected in kilns, or pits, which are usually constructed of stone, upon a hill-side, so that the limestone may be conveniently introduced at the top, and the lime, after calcination, removed from the opening at the bottom. Wood is very generally used for the fuel, but bituminous coal may be substituted for it, the pieces of limestone being placed so that the flames pass through it. When the calcination is finished, of which the experienced eye can easily judge, the fire is extinguished, and the lime, when cold, removed. The calcination of a large kiln usually require five or six days.

Sometimes the calcination is carried on in perpetual kilns, as they are called. The limestone is then introduced in successive layers, with layers of coal between them; and the fire, once kindled, is continned for an indefinite time, layer after layer of the coal being consumed, and the lime, after calcination, being removed from below. As the mass settles down in the kiln, new layers of limestone and coal are introduced at the top.

Lime is a brittle, white, earthy solid, the specific gravity of which is about 2·3. It phosphoresces powerfully when heated to full redness, and hence its use in the Drummond light (201). It is one of the most infusible bodies known; fusing with difficulty even by the heat of the oxyhydrogen blowpipe.

Exposed to the air, it gradually absorbs carbonic acid, and crumbles to powder. It has also a powerful affinity for water,

---

QUESTIONS.—How is the calcination usually effected? Describe the mode of calcining lime by the use of coal for fuel. Describe the properties of lime.

which is absorbed instantly on being poured upon it; and the combination is attended with great increase of temperature, and formation of a white bulky hydrate. The process of *slaking* lime consists in forming this hydrate, and the hydrate itself is called *slaked* lime. It differs from the hydrates of strontia and baryta, in parting with its water at a red heat. Recently-slaked lime dissolves sparingly in water, and has this singular property, that it is more soluble in cold than in hot water. The solution has a caustic, acrid taste, and acts upon vegetable colors like the alkalies. Exposed to the air, it absorbs carbonic acid; and if agitated, becomes milky, from the formation of insoluble carbonate of lime.

Mortar, for building, is prepared by mixing sand with recently-slaked lime. It becomes very hard by exposure to the air, in consequence of the absorption of carbonic acid by the lime. Combination seems also to take place, to some extent, between the silica of the sand and the lime. When the limestone from which the lime is made contains a considerable portion of silica, alumina, &c., it constitutes *hydraulic cement*, or *water-lime*. Mortar prepared from this, has the property of becoming hard under water, which is not the case with that prepared from pure lime.

405. **Chloride of Calcium,** CaCl.—This compound is formed by dissolving carbonate of lime in hydrochloric acid. It is much used in the operations of the laboratory, especially for removing moisture from gases, which it effects readily in consequence of its great affinity for water. It is often called muriate of lime.

Fluoride of Calcium, CaF, is the *fluor spar*, or *Derbyshire spar* of mineralogists. It is of great use to the chemist as affording the chief source of the element fluorine.

Salts of Lime.

406. Carbonate of Lime, CaO,CO_2.—This is one of the most abundant mineral productions known; it is found in every country

QUESTIONS.—In what consists the *slaking* of lime? Is lime soluble in water? How is mortar prepared? What is hydraulic cement, or water-lime? 405. How is chloride of calcium prepared? What use is made of it? 406. What varieties of carbonate of lime are mentioned?

as *limestone*, *chalk*, *Iceland spar*, *marble*, &c. It is decomposed by heat, and furnishes the *quicklime* used in preparing mortar.

The beautiful *stalactites*, frequently seen suspended from the roofs of caverns, are formed of this compound. Though insoluble in pure water, it is slightly soluble in water containing an excess of carbonic acid. Water permeating the soil above the caverns, first becomes charged with carbonic acid, and afterwards takes up a little carbonate of lime, which is again deposited as the water, drop after drop, hangs suspended for a time from the roof. This is occasioned partly from the evaporation of the water, and partly by the escape of the carbonic acid, when the water becomes exposed to the free air of the cavern. As a portion of the water falls to the bottom of the cavern, corresponding deposits of carbonate of lime, called *stalagmites*, are formed upon the floor, and gradually build themselves upward. Sometimes a stalactite from the roof is formed downward until it reaches the corresponding stalagmite from below, when the two, becoming connected, form

Stalactites.

a column or pillar, as shown at the left in the figure, which is from Knapp's Chemical Technology.

407. Sulphate of Lime, $CaO,SO_3 + 2HO$.—This compound is well known as *gypsum*, and *plaster of Paris*. Pure, crystalized

QUESTIONS.—How are stalactites in caverns formed? What are the corresponding deposits on the floor of the cavern called? 407. What varieties of sulphate of lime are mentioned?

specimens are sometimes called *selenite,* and compact varieties, *alabaster.* Common gypsum contains considerable water, which may be expelled by heat; but there is a variety destitute of water, called *anhydrite* by mineralogists. When powdered gypsum, the water of which has been expelled by a moderate heat, is again made into a paste with water, it soon becomes hard, or "sets," as the workmen say;—a property which adapts it admirably for many purposes in the arts. In stereotyping, a coat of this paste is spread carefully over a page of type, set in the ordinary manner, which, soon becoming hard, is removed, and a cast in common type-metal taken from it. This, after certain preparations, and the emendation of any broken letters that may be found, constitutes a *stereotype plate,* used in printing.

In a very similar manner it is used for preparing busts of living persons. The process is conducted as follows: The individual is prepared by removing the clothing from his neck and shoulders, coating the hair with paste, and applying a little oil or soap to the part which is to be covered by the plaster. He then places himself on his back on a table, his head being supported about an inch above the table by a small block of wood, and surrounded on three sides by a box prepared for the purpose, as represented in the figure. The operator, having his calcined plaster properly mixed with water, pours a portion of it into the box, so that it may fill up the space under the head; and continues to mix small portions at a time, and add it to the mass, until the whole head and face are inclosed, except a small opening at the nostrils. The whole soon becomes hard, attended by a considerable elevation of temperature, but not so much as to be uncomfortable to the subject of the operation.

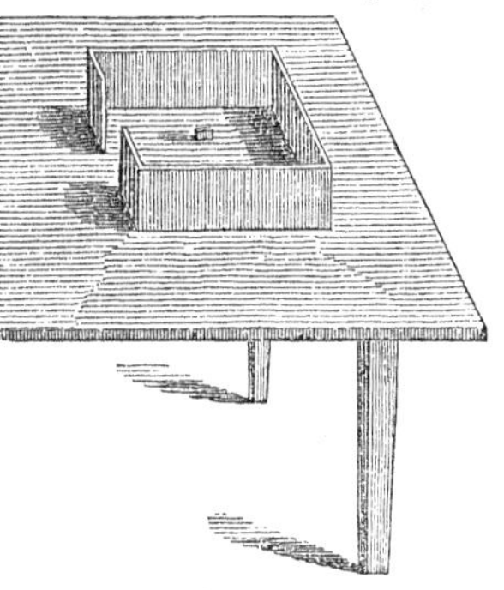

Formation of Busts.

In order to remove this plaster inclosure, the operator has taken the precaution, before applying the plaster, to draw around the head two pieces of thread or twine, one so that it shall come just below the ears, as the person lies upon the table, and the other just above them, bringing the ends of both around under the chin; and just as the plaster is about to set, taking one of these threads by the two ends, he pulls it out laterally so as to divide the mass into two parts, as if cut with a knife.

QUESTIONS.—What is the effect when powdered gypsum is exposed to a moderate heat? What now is the effect when it is mixed with water? Describe the mode of forming plaster busts of living persons.

When the second thread has been drawn out in this way, the plaster inclosure of the face and head will of course be divided into three parts, the upper part covering the face, and beneath this a ring covering the ears, and below this the third part covering the back of the head and neck. The part covering the face is now to be carefully removed, which may be done without difficulty; but the second piece which incloses the ears, will have to be divided below the chin and at the top of the head, and removed in two pieces. This being done, the subject of the operation is again at liberty to remove himself from the table, leaving the remaining piece in its place.

The four pieces being brought together, each in its proper place, it is evident that the operator has a perfect model or mould of the head and all the features of the face, from which the bust is to be prepared; but a further description of this part of the process will not be needed.

If a part of the breast is to be included with the bust, the arrangements must of course be made accordingly, in preparing the mould. The operation requires some labor, but is less disagreeable to the subject than might be supposed before making the trial.

Sulphate of lime is extensively used as a manure in many countries, with excellent effect. It is slightly soluble in water, and is often found in well and spring water, and gives it the property called hardness.

Phosphates of Lime.—There are several of these salts. One variety is found native, and is called *apatite;* it is an essential ingredient of all fertile soils, and is contained in all varieties of grain used for bread. It also constitutes the chief part of the solid matter of the bones of animals.

408. Hypochlorite of Lime, CaO,ClO.—This is the well-known *chloride of lime*, *bleaching-powder*, or *bleaching-salt* of commerce. It is formed by passing a current of chlorine through recently-slaked lime. It is a white powder, and emits a faint odor of chlorine. Great use is made of it in bleaching (229). For this purpose it is dissolved in water, and the articles to be bleached soaked in the solution, and then dipped in very dilute acid. The chlorine which is thus liberated produces the bleaching effect. The process is usually several times repeated. It is also used as a disinfecting agent, the chlorine, as it is gradually liberated, having the property of destroying deleterious gases present in the atmosphere.

The commercial value of bleaching-salt depends entirely upon the quantity of chlorine it is capable of evolving when used, and is usually determined by the quantity of indigo a given weight of it will bleach.

QUESTIONS.—What is said of the phosphates of lime? 408. What is hypochlorite of lime? What use is made of it?

Tests of Lime.—The proper test for lime is oxalic acid, which forms with it, in solution, an insoluble white precipitate. Oxalate of ammonia is generally used. A salt of lime dissolved in alcohol gives to the flame a red color, very similar to that communicated by strontia (402), but of a slightly different tint.

MAGNESIUM.

Symbol, Mg; *Equivalent*, 12; *Density*, 1·87

409. History, Etc.—Magnesium is obtained from its chloride by passing vapor of sodium or potassium over it when heated to redness in a glass tube; the alkaline chloride formed, and any undecomposed chloride of magnesium which may remain, are washed out with cold water, and the metallic magnesium subsides.

It is a white metal, of considerable brilliancy, and quite malleable. Heated in the open air, it readily takes fire, and burns with a brilliant flame, producing the protoxide of the metal. It is rapidly oxydized by boiling, but not by cold water.

Binary Compounds of Magnesium.

410. Protoxide of Magnesium—Magnesia, MgO.—Magnesia is best obtained by heating the carbonate to redness, by which the carbonic acid is expelled. It may also be prepared by decomposing nitrate of magnesia by heat. It is a soft, white powder, and is usually sold under the name of *calcined magnesia*. It is very slightly soluble in water, requiring for this purpose 5000 or 6000 times its own weight of water. Hydrate of magnesia, MgO,HO, is found native at Hoboken, New Jersey, and other places. Magnesia is very infusible, and communicates this property to minerals in which it predominates, as talc and soapstone.

Magnesia is extensively used in medicine as an antiacid. In

QUESTIONS.—What tests of lime are mentioned? 409. How is magnesium obtained? What is said of the metal? 410. Describe the protoxide of magnesium. What use is made of magnesia?

the state of hydrate it is said to be a good remedy in cases of poisoning with arsenic.

Chloride of Magnesium, MgCl.—Chloride of magnesium, which, as we have just seen, is made use of to obtain the metal, is best procured by dissolving magnesia in hydrochloric acid, and adding to the solution an excess of sal-ammoniac; and, after expelling the water, heating the residue in a platinum crucible, by which means the sal-ammoniac used will be driven off. Without the sal-ammoniac, the chloride of magnesium would be decomposed by the heat required to expel the water.

Salts of Magnesia.

411. **Carbonate of Magnesia**, MgO,CO_2.—Carbonate of magnesia is found native, in the *magnesite* of mineralogists, and may also be formed from the native sulphate by precipitation with an alkaline carbonate. It is nearly insoluble in pure water, but dissolves in water impregnated with carbonic acid, forming the liquid magnesia of the shops. When obtained by precipitation with an alkaline carbonate, it always contains a portion of hydrate of magnesia, and is usually seen in beautiful square blocks, which are remarkable for their lightness.

It is extensively used in the practice of medicine.

Sulphate of Magnesia, MgO,SO_3.—This is the well-known *Epsom salt*, used in medicine. It is not unfrequently found in the waters of mineral springs, as at Epsom, in England, and may readily be formed by dissolving magnesia, or its carbonate, in sulphuric acid, and by the action of this acid upon the mineral called dolomite, which is a double carbonate of magnesia and lime. Its crystals contain 7 eq. of water of crystalization.

It is very soluble, and has a bitter, saline taste. It may readily be distinguished from sulphate of soda by the form of its crystals, or by pouring into a solution of it some caustic potassa, which will cause a white precipitate. In sulphate of soda, no precipitate will be formed.

QUESTIONS.—Describe the mode of preparing chloride of magnesium. 411. Describe the carbonate of magnesia. What is the common name of sulphate of magnesia? Where is it sometimes found? How may it be distinguished from sulphate of soda?

Silicates of Magnesia, of which there are several, abound in nature, especially in the talcose and serpentine rocks.

There is no specific test of magnesia, but it is distinguished from other substances by different tests; and from most of its soluble salts phosphate of soda, with ammonia, separates a white precipitate, which is a double phosphate of magnesia and ammonia.

Group III.

Metals	
Aluminum Glucinum Zirconium Thorium Yttrium Erbium Terbium Cerium Lanthanum Didymium	*Metals, the protoxides or sesquioxides of which are earths.*—The oxides of these metals, which constitute the earths, are called *alumina, glucina, zirconia, thorina, yttria,* &c. They are distinguished from both the alkalies and alkaline earths by being quite insoluble in water, and, of course, destitute of any alkaline reaction. They, however, combine readily with, and neutralize the most powerful acids.

ALUMINUM.

Symbol, Al; *Equivalent,* 13·7; *Density,* 3·7.

412. History, Etc.—The metal aluminum is obtained by decomposing chloride of aluminum by the action of sodium or potassium, in the same manner as magnesium is prepared from its chloride. It is found, after the process, in small globules, which may be brought into a single mass by melting it in a close crucible, or under dry chloride of sodium. Instead of chloride of aluminum, the mineral called *cryolite,* which is a double fluoride of aluminum and sodium, may be used in its preparation.

Aluminum is very malleable, and has the brilliant lustre and white color of silver. It is not oxydized in the atmosphere at ordinary temperatures, but burns brilliantly when heated to redness. Cold water does not affect it. A small piece that had been rolled was found to have a density of 3·7, as given above.

The name of this metal, *aluminum,* is derived from *alum,* a double sulphate of alumina and potassa, from which the earth is very readily obtained.

Questions.—What metals belong to Group III.? How are they characterized? 412. How is metallic aluminum prepared? What native mineral may be used for the purpose, instead of the chloride? Describe the metal.

Binary Compounds of Aluminum.

413. **Sesquioxide of Aluminum,** Al_2O_3.—This is the earth *alumina*, and is one of the most abundant of nature's productions. Like silica, it is found in every soil, and in almost all rocks upon the face of the earth. Crystalized, it forms the *ruby* and the *sapphire*, two of the most valuable gems. *Emery*, also, so much used in the arts, is chiefly composed of this earth. It forms a large part of clay, and gives to it its tenacious character, fitting it for the use of the potter.

Pure alumina is a white powder, without taste or smell. It is easily prepared by pouring solution of caustic potash into a solution of alum, and washing and heating the soft mass that is precipitated. It contracts much in drying, and the dried mass adheres tenaciously to the tongue when applied to it.

Alumina, though usually serving as a base in the compounds which it forms, occasionally becomes the electro-negative element, and serves the part of an acid. Thus, it combines with potassa to form aluminate of potassa, and with baryta in like manner. The mineral called *spinelle* is a native aluminate of magnesia.

This earth is remarkable for its tendency to unite with organic substances. If a cotton cloth is immersed in a solution of acetate of alumina, the earth will deposit itself completely on the fibres of the cotton and leave the acetic acid free. On this principle depend some of the most important processes in calico-printing.

When heated with nitrate of cobalt, it forms a beautiful blue compound.

Native hydrate of alumina is occasionally found, as *gibbsite* and *diaspore*.

Chloride of Aluminum, Al_2Cl_3.—This compound is interesting as furnishing the means of obtaining the metal aluminum. For this purpose it must be anhydrous, and is obtained by passing a current of dry chlorine through a mixture of alumina and charcoal, heated to redness, in a porcelain tube.

QUESTIONS.—413. What is said of the abundance of alumina? What gems are mentioned as composed of this earth? Describe alumina. Does it, in combination, act as a base or an acid? What is said of its tendency to unite with organic substances? How is chloride of aluminum formed?

Salts of Alumina.

414. Sulphate of Alumina, $Al_2O_3,3SO_3$.—This salt, though containing 3 eq. of acid, is considered a neutral (349) sulphate. It is obtained by treating the purest clays with sulphuric acid, moderately diluted. It is very soluble in water, and may be obtained in small crystals, which contain 18 eq. of water. It is used in dyeing.

415. Sulphate of Alumina and Potash, $Al_2O_3,3SO_3+KO,SO_3$.—This double salt is the well-known *alum* of commerce. It is usually formed from a mineral substance called alum-slate, which is an argillaceous, slaty rock, containing iron pyrites. It is sometimes found naturally formed as an efflorescence upon the surface of the rock.

Alum is usually seen crystalized in octahedrons, which always contain 24 eq. of water; it is very soluble in boiling water, and has a sweetish, astringent taste. When the crystals are heated, they melt and froth up very much, in consequence of the large quantity of water they contain. Alum is much used in medicine and in the arts, especially in dyeing and calico-printing.

The *alumen ustum*, or *burnt alum*, used in medicine as a caustic, is alum that has been deprived of its water of crystalization by heat.

Common or potash alum is the type of a whole family of alums; as *soda alum*, *ammonia alum*, *iron alum*, *chromium alum*, &c. Soda and ammonia alums are produced by causing these substances to replace the potash in common alum; and iron and chromium alums, by replacing the alumina by the sesquioxides of iron and chromium. These alums are all exceedingly alike in their various properties, and all contain, when crystalized, 24 atoms of water.

The relation of these different alums (183) to each other in composition will best be seen by comparing their formulæ.

QUESTIONS.—414. How is sulphate of alumina formed? 415. What is the composition of alum? From what is it usually formed? What use is made of it? What are some of the different alums that are known?

I.	1. Potash, or common alum......	$KO,SO_3 + Al_2O_3,3SO_3 + 24HO$.
	2. Soda alum......................	$NaO,SO_3 + Al_2O_3.3SO_3 + 24HO$.
	3. Ammonia alum.................	$NH_4O,SO_3 + Al_2O_3.3SO_3 + 24HO$
II.	4. Ferio-potassa alum............	$KO,SO_3, + Fe_2O_3,3SO_3 + 24HO$.
	5. Ferio-soda alum................	$NaO,SO_3 + Fe_2O_3,3SO_3 + 24HO$.
	6. Ferio-ammonia alum..........	$NH_4O,SO_3 + Fe_2O_3,3SO_3 + 24HO$.
III.	7 Manganeso-potash alum.......	$KO,SO_3 + Mn_2O_3.3SO_3 + 24HO$.
IV.	8. Chromio-potash alum..........	$KO,SO_3 + Cr_2O_3,3SO_3 + 24HO$.

Both the manganeso and chromo series have a soda and an ammonia alum, the same as the other series.

Cubic alum, so called because its crystals usually are cubes, is formed by pouring solution of carbonate of potassa into a saturated solution of common alum at about 122°, constantly stirring the mixture for some time. Its composition is $KO,SO_3 + Al_2O_32SO + 9HO$.

416. Silicates of Alumina.—Several double silicates of alumina and other bases are found native, some of which are of great importance in the arts. *Feldspar* is a double silicate of alumina and potash, while *albite,* or *Cleavelandite,* is a double silicate of alumina and soda. *Spodumeme* and *petalite* are the same as the latter, except that they contain less soda, and a small portion of lithia.

Sometimes feldspar undergoes a natural decomposition, losing its potash and part of its silica, and is then called *kaolin.* This substance, which is essentially silicate of alumina, more or less pure, is the basis of all the varieties of *porcelain* or *China-ware.* The articles are made of this of the proper form, and, when dry, are exposed to a high temperature in a furnace, by which they become very compact, but do not fuse. They are then dipped in the glaze, which consists chiefly of feldspar, ground to a fine powder, and suspended in water, a coating of which adheres over the surface; and when the articles are dried and again subjected to the heat of the furnace, it fuses and forms a glassy envelope, which incorporates itself with the body previously formed, and increases its compactness and strength. The glaze also renders

QUESTIONS.—Of those alums mentioned in the table, what is the difference in composition between the first and second? The first and third? The first and fourth? First and seventh? Seventh and eighth? 416. What native silicates of alumina are mentioned? What is *kaolin?* What use is made of it? Describe the mode of manufacturing articles of porcelain.

them impervious to liquids, and even to the gases. For the finer kinds of porcelain, the kaolin, which constitutes the body, is mixed with some substance, as alkali or lime, by which it is rendered partially fusible, and the glaze therefore becomes more perfectly incorporated with it, so as to render articles made of it slightly translucent. The colors are applied after the first burning, and before the glaze, and are composed entirely of metallic oxides.

417. All the different kinds of *clay* are composed essentially of alumina and silica, both of which are very infusible, except when mixed with the alkalies or lime, or certain metallic oxides, especially the oxides of iron. Common clay always contains carbonate of lime and oxide of iron, and is therefore quite fusible. *Fire clay*, so called because of its infusibility, contains no lime; when free from metallic oxides it is called *pipe clay*, and articles made of it are uncolored when removed from the fire.

Stone-ware is made of an infusible clay; and when the articles are sufficiently heated in the furnace, common salt is thrown upon them, the soda of which, combining with the materials of the clay, forms a fusible compound that constitutes the glaze. Without this the articles would be porous, and water and other liquids would percolate through them.

Red *earthen-ware* is made of the most common kinds of clay, which contain lime and iron, and is so fusible that only a moderate heat is needed in baking articles made of it. The articles, after being shaped upon the potters' wheel, are thoroughly dried, and then coated over with litharge (oxide of lead), in fine powder, which, by the heat of the furnace afterwards applied, fuses and spreads over the surface so as to form a fine glaze. Such vessels, however, should never be used with acids, which would attack the oxide of lead and produce poisonous compounds.

Bricks are usually made of the same kind of clay as that just described. No glaze is required for them. The best bricks are now pressed when partially dry, to render them more solid, and

---

QUESTIONS.—417. Of what are all the different kinds of clay composed? What is *fire clay?* How is the glaze applied to articles of stone-ware? Describe red earthen-ware. How are bricks made?

to give them a smoother surface. The red color is occasioned by the oxide of iron in the clay, or the sand mixed with it, which, by the heat in burning, becomes peroxidized.

418. **Glucinum,** G; Eq., 4·7.—This metal receives its name from the circumstance that the salts of its oxide, *glucina*, are sweet (Greek, *glukus*, sweet,) to the taste. It is obtained from its chloride in the same manner as magnesium and aluminum.

Glucina is an oxide of the metal, but whether it is a protoxide, GO, or a sesquioxide, G_2O_3, has not been determined. It is obtained chiefly from the mineral species called beryl, and therefore the metal has by some been called *beryllum*. If we consider it as a sesquioxide, the equivalent of the metal will be 6·96, instead of the number given above.

419. **Zirconium,** Zr; Eq., 34.—Zirconium, in combination with oxygen, is found in the mineral species *zircon*, which is a silicate of zirconia. It is present also, in small quantity, in some few other minerals.

The earth *zirconia* is believed to be a sesquioxide of the metal, Zr_2O_3.

420. **Thorium,** Th; Eq., 59·6.—Thorium is found only in a few very rare minerals, as *thorite*, *pyrochlore*, and *monasite*.

The earth *thorina* is a protoxide, ThO.

Yttrium, Erbium, and **Terbium** are found associated in a very few rare minerals, especially in *gadolinite*, a mineral species obtained chiefly from Sweden.

Cerium, Lanthanum, and **Didymium** occur together in several mineral species, *cerite*, *allanite*, *orthite*, &c., but are obtained only in small quantity. Of these six metals last mentioned, little is really known;—except yttrium and cerium, they have but recently been discovered.

Group IV.

Metals	
Manganese Iron Zinc Chromium Cadmium Tin Cobalt Nickel	*Metals which decompose vapor of water at a red heat, but are not acted upon by liquid water.*

421. The metals of this group are not as well characterized as those of some of the other groups. All that we have named as included in it do indeed, at a red heat, decompose the vapor of

---

Questions.—What occasions the red color of bricks? 418. Describe glucinum, and the mode of preparing it. 419. In what is zirconium found? 420. What other metals are mentioned as belonging to this group? Name the metals of Group IV. How are they characterized? 421. Are the metals of this group as well characterized as those of some other groups?

water, but some of them also act slightly upon liquid water. This is the case more particularly with the first named, manganese; but some of the others, as iron and zinc, are oxydized by water, at ordinary temperatures, especially if atmospheric air be also present.

MANGANESE.

Symbol, Mn; *Equivalent*, 28; *Density*, 8·3.

422. History.—Manganese was first obtained by Gahn, from the substance then called *magnesia nigra*, which has since been found to be an oxide of this metal. It received its present name to distinguish it from magnesium, which has already been described. It is never found in nature in its metallic state, but its compounds are very generally diffused, and traces of it occur in both animal and vegetable substances.

423. Preparation.—Metallic manganese, in consequence of its great affinity for oxygen, is not obtained without considerable difficulty. To procure it, the black oxide, in fine powder, is mixed into a paste with oil and lampblack, and exposed, in a close crucible, to the highest heat of a powerful furnace. The oxide should first be heated alone in a covered crucible, to reduce it to the form of protoxide, and this then mixed with the lampblack and oil, as directed above. The last heating, which should be continued two hours, is best performed in a porcelain crucible, well closed by a cover, and then luted in a common Hessian crucible, as shown in the figure, which represents a section through the centres of both crucibles. A little powdered borax mixed with the materials facilitates the collection of the metallic globules into a mass.

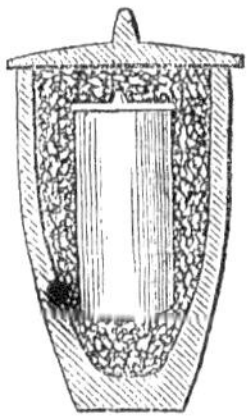

Preparation of Manganese.

The metal will be found in a button at the bottom of the crucible.

QUESTIONS.—422. Give the history of manganese. 323. How is the metal prepared?

424. Properties.—Manganese is a hard, brittle metal, of a grayish-white color, and granular texture. It is exceedingly infusible, requiring the highest heat of a wind-furnace for fusion. It soon tarnishes on exposure to the air, and absorbs oxygen with rapidity when heated to redness in open vessels. In water at ordinary temperatures it is slowly oxydized, and the action becomes more decided as the heat is raised.

The metal is best preserved in pieces of glass tube, hermetically ealed.

Binary Compounds of Manganese.

425. Oxides of Manganese.—There are six oxides of manganese, viz.:

The *protoxide*, MnO, which is a powerful base.
The *sesquioxide*, Mn_2O_3, which is a weak base.
The *red oxide*, Mn_3O_4, $= MnO + Mn_2O_3$, a saline oxide (344 : 4).
The *binoxide*, MnO_2, called also *peroxide* and *black oxide*.
Manganic acid, MnO_3, and
Permanganic acid, Mn_2O_7.

426. Protoxide of Manganese, MnO.—This compound, though existing frequently in combination, is obtained in a separate state with some difficulty, in consequence of its strong tendency to absorb oxygen from the air. The following is the best method of procuring it. A glass tube with a bulb, A, near the middle, is provided, and the bulb partly filled

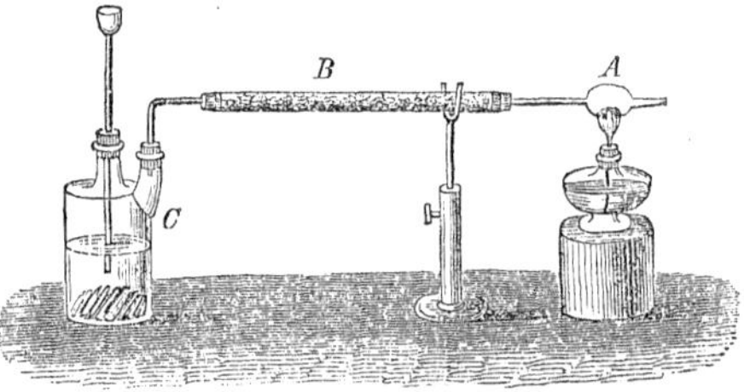

Preparation of MnO.

with powdered carbonate of manganese. A larger tube, B, is filled loosely with dry chloride of calcium, or pieces of pumice stone impregnated with strong oil of vitriol, and connected with the first tube, as

QUESTIONS.—424. What are the properties of manganese? 425. How many compounds of manganese and oxygen are there? 426. Describe the mode of preparing the protoxide by means of hydrogen gas.

shown in the figure. The two-necked bottle, C, contains pieces of zinc and water.

When the whole is arranged as represented, some oil of vitriol is poured into the bottle, C, and the whole interior of the apparatus is soon filled with hydrogen gas, which is deprived of its moisture by passing the tube, B, so that dry hydrogen only surrounds the manganese compound. A spirit-lamp is now placed under A, the heat of which decomposes the carbonate of manganese, leaving the protoxide as a fine powder; and the absorption of oxygen is prevented by the hydrogen.

After a proper time the tube, A, may be removed, and the extreme point closed hermetically by the lamp; the tube on the other side of the bulb may also be drawn out and closed in the same manner, which is accomplished the more readily if, after introducing the manganese compound, the tube has been a little reduced, as shown in the accompanying figure.

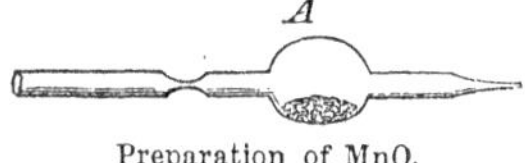

Preparation of MnO.

Prepared in this way, the protoxide is a powder of a delicate green color; and thus inclosed from the air, may of course be preserved for any length of time.

427. Peroxide of Manganese, MnO_2.—This is the common, or *black oxide*, of this metal. It is found in considerable abundance in the State of Vermont, and in other places in this country, and in Europe. Heated alone, it is reduced to the sesquioxide, Mn_2O_3, but when heated with sulphuric acid it is reduced to the protoxide, with which the acid combines. This last operation may be performed in glass vessels. Heated with hydrochloric acid, chloride of manganese is formed, and chlorine evolved. It is much employed in the arts, in the manufacture of glass, and bleaching-salt, and for other purposes. When crystalized, it is called by mineralogists *pyrolusite*. The sesquioxide very much resembles the peroxide in appearance, and is often fraudulently sold for it.

428. Manganic Acid, MnO_3.—This compound has never been obtained in a separate state, but only in combination with bases. It is interesting as the first *metallic acid* we are called to consider in the progress of our course. It is formed, in combination with potash, by heating equal weights of peroxide of manganese and nitrate of potash to dull redness in an open crucible. The compound, *manganate of potash*, is of a dark-green color, and has long been known by the name of *chameleon mineral*. Dissolved in cold water, it forms a beautiful green solution, which by continued dilution changes to blue, purple, and, finally, to a brilliant red.

QUESTIONS.—427. Is the black, or peroxide, found native? What use is made of it? 428. Describe the mode of preparing manganate of potash. What is it called?

Hence its name. The changes are much more rapid when hot water is used.

429. **Permanganic Acid,** Mn_2O_7.—This acid, in combination with potash, is formed when solution of the manganate of potash is made with hot water, as above described, by the absorption of oxygen from the air. It is to the formation of this compound that the red color is owing, which is finally obtained by solution of chameleon mineral.

From this red solution purple crystals of permanganate of potash may be obtained; but when an attempt is made to separate either this or the preceding acid from the bases with which they are united, they are decomposed.

Chloride of Manganese is formed by digesting the black oxide in hydrochloric acid. There are two known, the protochloride, MnCl, and sesquichloride, Mn_2Cl_3.

Salts of Manganese.

There are several salts of manganese, but the only one of importance in the arts is the following:

430. **Sulphate of Manganese,** MnO,SO_3.—This salt is formed by digesting the peroxide in strong sulphuric acid by the aid of heat, and filtering when it has become cold. The crystals of the salt are of a rose-red color, and are very soluble in water. They always contain some water of crystalization, the quantity depending upon the temperature of the solution in which they are formed.

Carbonate of Manganese, MnO,CO_2, is found native in the mineral called *diallogite.*

IRON.

Symbol, Fe (*Ferrum*); *Equivalent,* 28; *Density,* 7·7 to 7·9.

431. **History.**—Iron, the most abundant and most useful of all the metals, has been known from the remotest antiquity. The ores of the metal, as well as the metal itself, and some of its manufactures, are mentioned in the writings of Moses, and it is well known the ancient Greeks and Romans were acquainted with it.

Iron has been found native in small quantities in different

QUESTIONS.—429. How is permanganate of potash formed? 430. Describe the sulphate of manganese. 431. Has iron been long known? Has it been found native?

countries, but recently a real mine of very pure iron seems to have been discovered in the territory of Liberia, on the western coast of Africa.

The occurrence of iron of meteoric origin, associated with nickel, and sometimes with cobalt and other metals, is not uncommon. One mass, at least, of this kind was actually seen to fall from the atmosphere, which was subsequently examined; but many others have been found in such situations as to leave no doubt that they originated in the same manner. As no such compound has ever been found in proper iron-mines, it is believed that these bodies must have their origin in some region foreign to the earth.

Common meteorites, which have often been seen to fall from the atmosphere, are of a different composition, containing usually no metallic iron, but various compounds of iron, manganese, sulphur, &c.

There are many ores of iron, but the most important are the hydrated peroxide, called by mineralogists *brown hematite;* the peroxide (*specular iron*, or *red hematite*), and the black or magnetic oxide, which is a compound of the two preceding. The last is the natural magnet, or *loadstone.* English iron is obtained chiefly from the *clay-iron stone*, which is an impure carbonate of iron, found abundantly in connection with the coal-measures; but in this country the metal is extracted almost entirely from the ores previously mentioned.

432. Preparation.—The preparation of perfectly pure iron is a matter of considerable difficulty; but the ordinary method of reducing it from its ores, is, to heat them intensely, after having been reduced to small fragments, with charcoal or coke, and lime or siliceous sand, as the nature of the particular ore may require. The oxygen of the oxide of iron is absorbed by the heated carbon and carried off as carbonic acid, while the flux — for so the lime or sand is called when used for this purpose — unites with the earthy part of the ore and forms a fusible compound, that remains

---

QUESTIONS.—What is said of meteoric iron? What metals are usually found in combination with meteoric iron? What are some of the most important ores of iron? 432. Describe the ordinary modes of reducing the ores of iron. Why are fluxes used in the operation?

upon the surface; the melted iron, by its superior weight, falling to the bottom.

The nature of the flux used must depend, in any particular case, upon the character of the ore that is used. Most iron ores occur in the granite rocks, and therefore the *gangue* (that is, the earthy matter of the ore,) is mostly silica, which of itself is very infusible (323), but forms a fusible mass when mixed with lime. Limestone, in fragments, is therefore introduced with the ore, which is first by the heat changed to caustic lime, by expulsion of the carbonic acid; and the lime then unites with the silica, producing the fusible silicate of lime, which, at the high temperature, flows freely, and allows the reduced particles of iron to unite in a mass.

Occasionally, an iron ore occurs in a limestone region; and then, the lime alone being also infusible, a flux of silica (sand) is required.

When a sufficient quantity of melted iron has accumulated in the furnace, it is drawn off by an aperture at the bottom, which is opened for the purpose. After the iron has been removed, the slag, formed by the union of the flux with the earthy matter of the ore, is also drawn off.

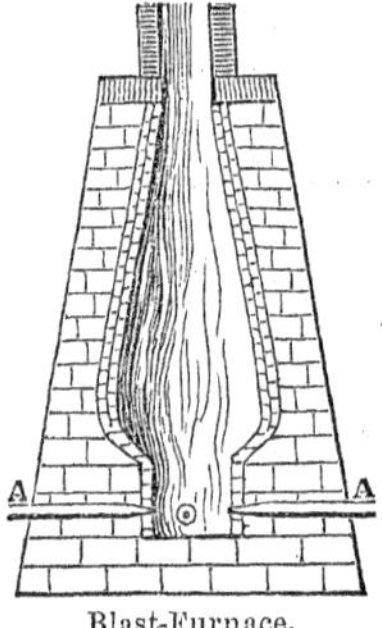

Blast-Furnace.

The accompanying figure represents a section of a *blast-furnace*, which is used for the reduction of iron ores. It is usually built of stone, thirty or forty feet high, and lined inside with fire-brick. The charcoal, or coke, and the ore, in proper proportion, with the necessary flux, are thrown in at top; and A A are pipes leading from a powerful bellows, worked by water or steam-power, for supplying the blast of air. This air thus forced in, will, of course, be of the same temperature as the external atmosphere, and constitute the cold blast; but sometimes the hot blast is used, in which case the tubes are so arranged over the top of the furnace that they are kept hot by the blaze from the burning mass, and the air, passing through them, becomes heated, and enters the fire at a temperature of 400° or 600°.

When once put in operation, a furnace is usually kept in full employment for many months, until repairs are needed; the fuel,

---

QUESTIONS.—What is used for flux when the ore is dug from siliceous rocks? Describe the blast-furnace used for reducing ores of iron. How is the air sometimes heated before being forced into the furnace? Is a furnace, when once heated, kept long in operation?

with a proper proportion of ore and flux, being regularly supplied at the top.

The iron obtained by this process is the *cast* or *pig-iron* of commerce, and contains a considerable quantity of carbon and other substances, by which it is rendered much more fusible than pure iron, but is at the same time harder and more brittle.

It receives it name from the fact that it may be *cast*, or *melted*, and poured into moulds, so as to form articles of any desired figure For this purpose, a pattern of the desired article is formed of wood, and moulded in sand; and the melted metal is poured into the cavity thus produced. The iron is usually melted in a cupola furnace, and drawn off in vessels lined with clay (see figure), from which it is conveniently poured.

Pouring Cast-iron.

To convert cast into *malleable iron*, it is exposed, in a melted state, to a current of air, which plays over its surface, or is forced through it. The process is usually conducted in a reverberatory furnace, a section of which is shown in the figure in the margin. G is the grate upon which the fire is kindled, and H H the hearth which contains the melted metal. As the blaze from the fuel passes to the chimney, C, it comes in contact with the melted iron, and the carbon it contains—and perhaps other impurities—is gradually burnt out, and the iron becomes malleable. This process is called *puddling*.

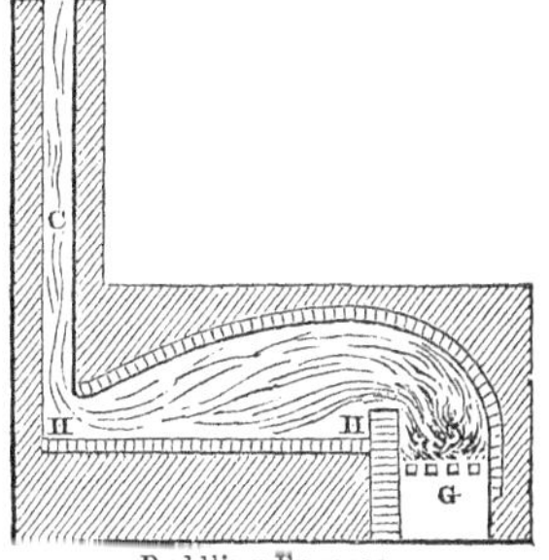

Puddling Furnace.

But it is not absolutely essential that cast-iron should be fused in order to be changed into the malleable state. When small

QUESTIONS.—What is the kind of iron obtained by the process described? Why is it called cast-iron? How is it fashioned into articles of any desired form? How is cast-iron converted into malleable iron? May small articles made of cast-iron be converted into malleable iron without being fused? Describe the process.

articles made of cast-iron are heated for a time in contact with powdered oxide of iron in close vessels, they are converted into malleable iron, and still retain their form perfectly. The carbon is gradually extracted by the oxygen of the oxide of iron used.

433. Properties.—We have already, under the last head, in part described the properties of iron. It is a hard metal, of a peculiar gray color, and strong metallic lustre, which is susceptible of being considerably heightened by polishing. Heated to redness, it becomes very soft and pliable, and is easily worked under the hammer, which gives it a decidedly fibrous structure. When malleable iron is strongly heated, it does not at once change to the liquid state, like most of the other metals, but becomes soft and pasty at the surface; so that two pieces in this state, upon being hammered or firmly pressed together, unite in one piece, or are *welded* together. Iron is, perhaps, all things considered, the most useful metal known, and it is owing in a great measure to this property, in which it is peculiar; only a few other metals possessing it, in an inferior degree.

Iron has a strong affinity for oxygen, and exposed to the air and moisture, it *rusts*, or oxydizes. Heated intensely in the blacksmith's forge, or in the flame of the compound blowpipe, it burns with brilliancy. The same effect is produced by igniting a wire in a receiver of oxygen gas (192), or dropping iron-filings into the flame of a spirit-lamp. It has the property also of becoming magnetic, under the influence of another magnet (121), or of the galvanic current (136). But pure iron loses its magnetism when the influence of the current is removed. Some of the compounds of iron, especially steel and the black oxide, however, retain their magnetism permanently.

The iron of commerce is never pure, but contains in combination more or less carbon, and other substances.

Perfectly pure iron can be obtained only by heating one of the oxides,

---

QUESTIONS.—433. Describe the properties of iron. How is malleable iron affected when heated to redness? Describe the process of welding. Are there other metals capable of being welded? What is said of the affinity of iron for oxygen? How is it affected when exposed to air and moisture? How may iron be made to burn? How may pure iron be procured?

or the protochloride, in an atmosphere of hydrogen. For this purpose, the same apparatus may be used as figured in paragraph 426 for forming protoxide of manganese. The oxide or chloride being heated by a lamp, is decomposed by the current of dry hydrogen, and the iron remains as a grayish-black powder, which may take fire spontaneously if the air be admitted, but may be preserved any length of time by sealing the tube hermetically. By reducing the iron in a porcelain crucible, at a very high temperature, it may be obtained in a solid mass, with a metallic lustre.

434. *Steel* is a carbonide of iron. It is formed by heating bars of malleable iron in close vessels in contact with charcoal, by which process a small quantity of carbon is absorbed and incorporated with the iron. This process is called *cementation;* and, although the proportion of carbon absorbed by the iron is small, yet very important changes are produced in its properties. It becomes more fusible, and may now be melted like cast-iron, and at the same time has become harder, and is capable of being *tempered*, that is, of being made hard or soft, at pleasure. This is done by first heating the article to redness, and then cooling it suddenly by plunging it into cold water, or oil, by which it is made very hard and brittle, and subsequently heating it moderately, and allowing it to cool slowly. By the last heating, a partial annealing (336) is effected, and, if the operation has been well conducted, a proper degree of hardness is produced.

Bars of steel, from the cementation process, always present a blistered surface, occasioned by the liberation of gaseous matter—probably carbonic oxide—within their substance; but when the metal has been melted and cast in moulds, it has a perfectly uniform texture, and is called *cast-steel.*

The melting of blistered steel, to form cast-steel, requires a very high temperature, and is performed in a furnace, a vertical section of which is represented by the first figure on next page. A is a small rectangular chamber, containing the crucible and fuel, and connects with the chimney, C, by a horizontal flue, B. The top of this chamber is covered with a lid, E F, which may be removed, at pleasure, to allow the introduction of the fuel and the crucibles, and the current of air is regulated by a damper, R. The crucibles are made of the most refractory clays, and of the form represented in the second figure. When the steel is perfectly

---

QUESTIONS.—434. What is steel? How is it formed? In what does it differ from iron in its properties? How are bars of blistered steel converted into cast-steel? How are articles made of steel tempered? What is meant by this?

fused, the crucible is withdrawn, and the melted metal poured into iron ingot-moulds prepared for the purpose.

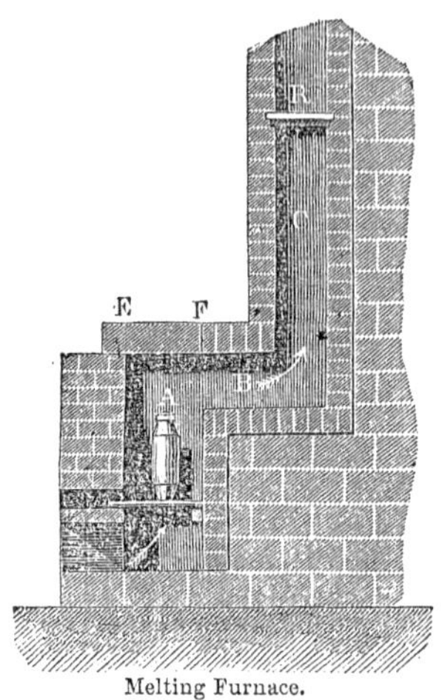

Melting Furnace.

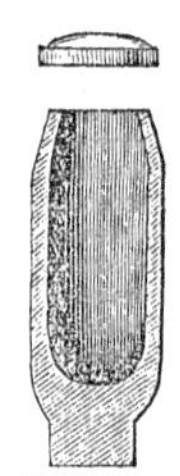
Crucible.

Binary Compounds of Iron.

435. Oxides of Iron.—There are four oxides of iron, as follows, viz:—

The *protoxide*, FeO, which is a powerful base.

The *sesquioxide*, Fe_2O_3, which is a feeble base, isomorphous with alumina, Al_2O_3. It is often called the *peroxide.*

The *oxide*, Fe_3O_4, called also *black* or *magnetic oxide*, which is considered a compound of the *proto* and *sesquioxides*;—thus, $FeO + Fe_2O_3 = Fe_3O_4$. It belongs to the class of "saline oxides" (344 : 4).

Ferric acid, FeO_3.

The red, or sesquioxide of iron, is obtained by heating green vitriol (sulphate of the protoxide of iron) to redness in the open air, and is used as a polishing-powder, under the name of *rouge*

QUESTIONS.—435. Describe the oxides of iron. How is the red or sesquioxide obtained?

or *colcothar*. Earth or clay highly impregnated with it forms the *red ochre* used by painters. The *black oxide* constitutes the scale that always forms upon the surface of iron when heated to redness in the open air. Large accumulations of it are often seen by the side of the smith's anvil. This oxide, found native, constitutes the *native magnet*, or *loadstone*, as stated above. It is found crystalized in beautiful octahedrons, and in masses, and is one of the best ores of the metal.

Ferric acid, FeO_3, in combination with potash, as ferrate of potash, is formed by heating a mixture of iron-filings and nitrate of potash in an iron crucible; and the salt may be separated from the mass by treating it with cold water. It cannot be obtained in a separate state.

436. **Chlorides of Iron.**—There are two chlorides of iron, the *protochloride*, $FeCl$, and the *sesquichloride*, Fe_2Cl_3; the first may be obtained by dissolving iron in hydrochloric acid, and the second by treating the sesquioxide in the same manner. They possess no particular interest.

437. **Sulphides of Iron.**—The *protosulphide*, FeS, of iron is readily formed by heating a mixture of iron-filings and sulphur, or by pressing a bar of iron heated to whiteness into a mass of sulphur. The sulphur and iron combine with great energy, and the liquid sulphide collects in a mass at the bottom, forming, on cooling, a hard, brittle solid. It is used in preparing hydrosulphuric acid (263).

Bisulphide of iron, FeS_2, is the *iron pyrites* of mineralogists. It is of a beautiful yellow color, resembling gold, for which it has often been mistaken. It is usually crystalized in cubes. It is used in the manufacture of green vitriol and sulphuric acid, and sometimes for extracting its sulphur.

Though the sulphides of iron are very abundant in nature, they cannot be used for the extraction of the metal, because of the great expense it requires, and the iron obtained is of an inferior quality.

438. **Carbonides of Iron.**—Both cast-iron and steel, as we have seen, are considered as carbonides of iron, but the ingredients do

---

QUESTIONS.—What is the red ochre of painters? What is the natural magnet, or *loadstone*? 436. Describe the chlorides of iron. 437. Describe the sulphides of iron. How may the protosulphide be formed? For what is the bisulphide sometimes mistaken? 438. What is said of the carbonides of iron?

not seem to be united exactly in definite proportions. Cast-iron usually contains from 2 to 5 per cent. of carbon, while steel seldom contains as much as 2 per cent. *Graphite*, called also *plumbago*, and *black lead*, has been considered as a carbonide of iron, but probably it is only a particular form of carbon, usually containing a portion of iron as an accidental impurity. It is found abundantly in nature, and is used in the manufacture of pencils and crucibles, and as a polishing-powder for stoves and other articles made of iron.

Protiodide of iron, FeI, is formed by digesting iron-filings or wire in water containing iodine, and evaporating the solution obtained. It is used in medicine.

Salts of Iron.

439. Carbonate of Iron, FeO,CO_2, occurs native and is called *spathic iron*, or *steel ore*, by mineralogists. An impure variety is called *clay-iron stone*. It may also be obtained by adding solution of carbonate of soda to solution of green vitriol. Carbonate of iron, though insoluble in pure water, is dissolved by water charged with carbonic acid, and is thus contained in the water of *chalybeate springs*.

440. Sulphate of iron, FeO,SO_3.—This salt is the *green vitriol* or *copperas* of commerce, so extensively used in the arts In crystals it contains 7 eq. of water, and its formula of course is $FeO,SO_3 + 7HO$. The crystals belong to the monoclinic system. It is a sulphate of the protoxide, and is prepared on a large scale at Strafford, Vermont, and other places, from iron pyrites, which is first roasted slightly, and then exposed in heaps to the atmosphere, from which oxygen is absorbed, converting the sulphur into sulphuric acid, and the iron into oxide of iron; the two together then uniting to form the salt in question. For use in the laboratory, it is readily formed by the action of dilute oil of vitriol upon metallic iron. It often forms upon specimens of

---

QUESTIONS.—What is graphite? 439. Is carbonate of iron found native? What is it called by mineralogists? 440. What is the green vitriol or copperas of commerce? How is it manufactured in Vermont? What use is made of it?

iron ores contained in cabinets, by the action of the atmosphere, and is seen as a yellowish-white powder.

It is used in coloring black, and in the manufacture of writing-ink, and for other purposes.

When this salt is heated it first parts with its water, and at a full red heat, with its sulphuric acid, a portion of which is decomposed into sulphurous acid and oxygen, but the rest passes off unchanged. By the oxygen of the decomposed acid, the iron is peroxydized, forming the colcothar of commence, before mentioned.

441. Nitrate of Iron, FeO,NO_5.—Nitric acid acts readily upon iron, and at the same time with the nitrate of iron there is formed also nitrate of ammonia, the ammonia being produced by the union of a portion of the nitrogen of the acid with the hydrogen of the water. The nitric acid should be considerably diluted.

442. Tests of Iron.—The usual test for iron is solution of yellow prussiate of potash, which forms with it a beautiful blue. For an experiment, let a small quantity of common hydrochloric acid be largely diluted with water, and then pour into it a few drops of solution of this prussiate, which will instantly give a fine sky-blue color if iron be present in the acid, as is almost certain to be the case. The iron is derived from the utensils made use of in the manufacture of the acid.

With tincture of nutgalls, the soluble salts of iron form at first a dark-blue precipitate, which finally becomes black.

CHROMIUM.

Symbol, Cr; *Equivalent*, 26·7; *Density*, 6.

443. History, Etc.—Chromium was discovered in 1797. It is found in considerable abundance in Massachusetts, Pennsylvania, and other States, in the mineral called *chrome iron*, and also in combination with oxide of lead. Its name comes from the Greek *chroma*, color, in allusion to the splendid color

QUESTIONS.—441. How is nitrate of iron formed? 442. What tests of iron are mentioned? 443. What is said of chromium? From what is the name derived?

of many of its compounds. It is prepared by heating its oxides mixed with charcoal, but not without difficulty. The metal has a white color, and distinct metallic lustre. It is very brittle, and with difficulty fusible.

Metallic chromium is not used in the arts, but several of its compounds are important.

Binary Compounds of Chromium.

444. Oxides of Chromium.—Oxygen forms several compounds with chromium, which, in composition and many of their properties, are analogous to the oxides of iron and manganese; so that these three metals form a kind of natural family, in the same manner as potassium, sodium, and lithium. Only two of these oxides, the sesquioxide, Cr_2O_3, and the teroxide, or chromic acid, CrO_3, are of sufficient importance to claim attention here.

The former of these, Cr_2O_3, in combination with protoxide of iron, forming the compound, FeO,Cr_2O_3, constitutes the native *chromic iron*, which is the most abundant ore of the metal. It may be prepared by various modes, but the following is perhaps the best, as it leaves the oxide in a proper state for use in the arts. Heat in a crucible an intimate mixture of 4 parts of bichromate of potash and 1 part of starch, and wash the mass thoroughly with hot water, to separate the carbonate of potash that has been formed. The residue, after being dried, is again heated, and the pure sesquioxide remains.

Sesquioxide of chromium is not decomposed by heat, and when fused with fluxes, it imparts to them a green color. It is used in staining glass and porcelain. It replaces alumina (415) in the chromic alums.

Chromic acid CrO_3, is prepared by decomposing a saturated solution of bichromate of potash with $1\frac{1}{2}$ times its volume of oil of vitriol; bisulphate of potash is formed, and the chromic acid separates in brilliant red crystals, which are very soluble in water, and deliquescent in the air. It is a powerful oxydizing agent, and strong alcohol thrown upon the dry crystals is very soon inflamed. With bases it forms important salts.

The chlorides and sulphides of chromium are not important.

Salts of Chromic Acid.

445. The Chromates of Potash are formed by heating a mixture of native chrome iron in powder with nitre, digesting the

---

QUESTIONS.—444. What is said of the oxides of chromium? How may the sesquioxide be formed? How is chromic acid prepared? What is said of the action of alcohol upon it? 445. How is bichromate of potash formed?

mass obtained in hot water, and saturating the solution with dilute sulphuric acid. After a little time the clear red liquid is poured off, and evaporated, when crystals of *bichromate of potash*, $KO, 2CrO_3$, are separated, from the excess of sulphuric acid present, and the sulphates of iron and alumina.

The *neutral chromate of potash*, KO,CrO_3, is now easily procured by neutralizing a solution of the bichromate with carbonate of potash, and crystalizing.

The *bichromate*, which is of a deep red color, is much used in dyeing, and in calico-printing, and for many purposes in the chemical laboratory.

446. Chromate of Lead, PbO,CrO_3.—This is the beautiful *chrome yellow*, used as a paint. It is formed by mixing solutions of one of the chromates of potash and acetate or nitrate of lead. The beauty of the color will depend somewhat on the strength of the solutions used, and other circumstances. *Chrome green* is formed by mixing the chromate of lead with Prussian blue, in a particular stage of the process of manufacture.

ZINC.

Symbol, Zn; *Equivalent*, 32·5; *Density*, 7.

447. History.—This metal has been known several centuries, but was not, for many years, much used. Its chief ores are *calamine*, which is a native carbonate, and *blende*, which is a sulphide; but several others are known. These ores are found in considerable abundance in New Jersey, and other parts of this country.

448. Preparation.—The ores of zinc are reduced by first roasting them in the open air, and then distilling them with charcoal in close crucibles, from which a tube descends directly

QUESTIONS.—How is the neutral chromate of potash formed? 446. Describe the mode of preparing chromate of lead. By what name is it familiarly known? What use is made of it? How is chrome green prepared? 447. Has zinc been long known? What ores of it are mentioned? 448. Describe the mode of reducing the metal from its ores.

through the bottom. The metal is volatilized by the heat, and descends through the tube, from which it is received into a vessel of water. The necessity of excluding the air perfectly, arises from the fact that vapor of zinc, on coming in contact with the oxygen of the air, is at once oxydized; and the object of having the tube which conveys away the volatilized metal pass directly downward, is, to preserve its temperature so high that it shall not become clogged by the condensed metal.

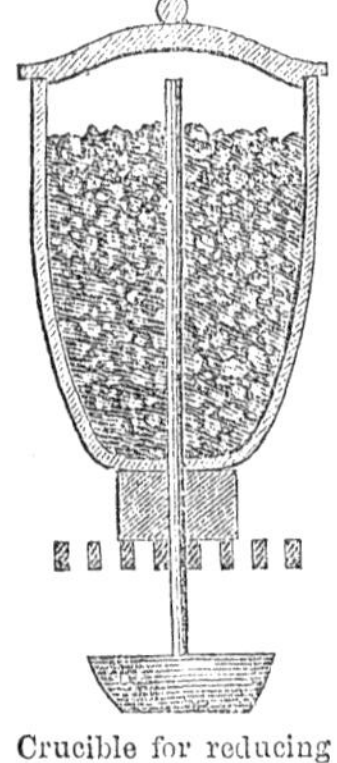

Crucible for reducing Zinc

The figure in the margin represents the section of a crucible used for this purpose, with its charge of ore and charcoal, and tube made of fire-clay passing downward through the bottom. The cover is carefully luted on, and it is supported a little above the grate of the furnace by a fire-brick. The zinc, as it is separated from the ore, taking the gaseous state, passes downward through the tube into a basin of water placed below to receive it. The part of the tube below the bottom of the crucible, not being surrounded by the fire, remains comparatively cold; and the metal, before it reaches the water, is condensed to the liquid state, and falls in drops into the water.

Commercial zinc usually contains iron, lead, and other impurities, from most of which it may be separated by distillation in this manner.

449. Properties.—Zinc has a bluish-white color and a strong metallic lustre. In masses, it always has a highly crystaline structure, and in commerce it is called *spelter*. When cold it is quite brittle; but heated to about 300°, it becomes malleable, and may be rolled into thin sheets. Heated to 773°, it melts; at a little higher temperature, in the open air, it takes fire and burns with a brilliant white flame, producing the protoxide, which assumes an exceedingly delicate gossamer appearance, and has been called *nihil album*, *philosophers' wool*, and by other names.

Zinc is much used in the arts, in the preparation of brass, in the construction of galvanic batteries, and, rolled in thin sheets, as a substitute for sheet-iron and tin-plate. Recently it has been

---

QUESTIONS.—Why is it necessary to exclude the air? Describe the form of crucible used in reducing these ores. 449. Describe the properties of zinc. What is its melting point? What is produced when it is highly heated in the open air? What use is made of zinc?

considerably used as a coating for sheet-iron, in the same manner as tin, to protect it from oxydation. The iron thus prepared is known in the arts as *galvanized iron.*

Binary Compounds of Zinc.

450. Protoxide of Zinc, ZnO.—This is the only oxide of zinc known. It is of a yellowish-white color, and may be prepared, as above described, by heating zinc in the open air, or by precipitating it from solution of white vitriol (sulphate of zinc) by carbonate of ammonia. It is much used in painting as a substitute for white lead; and is manufactured on a large scale directly from the ores of zinc, in France, and in New Jersey.

451. Chloride of Zinc, $ZnCl$.—Chloride of zinc is formed by dissolving commercial zinc in hydrochloric acid, or by burning zinc in chlorine gas. By evaporation it may be obtained as a white solid, but is very deliquescent in the air. Mixed with sal-ammoniac, it serves an excellent purpose in tinning articles of copper, brass, and other metals.

Sulphide of Zinc, ZnS.—This compound of zinc is found native, and constitutes, as has been stated, one of its ores. It may be prepared artificially by heating the metal with sulphur.

Salts of Zinc.

There are several salts of zinc, but the most important is the following:—

452. Sulphate of Zinc, ZnO,SO_3.—Sulphate of zinc is a white salt, which, in commerce, is called *white vitriol.* It is very soluble in water; and is sometimes used in medicine as a powerful emetic.

The ordinary salt contains 7 atoms of water of crystalization, and its formula is $ZnO,SO_3 + 7HO$; but by using the proper means, it may be crystalized with 5, 2, or even 1 atom of water.

QUESTIONS.—450. What is the only oxide of zinc that is known? What use is made of it? 451. How is chloride of zinc formed? To what use is it applied? 452. What is sulphate of zinc called in commerce?

Oxide of zinc is precipitated from its soluble salts by the caustic alkalies, but the precipitate is again dissolved by an excess of the alkali.

Carbonate of zinc is found native, and called *calamine.*

CADMIUM.

Symbol, Cd; *Equivalent,* 56; *Density,* 8·6.

453. History, Etc.—Cadmium is a very volatile metal, usually found associated with zinc. In appearance it can scarcely be distinguished by the eye from tin, but is rather harder. It melts at 442°, —the melting point of tin,—and sublimes at a temperature but little above the boiling point of mercury. Its properties indicate that it would be a useful metal in the arts, but it has hitherto been found only in small quantities.

Protoxide of Cadmium, CdO, is the only compound of these two elements that is known.

The protosulphide, CdS, is found native, as the *grenockite* of mineralogists.

TIN.

Symbol, Sn (*Stannum*); *Equivalent,* 58; *Density,* 7·3.

454. History. — Tin has been known from the most remote antiquity, and was in common use in the time of Moses. It is supposed the ancients obtained it chiefly from Cornwall, England, the mines of which still yield a large part of the tin of commerce. It is found also in India, Germany, Chili, and Mexico; but has not yet been discovered in the United States, except a few small crystals of the oxide in Chesterfield, Massachusetts, and at Middletown, Connecticut, and a small vein of the same ore in the town of Jackson, New Hampshire. The chief ores are the oxide and sulphide.

455. Preparation and Properties. — Most of the tin of commerce is obtained from the oxide, which is reduced by the action

---

QUESTIONS.—453. What is said of cadmium? What is its melting point? 454. Has tin been long known? Where was it obtained in ancient times? What countries now produce tin? Is it found in the United States? What are the chief ores of tin? 455. Mention some of its more important properties.

of charcoal at a high temperature. It is a brilliant white metal, like silver, but is less hard than the latter It is very malleable, and is rolled into very thin leaves, called *tin-foil*. It is inelastic, and when a rod of it is bent, a peculiar crackling noise, called the *cry of tin*, is produced, occasioned by its crystaline structure. It melts at 442°, and at a red heat is rapidly oxydized; at a white heat it burns with flame.

Tin is not readily acted on by the atmosphere or by moisture, but is dissolved slowly by hydrochloric acid, forming chloride of tin, and by hot sulphuric acid, forming sulphate of the protoxide; by dilute nitric acid it is converted into the binoxide, which is insoluble in the acids.

Uses.—Tin is used for many purposes in the arts, both alone and combined with other metals, as in *Britannia metal*, which is an alloy of tin and antimony, with a small proportion of copper. It is also used extensively to coat sheets of copper and iron, to prevent the oxydizing influence of the air and moisture, and other agents. Thin sheets of iron, coated over with tin, constitute the well-known and highly useful *tin-plate*.

Binary Compounds of Tin.

456. **Oxides of Tin.**—There are two well-defined oxides of tin, the *protoxide*, SnO, and the *peroxide*, SnO_2; but the latter alone possesses sufficient importance to require a description.

Peroxide of Tin is formed by exposing the metal to the action of nitric acid a little diluted. It may also be precipitated from a solution of the perchloride of tin (soon to be described) by an alkali. As obtained by the mode last mentioned, it is soluble in acids, but not as procured by the other mode. It is of a yellowish-gray color, and from the circumstance that it is capable of combining with bases in the manner of an acid, it has been called *stannic acid*. It is much used as a polishing-powder, under the name of *putty of tin*. Melted with ingredients for forming glass, it produces a white enamel.

457. **Chlorides of Tin.**—*Protochloride of Tin*, SnCl, is formed by dissolving the metal in boiling hydrochloric acid. By evaporating the solution it may be obtained in crystals, which contain 2 eq. of water. It is used as a mordant in dyeing, and as a powerful deoxydizing agent in the laboratory. The *perchloride*, $SnCl_2$, is formed by distilling a mixture of 1 part of tin-filings and 3 parts of corrosive sublimate, or by cau-

---

QUESTIONS.—To what use is tin applied? What is the *tin-plate* of commerce? 456. What oxides of tin are known? What is the substance called *putty of tin?* To what use is it applied? 457. What chlorides of tin are there?

tiously dissolving the metal in nitro-hydrochloric acid. It is much used as a mordant in dyeing, and as a disinfecting agent.

There are two *sulphides* of tin;—the *persulphide*, SnS_2, sometimes called *aurum musivum*, or *mosaic gold*, has a yellow color, and metallic lustre, and is used as a paint, and also instead of the zinc amalgam (90) for exciting electrical machines.

There are no important salts of tin.

COBALT.

Symbol, Co; *Equivalent*, 29·5; *Density*, 8·5.

458. History, Etc.—Cobalt is almost always found associated with nickel, the metal next to be described;—as found in mines both are usually in combination with arsenic. The pure metal is obtained with some difficulty. The best mode is to heat oxalate of cobalt in a small crucible, on which a cover is closely luted. It is of a reddish-white color, and is hard and brittle, and difficult to fuse. It is capable of becoming magnetic.

Cobalt is comparatively a rare metal. In combination with arsenic, it is found at Chatham, in Connecticut, in the minerals called *smaltine* and *chloanthite*.

Binary and Other Compounds of Cobalt.

459. Oxides of Cobalt.—There are two oxides of cobalt, the *protoxide*, CoO, and the *sesquioxide*, Co_2O_3, the former of which, mixed with some impurities, is sold in commerce as a gray powder, under the name of *zaffre*. Fused with silica and potash, it forms *smalt*, which is used for coloring glass, porcelain, &c., blue. The *sesqui* or *peroxide* is unimportant.

460. Chloride of Cobalt, CoCl, is formed by dissolving zaffre in hydrochloric acid. The solution has a pink color, and yields by evaporation small crystals of the same tint. Writing made with a diluted solution of it is nearly invisible, but becomes of a beautiful but pale blue color when the paper is warmed by the fire, and again disappears as the paper cools. It has been called *Hellot's Sympathetic Ink*. The addition of a salt of nickel gives the writing a green color.

The salts of cobalt possess no especial interest. The subcarbonate is a fine powder of a very delicate pink tint.

QUESTIONS.—What sulphides of tin are there? 458. With what other metal is cobalt usually found associated? Where is this metal found? 459. What oxides of cobalt are there? 460. How is chloride of cobalt formed? Describe the mode of using Hellot's Sympathetic Ink

NICKEL.

Symbol, Ni; *Equivalent*, 29·6; *Density*, 8·8.

461. History, Etc.—Nickel and cobalt, as before intimated, are in nature almost inseparable companions. Arsenical nickel and cobalt are found at Chatham, in Connecticut, and also in Missouri, and in various places in Europe; but mines of these metals are not common. Nickel is generally, if not always, combined with meteoric iron (431), of the mass of which it sometimes constitutes as much as ten per cent.

Pure nickel has a grayish-white color, and strong metallic lustre. It is quite hard, but malleable, and is nearly as difficult to melt as iron. Like iron and cobalt, it is capable of becoming magnetic. It is not readily acted upon by the air or by moisture. It is much used in the arts to form the alloy called *German silver*, which is composed of copper 10 parts, zinc 6, and nickel 4.

None of the compounds of nickel possess sufficient interest to require description here.

Group V.

Antimony Bismuth Lead Copper Vanadium Molybdenum Tungsten Titanium Uranium Columbium Pelopium	*Metals which are incapable of decomposing water, and whose oxides are not reduced by the mere action of heat.* All of them, except lead and copper, are brittle; and all, except antimony, bismuth, lead, and copper, are fusible only at very high temperatures.

ANTIMONY.

Symbol, Sb (*Stibium*); *Equivalent*, 129; *Density*, 6·7.

462. History.—Antimony is remarkable as having been the first metal discovered after the seven metals (332) known to the

---

Questions.—461. What is said of the metal nickel? In what is it almost always found? Describe the metal. For what is it used? How are the metals of Group V. characterized? Name the metals of this group. 462. For what is antimony remarkable?

ancients. It has been found in the metallic state, but is prepared chiefly from the sulphide, which is not of unfrequent occurrence.

463. Preparation.—The metal, called also *regulus of antimony*, is obtained by heating the sulphide with iron-filings, or carbonate of potash. When iron-filings are used, the sulphur is simply transferred from the antimony to the iron, and, of course, sulphide of iron is formed; but the metal thus obtained always contains a portion of iron, and is otherwise impure. By the other process, the sulphur is converted into sulphuric acid, which combines with the potash, forming sulphate of potash; and the metallic antimony, being thus set free, falls to the bottom of the crucible.

464. Properties.—Antimony is a very brittle metal, of a white color, and brilliant lustre. It always has a highly crystaline structure, and melts at a temperature a little below redness; and may be slowly distilled in a current of hydrogen gas. A small fragment, heated on a piece of charcoal, before the blowpipe, takes fire; and when thrown upon the floor breaks into numerous globules, which continue to burn as they are scattered, leaving a train to mark their path, and filling the air with fumes of the oxide.

465. Alloys of Antimony.—Antimony, being very brittle, is not adapted for use alone in the arts, but with other metals it forms several very useful alloys. One of these, called *Britannia metal*, is composed, of tin 88 or 90 parts, and antimony 12 to 10 parts. Sometimes a small proportion of copper is added, not exceeding 3 or 4 per cent. of the whole.

Its alloy with lead will be described hereafter.

Binary Compounds of Antimony.

466. Oxides of Antimony.—Antimony forms with oxygen two compounds which are well-defined, viz., oxide of antimony, SbO_3, and antimonic acid, SbO_5; but these are capable of combining with each other so as to form one or two other intermediate compounds.

QUESTIONS.—463. How is antimony obtained from the native sulphide? Describe the two processes. 464. Describe the metal. How is it affected when intensely heated in the air? 465. What alloys of antimony are used in the arts? 466. What oxides of antimony are mentioned?

467. Chlorides of Antimony.—There are two chlorides of antimony, $SbCl_3$, and $SbCl_5$, corresponding in composition, it will be seen, with the oxides. The terchloride, $SbCl_3$, is sometimes called *butter of antimony.* It is soluble in a small quantity of water, but if the solution be diluted, a white powder is precipitated, called *powder of algoroth.* It is an oxychloride, $SbCl_3,2SbO_3 + HO$.

468. Sulphides of Antimony.—Two sulphides of antimony are known, corresponding in composition with the oxides and chlorides, and having the formulæ SbS_3, and SbS_5. The first of these is the native sulphide;—when this is roasted in the air, oxygen is absorbed, and oxysulphid of antimony, $SbS_3 + SbO_3$, is formed, often called *glass of antimony*, or *liver of antimony.*

When the native sulphide is boiled in a solution of potassa or soda, a liquid is obtained, from which, on cooling, an orange-red matter is deposited, called *Kermes mineral.* On subsequently neutralizing the cold solution with an acid, the *golden sulphide* of the pharmacopœia is obtained. These compounds now possess little interest, but formerly were much used in medicine.

The salts of antimony are unimportant, except *tartar emetic,* used in medicine, which will be described hereafter.

BISMUTH.

Symbol, Bi; *Equivalent,* 213; *Density,* 9·8.

469. History, Etc.—Metallic bismuth is found in small quantities in Monroe, Connecticut, and other places, but is chiefly obtained from the native sulphide. In mass, it much resembles antimony in its crystaline structure, but has less lustre, and is of a reddish color. It is brittle when cold, but may be hammered when moderately heated. It melts at about 477°, and sublimes at a high temperature. By melting a considerable quantity in a large crucible, placing it in a situation to cool very slowly, and pouring off all that remains liquid, as soon as it begins to solidify at the surface, crystals of considerable size may be obtained.

The metal is unchanged in the atmosphere, but tarnishes slightly by moisture. Heated in the open air it burns with a bluish flame.

It is too brittle for use by itself in the arts, but with some other metals

QUESTIONS.—467. What is said of the chlorides of antimony? 468. What sulphides of antimony are known? 469. Is bismuth found native? Describe the metal. May it be obtained in crystals? What is said of its use in the arts?

it forms important alloys. It may be substituted, in whole or in part, for antimony in Britannia metal (465).

In some of their chemical properties, antimony and bismuth are analogous to phosphorus and arsenic, and also to nitrogen.

Binary and other Compounds of Bismuth.

There are several oxides, chlorides, &c.; but they possess no special interest.

470. Nitrate of Bismuth, BiO,NO_5.—This salt is formed by dissolving the metal in nitric acid. It is soluble in water; but if the solution is largely diluted, the salt is decomposed, and *subnitrate of bismuth* precipitated. This last salt is used in medicine, and also as a cosmetic.

LEAD.

Symbol, Pb (*Plumbum*); *Equivalent,* 103·7; *Density,* 11·44.

471. History.—Lead is one of the seven metals known to the ancients. Its most important and most abundant ore, from which all the lead of commerce is extracted, is the sulphide, the *galena* of mineralogists; but it is found in many other forms, as carbonate, sulphate, phosphate, &c. It is very abundant in different parts of this country.

472. Preparation.—The metal is reduced from the native sulphide by first roasting it in the open air, and subsequently heating it with lime in a charcoal fire. The ore usually contains a little silver, which remains in combination with the lead.

The metal is also easily reduced by heating the native sulphide, in fine powder, with iron nails, in a close crucible. The sulphur of the ore, in this case, passes directly to the iron, forming sulphide of iron.

473. Properties. — Lead is a bluish-gray metal, and when recently cut has a strong metallic lustre; but the surface soon tarnishes on exposure to the air. It is soft and malleable, but

---

QUESTIONS.—470. Describe the nitrate of bismuth. 471. What is the most common ore of lead? What other ores are mentioned? 472. How is the metal reduced from the native sulphide? What is the second mode mentioned? 473. Describe the properties of lead.

not very tenacious. Heated to about 635° it melts, and at very high temperatures is slightly volatilized. Exposed to the air and moisture, it is gradually corroded; and a crust, the white carbonate, is formed upon its surface.

The peculiar properties of lead fit it for use in the arts for many purposes; but one of its most important uses is in constructing pipes for conveying water. But when the water is to be used for drinking, care should always be taken to have the tubes kept constantly filled with water, as the introduction of air tends to form the highly poisonous carbonate; and even then the water should not be used until it has been proved by experiment that the particular water to be discharged by the tube is not capable of acting upon the lead.

Lead is gradually acted on by boiling sulphuric acid; but its only proper solvent among mineral acids is the nitric, which forms with it a soluble salt.

474. Alloys of Lead.—Lead forms with other metals many useful alloys. Two parts of tin and one of lead, fused together, form *soft solder*, which is much used in cementing together the different pieces of articles made of tin-plate, Britannia metal, &c. A coarser kind, which requires a higher temperature to melt it, is composed of lead 3 parts and tin 1 part. *Pewter* is an alloy of lead and antimony, and sometimes a little copper. *Type-metal* is an alloy of 3 parts of lead to 1 of antimony. Tin, lead, and bismuth, form a very fusible alloy; when made of 8 parts of bismuth, 5 of lead, and 3 of tin, it will melt in boiling water.

Binary Compounds of Lead.

475. Protoxide of Lead, PbO.—Protoxide of lead is formed by heating lead, in the open air, a little above its melting point; oxygen is gradually absorbed, and a yellow powder formed, which was formerly used as a paint, and called *massicot*. As usually manufactured, it is in the form of reddish-yellow scales, occa-

---

QUESTIONS.—What is said of the use of lead for water-pipes? What is the only mineral acid that dissolves lead? 474. What alloys of lead are mentioned? What is pewter? Type-metal? What alloy of lead melts in boiling water? 475. How is protoxide of lead formed?

sioned by its having been partially fused. It is much used in painting as a dryer, and is called *litharge.* It fuses readily at high temperatures, and enters into combination with several of the earths and alkalies, producing a transparent glass, which renders it an excellent substance for glazing some kinds of earthenware. With powerful bases it is capable of combining as an acid.

476. Peroxide of Lead, PbO_2.—This oxide is obtained by digesting red lead (the compound next to be described) in nitric acid, which dissolves out the protoxide, leaving the peroxide quite pure, in the form of powder. It is also called *plumbic acid.*

477. Red Oxide of Lead, Pb_3O_4.—This compound is the *red lead,* or *minium,* of commerce, much used in the arts as a paint, and as one of the ingredients in the manufacture of flint glass. It is formed by heating metallic lead to a temperature of 500° or 600°, in the open air, so as to oxydize it without fusing the oxide, and continuing the heat for some time. Heated to redness, it gives up a portion of its oxygen and is reduced to the protoxide. Minium is a compound of the proto and peroxides ($2PbO,PbO_2$), and is therefore a plumbate of the protoxide of lead.

478. Sulphide of Lead, PbS.—This compound has already been mentioned as constituting the only important ore of lead, called galena. It is also formed by passing a current of sulphuretted hydrogen through a solution of any salt of lead. As it occurs native, its color very much resembles that of metallic lead, and its structure is always crystaline.

Salts of Lead.

479. Carbonate of Lead, PbO,CO_2.—This is the *white lead* of commerce, so extensively used in painting. It is formed by several different modes, and is also found as a natural production. Nearly all the white lead of commerce, at the present time, is adulterated by mixing sulphate of baryta (399) with it in fine powder. By treating the mixture with nitric acid, the carbonate

QUESTIONS.—What use is made of the protoxide of lead? What is said of its fusibility? 476. Describe the mode of preparing the peroxide. 477. What is the red lead or minium of commerce? What use is made of it? 478. Describe sulphide of lead. 479. What is the white lead used by painters? With what is it usually adulterated? How may the fraud be detected?

of lead will be dissolved, and the baryta left as a white powder. Taken into the system, white lead acts as a violent poison, producing the disease called *painters' cholic.*

480. Sulphate of Lead, PbO,SO_3.—Sulphate of lead is found native, and called *anglesite* by mineralogists. It is also formed when solution of any sulphate, as sulphate of soda, is mixed with solution of any salt of lead. It is sometimes used in painting as a substitute for white lead.

Nitrate of Lead, PbO,NO_5.—Nitrate of lead is easily formed by dissolving metallic lead, or its protoxide, or carbonate, in nitric acid.

Zinc, iron, and tin precipitate lead from all its soluble salts in its metallic state. Make a solution of 1 part of nitrate or acetate of lead in 24 parts of distilled water, acidulated with acetic acid, and suspend in it, near the top, a piece of clean zinc; the precipitation of the lead will immediately commence, and in the course of 24 or 48 hours, the metal will appear in the form of large thin leaves, called sometimes *arbor Saturni.*

481. Test of Lead.—Hydrosulphuric acid precipitates lead as a dark-colored sulphide from all its soluble salts, and even blackens those that are insoluble, as the carbonate, when suspended as a fine powder in water.

COPPER.

Symbol, Cu (*Cuprum*); *Equivalent,* 31·7; *Density,* 8·9.

482. History.—Copper has been known from the earliest ages, and is often found in the earth in its metallic state. Masses of immense size of nearly pure metal have been found in the region of Lake Superior. One, 40 feet long, was estimated to weigh 200 tons. It is also found as a carbonate, sulphide, and oxide, as well as in other combinations. The ores of copper are very generally diffused, some of them being found in almost every country.

483. Preparation.—Metallic copper is obtained from the oxides and the native carbonates, simply by heating these ores with charcoal; but the sulphide, especially when mixed with iron, is

QUESTIONS.—480. What is said of the sulphate of lead? Describe the mode of forming the arbor Saturni. 481. What test of lead when in solution is mentioned? 482. Has copper been long known? What is said of masses that have been found native? What ores of copper are mentioned? 483. How may the metal be reduced from its oxides and carbonates?

reduced with more difficulty, and the process is too complicated to be given here in detail.

484. Properties.—Copper is distinguished by its peculiar red color. It is very ductile and malleable, but less tenacious than iron. To fuse it a bright red heat is required, but its melting point is much below that of cast-iron. It forms crystals belonging to the monometric system. It is less liable to be corroded by air and moisture than iron, but is gradually acted on by the joint agency of these elements, and becomes coated with a green crust, which is carbonate of copper. Heated to redness in the air, it becomes oxydized, and nitric acid readily dissolves it. It is used extensively in the arts, both alone and in combination with other metals.

485. Alloys of Copper.—Copper forms with other metals many very useful alloys. Three parts of copper with one of zinc constitutes *brass;* and by a variation of the proportions, the alloys called *tombac*, *Dutch gold*, and *pinchbeck*, are produced. *Bronze* is an alloy of copper and about 10 per cent. of its weight of tin; and *bell-metal*, an alloy of 4 parts of copper with 1 of tin; while *speculum metal*, used for the mirrors of reflecting telescopes, contains about 2 parts of this metal to 1 of tin. Equal parts of copper and zinc form *hard solder*, which is used in soldering articles of brass.

Binary Compounds of Copper.

486. Oxides of Copper.—*Red oxide* of copper, Cu_2O, is a suboxide; it is found native. The black or protoxide, CuO, is also found native, and is produced when copper is heated to redness in the open air. It is also easily procured by heating to redness the sulphate or nitrate of copper. There are two other oxides.

Chlorides of Copper.—Of these there are two, the subchloride, Cu_2Cl, and the protochloride, $CuCl$, but they possess no special interest.

Sulphides of Copper.—There are two sulphides of copper, corresponding in composition to the chlorides, viz., the subsulphide, Cu_2S, and the protosulphide, CuS. *Copper pyrites*, *vitreous copper*, and *variegated copper* are native sulphides of copper, or copper and iron.

QUESTIONS.—484. How is copper readily distinguished from the other metals? Describe its properties. 485. What alloys of copper are used in the arts? 486. What oxides of copper are there? What chlorides? What sulphides?

Salts of Copper.

487. Sulphate of Copper, CuO,SO_3, or, in crystals, $CuO,SO_3 + 5HO$.—This is the *blue vitriol* of commerce; and is formed by dissolving the protoxide in sulphuric acid. It is very soluble in water, and is extensively used in the arts. Its color is a fine blue.

488. Nitrate of Copper, CuO,NO_5, or, in crystals, $CuO,NO_5 + 3HO$, is easily formed by dissolving metallic copper in dilute nitric acid. It does not crystalize readily; and is deliquescent in the air, and exceedingly corrosive.

489. Carbonates of Copper.—There are several carbonates of copper, as *azurite* and *green malachite*, which are found native, and are made use of as ores of copper. They also contain water. *Green verditer*, or *mineral green*, is a hydrated subcarbonate, obtained by precipitating a solution of blue vitriol with carbonate of potash or soda.

Silicates of Copper, more or less hydrated, are found native in the mineral species called *dioptase* and *chrysocolla.*

490. *Arsenite of Copper*, or *Scheele's green*, is prepared by first dissolving arsenious acid and pearlash together in warm water, and then pouring into it gradually warm solution of sulphate of copper. It is of a beautiful green color, and is much used in painting. In commerce, it is called *Paris green.*

Test of Copper.—Ammonia produces a deep blue color in diluted solutions of any of the salts of copper, by which they may always be distinguished.

491. Vanadium, V; Eq., 68·6.—This is a metal of rare occurrence, being found only in some iron ores in Sweden, some lead ores in Scotland and Mexico, and recently in some copper ores from Lake Superior.

Molybdenum, Mo; Eq., 46.—Molybdenum is found as a sulphide in the mineral species called *molybdenite*, and also in other combinations. It is a white metal, and has a density of 8·6.

Tungsten, W, (*Wolfram*); Eq., 95.—This metal is found in several mineral species, especially the species called *wolfram*, which is a tungstate of iron and manganese. It forms several oxides, chlorides, sulphides, &c.

Titanium, Ti; Eq., 25.—Titanium is found in several mineral species, as *rutile, anatase*, and *brookite*. It is of a red color, and resembles cop-

---

QUESTIONS.—487. What is the commercial name for sulphate of copper? 488. How is nitrate of copper formed? 489. What is said of the carbonates of copper? 490. What use is made of arsenite of copper? What test of copper is mentioned? 491. What other metals are mentioned as belonging to this group?

per. Its density is about 5·3. The native oxide is used in coloring mineral teeth, so as to imitate the color of natural teeth.

Uranium, U; Eq., 60.—Uranium occurs in several species as *pitch-blende*, *uranite*, &c. Many binary compounds of it are known, as well as some of its salts.

Columbium, Cb, is the name given to a metal found in the mineral species called *columbite*, which is obtained at Middletown and Haddam, in the State of Connecticut.

Tantalum, Ta, is a metal very similar to the preceding, which is obtained from the European mineral, *tantalite*. Little is really known of these two metals last mentioned. The existence of the two metals named *pelopium* and *norium* (Table, p. 145), is doubtful.

GROUP VI.

MERCURY SILVER GOLD PLATINUM OSMIUM IRIDIUM PALLADIUM RHODIUM RUTHENIUM	*Noble metals, whose oxides are reduced by a red heat.*—No one of them, except osmium, is capable of decomposing water under any circumstances. The oxides of iridium and ruthenium, and some of those of osmium, are not perfectly reduced by heat, without the presence of carbon, or some deoxydizing agent.

MERCURY.

Symbol, Hg (*Hydrargyrum*); *Equivalent*, 100; *Density*, 13·6.

492. History and Preparation.—Mercury, or *quicksilver*, is one of the seven metals of the ancients. It is sometimes found in its metallic state; but most of the mercury of commerce is reduced from the native sulphide, called *cinnabar*. It is not very generally diffused, there being but few mines that afford it in any considerable quantity. Most of the mercury used in this country comes from Spain; but it is obtained also in Germany, Siberia, in Southern Asia, and in California.

The metal is extracted from the native sulphide either by roasting it alone, so as to oxydize the sulphur and sublime the mercury; or by heating it with lime.

QUESTIONS.—What use is made of the native oxide of titanium? How are the metals of Group VI. characterized? Name the metals of this group. 492. Was mercury known to the ancients? By what other name is it known? From what ore is it obtained? What is the mode of extracting the metal from the native sulphide?

Besides the native compound above mentioned, there are other ores of the metal, but they are found only in small quantities.

493. Properties.— Mercury is distinguished from all other metals by being liquid at ordinary temperatures. It is white as silver, and has a brilliant lustre. Cooled to —40°, it freezes or becomes solid, and is then very malleable, and nearly the color of lead; at 662°, it boils and forms a colorless vapor, the density of which (air being 1) is 6·976. At lower temperatures, even as low as 70°, it gives off vapor, as is shown by the whitening of pieces of gold-leaf suspended above it in a close glass bottle, and by its action upon the iodized silver plates in the Daguerreotype process.

The best method to obtain solid mercury is by means of solidified carbonic acid (58). A portion of the solid acid is made in the form of a ball, with a cavity in the upper side to receive the mercury, and the whole enveloped in cotton to protect it from the atmosphere. In a few minutes the mercury will be solid, and may be preserved in this condition for several hours with a very small quantity of the solid acid. The metal is also readily frozen by a mixture of the solid acid and ether.

By slow cooling, mercury forms crystals belonging to the monometric system.

494. Mercury is usually imported into this country in strong iron bottles, and generally is very pure, and serves for every purpose without purification. In this state it is scarcely oxydized by the atmosphere, but as it is capable of dissolving other metals, as tin, lead, silver, and gold, by use in the laboratory it often becomes contaminated with them, and thus is more liable to become oxydized, as will be shown by a pellicle of oxide floating upon its surface. In this state, a portion will adhere to the fingers, or other substance, when dipped in it; and it does not roll in perfect globules over the surface of other bodies, like pure mercury.

To purify mercury, several processes are resorted to;—one method is to pour it into a bottle with some sulphuric acid, or dilute nitric acid, which oxydizes and dissolves the foreign metals. It should stand in the bottle several days, and be frequently shaken to expose all the metal to the action of the acid. At the proper time, the mercury is to be removed and washed with water.

QUESTIONS.—493. Describe the properties of mercury. Give its freezing and boiling points. Does it evaporate at temperatures below its boiling point? How may it be solidified? 494. What other metals will mercury dissolve? How may it be known when other metals are held in solution in it? Describe the method of purifying mercury.

Distilling Mercury.

Another method is to distil the mercury, by which it is separated perfectly from silver, gold, and tin, but not from arsenic, zinc, or cadmium, which are also volatile. To distil mercury, a cast-iron vessel is used of the form represented in the figure. When using it, the cover should be well luted on, and the tube kept cold by a constant stream of cold water.

To obtain mercury of sufficient purity for use in thermometers and barometers, and especially the latter, it will generally be found necessary to subject it to both of these modes of purification in succession, the distillation being first in order.

495. The only acids that act on mercury are the sulphuric and nitric acids. The former has no action whatever in the cold; but on the application of heat, the mercury is oxydized at the expense of the acid, pure sulphurous acid gas is disengaged, and a sulphate of mercury is generated. Nitric acid acts energetically upon mercury, both with and without the aid of heat, oxydizing and dissolving it with evolution of binoxide of nitrogen.

496. Uses.—Mercury is made use of for many important purposes, in medicine, in the laboratory, and in the arts. In the construction of thermometers and barometers it is absolutely essential, and for the mercurial bath, to enable the chemist to collect and transfer gases that are absorbed by water. In union with various other substances it constitutes the bases of many important medicines. Several of these preparations will be described in their proper places.

497. Amalgams.—This term is exclusively applied to the alloys of mercury with the other metals.

Quicksilver unites with potassium and sodium when agitated in a glass tube with that metal, forming a solid amalgam. When the amalgam is put into water, the alkaline metal is gradually

---

QUESTIONS.—What will be necessary, in order to obtain mercury of sufficient purity for thermometers and barometers? 495. What acids only act upon mercury? Explain the action of sulphuric acid. Of nitric acid. 496. To what uses is mercury applied? 497. To what is the name amalgam given?

oxydized, hydrogen gas is disengaged, and the mercury resumes its liquid form.

A solid amalgam of tin constitutes the silvering of looking-glasses; and an amalgam made of 1 part of lead, 1 of tin, 2 of bismuth, and 4 of mercury, is used for silvering the inside of hollow glass globes. This amalgam is solid at common temperatures; but it is fused by a slight degree of heat.

The amalgam of zinc and tin (90) is used for promoting the action of the electrical machine.

Binary Compounds of Mercury.

498. There are two oxides of mercury, the gray oxide, which is considered a suboxide, Hg_2O, and the protoxide, HgO, which is of a red color. The *suboxide,* Hg_2O, is readily prepared by mixing calomel briskly in a mortar with pure potassa in excess, so as to effect its decomposition as rapidly as possible, and then washing the precipitate formed in cold water, and drying spontaneously in a dark place. This oxide is a black powder, and is insoluble in water, but unites with the acids as a weak base.

The *protoxide,* HgO, is the *red precipitate* used in medicine. It is formed either by heating mercury nearly to its boiling point in a vessel to which the air has access, or by cautiously heating the nitrate so as to expel the nitric acid. The latter mode is the one usually adopted. It is usually seen in very small, shining, crystaline scales, which have a brick-red color. Heated to redness, it is decomposed, yielding mercury in the gaseous state, and oxygen. Red oxide of mercury is slightly soluble in water, and gives it an alkaline reaction.

499. Subchloride of Mercury, Hg_2Cl, eq., (200 + 35·4 =) 235·4.—This compound, the well-known *calomel* used in medicine, is easily prepared by pouring a solution of nitrate of the suboxide of mercury into a dilute solution of common salt, when

---

QUESTIONS.—What is it that forms the silvering of mirrors? 498. What oxides of mercury are there? How is the suboxide prepared? How is the protoxide prepared? What is it called? How is it affected when heated? Is it soluble in water? 499. What is *calomel?* Describe the mode of preparing it. To what use is it applied?

the calomel is precipitated as a white powder, which is insoluble in water and the acids when cold. It is sometimes found native, and is called *horn-quicksilver*, or *native calomel*. Its density is about 7.

Calomel is sublimed without change by a moderate heat, and may be obtained in small crystals. It is usually seen as a white powder, with a slight yellowish tinge; and may always be known by instantly turning black as ink, when touched with a drop of aqua ammoniæ, or solution of any caustic alkali.

This compound of mercury has long been extensively used in medicine, and the estimation in which it has been held may, perhaps, be inferred from the names by which it has been known at different times, and in different countries. The following are some of them:—*Mercurius dulcis, draco mitigatus, sublimatum dulce, aquila alba, aquila mitigata, manna metallorum, panchymogogum minerale, panchymogogum quercetanus!*

Calomel vapor has a density of nearly 8·2, and is composed of 1 vol. of mercury vapor, ½ of a vol. of chlorine condensed to 1 vol. Thus,

1 vol. of mercury vapor weighs	6·976
½ " chlorine " "	1.220
1 vol. subchloride vapor weighs	8·196

500. Chloride of Mercury, HgCl; eq., (100+35·4=) 135·4.—Chloride of mercury—the *corrosive sublimate* of the pharmacopœia—is obtained either by acting on mercury by nitrohydrochloric acid, or by sublimation from a mixture of sulphate of the protoxide of mercury, and common salt. The latter is the mode usually adopted in practice. Both the sulphate and the salt should be perfectly dry and well mixed;—the reactions will then be as follows:

$$HgO,SO_3 + NaCl = NaO,SO_3 + HgCl.$$

The sublimation may be effected in a retort of hard glass over a charcoal fire; and the operation should be conducted in such a situation that all the fumes escaping may be immediately conveyed away by a strong draft of air.

QUESTIONS.—May calomel be sublimed? How many volumes of its constituents does one volume of it contain? 500. How is chloride of mercury prepared? By what name is it known in pharmacy? Describe the reactions when sulphate of mercury and common salt are used in its preparation.

As corrosive sublimate is usually seen, it is colorless, semi-transparent, and has a crystaline structure. It has an acrid, burning taste, and leaves a nauseous metallic flavor on the tongue. It has a density of 6·5, fuses at 509°, and boils at 563°. It is soluble in 16 times its weight of water, and in alcohol and ether. When its solution in water is agitated with ether, the latter abstracts the bichloride, and rises with it to the surface of the former; and it may thus be separated from many other substances when contained with them in solution. Its aqueous solution is gradually decomposed by light, calomel being deposited.

Applied externally, it is highly corrosive to the flesh; and taken internally is a most deadly poison. Albumen precipitates it as an inert compound, and is therefore indicated as a proper remedy in cases of poisoning with it.

501. The *kyanizing* (from the name of the inventor, Mr. Kyan,) of timber consists in soaking it for a time in a solution of this substance, which protects it from the attacks of worms and insects; and also, by combining with the albumen contained in it, preserves it from decay.

The method given above for the preparation of calomel, though very easy, is not the one usually adopted in practice; a better result is obtained by mixing 100 parts (1 eq.) of mercury and 135·4 parts (1 eq.) of corrosive sublimate, and subliming with a moderate heat, the whole being converted into calomel. Thus,

$$Hg + HgCl = Hg_2Cl.$$

502. Iodides of Mercury.—There are two iodides of mercury, corresponding in composition with the oxides and chlorides. The protiodide is precipitated as a beautiful red powder by mixing together solutions of iodide of potassium and chloride of mercury. The powder, after washing and drying, may be sublimed by a moderate heat, but it then becomes yellow. After cooling, the color gradually changes to red; and the change is instantaneous if it is rubbed in a mortar.

Its vapor has the highest density of any known gaseous substance, being 15·69. It appears to be composed of equal volumes of mercury vapor and iodine vapor, condensed to one volume.

1 vol. iodine vapor weighs	8·716
1 " mercury " "	6·976
1 vol. vapor of HgI weighs	15·692

QUESTIONS.—Describe chloride of mercury. Is it poisonous? How is a solution of it affected by albumen? 501. In what does the *kyanizing* of timber consist? Describe the mode of preparing calomel (or sub-chloride) by using the chloride and metallic mercury. 502. What iodides of mercury are there? Describe the protiodide, and the mode of preparing it. What is said of the density of its vapor? Of what is a volume of the vapor composed?

503. Sulphides of Mercury.—There are two sulphides of mercury, which are exactly analogous in composition to the preceding compounds. The protosulphide, HgS, as we have seen, constitutes the chief ore of mercury, being found native, as *cinnabar*. It may also be prepared by art in several ways, and then forms the brilliant red pigment called *vermillion*.

In vapor the density of this substance is about 5·4; and 1 volume of it contains $\frac{2}{3}$ of a volume of mercury vapor, and $\frac{1}{9}$ of a volume of sulphur vapor, as shown below:

$\frac{2}{3}$ vol. mercury vapor weighs	$(6{\cdot}976 \times \frac{2}{3} =)$	4·651
$\frac{1}{9}$ vol. sulphur " "	$(\frac{6{\cdot}654}{9} =)$	0·739
1 vol. vapor of HgS weighs		5·390

As $\frac{2}{3} + \frac{1}{9} = \frac{6}{9} + \frac{1}{9} = \frac{7}{9}$, it is evident that the union of these two substances is accompanied by an expansion;—it is the only instance of the kind yet known.

The other binary compounds of mercury possess no special interest.

Salts of Mercury.

504. Nitrates of Mercury.—Nitric acid acts violently upon mercury, and forms with its oxides several salts, differing from each other according as the temperature may be more or less elevated, or the acid more or less diluted. Cold, dilute nitric acid acts slowly upon mercury, and forms with it a salt of the suboxide, which crystalizes with two atoms of water, and has for its formula Hg_2O,NO_5+2HO; but if the acid is hot, and of the usual strength, a salt of the protoxide is formed, which, when crystalized, has the formula, $2HgO,NO_5 + 2HO$. Still other nitrates of this metal may be formed, but these are the most important.

505. Sulphates of Mercury.—Sulphuric acid acts but slightly upon mercury when cold; but if heated, a sulphate of either the suboxide or protoxide is formed, according to the temperature. If 5 parts of sulphuric acid are boiled upon 4 parts of mercury, sulphate of the protoxide, HgO,SO_3, is formed—the compound used (500) in the manufacture of corrosive sublimate. Boiling water decomposes this sulphate, forming a yellow basic salt, $3HgO,SO_3$, called *turpeth mineral*.

506. Chlorosalts of Mercury.—Protochloride of mercury combines readily with other metallic chlorides, forming numerous crystalizable salts, which by some have been called *Hydrargo-chlorides*. Compounds

---

QUESTIONS.—503. What is said of the sulphides of mercury? What is *vermillion?* Of what is one volume of vapor of HgS composed? What is said of the union of these two substances? 504. What is said of the nitrates of mercury? 505. What is said of the action of sulphuric acid upon mercury? 506. Does chloride of mercury combine with other metallic chlorides? What are the compounds called?

of this character, with the chlorides of potassium, sodium, lithium, ammonium, barium, &c., are made simply by mixing their solutions. The double chloride of mercury and ammonium was formerly used in medical practice, under the name of *salt of alembroth*.

SILVER.

Symbol, Ag (*Argentum*); *Equivalent*, 108; *Density*, 10·5.

507. History.—This metal was known to the ancients. It frequently occurs native in silver mines, both massive and in octahedral or cubic crystals. It is also found in combination with gold, tellurium, antimony, copper, arsenic, and sulphur. In the state of sulphide, it so frequently accompanies galena, the chief ore of lead, that the lead of commerce is rarely quite free from traces of silver.

Most of the silver mines, which afford the metal at the present day, are in South America and Mexico; but in smaller quantities it is obtained in several countries of Europe, and other parts of the world.

Nearly all the silver obtained from the mines at the present time is found native, or is extracted from the sulphide; but there are many other mineral species which contain the metal.

508. Preparation.—When metallic silver is contained in small particles, disseminated through the ore, it is extracted by triturating the ore in fine powder with mercury, which dissolves the silver, and separates it at once from the mass, as an amalgam. The amalgam is then subjected to pressure in leather bags, by which a portion of the mercury is separated, and the remainder is expelled by heat. This is called the process of *amalgamation*.

Other ores of silver require to be treated differently, but the numerous processes adopted cannot be here described.

Silver is purified from small quantities of other metals present, except gold, by the process of *cupellation*, which has been practised from a very remote age. The process depends upon the

---

QUESTIONS.—What is *salt of alembroth?* 507. Has silver been long known? With what other metals is it found associated? From what ores is most of the silver of commerce extracted? 508. Describe the mode of extracting silver from its ores by the amalgamating process.

fact that the metals (except gold) which are usually in combination with silver, as copper, lead, tin, &c., are more oxidable than silver, and therefore when the alloy is kept for a time at a high temperature in the open air, these metals are oxydized, and the silver left perfectly pure. In large refineries, the oxide, as it is formed, is blown off by a bellows, so as to expose constantly a fresh surface to the air; but in small operations, as in the testing of coin or plate, the *cupel* proper is used, which is made of bone-earth, of the form of a small cup, as represented in the figure. This has the peculiar property of absorbing the mixed oxides as they are formed, and thus the necessity of using the bellows is avoided. For small operations, cupels are made about an inch in diameter, and half an inch deep. A vertical section is shown in the annexed figure.

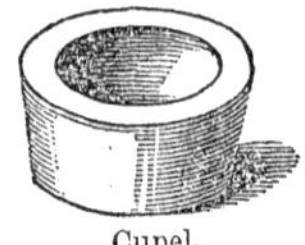
Cupel.

Section of Cupel.

509. We will suppose the object is to test some coin or plate, which contains with the silver some copper, and perhaps a little tin or lead. A small piece of the alloy is accurately weighed, and placed in the cupel with 10 or 20 times its weight of pure lead, and subjected to a strong heat, in such a manner that the air shall have constant access to it. When it becomes sufficiently heated, oxydation of the lead and other metals commences, and the mixed oxides being absorbed as they are formed, after a little time the silver is left perfectly pure. The cupel with the silver is then removed, and when cold the button of silver is detached and again carefully weighed.

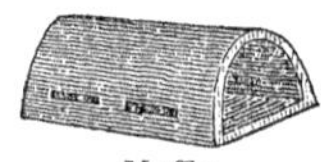
Muffle.

The heating is usually conducted in a muffle, placed in a proper furnace, so as to exclude dust and smoke, but allow the access of air. The muffle (see figure) is in fact a small oven, made of fire-clay, and is placed in the furnace so as to be entirely surrounded by the burning fuel, as shown in the figure on next page, which represents a vertical section of the furnace through its centre, with the muffle in its place. M is the muffle, with several cupels in it; A, B and C openings, which may be closed at pleasure by means of fire-brick doors, A, B, C, prepared for the purpose. To insure perfect success in the

---

QUESTIONS.—509. Describe the process of cupellation. Of what is the cupel formed? What purpose does it serve? Describe the muffle.

operation, attention is required to various particulars, which cannot be here given in detail.

Alloys of gold, subjected to the same process, are purified from all other metals except silver and platinum.

510. Another mode of testing alloys of silver, called the *wet* method, is to dissolve the silver alloy in nitric acid, and precipitate the silver by a standard solution of common salt, previously prepared with accuracy. This solution is made of such a strength, that 1000 parts of it, by measure, will precipitate exactly 1000 parts by weight of pure silver; and the proportion of silver in the alloy will of course be determined by the number of parts of the solution required to precipitate it. This is the mode practised at the United States mints, to determine the fineness of coin.

This method cannot be used when the alloy contains a metal, as mercury, whose chloride is insoluble.

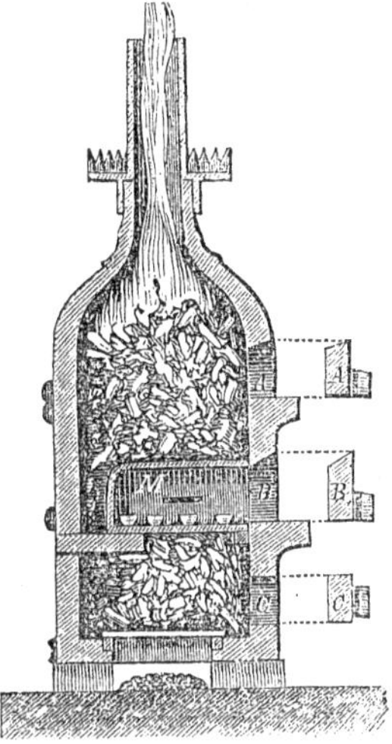

Section of Furnace and Muffle.

511. Properties.—Silver is a soft, white metal, and is very malleable and ductile. It has a brilliant lustre, and is susceptible of receiving a very fine polish. It is not acted upon by the atmosphere or by moisture, but is readily blackened by sulphur. Articles of silver often become tarnished, merely by the sulphurous gases which are diffused from the fires in houses in which mineral coal is used for fuel.

Its melting point is about 1873°, and if kept some time in fusion it absorbs oxygen in large quantity, which is given off again when the metal cools.

Silver is attacked by sulphuric acid only by the aid of heat, but is freely dissolved by nitric acid, even when cold.

Silver is used in every country for many important purposes—for coin, and for manufacture into various articles of utility or ornament; but to render it more stiff and hard it is always alloyed with a portion of copper.

QUESTIONS.—510. Describe the wet method of testing alloys of silver. 511. Describe the properties of silver. What acids act upon it? For what purposes is silver used?

The standard of purity for silver coin varies in different countries; but the coin of the United States contains 900 parts of pure silver and 100 parts of copper;—that is, one-tenth part of the weight of the coin is copper, except the three-cent piece, of which $\frac{3}{4}$ths are silver and $\frac{1}{4}$th copper.

At the present time, the largest silver coin issued from the United States Mint is the half-dollar, which is required to weigh 192 grains; the weight of the smaller coins being in the same proportion, except that of the three-cent piece, which is $12\frac{3}{8}$ grains. The quantity of pure silver in the half-dollar is, according to the above, 172·8 grs. = 192 — 19·2. This gives as the value of pure silver $1,38.8 per ounce.

The value of coin is always estimated in proportion to the amount of pure metal it contains, no attention being paid to the alloy. It is remarkable that the addition of copper scarcely produces any change in the brilliant white color of silver, provided it does not exceed about one-eighth of the latter metal.

Silver combines with other metals, forming alloys, which, however, possess no particular interest.

Binary Compounds of Silver.

512. Oxides of Silver.—Silver forms with oxygen three compounds, a *suboxide*, Ag_2O, *protoxide*, AgO, and *binoxide*, AgO_2; but only the protoxide is important. This is thrown down as a dark-colored powder, when solution of caustic potash is poured into a solution of nitrate of silver. Heated to redness it gives up all its oxygen, and pure silver is obtained. It forms the base of all the oxysalts of silver;—is slightly soluble in water, but very soluble in aqua ammoniæ.

By digesting this oxide for a time in concentrated aqua ammoniæ, a black compound is formed, which is highly explosive, sometimes called *fulminating silver*. Its composition has not been satisfactorily determined. The most probable opinion is, that it is a nitride of silver, NAg_3, formed by the reaction:

$$3AgO + NH_3 = NAg_3 + 3HO.$$

513. Chloride of Silver, $AgCl$; eq., (108 + 35·4 =) 143·4.—This compound is occasionally found as a natural production, and called *horn silver*, and is easily formed artificially, by pouring solution of common salt into a solution of nitrate of silver. Formed by this mode, it is a beautiful white powder, which, however, soon becomes purple in diffused light, or black if exposed to the direct light of the sun, or if warmed before a fire. It is insoluble in water, but soluble in ammonia and hyposulphurous acid.

QUESTIONS.—What proportion of the silver coin of this country is silver? What is the alloy? What is the largest silver coin now issued from the United States Mint? What does this give as the value per ounce of pure silver? In estimating the value of coin, is any allowance made for the value of the alloy used? 512. What oxides of silver are there? Which of these constitutes the base of the salts of silver? 513. Describe the chloride of silver. How is it prepared artificially?

Nascent hydrogen reduces it to the metallic state by absorbing the chlorine to form hydrochloric acid. To effect the reduction, the chloride is covered with water acidulated with sulphuric acid, and pieces of zinc introduced, and the whole occasionally stirred. The metallic silver is obtained in fine grains.

Iodide of Silver, AgI, is found as a natural production, and may also be formed by precipitation from solution of nitrate of silver by iodide of potassium.

Sulphide of Silver, AgS.—This compound is found in nature, alone, and in combination with other metallic sulphides, particularly sulphide of lead, in the ore called *argentiferous galena.* Sulphide of silver may also be formed artificially by several processes.

Salts of Silver.

514. Nitrate of Silver, AgO,NO_5.—This is the only salt of silver of any practical importance, and is well known by the name of *lunar caustic.* It is usually sold in small sticks, which are wrapped in paper, but may also be obtained in beautiful white, tubular crystals. The sticks usually contain a portion of nitre, which has been melted with it when cast in the moulds. Nitrate of silver is formed by dissolving silver in nitric acid, diluted with twice its weight of distilled water. It is soluble in one-half its weight of boiling water, and its own weight of cold water. It is the basis of *indelible ink,* as it is called; but writing done with it, and stains made by it, may be removed by solution of cyanide of potassium. Nitrate of silver is very caustic to the flesh, and is used in medicine as a cautery. From its solution, metallic copper precipitates the silver as a fine powder; by mercury, the silver is thrown down in an arborescent form, which has been called the *arbor Dianæ.*

Sulphate of Silver, AgO,SO_3, may be formed by boiling sulphuric acid upon metallic silver. It is a colorless salt, slightly soluble in boiling water.

QUESTIONS.—Describe the mode of reducing the chloride by means of zinc and oil of vitriol. What other binary compounds of silver are mentioned? 514. What is lunar caustic? Describe its properties What use is made of it?

GOLD.

Symbol, Au (*Aurum*); *Equivalent*, 198; *Density*, 19·26.

515. History. — Gold appears to have been known to the earliest races of men, and to have been esteemed by them as much as by the moderns. With the exception of the rare mineral *telluride of gold*, it has hitherto been found only in the metallic state, either pure, or in combination with other metals. It is sometimes found in quartzose rocks, but more frequently in alluvial depositions, especially among sand in the beds of rivers, having been washed by water out of disintegrated rocks in which it originally existed.

Though usually found in irregular rounded lumps and grains, it is sometimes obtained in crystals of the monometric system as

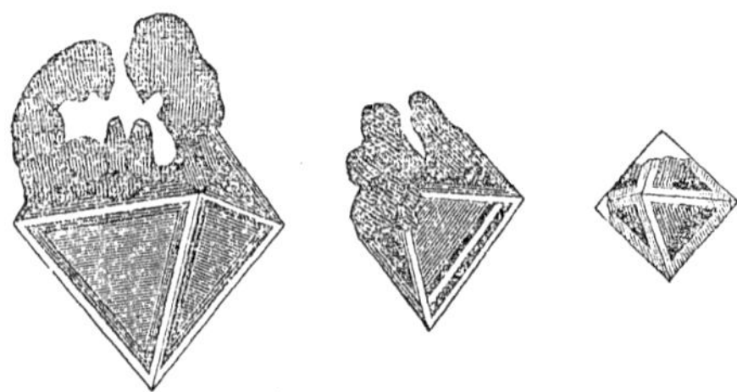

Crystals of Gold.

represented by the accompanying figures, taken from the American Journal of Science.

Gold is obtained at the present day in large quantities in California, Australia, and some parts of the Ural Mountains, and less abundantly in Hungary, and other countries of Europe. In small quantities, it occurs in Georgia, North Carolina, and Virginia.

516. Preparation.—As gold exists in its ores in the metallic state, it is generally separated from them by the process of amal-

---

QUESTIONS.—515. Give the history of gold. In what situations is it found? Is it occasionally found in crystals? 516. Describe the mode of separating silver from gold, called *quartation*.

gamation, similar to that already described for obtaining silver, by which means it is separated from all other metals except silver. To remove this, so much silver must be added to it that the gold shall constitute but a fourth of the whole, and the mass boiled in nitric acid, which then readily acts upon it, dissolving out all the silver, and leaving the gold in a state of purity. This process has been called *quartation*, from the circumstance that the proportion of gold, in order that the nitric acid shall dissolve out all the silver, must not exceed a *quarter* of the whole mass. Other metals, except silver, may also be separated from it by cupellation (508).

To prepare absolutely pure gold, a piece of coin may be dissolved in aqua regia, and precipitated with solution of sulphate of protoxide of iron. The reactions are as follows:

$$AuCl_3 + 6FeO,SO_3 = Au + Fe_2Cl_3 + 2(Fe_2O_3,3SO_3).$$

The gold thus obtained is in a minutely divided state, and is of a purplish brown color. It is to be collected on a filter, and washed with very dilute hydrochloric acid, and fused with a little borax and saltpetre.

517. Properties.—Gold is readily distinguished from all other metals by its brilliant yellow color, and by its great malleability and ductility. It is capable of being beaten out into leaves so thin that light may be transmitted through them, which then appears of a greenish yellow color. It is not acted upon by air or moisture, though exposed to their influence for ages; nor is it oxydized by being kept in a state of fusion for any length of time. When intensely heated by the galvanic current, or by means of the compound blowpipe, it burns with a greenish blue flame, and is dissipated in the form of a purple powder, which is supposed to be an oxide. Selenic acid, aided by heat, dissolves it, and a mixture of selenic and hydrochloric acids, without heat; but its proper solvent is *aqua regia*, which is a mixture of 1 part of nitric and 2 parts of hydrochloric acid. It fuses at about 2016°.

QUESTIONS.—Why does the process for separating silver from gold here described receive the name *quartation?* How may absolutely pure gold be prepared? 517. How is gold distinguished from the other metals? Describe some of its properties. What is the only proper solvent of gold?

518. Gold and silver, from the estimation in which they have been held, have been long known as the "precious metals;" and it is usual to estimate their purity in carats. A carat is to be understood as $\frac{1}{24}$th part of the mass; and a piece of gold or silver is 14, 18, or 20 carats fine, when so many 24ths of the whole are fine metal, the rest being alloy. But in the Mint of the United States, their fineness is estimated in thousandths: thus, gold or silver is said to be of the fineness 654, 789, 921, or 994 thousandths, when so many thousandths of the whole mass consist of pure metal, the rest being alloy. The alloy of silver is always copper, but the alloy of gold may be either copper or silver, or a mixture of the two. Pure gold is so soft that some alloy is always needed to give it the proper stiffness, and to prevent too rapid wearing. In the gold coins of this country, one-tenth part is alloy, which is a mixture of silver and copper. The gold eagle of the United States weighs 258 grains, of which 25·8 grains are alloy. The value of standard gold is therefore estimated at $18.604 per ounce, and that of pure gold at $20.671 per ounce.

Native gold is almost always alloyed with silver, but the proportion of this metal is very variable.

Gilding consists in coating over the surfaces of bodies with a thin film of gold. Articles made of metal were formerly gilded by applying to their surface, properly cleaned and smeared with nitrate of mercury, amalgam of gold, and then expelling the mercury by heat; but this mode has been entirely superseded by the electrotype process, heretofore (116) described. In both cases the surface of gold requires to be burnished.

Articles made of non-metallic substances are gilded by a coating of gold leaf.

Binary Compounds of Gold.

519. **Oxides of Gold.**—There are two, and perhaps three, oxides of gold; but the *teroxide*, AuO_3, alone possesses any special importance. It is of a yellow color when first formed, but becomes black when all the water is expelled. It is used in coloring glass and porcelain purple. In some cases it seems to act the part of a feeble acid, and has been called *auric acid.*

520. **Chlorides of Gold.**—There are two chlorides of gold; the *terchloride*, $AuCl_3$, the one usually seen, is formed when gold is dissolved in aqua regia. By evaporating the solution carefully, the chloride may be obtained as a solid, which is very soluble in water, alcohol, and ether. Solution of chloride of gold is very easily decomposed by green vitriol, and by organic

---

QUESTIONS.—518. Why have gold and silver been called the precious metals? How has the fineness of these metals, or rather alloys of them, usually been estimated? What is meant when an article of gold or silver is said to be 18 carats fine? How is the fineness of these metals estimated at the United States Mint? What is the alloy used for the gold coins of this country? What is the weight of the United States eagle? What does this give as the value per ounce of standard gold? Of pure gold? How is gilding performed? 519. What oxides of gold are there? 520. What chlorides?

substances. Protochloride of tin forms with it a beautiful purple powder, called *purple of Cassius.*

Aqua ammonia precipitates, from solutions of terchloride of gold, a fulminating compound called *fulminating gold,* of uncertain composition.

Salts of Gold.

521. The *oxides* of gold do not unite as bases with the acids, but the teroxide, as an acid, combines with bases, forming some unimportant salts, which are called *aurates.*

Chlorosalts of Gold.—Terchloride of gold combines with many other metallic chlorides, like the chloride of mercury, forming a series of chlorosalts, sometimes called *auro-chlorides.*

PLATINUM.

Symbol, Pt; *Equivalent,* 99; *Density,* 21·5.

522. **History.**—Platinum was first recognised as a distinct metal in 1741, but was not described until 1749. It has hitherto been obtained chiefly from Brazil, Peru, and some other parts of South America, and from the Ural Mountains. It occurs only in the metallic state, associated with other metals, as gold, silver, lead, palladium, osmium, iridium, and rhodium.

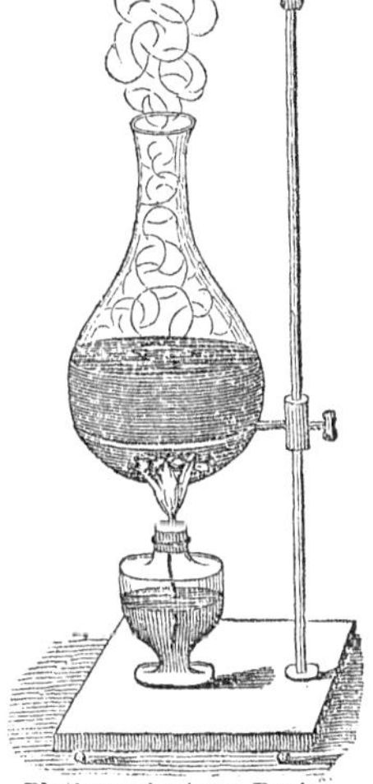

Platinum in Aqua Regia.

523. **Preparation.**—To prepare pure platinum, the native grains, or commercial platinum, are dissolved in boiling aqua regia, and, after standing for a time, the clear solution is poured off, and the platinum precipitated by solution of sal-ammoniac, as a double chloride of platinum and ammonium. By heating this precipitate, both the chlorine and ammonia are expelled, and the metallic

QUESTIONS.—How is purple of Cassius formed? 521. What is said of the chlorosalts of gold? 522. Give the history of platinum. What is the state in which it occurs? 523. Describe the mode of preparing pure platinum.

platinum remains as platinum sponge, which, when treated with hot water, appears as a dark gray mud.

This is now pressed in a hollow cylinder, by which means the particles of metal are made to adhere, so that the disc thus obtained may be carefully heated, and hammered on an anvil. The metallic grains now become firmly welded together, and the mass may be worked in any desired form.

524. **Properties.**—Platinum is a white metal, much resembling silver, but of a darker color, and of inferior lustre. When pure, it is very malleable and ductile. It is the most dense substance known to man (except, perhaps, iridium), but is quite soft; and pieces of it, when heated, may be welded like iron, though not so easily. No single acid attacks it, but it is soluble in heated aqua regia. By heated nitre, or potassa, or soda, it is oxydized. It cannot be melted by the most intense heat of the hottest furnace; but may be fused by the compound blowpipe. When a large surface of the metal is exposed to a mixture of oxygen and hydrogen, it has the singular property of causing them to combine, either silently or by an explosion. It acts in this way more readily when used in the spongy form (200, 523), as precipitated from its solution by sal-ammoniac.

Platinum black, in which the metal is in a still more finely divided state, acts energetically in the same manner, causing the rapid union of various gases besides oxygen and hydrogen, when submitted to its influence. This form of platinum is prepared by electrolizing a dilute solution of chloride of the metal, or by boiling a solution of the chloride mixed with carbonate of soda and sugar.

Wire of Platinum over Ether.

If a coil of platinum wire, recently ignited, be suspended in a deep glass, containing a little ether at the bottom, it will instantly become incandescent, and glow with a red heat, until the ether is entirely dissipated. The same effect may be produced by placing a coil of

QUESTIONS.—How are the particles of finely divided metal made to unite so as to form a solid mass? 524. Describe the properties of platinum. By what alone is it dissolved? How is it affected by heated nitre or potash? How is a mixture of oxygen and hydrogen affected when exposed to a large surface of this metal? What is said of *platinum black* in this connection? Describe the experiment with a coil of platinum wire and ether.

small platinum wire over the wick of a spirit-lamp, and after lighting it, suddenly extinguishing the flame. The wire will continue at a red heat until all the alcohol is consumed. Such a lamp (called a *flameless lamp*) is represented in the accompanying figure.

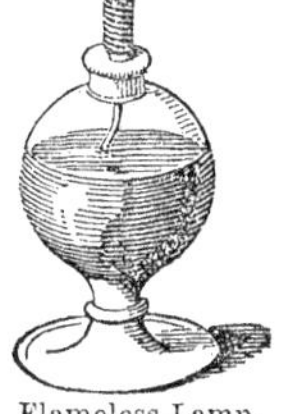
Flameless Lamp.

525. Uses.—Platinum is of great importance in the laboratory, and is much used in the arts, especially for retorts for condensing (261) sulphuric acid. Its present value in the market is about half that of gold. It was formerly issued as coin by the Russian government, but the practice has been discontinued.

Binary Compounds of Platinum.

526. Oxides of Platinum.—Platinum forms with oxygen two compounds, the *protoxide*, PtO, and the *peroxide*, PtO_2, both of which are feeble bases, and unite with some of the acids to form salts.

Chlorides of Platinum.—Two chlorides of platinum are known, analogous in composition to the oxides. The bichloride, $PtCl_2$, is the most important of all the compounds of this metal, and is obtained by treating the metal with boiling aqua regia, and carefully evaporating, to expel the excess of acid. It is much used in the laboratory.

Sulphide of Platinum, PtS, may be formed by heating platinum-filings in vapor of sulphur.

527. Salts of Platinum.—The salts of platinum, at least the oxysalts, are not important. Oxalate of the protoxide, and sulphate and nitrate of the binoxide, are known.

Chlorosalts of Platinum.—The bichloride of platinum combines with other metallic chlorides, forming a series of chlorosalts, called also *platino-chlorides.*

528. Osmium, Os; Eq., 99·5.—This rare metal is in combination with platinum and iridium;—with the former in the so called native platinum grains, and with the latter in the mineral species called *iridosmine.* It is of a grayish color, and metallic lustre; is slightly malleable, and has a density of about 10. It combines readily with oxygen when heated, and is attacked by nitric acid, which converts it into osmic acid.

Small grains of the native iridosmine are used for the tips of gold pens, because of their hardness and capability to resist the corrosive action of the ink.

QUESTIONS.—Describe the flameless lamp. 525. To what uses is platinum applied? 526. What oxides of platinum are known? What chlorides? How may sulphide of platinum be formed? 527. Are there any important salts of platinum? 528. What is said of osmium? What use is made of the native grains of osmium and iridium?

There are known no less than five *oxides* of osmium, viz., OsO, Os_2O_3, OsO_2, OsO_3, and OsO_4; of which the last two possess acid properties, and are called the osmious and osmic acids.

Two *chlorides* only of osmium are known, a *protochloride* and a *bichloride.*

529. **Iridium,** Ir; Eq., 99.—Iridium, as stated above, is found chiefly in combination with osmium. It has not been obtained in a malleable state, but only as a hard compact mass; and its density cannot therefore be well determined, but by some it is believed to exceed that of platinum. It is not attacked by nitric acid, nor even by aqua regia when pure, but at a red heat enters into combination with chlorine, with which it forms two compounds, the protochloride, IrCl, and the bichloride, $IrCl_2$.

530. **Palladium,** Pd; Eq., 53·3.—This metal, often contained in platinum ores, is obtained chiefly from a native compound of this metal with gold, found in some parts of South America. It is nearly as white as silver, and scarcely less fusible than platinum. It is malleable and ductile, and receives a fine polish under the burnisher. Its density is 11·8.

Palladium is used for the construction of the beams for delicate balances, and also for the graduated scales of astronomical instruments.

531. **Rhodium,** Rh; Eq., 52·2.—Rhodium, which receives its name from the *rose-color* of some of its compounds, is contained in small quantities in most platinum ores, and is sometimes found in combination with gold. It is of a gray color, and even more infusible than platinum. It is attacked by aqua regia only when alloyed with platinum, or some other metal. Its density is 10·6.

Ruthenium, Ru, is the name given to a metal recently discovered in some platinum ores. In its general properties it resembles iridium.

QUESTIONS.—What oxides of osmium are known? 529. In what is iridium found? 530. Describe palladium. What use is made of it? 531. Describe rhodium. What is said of ruthenium?

PART IV.

SPECIAL CHEMISTRY—ORGANIC.

GENERAL PROPERTIES OF ORGANIC BODIES.

532. Introduction.—In the preceding part of this work, we have traced the chemical history of all the elementary substances at present known, and that of many of their most important compounds, as they are produced by the action of their affinities, uncontrolled except by the agencies of heat, light, and electricity; but we have now to discuss altogether another class of compounds, called *organic*, because produced almost exclusively by the *organs* of plants and animals, or derived from substances so produced.

Organic Chemistry, therefore, is that branch of the general science of Chemistry which treats of the history, properties, and transformations of animal and vegetable substances.

These are always compound, and differ from inorganic or mineral compounds chiefly in their origin, and in the circumstance that most of them are of a more complex composition.

533. Organic and Organized Bodies.—There is, however, a certain class of organic bodies,—called *organized* bodies,—whose essential physical properties are altogether peculiar. These are always insoluble, and incapable of crystalization, and exhibit an organized structure, often visible to the naked eye, and always apparent under the microscope. To this class belong all the proper tissues of the animal and vegetable systems, constituting the organs by which all their various functions are performed. Sugar, gum, alcohol, and urea are organic substances, the two

---

QUESTIONS.—532. What has been treated of in the preceding parts of this work? What are now to be treated of? Define Organic Chemistry. How do organic bodies differ from inorganic? 533. What are *organized* bodies? What is said of sugar, gum, &c ?

former being found ready formed in plants and sometimes in animals, and the two latter, being derived usually from substances so produced, but they are not organized. On the other hand, the cellular tissue of wood, the pith of an elder tree, and the skin of an animal, are organized; and their peculiar structure is apparent to the eye, at least when aided by the microscope. The figure in the margin, from Regnault's Chemistry, represents a cross section of the pith of the elder, as seen under the microscope. This exhibits a peculiar regularity of structure, which, however, is not common;—the structure of two different organized bodies seldom presents any striking similarity. Of the next two figures, the first represents a microscopic view of a longitudinal section of a stalk of asparagus, and the second a cross section of the same.

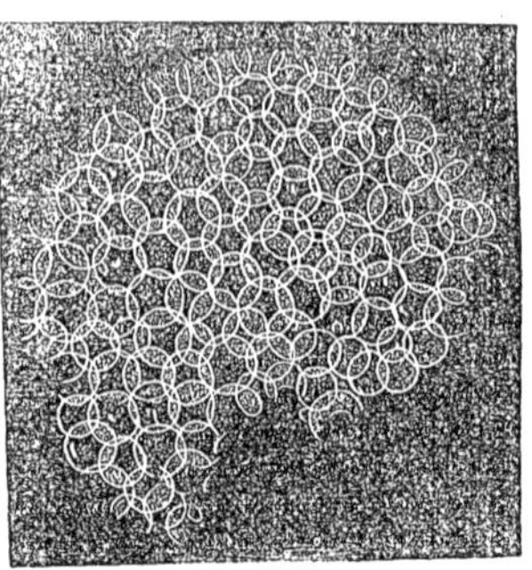

Pith of Elder.

Stalk of Asparagus.

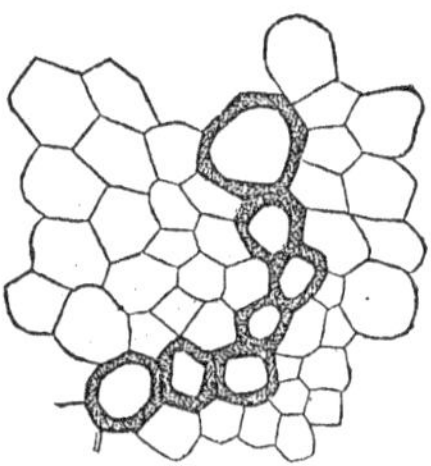

Cross Section of Same.

534. Vitality.—In the production of organic bodies, the simple affinities of the particles of which they are composed are over-

QUESTIONS.—What is said of the various tissues of the system? How may the organized structure always be seen? What is represented by the figure in the margin? What by the next two figures? 534. By what are the affinities of matter controlled in the production of organic compounds?

ruled or controlled by another power or force, called the *vital power*, *vitality*, or *the principle of life*. Its influence is absolutely essential for the production of most organic compounds, which, it is believed, can never be formed by the simple operation of the ordinary affinities of the elements of these compounds. And it follows, as a necessary consequence, that after death, that is, when this principle has ceased its influence, and the simple affinities of the elements of an organic compound are left uncontrolled, they will show a disposition to break up, and re-arrange themselves in a new order. In this way the old compound is destroyed, and perhaps several new ones formed; or the simple elements may be entirely set free. This is what is termed the spontaneous decay, or putrefaction of a substance. Often this process is affected by the presence of the atmosphere, and is always much influenced by the temperature, and other circumstances.

Although many organic compounds are found, as such, in the bodies of plants and animals, by far the greater number are produced by the actions of the various reagents, as the acids, alkalies, and salts, either cold or aided by heat, upon these compounds. Thus, sugar and starch occur abundantly in plants, but from them, by the action of reagents, a long list of other compounds are produced, which are never found in the plants themselves. Alcohol, the almost innumerable ethers, and many acids, are of this kind. Some compounds, as acetic and oxalic acid, are found ready formed in organic bodies, and may also be produced by the action of reagents upon other organic substances.

But it is not to be understood that all organic substances are equally liable to decay; some of them, as sugar, wood, and gum, of vegetable origin, and gelatine, of animal origin, if kept dry, may be preserved apparently for any length of time.

The chief elements of organic bodies are carbon, hydrogen, oxygen, and nitrogen, which are combined in different modes, and in different proportions; but besides these some organic substances also contain sulphur, phosphorus, chlorine, calcium, potassium, sodium, magnesium, iron, silicon, &c.

Some few organic compounds have been formed artificially, that is, without the aid of the vital principle; but not any *organized* body.

QUESTIONS.—Is the influence of this principle of vitality essential to the production of organic compounds? Why, after death, do most organic substances spontaneously decay? Are most organic compounds found ready formed in the bodies of plants and animals? What is said of alcohol and the ethers? May some organic bodies be preserved for a long period? What are the chief elements of organic bodies?

535. Molecular Structure of Compounds.—It is well known that two compounds, exactly the same in composition (172), often differ very considerably in their properties. This difference is believed to be occasioned by differences in the mode of arrangement of the particles in the compounds, or, in other words, in their molecular structure. In general, while we understand well the elements of many compounds, we really know little of the mode in which these elements are grouped together in any particular case. We know, for instance, that the equivalent of sulphuric acid, SO_3, contains 1 atom of sulphur and 3 atoms of oxygen, but we cannot say whether these are grouped as $S+3O$, $SO+2O$, or SO_2+O, for these three modes are all alike possible. Here it would seem that there can be only three modes of combination, but in more complex compounds the number of possible modes may be greatly increased.

This supposed difference in the mode of combination or grouping of the atoms of a compound, may be represented to the eye by means of a diagram. Thus, let us suppose that we have a compound of two simple

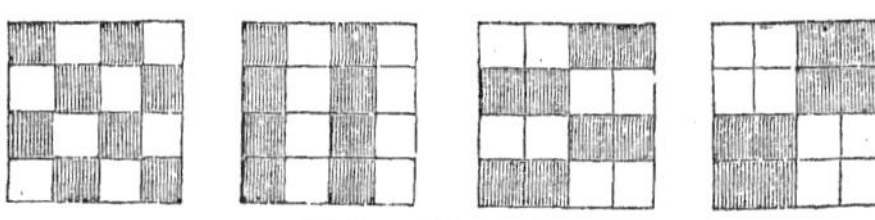

Modes of Grouping.

substances, and that in the atom of the compound there are 8 atoms of each of the elements;—representing the particles of one kind of matter by the dark squares, and the other by the light ones, the four figures show as many independent modes of grouping. Now, as every independent mode of grouping of the particles may produce a new substance, it is plain that we may thus have four substances, all having the same ultimate composition, yet possessing properties altogether different.

536. Compound Radicals — Nomenclature.—Compound radicals are chemical compounds which are capable of performing the functions of simple substances; they combine with elementary bodies and with other compounds in the same manner as simple

---

QUESTIONS.—535. May substances having the same composition differ in their properties? How is this difference believed to be occasioned? What is said of sulphuric acid in this connection? How is the supposed difference in the mode of grouping among the atoms of a compound illustrated by the figure? 536. What are compound radicals?

substances, and may often be substituted for these in the compounds of which they form a part.

One of the best known of these compound radicals is cyanogen, C_2N (316), which, as is shown by the formula, is a bicarbonide of nitrogen. This group of atoms, it is well determined, is capable of combining with the metals and other substances in the same manner as oxygen, chlorine, sulphur, &c., producing compounds similar to the oxides, chlorides, &c., which are called *cyanides.* United with oxygen it forms cyanic acid, with hydrogen it forms hydrocyanic acid, &c. So through a long series of other compounds, this group of atoms is found everywhere performing the functions of a simple substance.

Other compound radicals which have been determined are *methyle,* C_2H_3; *ethyle,* C_4H_5; *butyryle,* C_6H_7; *valyle,* C_8H_9; *amyle,* $C_{10}H_{11}$, &c. Still other compounds of the same character, but having a more complex composition, are *benzyle,* $C_{14}H_5O_2$; *cacodyle,* C_4H_6As; *stanmethyle,* C_4H_5Sn; *zincmethyle,* C_2H_3Zn; *stibmethyle,* $(C_2H_3)_3Sb$, &c.

537. That these groups of atoms, and many others, do enter, as such, into composition with the simple substances, and with each other, and are capable of being transferred by single or double decomposition from one compound to another, are facts too well established to be controverted. All those mentioned above, except perhaps benzyle, have also been obtained in a separate state; and many that have not been thus obtained may, with great probability, be assumed as having a real existence.

Considering the existence of these groups denominated compound radicals as fully established, it will be seen at once that it furnishes a ready mode of classifying organic compounds, and also for forming for them a systematic nomenclature. Thus, if we take ethyle, C_4H_5, as the basis or starting point of a series, we have for its binary compounds the oxide, chloride, iodide, sulphide, &c.; and for salts of its oxide, the hyponitrite, nitrate, acetate, &c., as follows, viz.:

Binary Compounds.

1. Ethyle.............	C_4H_5.		
2. Oxide of ethyle,	C_4H_5O,	common sulphuric	ether.
3. Chloride "	C_4H_5Cl,	hydrochloric	"
4. Iodide "	C_4H_5I,	hydriodic	"
5. Sulphide "	C_4H_5S,	hydrosulphuric	"

Salts of Oxide of Ethyle.

1. Nitrite of oxide of ethyle,	C_4H_5O,NO_4,	hyponitrous	ether.
2. Nitrate " "	C_5H_5O,NO_5,	nitric	"
3. Acetate " "	$C_4H_5O,C_4H_3O_3$,	acetic	"

QUESTIONS.—What compound radicals are mentioned? 537. Have many of these groups been obtained in a separate state?

The foregoing are given as examples only; — this series might be extended much further, and many other series might be introduced, as the methyle series, which would include the compounds of methyle, C_2H_3, analogous to the above, the acetyle series, the cacodyle series, &c. A nomenclature of organic compounds constructed on this principle has been adopted by able and judicious chemists; but as there is no proof that such a nomenclature truly represents the real molecular structure of these compounds, we do not make use of it in this work.

538. A slight examination only is needed to show that most organic compounds may be arranged in series in a variety of ways, each new mode requiring or supposing a new molecular arrangement. Thus in the compounds represented in the table above, if we commence with olefiant gas, or ethylene, C_4H_4, instead of ethyle, C_4H_5, then we may consider the latter, $C_4H_5 = C_4H_4,H$, as the hydride of ethylene, and ether, $C_4H_5O = C_4H_4,HO$, as hydrate of ethylene, hydrochloric ether, $C_4H_5Cl = C_4H_4,HCl$, as hydrochlorate of ethylene, &c.

539. The real mode in which the particles of organic compounds are arranged (535), at least in most cases, has not yet been satisfactorily determined; and it is therefore impossible, in the present state of our knowledge, to fix upon a proper systematic nomenclature. The same may also be said of the formulæ used to represent these compounds. A distinction is often made between *empirical* and *rational* formulæ, the former representing the composition as determined by ordinary analysis, without any attempt to indicate the arrangement of the particles; the latter, on the other hand, representing not only the composition of the compound, but also its molecular structure. Thus, the composition of acetic acid, as determined by analysis, is $C_4H_4O_4$; but when this acid combines with a base it always parts with one equivalent of water (or its elements), so that its salts, the acetates, have the general formula, $RO,C_4H_3O_3$. This would seem to indicate that the formula for the acid should be written $C_4H_3O_3,HO$. But certain considerations have led us to the belief that in anhydrous acetic acid, $C_4H_3O_3$, 1 equivalent of the oxygen is held in a different state from the rest, and therefore its formula should be $C_4H_3O_2,O$. Adopting these views, then, while it is admitted that the empirical formula, $C_4H_4O_4$, represents correctly the elements of acetic acid, its rational formula will be, $C_4H_3O_2.O + HO$. In other words, it is a compound having for its primary radical, acetyle, C_4H_3, and for its secondary radical, $C_4H_3O_2$.—(*Dr. Gibbs' Report, p.* 43.)

540. In the following pages, a systematic nomenclature, according to any particular theory, is not attempted; — the names of compounds adopted are those in general use, with a few unimportant modifications.

QUESTIONS.—Why is not the nomenclature of the compound radical theory made use of in this work? 538. May most of the organic compounds be arranged in series in a variety of ways? Give the illustration in the text. 539. Has the real molecular structure of organic compounds been determined? What are *empirical*, and what *rational* formulæ? Give the illustration by reference to acetic acid. 540. What is said of the nomenclature used in the remaining part of this work?

In many cases, compounds are named from some one of the natural productions in which they are found, as *malic acid* from *malum*, an apple; *citric acid*, from *citron*, a lemon; *valerianic acid*, found in the root of the plant called *valeriana officinalis; cinchonia*, obtained from the bark of the *cinchonia condaminea*, &c. The formulæ, in general, are to be considered only as empirical; frequently, two formulæ are given for the same compound, with the sign = between them. In such cases, the first is always the empirical formula, while the second indicates something further, as it regards the supposed constitution of the compound, or the mode of its reactions with other substances. The formula for cane sugar, for instance, is written $C_{12}H_{11}O_{11} = C_{12}H_9O_9 + 2HO$, by which it is indicated that 2 equivalents of water (or the elements of water) sustain a relation to the compound different from that of the other atoms of these elements.

Laws of Combination and Transformation.

541. We have seen (536), that in organic compounds certain groups of atoms, called compound radicals, frequently are found to perform the functions of simple substances. Many of these groups are already known, and many more will probably be hereafter discovered. If all the groups capable of acting in this manner, with their properties and relations, were known, it is very probable that the transformations of organic bodies would not differ essentially from those of inorganic matter, except as they are affected by this circumstance. In other words, all cases of chemical action would be reduced to instances of direct union, or of single or double elective affinity.

When potassium is burned in oxygen gas, direct union takes place between the two elements, and potassa (oxide of potassium) is formed—$K + O = KO$. But when potassium is thrown into water, we have a case of single elective affinity, as shown by the equation representing the reaction;—thus, $K + HO = KO + H$. We may say in this case, either that the oxygen is transferred from the hydrogen to the potassium, or that the potassium has replaced the hydrogen of the water. When solutions of nitrate of baryta and sulphate of soda are mixed together, we have an instance of what is called double elective affinity; and by a double transfer of elements we have formed the two new compounds, sulphate of baryta and nitrate of soda. Thus,

$$BaO,NO_5 + NaO,SO_3 = BaO,SO_3 + NaO,NO_5.$$

QUESTIONS.—How are compounds often named? Are the formulæ given to be considered as empirical or rational? 541. What is said of the transformations of organic compounds as compared with those of inorganic matter? What illustrations are given from inorganic chemistry?

542. So in organic chemistry, most if not all reactions will be similar to one or another of the above cases. When olefiant gas, C_4H_4 (308), which is properly an organic product, and chlorine, Cl, are brought together, they combine to form an oil-like liquid, $C_4H_4Cl_2$. But often when two compounds have in this way combined, the new and more complex compound that is formed may, by a slight change of circumstances, or even spontaneously, break up into compounds of less complex constitution. We have a case of this kind in sulpho-vinic acid, which is formed by the union of alcohol, $C_4H_6O_2$, with sulphuric acid. Thus,

$$C_4H_6O_2+2(SO_3,HO)=C_4H_6O_2,2(SO_3HO)=C_4H_5O,HO,\ 2(SO_3HO).$$

This last substance, called monohydrated sulpho-vinic acid, when heated, breaks up, not into alcohol and monohydrated sulphuric acid, but into ether, C_4H_5O, and $2SO_3,3HO$.

But most of the transformations in organic bodies may be considered as instances of double decomposition, or double elective affinity. The action of chlorine upon acetic acid is properly of this kind, as is shown by the following equation. Thus,

$$\underset{\text{Acetic Acid.}}{C_4H_4O_4}+6Cl=\underset{\text{Chloracetic Acid.}}{C_4HCl_3O_4}+3HCl.$$

It is, indeed, true that chlorine, one of the substances used, is not a compound, but we are to consider that the action is the same as if each of the three atoms of hydrogen were successively replaced, giving for the first step in the process the reaction, $C_4H_4O_4+2Cl=C_4H_3ClO_4+HCl$. One atom of the chlorine is transferred to the acetic acid, which at the same time gives up 1 atom of its hydrogen to combine with the second atom of chlorine, to form hydrochloric acid. This reaction three times repeated results as given above, in the replacement of 3 atoms of hydrogen by as many atoms of chlorine.

543. Substitution or Metalepsy.—We have seen above that by the action of chlorine upon acetic acid, $C_4H_4O_4$, the latter loses 3 atoms of hydrogen, and takes in their place 3 atoms of chlorine; in other words, 3 atoms of chlorine are *substituted* in the acid for an equal number of atoms of hydrogen. The new compound, $C_4HCl_3O_4$, is called chloracetic acid, and in most of its properties closely resembles acetic acid, from which it has been formed. Transformations of this kind are of very frequent occurrence;

QUESTIONS.—542. What is the effect when olefiant gas and chlorine are brought together? When two substances combine so as to form a more complex compound, will they sometimes break up in such a manner as to form compounds different from those which at first united? Give the illustration by reference to alcohol and sulphuric acid. What is said of acetic acid and chlorine in this connection? 543. What is the change produced in acetic acid by the action of chlorine?

and the exchange or substitution may take place between atoms or groups of atoms (compound radicals) which are similar to each other in their chemical relations.

Hydrogen seems to be more frequently replaced than any other element; and there may be substituted for it chlorine, iodine, bromine, or a metal, or even a compound radical, as methyle, ethyle, &c. Instances of the latter kind are seen in ethylamine, diethylamine, &c., in which one or more atoms of hydrogen in ammonia are replaced by ethyle. Thus, ammonia $= NHHH$, ethylamine $= NHHC_4H_5$, diethylamine $= NH(C_4H_5)(C_4H_5) = NH(C_4H_5)_2$, &c.

As hydrogen, chlorine, iodine, &c., form a natural family, one of which may replace another in the compounds they form; so nitrogen, phosphorus, arsenic, antimony, and bismuth form another family, the individuals of which sustain a similar relation to each other. Arsenide of hydrogen, AsH_3 (292), corresponds to ammonia in which the nitrogen is replaced by arsenic; and the compound, $Sb(C_4H_5)_3$, may be regarded as ammonia in which the nitrogen is replaced by antimony, and the hydrogen by ethyle.

Still another natural family is formed by oxygen, sulphur, selenium, and tellurium. Alcohol, $C_4H_6O_2$, by a substitution of sulphur for its oxygen, forms mercaptan, or sulphur alcohol, $C_4H_6S_2$.

544. Conjugated or Coupled Compounds.—Many of the compounds produced by transformations in accordance with the above principles, are frequently called conjugated or coupled compounds. They may be radicals only, or acids, or bases. As the name implies, they are supposed to be formed by the union of other compounds; and usually the characteristic properties of one or the other of the coupling compounds will be more or less preserved in the new or coupled compound. Thus, ethylamine, $NHHC_4H_5$, is a coupled or conjugated ammonia;—it is a substance formed on the *type* of ammonia, and possessing nearly the same properties, ethyle, C_4H_5, being the couplet. So the compounds, $(C_4H_5)_3Sb$, C_4H_5Zn, C_2H_3Bi, &c., are called conjugate metals.

545. Homologous Bodies. — Homologous bodies are bodies which may be arranged in series, all the members of which are similar in their general properties and chemical relations,

---

QUESTIONS.—Between what may substitutions take place? What other elements or compound radicals may be substituted for hydrogen? What elements constitute a natural family with nitrogen, capable of replacing each other? 544. What are conjugated or coupled compounds? What examples are given? 545. What are homologous bodies?

and are composed of the same elements, but differ from each other in composition by the addition or subtraction of the elements, C_2H_2, or some multiple of this expression. Of this kind are the alcohols, all of which, though they differ greatly in some of their properties, still have many points of resemblance. The composition of all that are known will be seen by the following table:

	Names.	Formulæ.
1.	Methylic alcohol (wood spirit)	$C_2H_4O_2$
2.	Common, or wine alcohol	$C_4H_6O_2$.
3.	Propylic alcohol	$C_6H_8O_2$.
4.	Butyric "	$C_8H_{10}O_2$.
5.	Amylic " (fusel oil)	$C_{10}H_{12}O_2$.
6.	Caprylic "	$C_{16}H_{18}O_2$.
7.	Ethal (ethalic alcohol)	$C_{32}H_{34}O_2$
8.	Cerotine (cerotic alcohol)	$C_{54}H_{56}O_2$.
9.	Melissine (melissic alcohol)	$C_{60}H_{62}O_2$.

Each of these compounds, denominated alcohols, it will be seen, contains 2 atoms of oxygen;—the first, or methylic alcohol, contains C_2H_4; the second, $C_4H_6 = C_2H_4 + C_2H_2$; the third, C_6H_8, and so on for the others, by the addition in each case of C_2H_2, or some multiple of this expression. In the cases requiring a multiple of C_2H_2, it would seem that one or more intermediate compounds are wanting. These may hereafter be supplied;—as, for instance, between amylic and caprylic alcohol the compound, $C_{14}H_{16}O_2$, is not yet known, but when a compound having this constitution shall be discovered, it is safe to predict that it will have the general properties of this class of bodies.

We have, therefore, as a general expressive for the alcohols, the formula, $C_{2n}H_{2(n+1)}O_2$.

Besides the above series of alcohols, many other series are known, among which are the following, taken from Gibbs' Report:

	Hydrogens.	Acetenes.	Formic Acids.	Oleic Acids.
	H	C_2H_2	$C_2H_2O_4$	$C_6H_4O_4$
	C_2H_3	C_4H_4	$C_4H_4O_4$	$C_8H_6O_4$
	C_4H_5	C_6H_6	$C_6H_6O_4$	$C_{10}H_8O_4$
	C_6H_7	C_8H_8	$C_8H_8O_4$	$C_{12}H_{10}O_4$
Gen. Form.	$C_{2n}H_{2n+1}$	$C_{2n}H_{2n}$	$C_{2n}H_{2n}O_4$	$C_{2n}H_{2(n-1)}O_4$

546. The compound, C_2H_2, which in all these series sustains so important a relation, has been called the homologizing body. But it is

---

QUESTIONS.—In what do homologous bodies of the same series differ from each other in composition? What series of substances of this kind is mentioned? What is said of the composition of the alcohols? Give the general formula for the alcohols. 546. What is the compound, C_2H_2, here called?

probable, that in other series other compounds, or perhaps a certain number of atoms of an element, may serve as the homologizing body. Thus, the ethyle and phenyle compounds differ from each other by C_8, which may be considered as the homologizing body connecting the corresponding compounds of the two classes, so as to form two terms of a homologous series.

547. Homologous bodies of the same series are in general similarly affected by the action of a reagent; and the resulting compounds will be homologous. This is implied in the definition of those bodies given above. The alcohols, $C_{2n}H_2(n+1)O_2$, for instance, by the action of oxydizing reagents yield corresponding homologous acids, $C_{2n}H_{2n}O_4$;—and intermediate between the alcohol and acid, in several cases, another compound is known, called an aldehyde (from alcohol dehydrogenatus), and having the composition, $C_{2n}H_{2n}O_2$. Other compounds of this series, we may confidently anticipate, will hereafter be discovered.

The following table contains the formulæ of the alcohols, their corresponding aldehydes (when known), and acids:

	Alcohols.	Aldehydes.	Acids.
1. Methylic...	$C_2H_4O_2$	———	$C_2H_2O_4$.
2. Wine........	$C_4H_6O_2$	$C_4H_4O_2$	$C_4H_4O_4$.
3. Propylic....	$C_6H_8O_2$	$C_6H_6O_2$	$C_6H_6O_4$.
4. Butyric.....	$C_8H_{10}O_2$	$C_8H_8O_2$	$C_8H_8O_4$.
5. Amylic.....	$C_{10}H_{12}O_2$	$C_{10}H_{10}O_2$	$C_{10}H_{10}O_4$.
6. Caprylic ...	$C_{16}H_{18}O_2$	———	$C_{16}H_{16}O_4$.
7. Ethalic.....	$C_{32}H_{34}O_2$	$C_{32}H_{32}O_2$	$C_{32}H_{32}O_4$.
8. Cerotic.....	$C_{54}H_{56}O_2$	———	$C_{54}H_{54}O_4$.
9. Melissic....	$C_{60}H_{62}O_2$	———	$C_{60}H_{60}O_4$.

Besides the acids in the above list, several others are known of the same series, some of which will hereafter be mentioned; but the corresponding alcohols and aldehydes have not been discovered.

548. In every homologous series as yet known, containing oxygen, the number of atoms of this element is constant; and the same appears to be true of nitrogen, as will hereafter be shown.

In the alcohol series, $C_{2n}H_2(n+1)O_2$, it is evident the smallest value we can give to n is 1, and the formula then becomes $C_2H_4O_2$, which represents methylic alcohol, but what the extreme upper limit may be we are ignorant. If in the general formula for the alcohols we make $n = 0$, the formula becomes $H_2O_2 = 2HO$, which represents 2 atoms of water. Water is therefore said to be the type of the series. In like

QUESTIONS.—May there be other homologizing bodies? 547. What is said of the effect of reagents upon homologous bodies of the same series? Into what are the alcohols converted by oxydizing reagents? What intermediate product is obtained in some cases? What are contained in the table in this paragraph? 548. What is said of the oxygen and nitrogen in the homologous series now known? Why is water said to be the *type* of the series of alcohols?

manner the general formula for the oleic acids, $C_{2n}H_2(n-1)O_4$, when $n = 1$, becomes $C_2O_4 = 2CO_2$; the type of the series is therefore carbonic acid, CO_2.

549. Although all the compounds of any homologous series are essentially similar in their leading properties, yet we may often observe a gradual transition as we pass from one to another of the same series. The carbo-hydrogens (*Hydrogens* and *Acetenes*, p. 402,) having the smallest number of atoms are gaseous at ordinary temperatures, but as we ascend in the series, that is, as the number of atoms in the compound is increased, they become less and less volatile, until the higher members of the series, at the ordinary temperature, are liquid, or even solid. Methylic alcohol, the first in the alcohol series, boils at 152°, and the others have each a higher boiling point in proportion as the number of atoms is increased. The last three are solid at ordinary temperatures. In general, in any series, the boiling point becomes higher as the whole number of atoms in the compound is greater.

550. **Analysis of Organic Substances.**—The analysis of a compound has for its object to determine its composition; and in organic chemistry may be either *proximate* or *ultimate.* By the *proximate analysis* of a substance we separate and determine its *proximate* principles, or, in other words, the several organic compounds which may be contained in it, as sugar, gum, albumen, &c.; but by its *ultimate analysis* we determine its *ultimate* elements, as carbon, hydrogen, nitrogen, &c. The methods of separating the proximate principles will be described as we progress, but for general modes of analyses the intelligent student will consult works treating specially of Analytical Chemistry.

551. Proximate principles are the proper organic compounds, which cannot be separated into other kinds without evidently changing their nature; they are usually characterized by one or more of the following properties, viz.:

1. Capability of combining in definite proportions with other elementary substances, or well determined compounds.
2. Capability of crystalizing, when obtained in the solid state, or having a definite melting point.
3. Having a definite boiling point, and being capable of distillation, or sublimation, without decomposition, or a change of properties.

These compounds are seldom found uncombined in organized bodies, and their separation, or the proximate analysis of the substances containing them, becomes an important object to the chemist, and is often attended with no little difficulty.

QUESTIONS.—549. Do we observe a gradual transition in the properties of the members of a homologous series as we pass from one to another? Give an instance to illustrate. 550. Define what is meant by the proximate and what by the ultimate analyses of an organic compound. What is a proximate principle? 551. What are the characteristics by one or more of which an organic compound will usually be distinguished?

STARCH, SUGAR, GUM, LIGNINE.

552. These four organic bodies constitute a natural family, possessing this remarkable peculiarity, that each member is composed of twelve atoms of carbon, united with a certain number of atoms of water, or rather with the elements of water, oxygen and hydrogen. In general, they are nutritious substances, and do not possess any very active chemical affinities.

STARCH, OR FECULA, $C_{12}H_{10}O_{10}$.

553. **Sources.**—Starch, or fecula, is obtained from a variety of vegetable substances, as the different grains; and from many roots, as the potato; and also sometimes from the stems of plants. It is contained in the cavities of the vegetable tissues, in the form of small, white grains, which always have a rounded outline, but vary considerably in size and form, as obtained from different substances. Each grain is inclosed in a delicate envelope that is not readily acted upon by cold water, but is ruptured by the expansion of the inclosed substance, when the water is heated nearly to the boiling point.

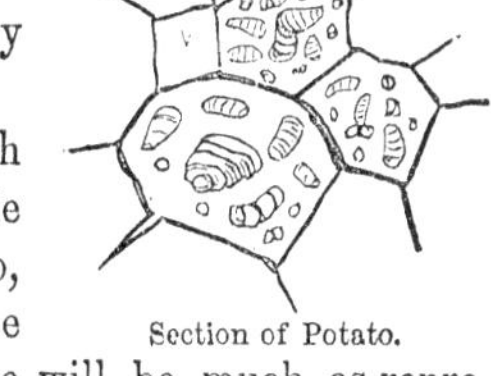

Section of Potato.

To show the appearance of the starch globules, in their cells in the vegetable tissue, cut a very thin slice of a potato, and examine it carefully by means of the compound microscope;—their appearance will be much as represented in the above figure.

If the slice before examination is moistened with a very dilute alcoholic solution of iodine, the starch globules will be colored blue, while the other parts remain uncolored.

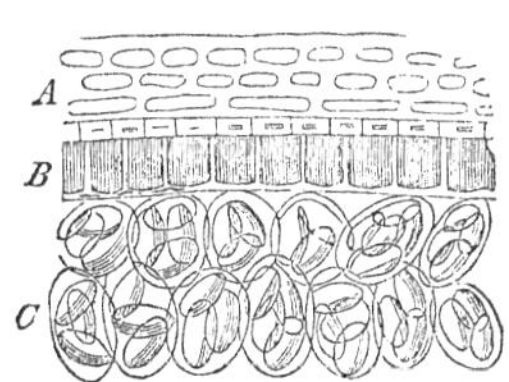

Starch Globules in a Grain of Rye.

The next figure shows the arrangement of the starch globules in a grain

QUESTIONS.—552. Of what are starch, sugar, gum, &c., composed? 553. From what is starch obtained? How is each grain inclosed? How may the grains of starch be shown in the potato?

of rye. A, the outer seed-coat, which constitutes the bran after grinding; B, the gluten (to be described hereafter); and C, the grains of starch.

The grains of starch from different sources vary greatly in size, and present different outlines. Of the following figures, A represents starch grains of the potato, B those of wheat, and C those of peas.

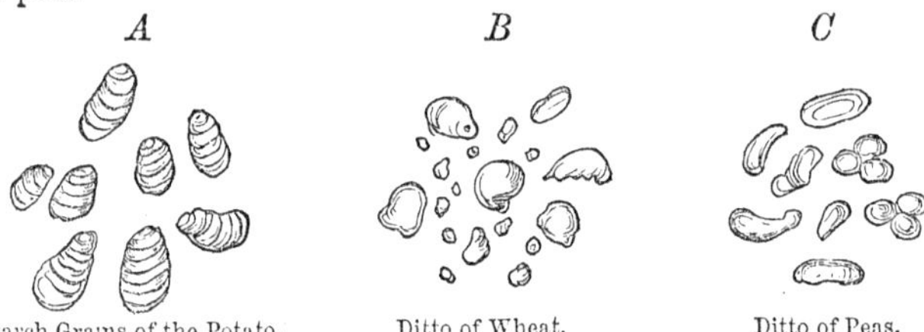

Starch Grains of the Potato. Ditto of Wheat. Ditto of Peas.

554. Preparation and Properties.—Starch is easily obtained from potatoes by mashing the tubers, and then inclosing the pulp in a piece of cloth, and washing it freely with cold water, at the same time pressing the mass between the hands.

Separating Starch from Wheat Flour.

From wheat or rye flour it is easily procured by placing the flour upon a piece of muslin, and working it with the hand while a small stream of water is poured upon it. The starch is washed through with the water, while the gluten remains as a tenacious mass upon the muslin. After a few hours the starch will all subside.

QUESTIONS.—What is said of the appearance of starch globules from different sources? 554. How may starch be obtained from the potato? How from wheat or rye flour?

Starch is an insipid white solid, quite insoluble in cold, but slightly soluble in boiling water. In the latter case, the granules are broken, and the broken envelopes floating in the solution give it the consistence of a jelly. In this state it is used for stiffening linen.

Though quite insipid to the taste, it forms a large part of many articles of food, as the different grains, rice, the potato, and other esculent roots. It also performs important functions in nearly all vegetables during their growth.

The substances known as *arrow-root*, *tapioca*, and *sago* are different varieties of starch.

Iodine forms with starch a beautiful blue compound, which is quite insoluble in water; it therefore serves as an excellent test for it.

555. When starch is kept for some time at a temperature between 300° and 400°, it undergoes a peculiar change, and becomes soluble in cold water, and is called *British gum*, or *leiocome*.

A substance very similar to the above, but called *dextrine*, is produced by gently heating starch in water acidulated with sulphuric acid, or containing infusion of malt. It has the same composition as starch, but is very soluble in cold water, and is not colored by iodine. If the mixture is boiled for some time, grape-sugar is formed, of which more will be said hereafter.

Diastase is a substance produced in small quantity in the process of malting grain, and is found in the potato soon after germination commences, in the parts near the young germs. It is noted for its specific action upon starch, converting it first into dextrine, in the same manner as diluted sulphuric acid, and afterwards into grape-sugar. Diastase is known to contain nitrogen, but its composition, further than this, has not been well determined.

556. The operation of *malting* consists in exposing grain (usually barley) to the proper degree of heat and moisture, with the free accession of atmospheric air, to produce incipient germination, and then suddenly checking it by elevating the temperature. This is done by first soaking the grain in water until it is fully swelled, and then placing it in heaps upon a floor until it begins to germinate, when the further progress of the vegetable process is arrested by quickly drying it at a moderately-elevated temperature. During the incipient germination, a portion of diastase is produced, by which, in the subsequent processes to which the grain is subjected, much of the starch of the grain is converted into dextrine and grape-sugar; and the grain (now called *malt*) becomes fitted for the use to which it is applied. It is chiefly used in the manufacture of beer.

QUESTIONS.—What is said of the properties of starch? What varieties of it are mentioned? 555. What test of starch is mentioned? How is starch affected by a heat of 300° or 400°? What is *dextrine?* What *diastase?* 556. What is *malt?* What use is made of malt?

SUGARS.

557. There are several varieties of sugar, but the most important are cane and grape-sugar,—names suggested by the sources from which they are respectively obtained. All the varieties of sugar possess a sweet taste, are soluble in water, and are susceptible of undergoing a peculiar change, called the *vinous fermentation*, by which alcohol is produced.

558. **Cane-Sugar,** $C_{12}H_{11}O_{11}$.—Most of the sugar of commerce is obtained from the sugar-cane *(arundo saccharifera)*, represented in the figure;—scale, one inch to four feet. But it is procured also in this country in large quantities from the sap of the sugar-maple *(acer saccharinum)*. Many plants contain it in their juices, as the common beet and other roots, and the stalks of Indian corn. Their juices, after being expressed from the plant, are evaporated until a dense syrup is obtained, from which a large portion of the sugar crystalizes on cooling; and the remaining liquid portion is then drained off, and constitutes *treacle*, or *molasses*.

Sugar-cane.

Pure sugar is a white, inodorous substance, of a very agreeable, sweet taste, which it imparts to its solutions. By slow evaporation, in a very warm room, it is obtained in large rhomboidal crystals, which are sold as *rock-candy*. It is very soluble in water, but is dissolved only in small quantity in alcohol. Its density is 1·6. It melts at about 356°, and on cooling forms a transparent, vitreous mass, called *barley-sugar*, which, however,

QUESTIONS.—557. How are the sugars characterized? 558. From what is cane-sugar obtained? Give some of the properties of sugar? What is *barley-sugar?*

after a time, becomes white and opaque, and is then found to be a mass of small crystals. Its composition remains without change.

By heating cane-sugar to 420° or 425°, a change of composition is effected; it then gives up 2 atoms of water, and a brown substance is formed, called *caramel*, which has the composition, $C_{12}H_9O_9$.

559. Grape-Sugar—Glucose, $C_{12}H_{14}O_{14} = C_{12}H_{12}O_{12} + 2HO$.—This substance, which much resembles the preceding, has for its composition, when crystalized, $C_{12}H_{14}O_{14}$; but by a boiling heat, two equivalents of water are expelled. It is more generally diffused in nature than cane-sugar, being found in the grape and most other sweet fruits. It constitutes also the solid part of honey. It may be obtained from grapes by expressing the juice, and neutralizing the free acid with chalk, and then clarifying and crystalizing in the same manner as with cane-sugar.

Grape-sugar is less soluble in water, and forms a less tenacious syrup, and is less sweet than cane-sugar. One ounce of water will dissolve three ounces of cane-sugar, but only about two-thirds of an ounce of grape-sugar; and it is estimated that one ounce of the former has an equal sweetening capacity with two and a half ounces of the latter. Grape-sugar does not crystalize as readily as cane-sugar, and is soluble in oil of vitriol, while cane-sugar is blackened by it.

A dilute solution of sulphate of copper, containing a little potassa, or tartrate of potassa, is at once rendered colorless by grape-sugar, at the ordinary temperature, but this effect is produced by cane-sugar only by boiling. This serves as an unfailing test to distinguish the two varieties of sugar. The color of the copper solution is destroyed by the precipitation of suboxide of copper, Cu_2O.

560. Grape-sugar may be prepared from several substances which have a similar composition, as starch, gum, and woody-fibre or lignine, and is occasionally found in the animal system, as in the disease called *diabetes*. It then makes its appearance in the urine. It is sometimes called *starch, sugar*, *diabetic sugar*, and *glucose*. The latter name is more properly applied to the

QUESTIONS.—What is *caramel?* 559. Describe grape-sugar. What is said of its solubility in water and its sweetness, as compared with cane-sugar? How may the two kinds be distinguished from each other? 560. Describe the mode of preparing grape-sugar, or glucose, from starch.

sugar prepared from starch, &c., but the identity of this substance with the sugar of grapes is generally admitted.

To prepare grape-sugar from starch, 1 part of sulphuric acid is mixed with 200 parts of water, and heat applied until the mixture begins to boil; 50 parts of starch are then added, and the boiling continued until the liquid becomes perfectly clear. Powdered chalk is then introduced, a little at a time, in order to neutralize the acid; and by a few hours standing the sulphate of lime formed will be precipitated, leaving the liquid clear, which is now solution of glucose. By evaporation of the water it may be obtained in crystals.

Wood, the composition of which, as we shall hereafter see, is nearly the same as that of starch, also yields grape-sugar, or glucose, by boiling with sulphuric acid. The process is essentially the same as the above, except that a large proportional quantity of the acid is used.

In both of these processes the acid remains unchanged, but by its presence, in some unexplained mode, it occasions the starch and the cellulose of the wood to unite with an additional quantity of water—or the elements of water—thus converting it into sugar. Thus,

$$\text{Starch (or cellulose)}, C_{12}H_{10}O_{10} + 4HO = C_{12}H_{14}O_{14}.$$

From clean linen, or cotton rags, more than their own weight of sugar may be formed.

Grape-sugar, prepared in this way, is used to adulterate cane-sugar, and in the manufacture of beer and alcohol.

Sugar of sour fruits, as currants, cherries, plums, &c., possesses the composition, $C_{12}H_{12}O_{12}$, and readily ferments, producing alcohol, but is uncrystalizable.

561. Milk Sugar, Lactine, $C_{24}H_{24}O_{24} = C_{24}H_{19}O_{19} + 5HO$.—This sweet principle is obtained by evaporating the whey of milk, purifying with animal charcoal, and crystalizing. It is less soluble in water than either of the other varieties of sugar, and less sweet to the taste. Under certain circumstances, it is capable of undergoing the alcoholic fermentation, like the other varieties of sugar; and in some countries, it is well known that an intoxicating drink is made from camels' milk. But when in solution it is allowed to stand in the open air, at ordinary temperatures, *lactic* fermentation takes place, and *lactic acid*, $C_6H_6O_6 = C_6H_5O_5,HO$, is formed. This change takes place in the ordinary souring of milk. By the action of dilute acids at 212°, lactine is converted into grape-sugar.

562. Mannite, $C_6H_7O_6$, though not analogous to sugar in composition, is similar to it in some of its properties. It is found in many plants, but chiefly in a substance, called *manna*, obtained from certain species of the ash.

QUESTIONS.—Describe the mode of preparing grape-sugar from wood. Does the acid remain unchanged in the operation? What is said of the sugar of sour fruits? 561. What is milk-sugar, or lactine? May it undergo the alcoholic fermentation? 562. Describe mannite.

GUMS, $C_{12}H_{10}O_{10}$.

563. We designate by the name of *gum* a variety of vegetable substances, which are very soluble in water, but insoluble in alcohol, and uncrystalizable. Their composition is the same as that of starch, but they differ from this substance in several of their properties.

The gums generally exude from the bark of trees, as the cherry and peach trees, and are found on the outside in transparent masses, which are more or less globular in form.

A gum is a colorless, transparent, insipid, inodorous solid, and when perfectly dry is very brittle, and has a vitreous fracture. When put into water, it first softens and swells up considerably, and then dissolves, forming a mucilage which is often used as a substitute for paste, for which it answers well. Its solubility is increased both by acids and alkalies. *Gum Arabic*, *gum Senegal*, and *gum tragacanth* are the most common varieties of this substance.

Solutions of gum Arabic, and probably also those of the other gums, yield sugar by boiling with sulphuric acid;—boiled with nitric acid, *mucic acid* is formed.

564. *Pectine* is a substance closely resembling gum, which is found in many fruits and in certain roots; it is the substance contained in currants, cherries, apples, &c., which they yield on being boiled, and especially when boiled with sugar. It is a kind of vegetable mucus, and found in small quantity in many vegetables. By the action of an alkali, or almost any base, it is converted into an acid, called *pectic acid.*

WOODY FIBRE, LIGNINE, CELLULOSE.

565. Wood from a growing tree is of a very complex composition. By examination with the microscope, it is found to possess a highly organized structure (533), consisting of vascular tissue, having its cells filled with a variety of substances, as starch, solution of sugar and mineral salts, albuminous com-

---

QUESTIONS.—563. Describe the gums. From what are they obtained? What are some of the varieties of gum? How are the gums affected by sulphuric acid? By nitric acid? 564. Describe pectine. 565. What is said of the composition of growing trees?

pounds, oils, and resins, depending upon the natural family and species to which the tree belongs.

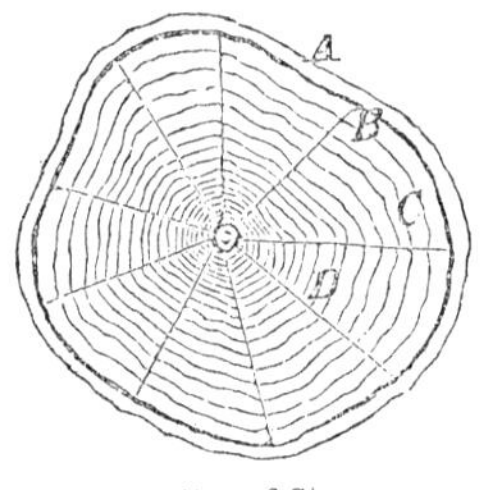

Section of Stem.

The first figure in the margin represents a transverse section of the stem of a tree. A is the outer bark, which is usually rough, and has lost its vitality, and serves only as a covering to the parts within. B is the inner fibrous bark, in which the sap-vessels are found, serving the purpose of the veins of animals. Inside of this is the wood, consisting of the part C, which is usually whiter than the rest, and is therefore called the *alburnum*, or *sap-wood*; and the *heart-wood*, D, which is more solid and durable than the sap-wood, and of a darker color. Both the alburnum and the heart-wood are composed of concentric layers, an addition of a layer being made to the alburnum each year, immediately beneath the bark. The next figure represents a transverse section, magnified, of a piece of pine, showing the two kinds of wood—A, the alburnum; B, the heart-wood. The inner rings of the alburnum are gradually converted into firm heart-wood, and seem then no longer to partake of the vitality of the tree. In annual plants, the woody part corresponds to the alburnum of trees.

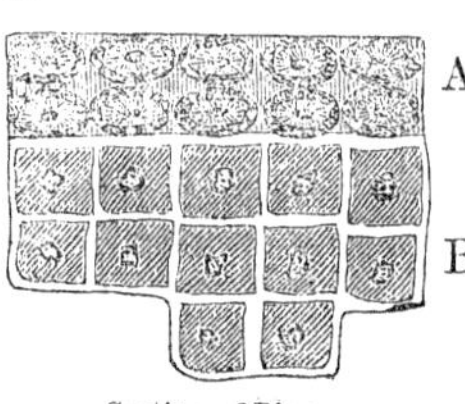

Section of Pine.

566. **Cellulose,** $C_{12}H_{10}O_{10}$.—The vascular tissue of which we have spoken is composed chiefly of cellulose, the composition of which, it will be observed, is the same as that of starch, though it differs essentially from this substance in some of its properties.

Cellulose constitutes the basis of wood, and is obtained by digesting saw-dust, paper, or linen or cotton rags, successively in alcohol, ether, diluted acid, diluted alkaline solutions, and water, so as to remove everything which is soluble in these menstrua.

It is found in very different states; as indigestible and hard, in wood, and in the shells of nuts; as tender and easily digestible, in the esculent

QUESTIONS.—What is represented by the first figure on this page? What by the second? 566. In what is cellulose found? What is said of its composition?

roots, and in the pulp of fruits, as the apple, pear, &c.; as light and porous, in the pith of the elder, and in cork; and as soft and flexible, in the fibres of cotton, flax and hemp.

567. **Lignine, Woody Fibre.**—The vascular tissue of plants, when first formed, is composed of nearly pure cellulose, but afterwards the cells become lined with a hard incrusting substance, which, in trees and shrubs, for years increases in thickness and solidity. This is called lignine, or sometimes woody fibre, though the latter term is more properly applied to wood as found in the tree, and composed of both cellulose and lignine. Lignine is found more abundant in the heart-wood than in the alburnum.

The composition of lignine is believed to be essentially the same as that of cellulose, but it has not been fully determined.

We have seen above (560), that wood treated by sulphuric acid is converted into grape-sugar, but it is only the cellulose that is capable of this change; lignine is not affected by sulphuric acid, or only charred.

The mutual relations of starch, sugar, and woody fibre, are singular and important. Their composition, we have seen, is nearly the same; and they are convertible into each other by easy processes; indeed, we are able to recognise this conversion as really taking place, in certain cases, in the natural process of vegetation, as shown in the malting of grain. The same change takes place in the ripening of many fruits, as the apple and pear, which are acid until they approach maturity, when they become more or less sweet. The sap of the maple and other trees contain sugar, which subsequently becomes changed into woody fibre, and thus contributes to the enlargement of the tree.

Changes produced upon Woody Fibre by Acids.

568. Wood, plunged into strong sulphuric acid, especially if a little warm, is instantly charred, as if held near a hot fire. This is occasioned, it is believed, by the strong affinity of the acid for water, the elements of which, oxygen and hydrogen, are abstracted by it from the wood, leaving the black carbon.

If the sulphuric acid be diluted, or if added in small quantities, so as entirely to avoid any rise of temperature, the effect is to form sugar, as we have already seen.

QUESTIONS.—567. In what is lignine or woody fibre found? What is said of the mutual relations of starch, sugar, and woody fibre? Are they capable of conversion one into another? 568. What is the effect of strong sulphuric acid upon wood? What if the acid is diluted?

569. Gun Cotton — Pyroxyline.—This substance, which is noted for its explosive property, is formed by the action of very strong nitric acid, or better, by a mixture of the most concentrated nitric and sulphuric acids, upon cotton, flax, paper, or fine saw-dust.

To prepare it, make a mixture of equal parts (by volume) of the strongest nitric and sulphuric acids, and then press into it as much cotton as can be moistened with it; and, after standing five or ten minutes, press out as much of the acid as possible, and wash thoroughly with a large supply of pure water, and dry carefully without artificial heat. It will be found that two ounces of each of the mixed acids will be sufficient for 75 or 100 grains of cotton.

When thus prepared, the cotton appears much as before the process, but has a harsh feeling, and the fibres are less tenacious than in the original cotton. It also gains considerably in weight during the process, so that from 100 grains of cotton as much as 175 grains of gun-cotton will often be obtained. It takes fire very readily, often at a temperature even below 212°, especially if the heat is suddenly applied; and burns with an immense volume of flame. Placed on a plate of metal, and very gradually heated, it may sometimes be completely decomposed, without igniting, leaving behind a residue of carbon. When properly prepared, it explodes with great violence, and is entirely consumed. Its power to propel balls is much greater than that of the best gunpowder, which is still further increased by soaking it in a solution of chlorate of potash before drying.

The composition of pyroxyline is is uncertain; but it is known that, by the action of the acids, oxygen and hydrogen (in the form of water) are separated from the cotton, and, at the same time, nitric acid combines with it. The most probable opinion is that 2 equivalents of cellulose combine with 5 equivalents of nitric acid, giving up at the time 3 equivalents of water. Thus,

$$C_{24}H_{20}O_{20} + 5NO_5 = C_{24}H_{17}O_{17},5NO_5 + 3HO.$$

Gun-cotton, though insoluble in water or alcohol, is usually found quite soluble in sulphuric ether containing a little alcohol. But this is not always the case; and it is believed there are at least two different compounds formed in the process, one of them being soluble in alcoholic ether, and the other insoluble. The insoluble variety appears also to explode with more violence than the other.

The gelatinous ethereal solution of gun-cotton is used in surgery, as a substitute for sticking-plaster, or court-plaster, under the names of *collodion* and *liquid cuticle.*

Xyloidine is an explosive compound similar to pyroxyline, produced by the action of strong nitric acid upon starch.

QUESTIONS.—569. Describe the process of preparing gun-cotton. Give its properties. What is the most probable opinion concerning the composition of pyroxyline, or gun-cotton? In what is it soluble? What is xyloidine?

Changes of Woody Fibre, by Air and Moisture.

570. When lignine is kept perfectly dry, or constantly immersed in water, it may be preserved for any length of time; but exposed to air and moisture, it undergoes a slow decay, called *eremacausis* (from *erema*, slow, and *kausis*, combustion), by the absorption of oxygen, and the evolution of water, or its elements, and carbonic acid.

The chemical changes which in this case occur, are very nearly the same as in combustion, except that they take place more slowly. In both cases, the constituents of the wood, with the addition of oxygen from the air, are converted into carbonic acid and water; in both cases, also, the hydrogen is oxydized more rapidly than the carbon, as is shown by the black color during combustion, and the dark brown during the slow decay of vegetable matter. The flame which appears during the combustion of wood and other vegetable substances, is occasioned by the burning of the gaseous hydro-carbons, evolved as the first effects of the application of heat.

571. **Humus, Geine.**—Every fertile soil contains more or less organic matter in a state of decay, to which its fertility is, in a great measure, owing, and which has received various names, as *humus*, *geine*, *ulmine*, *vegetable mould*, *humic*, *geic*, and *ulmic* acids, &c. This decaying matter, which we will call *vegetable mould*, is ever changing, as the carbon and hydrogen are oxydized and separated; and from this and the carbonic acid, water, and ammonia, formed from them in the soil, the plants derive their chief nourishment, by means of their roots, which are extended in every direction.

572. **Peat.**—Peat consists of partially decayed vegetable matter, which is found in beds, in moist places, in every country, and is usually mixed with more or less matter of mineral origin. It is formed from vegetable matter when slowly decaying under water, and of course free from the influence of the oxygen of the air. Water is decomposed, yielding its oxygen to one portion of carbon, to produce carbonic acid, while another portion of carbon unites with the hydrogen, forming light carburetted hydrogen (307), which often issues from the soil in large quantities, as the *fire-damp* of coal mines; but most of the carbon remains behind as peat. Recently formed peat is also usually found interlaced with the

QUESTIONS.—570. May wood be preserved if kept perfectly dry? What is the effect when exposed to air and moisture? What is said of the chemical changes which take place in eremacausis? 571. What is contained in every fertile soil? 572. Describe the formation of peat.

fibrous roots of growing plants, and retains, more or less distinctly, the forms of many of the plants of which it has been composed; but old peat is more homogeneous, and may be cut into the form of bricks, like moist clay. In many countries it is used extensively for fuel, being first thoroughly dried.

573. Coal.—A sufficient description of the different kinds of mineral coal has already (298) been given. Immense deposits of this mineral are found in many countries, which have no doubt been formed from vegetable matter, at a comparatively early period in the world's history. They are usually not very distant beneath the surface; and are contained between rocky strata of great extent. Occasionally these strata are covered with many feet of rock and earth;—everything indicating that vast changes have taken place in the earth's surface since their formation.

That all the varieties of mineral coal have been produced from the matter of plants, which in former periods grew up and flourished precisely as plants now do, is universally believed by men of science. The evidence is found in the position of the coal, which always occurs in the form of beds interstratified above and below with solid rock formed at the same time with itself;—in the vegetable impressions which abound in the rocky strata of every coal formation;—and in the organized structure often exhibited by the coal itself.

We may suppose that, by the decay of the vegetable matter, peat was first produced, which was subsequently converted into proper coal, in a manner not fully understood.

Mines of coal are limited to the temperate zone, none of it being found in very warm or very cold climates.

574. Petroleum.—Petroleum, or *rock-oil*, is an odoriferous, oily liquid, which, in many countries, exudes from the rocks and ground, being formed, in all probability, from vegetable matter at the same time with peat and coal. It is often found upon the surface of lakes and springs, as at Seneca Lake in New York, where it is called *Seneca* oil. By distillation, petroleum yields a yellowish liquid, lighter than water, called *naphtha*, or *oil of naphtha;* which is probably a mixture of several compounds that have not yet been separated. It is the liquid used to preserve the alkaline metals. *Asphaltum*, or *mineral pitch*, is a substance closely allied to petroleum. When cold it is solid, but becomes soft as it is heated, and melts at about 212°. Dissolved in oil of turpentine, it forms a varnish which is used for certain purposes.

575. Distillation of Coal and Wood.—Illuminating Gas.—Coal and wood are subjected to distillation by heating them to redness in large cast-iron retorts, prepared for the purpose. Usually the retorts are kept at a red heat, and are capable of being opened at one end, so that the charge may be changed when necessary.

QUESTIONS.—What use is made of peat? 573. How have the deposits of mineral coal been formed? What evidence of this is there? 574. Describe petroleum. How is naphtha obtained from it? What is asphaltum, or mineral pitch? 575. How are coal and wood distilled?

Coal subjected to this process gives off large quantities of gaseous carburetted hydrogens (309), with carbonic acid and sulphuretted hydrogen, and a resinous substance not unlike *tar*, called *coal-tar*.

This last substance is a mixture of several compounds, as *naphthalene*, *phenol*, salts of ammonia, &c.

A gas, not unlike that prepared by the distillation of coal, is often found to issue, ready formed, from the earth, and may be collected and used like that formed by art. The village of Fredonia, in the State of New York, is lighted by natural gas, in this way; and in some salt-works in Virginia, it is said, no other fuel is used for boiling the brine.

576. Wood by distillation yields illuminating gas with *water*, *acetic acid*, *creosote*, *pyroxylic spirit* (*methylic alcohol*), *tar*, &c. The acetic acid, as first obtained, is mixed with creosote and other substances, and is called *pyroligneous acid*.

Creosote, $C_{28}H_{16}O_4$, which passes over, with other products, in the distillation of wood, is a colorless, transparent liquid, of an oily consistence, and retains its fluidity at —17°. It has a specific gravity of 1·04; boils at 397°, and is a non-conductor of electricity. It has a burning taste, and its odor is like that of wood-smoke, or rather of smoked meat. It is highly antiseptic to meat: the antiseptic virtue of tar, smoke, and crude pyroligneous acid, seems owing to the presence of creosote. Its name (from *kreas*, *flesh*, and *sozein*, *to save*) was suggested by this property.

Creosote requires about 80 parts of water for solution, but is soluble in every proportion in alcohol and ether. It has neither an acid nor alkaline reaction with test-paper, but combines both with acids and alkalies. It is used both internally and externally in medical practice.

Naphthaline, $C_{20}H_8$, is a solid substance, obtained by distillation from coal-tar. It is a white, crystaline solid, which melts at 175°, and boils at about 422°. It has a specific gravity of 1·05, and is insoluble in water, but dissolves in alcohol and ether.

By the action of chlorine and bromine, and also by the action of acids, as the sulphuric and nitric, upon this substance, various interesting compounds are formed.

Paraffine is sometimes found in petroleum, but is generally procured by the distillation of wood-tar. It is a white, waxlike solid, of specific gravity 0·87, which melts at about 110°. It may be distilled without alteration. It received its name (*parum affinis*) from the circumstance that it manifests little affinity for other substances.

Eupione, C_6H_6, is a fragrant, colorless liquid, obtained by the distillation of wood. Its specific gravity is 0·655, and its boiling point about 340°.

Phenol is obtained from coal-tar by distillation. Its composition is $C_{12}H_6O_2$, in which, as well as in many of its chemical relations, it re-

---

QUESTIONS.—What are formed when coal is distilled? 576. What compounds are mentioned as being obtained by the distillation of wood? Describe creosote. What is the derivation of the name, creosote? Describe naphthaline. Paraffine. Eupione. What is said of phenol?

sembles the alcohols (547), and has therefore been called *phenylic alcohol* By the action of reagents, many compounds are formed from it, analogous to those obtained from the alcohols.

ALCOHOLS AND SUBSTANCES DERIVED FROM THEM.

577. The name alcohol has long been applied to a well known liquid, obtained by fermentation and distillation from solution of sugar, and starch, or substances containing one or the other of these compounds. But in the progress of the science, several other compounds, analogous to common alcohol, both in composition and properties, have been discovered, to which the name is also given. The general formula, $C_{2n}H_{2(n+1)}O_2$ (545), may represent their composition.

The three most important of the alcohols are, common, or *wine alcohol*, *methylic alcohol*, and *amylic alcohol*, which, with their derivatives, will be here described; the others will be attended to in their proper places.

WINE ALCOHOL, $C_4H_6O_2$.

578. Common, or wine alcohol, is always produced by the fermentation of sugar or starch, or compounds containing one or the other of these substances.

579. Fermentation.—By the fermentation of a substance, we mean a change or modification which takes place in its constitution under the influence of another substance, called a *ferment*, which acts simply by its presence.

Several different fermentations have received separate names, as the *vinous* or *alcoholic* fermentation, by which alcohol is produced from starch or sugar; the *viscous* fermentation, by which sugar is converted into a viscous, or mucilaginous substance; the *acetic* fermentation, by which diluted alcohol and other substances are changed to acetic acid, &c.

The most important ferment used for inducing the vinous fermentation is *yeast*, or *leaven;* but blood, albumen, caseine, and the juices of many

---

QUESTIONS.—577. Give the general formula for the alcohols. Name the three most important alcohols. 578. How is common, or wine alcohol, always produced? 579. What is meant by fermentation? What different fermentations are mentioned? What ferment is generally used to produce the vinous fermentation?

fruits, as currants, apples, grapes, &c., are capable, after being exposed for a time to the air, of producing the same effect.

Yeast is a substance which collects as a froth upon the surface of beer, and other liquids, in the process of fermentation. It always contains a portion of nitrogen, which seems essential to it, but the mode of its action has not been determined.

Active yeast is believed by many to contain a kind of microscopic vegetable, which is developed spontaneously in the organs of plants, and in many nitrogenized substances when left to putrefy. It can be observed only by use of the microscope. See figure in the margin.

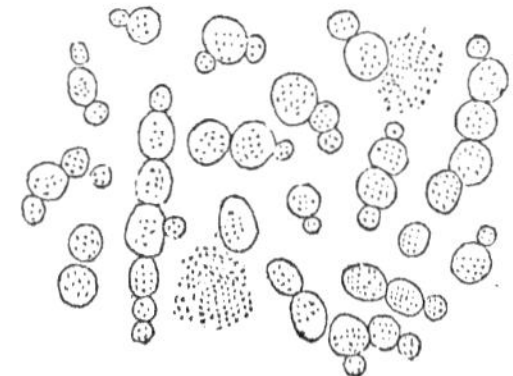
Yeast Plant.

To induce the vinous fermentation, a quantity of sugar is dissolved, or an equal quantity of starch diffused, in 4 or 6 times its weight of water, a small quantity of yeast stirred in, and the mixture placed in a situation where it may retain a uniform temperature of 70° to 85°. After a time the solution will be found in brisk effervescence, occasioned by the escape of gaseous matter, which may easily be collected over mercury (see figure), and on examination proves to be pure carbonic acid. When the action has ceased the liquid again becomes clear, and is found to contain, not sugar, but alcohol, which may be separated by distillation, and by other modes.

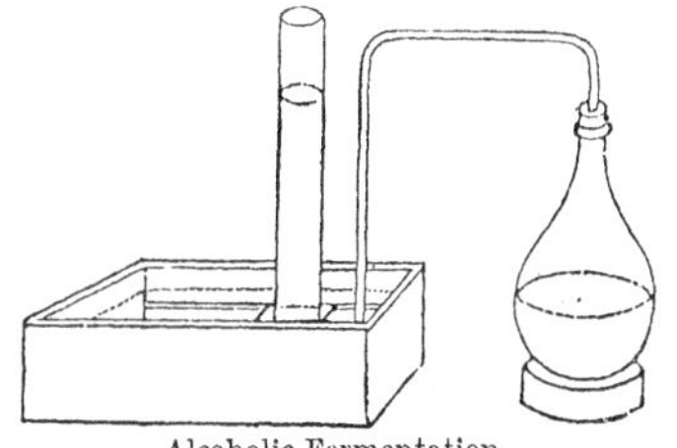
Alcoholic Fermentation.

In this way it is found that alcohol and carbonic acid alone are produced by the vinous fermentation, in the proportion of 2 equivalents of the former to 4 equivalents of the latter. It would not seem difficult, therefore, to understand the nature of the change that takes place when fruit-sugar (560) is used, for this sugar contains precisely the elements of these compounds in this proportion. Thus,

$$C_{12}H_{12}O_{12} = 2(C_4H_6O_2) + 4CO_2.$$

But when starch, $C_{12}H_{10}O_{10}$, cane-sugar, $C_{12}H_{11}O_{11}$, or grape-sugar, $C_{12}H_{14}O_{14}$, is fermented, the case is a little different, and the first effect of the ferment is to convert them into fruit-sugar, by an obvious change.

580. Preparation and Properties.—Alcohol obtained by distillation always contains a portion of water, even after several

---

QUESTIONS.—What is yeast? What is represented by the figure? What are the products of the vinous fermentation? How many equivalents of alcohol and carbonic acid are produced from an atom of fruit-sugar? 580. May pure alcohol be obtained by distillation?

successive distillations. When most highly rectified in this mode, it contains about 90 per cent. of pure alcohol, the rest being water. The alcohol, or *spirits of wine*, of commerce is seldom as pure as this. To obtain *absolute alcohol*, or alcohol in a state of purity, common alcohol is carefully distilled, by the heat of a water-bath, from pearlash or chloride of calcium, previously dried and mixed with it. The water combines with the salt and remains behind, while the pure alcohol distils over.

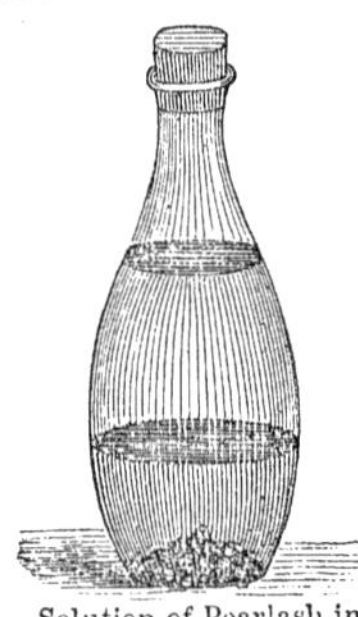

Solution of Pearlash in Proof Spirit.

Alcohol sufficiently pure for almost any purpose, may be obtained, without distillation, by means of well-dried pearlash. For this purpose, take common alcohol, or even proof-spirit, or whiskey, and pour into it half its weight or more of pearlash, which has previously been thoroughly dried by heat. The whole is then to be shaken together a few minutes, and allowed to stand several hours; there will then appear in the vessel two liquids entirely separate, one above the other, as shown in the figure. The upper portion is to be carefully drawn off and subjected to a second operation of the same kind, by means of a fresh portion of dried pearlash; and this operation is to be continued as long as the introduction of the pearlash produces any effect. Pearlash is soluble in water, but quite insoluble in alcohol; when, therefore, a portion of it, perfectly dry, is introduced into alcohol containing water, the latter dissolves it, and the dense solution settles to the bottom, while the alcohol remains above. After several repeated operations, all the water, or nearly all, is thus separated.

581. Pure alcohol is a colorless liquid, of a pungent taste and odor; and at 60° has a density of 0·795. It boils at 172°; and its vapor is highly inflammable, and burns with a pale yellowish flame, without smoke. It has never been frozen by any cold yet produced; but at a temperature of —146°, becomes thick and tenacious, like melted wax. It is a powerful solvent for many substances totally insoluble in water, especially many of the resins. Hydrate of potassa (and also of the other alkalies) dissolves readily in alcohol, forming a solution often used in the laboratory.

When potassium (or sodium) is immersed in absolute alcohol,

QUESTIONS.—How may pure or absolute alcohol be procured? Describe the mode by the use of pearlash. 581. Describe the properties of alcoho'.

hydrogen is given off, and a crystaline compound formed having the same composition as alcohol, except that 1 atom of hydrogen is replaced by 1 atom of potassium. Thus.

$$C_4H_6O_2 + K = C_4H_5KO_2.$$

By contact with water, this compound is converted into alcohol and hydrate of potassa.

Vapor of alcohol has a density, as determined by experiment, of 1·613; but, as calculated from the densities of its constituents, it is 1·596 Thus,

4 vols.	carbon vapor weigh	(·836 × 4)	3·344
12 "	hydrogen "	(·069 ×12)	·828
2 "	oxygen "	(1·106 × 2)	2·212
4 vols.	alcohol vapor weigh		6·384

One vol. alcohol vapor therefore weighs 1·596.

Alcohol exists in every kind of spirituous liquors, and may be separated from them by distillation. The different kinds of *brandy*, *rum*, *gin*, and *whiskey*, usually contain from 45 to 55 per cent. of pure alcohol; the stronger wines from 18 to 25 per cent.; and the weaker, not more than 12 or 15 per cent. In cider, ale, and porter, the quantity varies from 4 to 10 per cent.

Alcohol is extensively used in the arts and in medicine, chiefly in consequence of its powerful solvent qualities. Taken internally, it operates, as is well known, as a powerful stimulant; and various alarming diseases, often terminating in extreme moral degradation and death, attend its habitual use.

Used in lamps instead of oil, great heat is produced, and the combustion is unattended with smoke, for which reason it is much employed in the laboratory. A cover, *a*, fitting accurately upon *b*, prevents evaporation when not in use.

Spirit-lamp.

582. Bread-making.—The ordinary mode of raising bread by means of yeast, also furnishes an instance of the vinous fermentation. The yeast contained in the flour-paste, or *dough*, causes the fermentation to commence, as already explained, and both alcohol and carbonic acid are formed in the dough; and the latter being retained in the tenacious mass causes it to swell up, or *rise*. This effect is further increased by the expansion of the gas by the heat in the process of baking, giving the

QUESTIONS.—What is the effect when potassium is immersed in alcohol? In what is alcohol always contained? 582. What is said of bread-making?

bread its light, porous character. The small quantity of alcohol formed is dissipated by the heat of the oven.

Bread is also raised, as is well known, by the use of bicarbonate of soda, or potash, and an acid, as the hydrochloric or tartaric acid, or the acid tartrate of potash (*cream of tartar*). The *rising* is in this case produced by carbonic acid liberated in the dough from the alkaline carbonate; and the same light, spongy character given to the bread.

Products of the Oxydation of Wine Alcohol.

583, Aldehyde, $C_4H_4O_2$. — Aldehyde (from *alcohol dehydrogenatus*), as will be seen by an inspection of its formula, is dehydrogenated alcohol, — alcohol from which two atoms of hydrogen have been separated. The separation of the hydrogen is effected by distilling alcohol with highly oxydized substances, which readily give up a portion of their oxygen to form water with the eliminated hydrogen.

The best method for preparing it, is to mix in a retort equal parts of powdered bichromate of potash and strong alcohol, and then add, in successive portions, $1\frac{1}{3}$ parts of sulphuric acid. Energetic action commences at once, much carbonic acid is given off, and aldehyde, mixed with a small quantity of acetic acid and other substances, distils over into the receiver, which is kept surrounded with ice. To the liquid thus obtained, some ether is added, and it is then saturated with ammonia, which forms with aldehyde a crystaline compound, *aldehyde ammonia*, $NH_3,C_4H_4O_2$. This solid, distilled at a gentle heat with dilute sulphuric acid, and then re-distilled from chloride of calcium, previously fused, affords the pure aldehyde.

Aldehyde is a colorless liquid, of a peculiar suffocating odor; at 60° its density is 0·790, and it boils at the low temperature of 70° or 71°. It is soluble in water, alcohol and ether; and burns with a pale flame. It dissolves sulphur, phosphorus, and iodine; and is especially remarkable for its affinity for oxygen, in consequence of which it is capable of reducing many of the metallic salts. Its action upon nitrate of silver is remarkable. If a solution of this salt is poured into a clean glass vessel, with

---

QUESTIONS.—What occasions the *rising* which gives to bread its spongy character? 583. What is aldehyde? Describe the mode of preparing it. What are some of its properties?

a little ammonia, the addition of some water, mixed with aldehyde, causes an immediate precipitation of the silver upon the surface of the glass, which retains its perfect metallic lustre, like the coating of a looking-glass.

Aldehyde is instantly decomposed by contact with an alkali; and if kept only a short time, it changes spontaneously into one or two other compounds isomeric with itself, depending upon the temperature;—these are, *elaldehyde* and *metaldehyde*, the former of which is liquid and the latter solid. By combining with oxygen, aldehyde forms *aldehydic* or *acetous* acid, $C_4H_4O_3$.

584. Acetic Acid, $C_4H_4O_4 = C_4H_3O_3,HO$.—Acetic acid is well known as the acid contained in *vinegar* (from the French *vin*, wine, and *aigre*, sour), which is, in fact, a very dilute acetic acid, containing also much saccharine and mucilaginous matter. Acetic acid is one of the most important of the organic acids, and is found, in combination with bases, in many plants.

This acid is formed artificially by a variety of processes, but the usual method of preparing it is to subject liquids containing alcohol, as the weaker wines, cider, &c., to the action of some ferment, while exposed to the atmosphere. The process has sometimes been called the *acetic fermentation;* and the chemical changes which take place in it consist in the separation from the alcohol of 2 atoms of hydrogen by the oxygen of the air, forming aldehyde and water, and the subsequent absorption of 2 additional atoms of oxygen by the aldehyde. Thus,

$$C_4H_6O_2 + 2O = C_4H_4O_2 + 2HO, \text{ and}$$
$$C_4H_4O_2 + 2O = C_4H_4O_4.$$

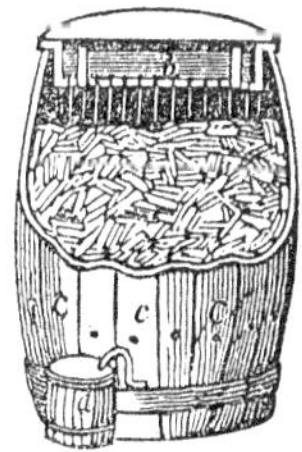
Prep. of Vinegar.

The access of atmospheric air is absolutely essential to the formation of vinegar by the ordinary process, as is well known; and its production is much facilitated by a method invented in Germany. A cask, as shown in the figure, is filled with wood shavings, and closed at top by a pan, *b*, the bottom of which is perforated with many small holes, through which small threads are passed to conduct the liquid downward. The shavings, being first well soaked in vinegar, are placed lightly in the cask; and below them are several small holes, *c c*, about half an inch in diameter, to admit the free

QUESTIONS.—Can aldehyde be kept without change? 584. What is vinegar? How is it formed artificially? Describe the German process.

accession of air. If now proof-spirit, diluted with four times its weight of water, and having mixed with it a very little honey or yeast, is poured into the pan above, it gradually trickles down upon the shavings, where, a large surface being exposed to the atmosphere, rapid absorption of oxygen takes place, the temperature is raised, and acetic acid is formed. As the liquid passes down, it is collected in the vessel *a;* and when passed through three or four times, which requires but about 36 hours, it is converted into vinegar.

Acetic acid cannot be separated from water by mere distillation; but the pure acid is obtained by distilling some acetate, as acetate of soda or lime, &c., with a proper quantity of sulphuric acid, and collecting the product in a cold receiver.

Wood-vinegar, or *pyroligneous acid*, is obtained by distilling wood in close vessels. It is a very impure acetic acid, having a disagreeable, smoky odor, and containing empyreumatic oils, and other substances derived from the wood. It is much used in calico-printing; and often the cloths, not having been properly cleansed, possess its disgusting odor.

585. Pure acetic acid, at 63°, is a colorless liquid, of a pungent, refreshing odor, and excessively sour to the taste. Applied to the skin for a time, it produces blisters. It boils at 248°; and cooled below 63°, it may be obtained in crystals. At 63°, the density of the liquid is 1·06. It mixes readily with water, ether, or alcohol.

586. Acetal, $C_{12}H_{14}O_3$.—This compound is a clear, colorless liquid, which boils at about 167°, and has a density of 0·844. It is formed by the slow action of moistened platinum black upon a mixture of vapor of alcohol and oxygen, in a large bell-glass. The bell-glass is to be open at top, and elevated a little from the bottom of the basin in which it is placed, as represented in the figure, so as to allow a gradual circulation of the air through it. A watch-glass, A, is placed in the centre of the basin, containing a little platinum black; and a small funnel, with a long neck, B, contains strong alcohol, which drops very slowly into the watch-glass. By the

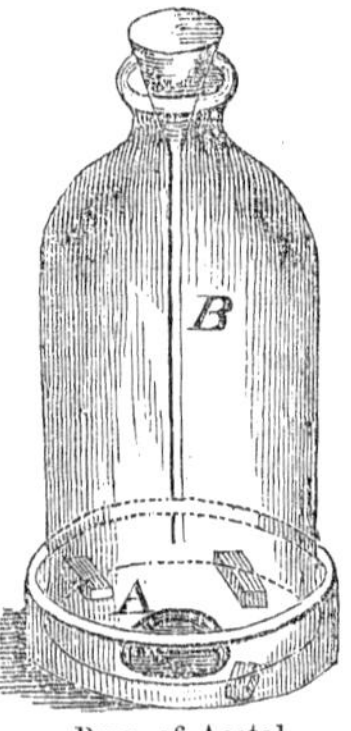

Prep. of Acetal.

QUESTIONS.—What is pyroligneous acid? 585. Describe the properties of pure acetic acid. 586. How is acetal formed?

catalytic action of the platinum, the alcohol is decomposed, and acetal, aldehyde, and acetic acid, are formed, and, condensing together upon the sides of the bell-glass, are collected in the basin. This liquid is now to be neutralized with powdered chalk, and carefully distilled from chloride of calcium.

Salts of Acetic Acid, and Other Allied Compounds.

587. Acetic acid combines with most bases, forming salts called acetates. In combination, its composition is $C_4H_3O_3$; from which it appears that in the process of combining, one atom of water—or the elements of water—is given up. Or, we may consider (with some chemists) the metal of the base as simply replacing 1 equivalent of the hydrogen of the acid (353). Thus, when acetic acid combines with soda, the acetate of soda formed has for its composition, $NaO, C_4H_3O_3$; and we may indicate the supposed relationship of the sodium as replacing 1 equivalent of the hydrogen of the acid by writing its formula thus, $C_4H_3(Na)O_4$.

Acetic acid is monobasic; that is, the salts it forms contain a single equivalent of the acid with one equivalent of base. These salts are also said to be monobasic (170, 349).

We shall describe only a few of the more important acetates.

588. Acetate of Lead, $PbO,C_4H_3O_3 + 3HO$. — This is the *sugar of lead* of commerce. It is formed by dissolving oxide of lead (litharge) in acetic acid. It is very soluble in pure water, and has a sweet, astringent taste. Taken internally, it is poisonous; but is used in various preparations in medicine. The $3HO$ is water of crystalization.

If we consider the metal as replacing an equivalent of the hydrogen of the acid, its formula may be written, $C_4H_3(Pb)O_4$.

Besides the above neutral acetate of lead, there are several other acetates of the same base, which are formed by the combination of this compound with additional equivalents of oxide of lead, as $2PbO,3C_4H_3O_3$, and $3PbO,C_4H_3O_3$. The latter of these, sometimes called tribasic acetate of lead, is used in medicine, and in the proximate analysis of organic

---

QUESTIONS. —587. What are the salts of acetic acid called? What is said of acetic acid as it combines with bases? Is this acid monobasic? What is meant by this? 588. Describe acetate of lead. What is it called in commerce?

compounds. It is formed by digesting 7 parts of litharge with [illegible] parts of the neutral acetate in 30 parts of water, and evaporating the solution until crystals are formed on cooling. The solution is called in pharmacy *Goulard's extract of lead.*

589. Acetate of Copper, $CuO,C_4H_3O_3$.—Acetate of copper is obtained by dissolving verdigris in hot acetic acid. On cooling, it is obtained in fine green crystals, which contain one atom of water, and is sometimes called *distilled verdigris.* This salt is capable of uniting with additional equivalents of oxide of copper, forming sub-acetates; and the *verdigris* of commerce, used as a paint, appears to be a mixture of these. It is prepared in large quantities, in the south of France, by covering copper with the refuse of grapes, after the juice has been extracted for making wine; the saccharine matter contained in the husks furnishes acetic acid by fermentation, and in four or six weeks the plates acquire a coating of the acetate. A purer and better article is prepared by covering copper plates with cloths soaked in pyroligneous acid

590. Acetate of Alumina, $Al_2O_3,3(C_4H_3O_3)$.—This salt is prepared by decomposing solution of sugar of lead by alum. It is used in dyeing, and calico-printing.

591. Acetate of Iron.—By treating iron-filings with dilute acetic acid, a mixture of the acetates of the proto and sesquioxides of iron is obtained, which is considerably used in the arts.

592. *Acetate of Ammonia,* $NH_3,C_4H_3O_3,HO = C_4H_3(NH_4)O_4$, is readily formed by neutralizing acetic acid with ammonia, or its carbonate. It has been used in medicine under the name of *spirit of Mindererus.*

593. Chloracetic Acid, $C_4HO_4Cl_3 = C_4Cl_3O_3,HO$, or $C_4H(Cl_3)O_4$.—This acid is formed from acetic acid, by the abstraction of 3 atoms of its hydrogen, and the substitution of 3 atoms of chlorine. It is prepared by placing some crystals of acetic acid under a large bell-glass filled with chlorine, and exposing it to the direct rays of the sun. Other compounds are formed at the same time, which are to be separated from it, and then it is obtained in colorless rhomboidal crystals.

QUESTIONS.—589. How is acetate of copper formed? What is the verdigris of commerce? How is verdigris prepared? 590. What other acetates are mentioned? 593. How is chloracetic acid formed?

Neglecting the secondary compounds formed, the reactions will be represented as follows, viz.:

$$C_4H_3O_3,HO + 6Cl = C_4Cl_3O_3,HO + 3HCl.$$

The crystals, when heated, melt at 113°, and the liquid boils at about 392°. It closely resembles acetic acid in its properties, neutralizes the same quantity of bases, and forms salts similar to the acetates.

594. Acetone, $C_6H_6O_2$.—Acetone, or *pyroacetic spirit*, is a limpid liquid, obtained by passing vapor of acetic acid through a red-hot tube, or by distilling a mixture of dry sugar of lead and powdered quicklime, and condensing the product in a cool receiver. Much uncondensable, gaseous matter passes over at the same time, which is allowed to escape. It has a density of 0·792, and boils at 132°. In some of its properties, acetone appears to be closely allied to the alcohols, but in its composition it is isomeric with the aldehyde of propylic alcohol (547).

In the preparation of acetone, another allied compound often makes its appearance, called *metacetone*, or *propione*, C_6H_5O. (We should prefer to call it *propylone.*) A corresponding acid is known, called the *metacetonic*, *propionic*, or *propylic* acid, $C_6H_6O_4$.

Action of Chlorine upon Alcohol.

595. When a current of dry chlorine is passed through anhydrous alcohol, the first effect is to produce aldehyde and hydrochloric acid. Thus,

$$C_4H_6O_2 + 2Cl = C_4H_4O_2 + 2HCl.$$

Two equivalents of chlorine, therefore, have separated 2 eq. of the hydrogen of the alcohol, forming 2 eq. of hydrochloric acid, the residue of the alcohol constituting aldehyde.

But if the action of the chlorine is continued, a new compound is formed, called *chloral*, $C_4HCl_3O_2$. It may be regarded as aldehyde in which 3 eq. of hydrogen have been replaced by 3 eq of chlorine. The reactions are as follows:

$$C_4H_4O_2 + 6Cl = C_4HCl_3O_2 + 3HCl.$$

Or, regarding it from the beginning of the process,

$$C_4H_6O_2 + 8Cl = C_4HCl_3O_2 + 5HCl.$$

QUESTIONS.—594. Describe the mode of preparing acetone. 595. What is the first effect when dry chlorine is passed through pure alcohol? What is formed when the process is continued? How may chloral be regarded?

Chloral is an oily liquid, of specific gravity 1·5, which boils at 201°, and has a strong affinity for water, being soluble in water, alcohol, or ether.

METHYLIC ALCOHOL, OR WOOD-SPIRIT, $C_2H_4O_2$.

596. Preparation and Properties.—When wood is subjected to distillation in a close vessel, besides the uncondensable gases which pass over, there is obtained an aqueous liquid, composed of pyroligneous acid (584), and other compounds; among which, is a volatile, inflammable liquid, long known as *wood-spirit*, or *pyroxylic spirit* (from *pur*, fire, and *xulon*, wood). But from its resemblance to alcohol, in composition (547) and many of its properties, it is more properly designated as *methylic alcohol*, (from *methu*, wine, and *xulon*, wood).

To separate it from the other compounds, the crude liquid is first neutralized with lime, and carefully distilled, only the part that passes over first being preserved, as this will contain nearly the whole of the wood-spirit. To obtain it pure, it must be several times re-distilled from chloride of calcium and lime.

Methylic alcohol is a colorless liquid, with an odor and taste somewhat resembling those of common alcohol. It has a density of 0·798, and boils at 152°, and burns freely in a lamp.

It mixes readily with water, alcohol, and ether, and, like common alcohol, dissolves freely most resinous substances. Some use has been made of it in the treatment of diseases, under the name of *wood naphtha*.

The formula of this compound, $C_2H_4O_2$, represents 4 vols. of its vapor, the density of which, by calculation, is 1·109, as follows:

2 vols. carbon vapor weigh	(·836 × 2)	1·672
8 vols. hydrogen,	(·069 × 8)	·552
2 vols. oxygen,	(1·106 × 2)	2·212
Giving for 4 vols. vapor of methylic alcohol,		4·436
The weight of 1 vol., or density, therefore, is		1·109

Products of the Oxydation of Methylic Alcohol.

597. Formic Acid, $C_2H_2O_4 = C_2HO_3,HO$.—This acid was first obtained, by distillation, from the bodies of red ants (*formica*

QUESTIONS.—596. From what is methylic alcohol, or wood-spirit, obtained? Describe methylic alcohol. 597. From what was formic acid first obtained?

rufa); and hence its name. Its relation to methylic alcohol is the same as that of acetic acid to common alcohol. It may be obtained by exposing the vapor of wood-spirit, mixed with air, to the action of platinum black, under a receiver, water and formic acid being produced at the same time. It appears that two atoms of oxygen combine with 2 atoms of the hydrogen of the spirit, forming 2 atoms of water; and then 2 atoms more of oxygen unite with the residue to form the acid. Thus,

$$C_2H_4O_2 + 2O = C_2H_2O_2 + 2HO, \text{ and}$$
$$C_2H_2O_2 + 2O = C_2H_2O_4.$$

Formic acid may also be produced by distilling a mixture of sugar, bichromate of potash, and oil of vitriol, and by other means.

Formic acid is a clear liquid, of specific gravity 1·24, has a strong acid taste and pungent odor, and quickly blisters the skin. It boils at 212°, producing an inflammable vapor, and solidifies at about 32°. It is contained in small quantities in many plants, as the nettle, and produces the pain occasioned by handling them. The figure represents the "sting" of the nettle, greatly magnified, with the sacks at the base containing the acid.

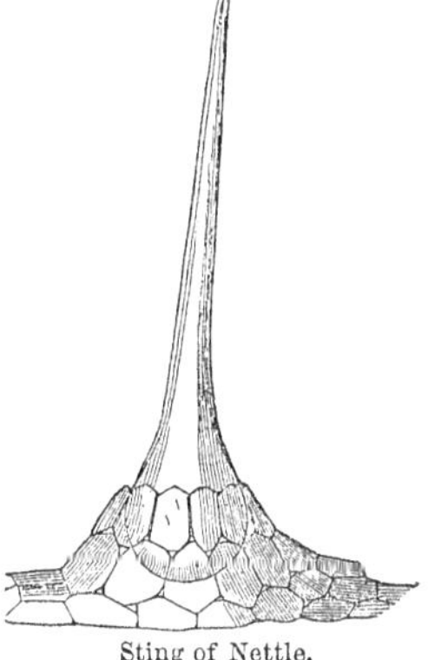
Sting of Nettle.

When this acid combines with bases, like acetic acid, it gives up the elements of an atom of water; and is therefore by many considered as a hydrate, and its formula written $C_2HO_3 + HO$. Or, as in the case of acetic acid (587), we may consider the metal of the base as replacing 1 equivalent of the hydrogen of the acid. Thus, the formiate of lead, PbO,C_2HO_3, on this supposition, will have for its formula, $C_2H(Pb)O_4$. It is monobasic, like acetic acid.

QUESTIONS.—What is said of the relation of formic acid to methylic alcohol? May it be procured from sugar? Describe the process. Describe its properties. Is it found in plants? What is said of the nettle?

598. The *formiates*, in their general properties, resemble the corresponding acetates. The alkaline formiates are used in the analyses of some minerals.

599. **Methylal**, $C_6H_8O_4$.—This substance is procured by distilling a mixture, in proper proportion, of methylic alcohol, dilute oil of vitriol, and peroxide of manganese, and purifying the product. It is a colorless liquid, of an agreeable aromatic odor, and burns with a yellow flame. It has a density of 0·85, and boils at 108°.

AMYLIC ALCOHOL, $C_{10}H_{12}O_2$.

600. Amylic alcohol is a peculiar oily substance, which is collected in the process of distilling spirit from potatoes. It is supposed to be formed from the starch (*amylum*) of the potato, and hence its name. It is also called *potato oil*, and *fusel oil*. It is found in some kinds of brandy, and in other spirituous liquors, being probably formed during the fermenting process, in circumstances not well understood.

In distilling spirit from potatoes, it is particularly abundant towards the close of the process; it is then mixed with water, to which it gives a milky appearance, but after standing for a time it rises to the surface.

Amylic alcohol is a clear, colorless liquid, insoluble in water, but very soluble in alcohol and ether. Its density is 0·82, and it boils at 270°. Its taste and smell are burning and pungent. If its vapor, mixed with air, is breathed for a little time, asthmatic pains and coughing are likely to ensue, and even vomiting.

It dissolves phosphorus, sulphur, and iodine, without change; and, unlike wine alcohol, is congealed by a temperature of —4°.

The formula of amylic alcohol, $C_{10}H_{12}O_2$, answers to 4 vols. of the substance in the gaseous state. The calculated density of its vapor is 3·057. Thus,

10 vols. carbon vapor weigh		(·836 × 10)	8·360
24 " hydrogen	"	(·069 × 24)	1·656
2 " oxygen	"	(1·106 × 2)	2·212
Thus, 4 vols. amylic alcohol vapor weigh			12·228
The weight of 1 vol., or the density, is therefore			3·057

QUESTIONS.—598. What do the formiates resemble in their general properties? 599. Describe methylal. 600. From what is amylic alcohol obtained? Describe its properties. When its vapor, mixed with air, is breathed, what is the effect?

Product of the Oxydation of Amylic Alcohol.

601. Valerianic Acid, $C_{10}H_{10}O_4 = C_{10}H_9O_3,HO$. — This acid sustains the same relation to amylic alcohol as acetic acid sustains to wine alcohol, or formic acid to methylic alcohol. It is contained in the valerian root (*valeriana officinalis*), which is extensively used in medicine, and is obtained by distilling the root with water. It is also formed artificially by dropping warm potato oil upon platinum black in contact with atmospheric air, or by distilling a mixture of amylic alcohol, sulphuric acid, and bichromate of potash.

In the process of its formation from amylic alcohol, the same chemical changes are required as in the formation of acetic acid from wine alcohol;—the alcohol loses 2 eq. of hydrogen, which combine with oxygen, forming water, and then 2 eq. additional of oxygen are absorbed. Thus,

$$C_{10}H_{12}O_2 + 2O = C_{10}H_{10}O_2 + 2HO, \text{ and}$$
$$C_{10}H_{10}O_2 + 2O = C_{10}H_{12}O_4.$$

Pure valerianic acid is a colorless, oily liquid, of specific gravity 0·937. Its taste is pungent and acid, and its odor like that of valerian. Its boiling point is 347°. It is little soluble in water, but dissolves freely in alcohol and ether. It combines with bases, forming salts in many respects similar to the acetates and formates. When it enters into combination, it gives up the elements of 1 equivalent of water, precisely like the acetic and formic acids, and is therefore monobasic.

Some of the valerianates, especially valerianate of zinc, are used in medical practice, as a substitute for the preparation of the valerian root.

602. *Amylic aldehyde,* $C_{10}H_{10}O_2$, analogous to the aldehyde of wine alcohol, is formed by distilling valerianate of baryta.

QUESTIONS.—601. From what is valerianic acid obtained? How may it be formed artificially? What is said of the chemical changes which take place in its formation? Describe its properties. What salt of this acid is used in medical practice? 602. How is amylic alcohol obtained?

The three alcohols above described are the most important of this homologous series (547), but six others are known, which will be hereafter described. These all have the formula, $C_{2n}H_{2(n+1)}O_2$, as we have before (545) seen.

Some other compounds, similar in their properties to the above, but not strictly homologous with them, have been denominated alcohols, as phenylic alcohol, $C_{12}H_6O_2$, and acrylic alcohol, $C_6H_6O_2$.

SULPHUR ALCOHOLS, OR MERCAPTANS.

603. These compounds are formed on the same type as the alcohols, but differ from them by containing sulphur instead of oxygen. They are called mercaptans (*mercurium captans*), because of their great affinity for mercury, with which they eagerly combine, on coming in contact with its oxide.

Wine-alcohol Mercaptan, $C_4H_6S_2 = C_4H_5S,HS$. — This compound is formed by saturating a solution of caustic potash, of density 1·3, with sulphuretted hydrogen, and distilling it with an equal measure of sulphovinate of lime (a substance to be hereafter described,) of the same density. It is a colorless liquid, which has a specific gravity of about 0·84, and boils at 97°. Its odor resembles that of onions.

Mercaptan vapor has a density of 2·152, as may be thus calculated:

4 vols. carbon vapor weigh	(·836 × 4)	3·344
12 vols. hydrogen " "	(·069 ×12)	·828
$\frac{2}{3}$ vol. sulphur " "	(6·654 × $\frac{2}{3}$)	4·436
4 vols. mercaptan vapor		8·608
The density therefore is, or the weight of 1 vol.,		2·152

By the action of this compound upon metallic oxides, 1 equivalent of its hydrogen unites with the oxygen of the oxide, forming water, and the metal takes the place of the hydrogen in the original compound. Thus, by the action of mercaptan on oxide of lead, PbO, we have—

$$C_4H_6S_2 + PbO = C_4H_5S,PbS + HO = C_4H_5PbS_2 + HO.$$

These metallic compounds, of which several more are known, are called *mercaptides*.

QUESTIONS.—Are there other alcohols besides the three described above? 603. What are the mercaptans, or sulphur alcohols? Describe wine alcohol mercaptan. What are the compounds of this substance with the metals called?

The compound, $C_4H_4S_2$, called *mercaptan aldehyde*, is known, as are also *methylic mercaptan*, $C_2H_4S_2$, and *amylic mercaptan*, $C_{10}H_{12}S_2$.

Selenium may be made to replace the sulphur in the common alcohol mercaptans, forming *seleno-mercaptans.*

ETHERS.—COUPLED, OR VINIC ACIDS.

604. The substance usually designated by the simple term, *ether*, is a compound obtained by the reciprocal action of alcohol and acids, as, the sulphuric, phosphoric, or arsenic, or the chlorides of zinc, tin, &c. It is often called *sulphuric ether*, from the circumstance that sulphuric acid is generally used in its preparation. The name *ether*, given to it, indicates its volatility.

But the term is now extended to a very numerous class of similar compounds, most of which are liquid at ordinary temperatures, and are very volatile, but differ from each other as to the temperature at which ebullition takes place. Some few of them are solid.

There are several families of ethers derived from the several alcohols; and in naming them, except in the case of those belonging to the common alcohol group, the family or group is usually indicated. Thus we have hydrochloric ether, methylic hydrochloric ether, and amylic hydrochloric ether; the first of which belongs to the wine alcohol group, but the others to the groups severally indicated.

And in each alcohol series, there are also two distinct classes of ethers, which we may call *simple ethers*, and *compound ethers.* The former, or simple ethers, have the general formula, $C_{2n}H_{2n+1}\Delta$ (the Greek letter Δ being used to indicate one equivalent of any one of the elements, oxygen, chlorine, bromine, iodine, fluorine, sulphur, selenium, tellurium, or a compound radical, cyanogen, bisulphide of carbon, &c.,) while the compound ethers always contain an acid in combination with a simple ether, $C_{2n}H_{2n+1}\Delta$. Thus, common sulphuric ether is a simple ether, C_4H_5O; but

---

QUESTIONS.—604. How is the compound usually called *ether* obtained? Why is it often called *sulphuric ether?* How is the name ether now used? Do the different alcohols afford ethers? What two classes of ethers are there in each alcohol series? In what do the individuals of the two classes differ?

nitric ether is a compound ether having the formula, $C_4H_5NO_6 = C_4H_5O,NO_5$.

There are some simple ethers which contain more than one equivalent of Δ; and in both the simple and compound ethers, one or more equivalents of the hydrogen may be replaced by chlorine (and possibly other elements), as in the chlorocarbonic ether, $C_4H_3Cl_2O,CO_2$.

Ethers of Wine Alcohol.

I. SIMPLE ETHERS.

605. Ether (Sulphuric Ether), C_4H_5O. — This ether may be formed by several modes; but the best is to distil a mixture of equal parts of alcohol and strong sulphuric acid in a glass retort, the process being discontinued as soon as the mixture begins to become colored. For distilling it, an alembic or other distilling apparatus (48) may be used.

The product should be washed with water, to separate a little alcohol and sulphurous acid that usually pass over with the ether. This is done by filling a bottle about half full with the impure ether, and then pouring in a quarter or one-half as much water, and shaking them well together. After standing a few minutes, the liquids separate, and the water may be drawn off by perforating the cork and inverting the bottle, taking care to notice when the water has all escaped.

Pure sulphuric ether is a colorless liquid, of a hot, pungent taste, and fragrant odor. At the temperature of 60°, its density is 0·72, and it boils at 96° or 98° in the atmosphere, and at about 40° below zero in a vacuum. In the open air, it evaporates with great rapidity, producing intense cold, so that water may easily be frozen by it. It is very combustible, and burns with a yellow flame. Exposed to the light, in vessels partly filled with air, it gradually absorbs oxygen, with the formation of acetic acid. The solvent powers of ether are not as extensive as those of alcohol, but it dissolves the essential oils, resins, and many of the fatty principles.

606. When the vapor of ether is inhaled, it first produces a species of intoxication, similar to that occasioned by exhilarating

---

QUESTIONS.—605. Describe the mode of preparing ether. What are its properties? 606. What is the effect when vapor of ether is inhaled?

gas; and afterwards a kind of stupor follows, during which the person is nearly insensible to pain; the effect being much the same as that produced by chloroform. When used for this purpose, it has been called *letheon*.

607. *Sulphovinic Acid.—Theory of the Formation of Ether.*—By comparing the formula of alcohol, $C_4H_6O_2 = C_4H_5O,HO$, with that of ether, C_4H_5O, it is evident that the former may be considered as a hydrate of the latter; and the effect of the acid upon the alcohol in producing ether is simply to abstract from it one equivalent of water, or its elements. But although this is the final result, other intermediate chemical changes take place.

When equal parts of sulphuric acid and alcohol are mixed, and slightly heated, *sulphovinic acid*, $C_4H_5O,2SO_3HO$, is first formed, and may be obtained in a free state, as a syrupy liquid. This acid is capable of combining with bases and forming salts, which are called *sulphovinates*, as sulphovinate of baryta, $BaO,C_4H_3O,2SO_3$, and sulphovinate of lime, $CaO,C_4H_5O,2SO_3$. These, and sulphovinates of other bases, are easily obtained in crystals.

Sulphovinic acid, though very stable at ordinary temperatures, is decomposed when heated to about 310°, and ether is separated in the gaseous state, and may be condensed, the sulphuric acid and water remaining behind. Thus we have, neglecting the water of the oil of vitriol,

$$C_4H_6O_2,2SO_3 = C_4H_5O + 2SO_3,HO.$$

608. By using alcohol nearly pure, and regulating the temperature constantly at about 300°, both the ether and the water formed from the alcohol may be made to distil over together, leaving unchanged the acid, which has produced the whole effect.

Let A (see figure on next page) be a flask, with rather a wide mouth, containing a mixture of five parts of alcohol and eight parts of sulphuric acid, so as to fill it about half full. In the cork insert a bent tube, B, through which the alcohol required in the process is to pass, and another, C, to connect with the condenser of the distilling apparatus (96);

QUESTIONS.—607. Why may alcohol be considered a hydrate of ether? What is the effect produced upon alcohol by sulphuric acid? What is first produced when equal parts of alcohol and sulphuric acid are mixed? What is the composition of this acid? Is it capable of forming salts with bases? What is the effect when sulphovinic acid is heated? 608. Describe the mode of forming ether by a continuous stream of alcohol.

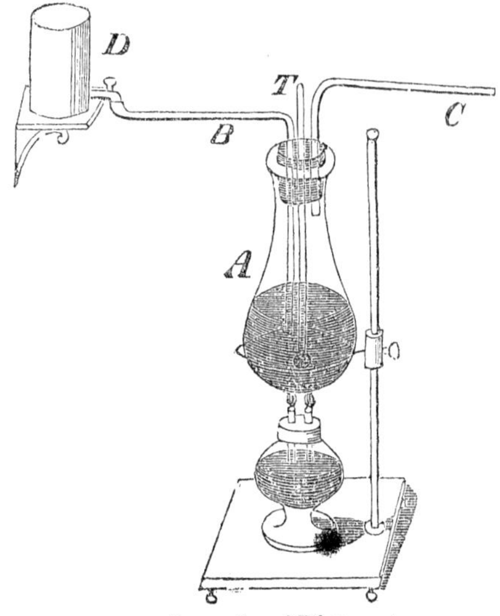

Preparation of Ether.

and between them place a thermometer, T, the bulb of which must extend to the mixture within, to determine the temperature. So also the alcohol tube, B, should extend a little below the surface of the liquid, as represented in the figure. When the mixture has become sufficiently heated, a continuous stream of alcohol is made to enter, just so as to keep the liquid in the flask at its original level. The supply of alcohol is regulated by the stopcock in the vessel D, which must be supported a little above the flask by a shelf or stand. The ether and water that pass over are, of course, collected together; but, as they do not mix, the ether is easily separated. The mixture, during the operation, should be kept continually boiling, and the heat very carefully regulated.

The ether, after being separated from the water, should be redistilled from a weak solution of caustic potash, to separate it from any acid which may have passed over with it.

609. Many other acids, as the phosphoric, arsenic, carbonic, oxalic, tartaric, &c., produce with alcohol *vinic acids*—called also *coupled acids*—which are entirely similar to the sulphovinic. Such acids are properly to be considered as bibasic (or tribasic).

To indicate the basic character of the water in sulphovinic acid, we may write its formulæ, $HO,C_4H_5O,2SO_3$;—now this water may be replaced by another equivalent of ether, C_4H_5O, and we then have the compound, $2(C_4H_5O,SO_3)$, which is perfectly neutral, and is called a *compound ether*.

QUESTIONS.—609. What other acids are mentioned as capable of forming vinic or coupled acids with alcohol? To what class of acids do they belong?

Some acids, as the sulphuric, oxalic, and carbonic, form both vinic acids and compound ethers, while others, as the phosphoric, form only vinic acids; and others still, as the nitric and acetic, form only compound ethers. Acids of this latter kind are strictly monobasic.

Besides the acids, several of the chlorides (604), as the chlorides of zinc and tin, and the fluoride of boron, effect the transformation of alcohol into ether, with the elimination always of an atom of water.

By distillation with an alkaline solution, the compound ethers and vinic acids are decomposed, and alcohol reproduced. In this case the acid of the compound ether, or vinic acid, combines with the alkali; and the ether, C_4H_5O, as it is set free, unites with an atom of water, thus reproducing alcohol, $C_4H_6O_2$.

610. Hydrochloric Ether, C_4H_5Cl.—This ether is formed by saturating common alcohol with hydrochloric acid gas, and then distilling with a very moderate heat. As the vapor distils over, it should be made to pass through warm water, to free it from impurities, and is then to be condensed in a receiver surrounded by ice.

It may also be formed by distilling a mixture of equal parts of alcohol and the hydrochloric acid of commerce, and by other means.

The reaction producing it is shown by the following equation:

$$C_4H_6O_2 + HCl = C_4H_5Cl + 2HO.$$

This ether is a colorless liquid, which has a density of 0·874, and boils at the low temperature of 52°. In the warm weather of summer, it can be preserved only in tubes hermetically sealed.

By its formula it will be seen that its composition is the same

---

QUESTIONS.—What acids are mentioned as forming both vinic acids and compound ethers? What as forming only compound ethers? What is their character? What other compounds are mentioned as being capable of transforming alcohol into ether? How are the compound ethers and vinic acids affected when distilled with a solution of alkali? 610. How is hydrochloric ether formed? What are the reactions by which it is produced when alcohol and hydrochloric acid are used? What are some of the properties of this ether?

as that of sulphuric ether, except that the oxygen of the latter substance is replaced by chlorine.

By passing a current of chlorine through a portion of this ether, in direct sunlight, 1 equivalent after another of the hydrogen is replaced by chlorine, until at length it all disappears, and we have only the compound C_4Cl_6. We have therefore the following series of compounds, viz.:

		Boiling Point.	Density.
Hydrochloric ether........	C_4H_5Cl	.... 52°	0·874
Bichlorinated do	$C_4H_4Cl_2$	.... 147°	1·174
Trichlorinated do	$C_4H_3Cl_3$	.... 167°	1·372
Quadrichlorinated ether	$C_4H_2Cl_4$	.... 215½°	1·539
Quinquechlorinated do ..	C_4HCl_5	.... 295°	1·640
Sesquichloride of carbon	$C_4H_6 = C_2H_3$	.... 356°	

611. **Hydrobromic Ether,** C_4H_5Br.—This ether is prepared from a mixture of alcohol, phosphorus, and bromine. It is a colorless liquid, having a density of 1·47, and boiling at 106°.

Hydriodic Ether, C_4H_5I.—Preparation similar to the preceding. It is a limpid liquid, of density 1·97, and boils at 158°.

Hydrosulphuric Ether, C_4H_5S.—This ether is very volatile, colorless, and of a penetrating, nauseous odor. Its density is 0·825, and its boiling point 163°.

Hydrocyanic Ether, C_4H_5Cy.—A poisonous, volatile liquid, of density ·787, and having its boiling point at 180°.

II. COMPOUND ETHERS.

612. **Sulphuric Ether (proper),** C_4H_5O,SO_3.—This is the only compound properly called *sulphuric ether*, on the principles adopted in naming the other compound ethers.

It is obtained by the action of anhydrous sulphuric acid upon ordinary ether. It is a neutral, oily liquid, having a density of 1·12, and a sharp, burning taste. Heated to 285° or 300°, it is decomposed, so that its boiling point cannot be ascertained.

613. **Hyponitrous Ether,** $C_4H_5NO_4 = C_4H_5O,NO_3$.—This ether may be formed by the direct action of nitric acid upon alcohol; but a better mode is to pass a current of nitrous acid vapor (obtained by the action of nitric acid upon starch) through dilute

---

QUESTIONS.—What is the effect when a current of chlorine is passed through hydrochloric ether? 611. What other simple ethers are mentioned? 612. What is the composition of sulphuric ether proper? How is it obtained? 613. Describe hyponitrous ether.

alcohol. Heat is generated by the process, and the vapor is condensed in a cold receiver. As the action of the acid upon the alcohol is often very tumultuous, the process should be conducted with caution. Hyponitrous ether is a liquid of a pale yellow color, and fragrant odor; its density is about 0·94, and it boils at 70°. In a pure state it cannot be kept long; but mixed with alcohol it is more permanent, and is extensively used in medicine, under the name of *sweet spirits of nitre.*

Nitric Ether, $C_4H_5NO_6 = C_4H_5O,NO_5$.—Nitric ether is formed by distilling equal parts of nitric acid and alcohol with a few grains of urea. It is a colorless liquid, of a sweet taste, and is heavier than water. It boils at about 185°, and its vapor explodes by heat.

614. Carbonic Ether, C_4H_5O,CO_2.—This ether is obtained by distilling oxalic ether (soon to be described) with potassium or sodium. It is a colorless liquid, very volatile, and having an aromatic odor and acrid taste. Its density is 0·975, and it boils at 259°. With chlorine it yields several chlorinated carbonic ethers, by the replacement of 1 or more equivalents of the hydrogen of the simple ether, C_4H_5O, by an equal number of equivalents of chlorine.

615. Silicic Ethers.—Silicic ethers, of which there are two (or perhaps three), are formed by the action of chloride of silicon upon absolute alcohol. They are volatile liquids, having a penetrating odor and a pungent taste. By water, or by long standing in bottles, from which the air is not perfectly excluded, they are gradually decomposed; and the hydrated silica is left in hard, vitreous masses, resembling quartz.

Acetic Ether, $C_8H_8O_4 = C_4H_5O,C_4H_3O_3$. — Acetic ether may be prepared by several modes, but the best is to distil a mixture of 3 parts of acetate of potash, 3 of absolute alcohol, and 2 of sulphuric acid. Acetic ether is a volatile liquid, of a fragrant odor, like that of strong vinegar. It boils at 165°, and has a density of about 0·89.

QUESTIONS.—What is said of nitric ether? 614. Of what is carbonic ether composed? 615. What is said of the silicic ethers? What of acetic ether?

616. Oxalic Ether, $C_6H_5O_4 = C_4H_5O,C_2O_3$.—This ether is prepared by distilling a mixture of 4 parts of binoxalate of potash, 5 parts of sulphuric acid, and 4 of strong alcohol, and thoroughly washing the product with water. It is a volatile liquid, which has an aromatic odor, and is a little heavier than water.

Oxalic ether is decomposed by contact with an alcoholic solution of potash, forming alcohol and oxalic acid. Ammonia acts upon it, producing two compounds, called *oxamide*, $C_2O_2NH_2$, and *oxamic acid*, $C_4O_5NH_2,HO$. By the action of chlorine upon it, several chlorinated compounds are formed.

617. Formic Ether, $C_6H_6O_4 = C_4H_5O,C_2HO_3$.—This ether is formed by distilling a mixture of dry formiate of soda, sulphuric acid, and strong alcohol. It is also produced in considerable quantity during the process for preparing fulminating mercury, to be hereafter described.

Formic ether is a colorless liquid, which boils at about 130°, and has a specific gravity of 0·915. It has a very penetrating, agreeable odor, and is soluble in about ten times its weight of water.

Many other compound ethers of the wine alcohol series, cannot be here described. Recent writers enumerate nearly a hundred, formed by the organic acids alone.

In their composition they evidently possess the character of salts, being formed by the union of ether, C_4H_5O, acting as a base, with the several acids; but in their properties they differ essentially from proper saline compounds. This appears in the fact that the acid cannot be detected by the ordinary tests; thus, oxalic ether is a compound of vinic ether, C_4H_5O, and oxalic acid, C_2O_3; but lime, which separates oxalic acid from its saline compounds, forming oxalate of lime, will not separate this acid from oxalic ether.

The same remark applies to the coupled, or vinic acids.

Ethers of Methylic Alcohol.

I. SIMPLE ETHERS.

618. Methylic Ether—Wood Ether, C_2H_3O. —This ether is formed from methylic alcohol, in the same manner as sulphuric

---

QUESTIONS.—616. How is oxalic ether prepared? 617. Formic ether? How are the compound ethers formed? Are they really salts? Do they differ in any respect from proper saline compounds? Illustrate with reference to oxalic ether. 618. How is methylic ether formed?

ether is prepared from common alcohol, by distilling methylic alcohol with an equal volume of sulphuric acid. It is a colorless gas, of a pungent odor and taste, and is rapidly absorbed by cold water, but given off again unchanged by boiling. It is condensed to the liquid form by a temperature of —3°; and is chiefly interesting as taking the place, in this series, which the ether first described (605), sometimes called sulphuric ether, occupies in the common alcohol series.

The theory of its formation corresponds in every respect with that already explained (607) in describing ordinary or sulphuric ether. When methylic alcohol and sulphuric acid are mixed, at a moderately elevated temperature, *sulpho-methylic acid* is formed, having the composition, $C_2H_4O_2,2SO_3 = C_2H_3O,2SO_3,HO$, which, by further elevation of temperature, is decomposed; the methylic ether being separated, and the sulphuric acid and water alone remaining.

619. Hydrochloric Methylic Ether, C_2H_3Cl.—This compound is formed by distilling a mixture of common salt, wood-spirit, and oil of vitriol. It is a colorless gas, of a peculiar odor, and corresponds to the ether of similar name in the common alcohol series; the oxygen of the preceding compound being replaced by chlorine.

It is not reduced to a liquid state by a temperature of zero.

By the action of chlorine upon this ether, in the direct rays of the sun, the several equivalents of hydrogen may be successively replaced by chlorine, as heretofore shown in another case (610), and important compounds formed,—the formulæ of which, with their densities and boiling points, are given in the following table:

		Boil. Point.	Density.
Hydrochloric methylic ether,	C_2H_3Cl	0·0°	1.74
Bichlorinated " "	$C_3H_2Cl_2$	87°	1·34
Trichlorinated " "	C_2HCl_3	142°	1·49
Sesquichloride of carbon,	C_2H_4	172°	1·60

620. Chloroform, C_2HCl_3, is the trichlorinated compound of this series. It is produced by the direct action of chlorine

---

QUESTIONS.—What is said of the relation methylic ether sustains to methylic alcohol? What is sulpho-methylic acid? 619. Describe hydrochloric methylic ether. What is the effect when a current of chlorine is made to pass through this ether? 620. What is chloroform?

upon hydrochloric methylic ether, but the best method of preparing it is to distil a mixture of hypochlorite of lime (common bleaching salt) and wine alcohol, or wood-spirit.

Chloroform is a dense, oily liquid, of an agreeable ethereal odor, and sweetish taste. It is not dissolved by water, but mixes readily with alcohol. Its density is 1·49, and it boils at about 142°. By breathing its vapor mixed with atmospheric air, a kind of intoxication is produced, much like that occasioned by exhilarating gas, or the vapor of sulphuric ether. If the vapor is breathed some time, total insensibility to pain is produced, and loss of consciousness, during which difficult surgical operations may often be performed, without even the knowledge of the patient.

It readily dissolves caoutchouc and gutta percha, and solid carbonic acid.

By alcoholic solution of potash, chloroform is changed into formiate of potash and chloride of potassium. It was this circumstance that suggested the name chloroform.

Iodoform, C_2HI_3, *bromoform*, C_2HBr_3, and *sulphoform*, C_2HS_3, are analogous compounds of iodine, bromine, and sulphur.

621. Hydriodic Methylic Ether, C_2H_3I, is a colorless liquid, obtained by distilling a mixture of iodine, methylic alcohol, and phosphorus. It is remarkable for its great density, which is 2·24.

Hydrobromic Methylic Ether, C_2H_3Br, is prepared in a manner similar to the above, only substituting bromine instead of iodine. It is liquid at temperatures below 55°.

Hydrosulphuric Methylic Ether, C_2H_3S, is a liquid, of a density 0·845, which boils at 106°.

Still other simple methylic ethers are known, but they cannot be here described.

II. COMPOUND METHYLIC ETHERS.

622. Sulphuric Methylic Ether, C_2H_3O,SO_3. — This ether is formed by distilling 1 part of methylic alcohol with 8 or 10 parts of sulphuric acid. It is a liquid, having a density of 1·32, and boils at 370°.

Nitric Methylic Ether, $C_2H_3NO_6 = C_2H_3O,NO_5$.—This ether corresponds to the ether of similar name in the common alcohol

---

QUESTIONS.—How is chloroform prepared? What use is made of it? 621. What other methylic ethers are mentioned? 622. How is sulphuric methylic ether formed? Describe nitric methylic ether?

series. It is formed by distilling a mixture of wood-spirit, nitrate of potash, and sulphuric acid. It is a dense, colorless liquid, which boils at about 150°. Its vapor, when heated to 248°, is decomposed with a violent explosion.

Oxalic Methylic Ether, $C_4H_3O_4 = C_2H_3OC_2O_3$.—Oxalic methylic ether is prepared by distilling a mixture, in proper proportions, of sulphuric acid, binoxalate of potassa, and methylic alcohol. At ordinary temperatures, it is a white crystaline solid, which melts at 124°, and boils at 322°.

Acetic Methylic Ether, $C_6H_6O_4 = C_2H_3O, C_4H_3O_3$.—This ether is formed by distilling an acetate with methylic alcohol, and sulphuric acid. It is always formed in the distillation of wood. It is a colorless liquid, of an agreeable ethereal odor, and has a specific gravity of 0·919. It boils at a temperature of 137°.

The close resemblance between the methylic ethers and those of the common alcohol group will be seen at once. There are many more of this group, which cannot be here described, late writers enumerating more than seventy.

Ethers of Amylic Alcohol.

623. Amylic alcohol, treated with sulphuric acid, produces sulphoamylic acid, similar to the sulphovinic and sulphomethylic acids, which is easily obtained in combination with bases, but with difficulty in a free state.

Treated with an excess of concentrated sulphuric acid, and heated to boiling, a carburetted hydrogen, $C_{10}H_{10}$, is obtained, called *amylene*, which is a volatile, colorless liquid, boiling at 102°. At the same time two other compounds, isomeric with it, are usually obtained, called *paramylene*, $C_{20}H_{20}$ and *metamylene*, $C_{40}H_{40}$.

624. Amylic Ether, $C_{10}H_{11}O$.—This ether is prepared from the hydrochloric amylic ether, to be next described, by the action of a strong solution of potassa. It is a liquid of an agreeable odor, the boiling point of which is about 234°.

QUESTIONS.—What is said of the resemblance between the methylic ethers and those of the common alcohol series? 623. What is said of the sulphoamylic acid? What is amylene? What two substances isomeric with it are mentioned? 624. What is amylic ether?

This ether, it will be seen, corresponds precisely to common ether in the wine alcohol group of ethers, and to methylic ether in the methylic alcohol group.

Hydrochloric Amylic Ether, $C_{10}H_{11}Cl$.—To obtain this ether, equal parts of amylic alcohol and perchloride of phosphorus are distilled in a glass retort, and the vapor condensed in the ordinary mode. It is a colorless liquid, with an aromatic odor, and boils at about 215°.

It may also be obtained by the action, long continued, of hydrochloric acid upon alcohol.

625. *Hydriodic amylic ether*, $C_{10}H_{11}I$, and *hydrosulphuric amylic ether*, $C_{10}H_{11}S$, are known. The latter has the smell of onions.

Acetic Amylic Ether, $C_{14}H_{14}O_4 = C_{10}H_{11}O,C_4H_3O_3$.—To prepare acetic amylic ether, equal parts of amylic alcohol and oil of vitriol are distilled with 2 parts of acetate of potash, and the resulting liquid purified. It is a limpid liquid, a little lighter than water, with an agreeable aromatic odor. Its boiling point is 257°. By some its composition is said to be $3(C_{10}H_{11}O),C_4H_3O_3$.

The odor of this ether closely resembles that of the ripe banana fruit, and it is used to give the banana flavor to sugar-candy. The preparation is sometimes called banana drops. The peculiar flavor of many ripe fruits may in this mode be very closely imitated by the use of ethers.

Oxalic Amylic Ether, $C_{12}H_{11}O_3 = C_{10}H_{11}O,C_2O_3$.—This compound is procured by distilling a mixture of amylic alcohol and oxalic acid. It is liquid at ordinary temperatures, and boils at 500°. Its odor is exceedingly offensive.

VOLATILE, OR ESSENTIAL OILS.

626. These oils constitute a very numerous class of compounds, and, as the name indicates, are distinguished by being volatile at ordinary or slightly elevated temperatures. Most of them are liquid at a temperature of 50°, or above; but a few are solid, as common camphor.

They are also usually distinguished by their powerful odor, which is often very agreeable. Most of them are obtained from

---

QUESTIONS.—What relation does amylic ether sustain to this series of ethers? What other ethers of this series are mentioned? 625. What is said of the odor of acetic amylic ether? 626. What is said of the volatile oils? Are any of them solid at ordinary temperatures? What is said of their odor? From what are they obtained?

the leaves and stalks of plants, but some are found in the flowers, fruit, bark, wood, or in the pericarp.

In general, these oils are obtained by distilling the part of the plant containing them, with water;—both the oil and the water pass over in vapor, and are condensed in the usual mode; and are afterwards separated by the oil rising to the surface. In a few cases only is the oil heavier than water, when, as a matter of course, it sinks to the bottom.

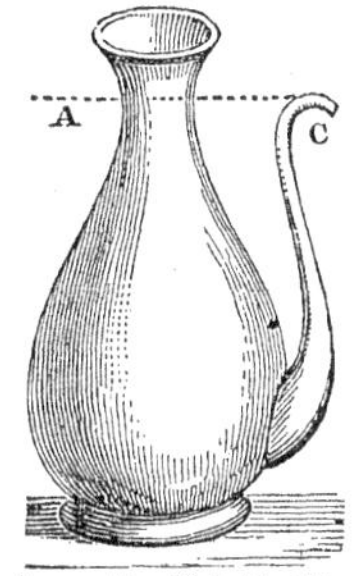

Separating Volatile Oils.

To separate such as are lighter than water, a vessel of the form represented in the margin answers well. It is simply a vessel of proper size, with a small spout which starts nearly from the bottom, and rises nearly as high as the mouth. It is first filled with water to the line A, and then the mixed liquids are poured into it in a small stream, the oil remaining at the surface, while the water escapes at C. This, it is evident, may be continued until the vessel is nearly filled with the pure oil; or until, if more were added, the oil itself would begin to escape by the spout C.

Water is necessary in the distilling process, to prevent the decomposition of a portion of the oil by the heat; and in most cases the distillation can be effected in this mode, even though the boiling point of the oil be considerably above that of water. In some cases, a higher temperature than 212° is required, when some soluble salt is added to the water, as common salt; a saturated solution of this compound boiling at about 230°.

Some few of the essential oils are obtained from the plant by pressure, or by the action of a solvent, as alcohol or ether.

627. These oils being volatile, if a small portion is droppod on a piece of white paper, the stain produced soon disappears, especially if the paper be warmed. In this way they may be distinguished from other oils, which produce upon paper a permanent stain.

Most of the essential oils consist of at least two principles; one of which is less fusible than the other, and may be separated by cold. Thus, exposing oil of peppermint to severe cold, a solid

QUESTIONS.—Why is water necessary in distilling these oils? 627. How may the volatile oils be distinguished? What is said of the composition of most of them?

not unlike camphor is crystalized out, and may be separated from the portion that remains liquid. These solids, sometimes called *stearoptens*, are different in the different oils; some being much more easily solidified than others. Their composition is also various; some being isomeric with the oil which yields them, and others, hydrates or oxides of the oil.

Most of them exist, as such, in the plant or part of the plant from which they are obtained, but some are formed in the process of distillation from materials contained in the substance distilled.

628. If we have regard to the elements of which these oils are composed, we may divide them into three classes, viz.:

1. Oils composed entirely of carbon and hydrogen.
2. Oils which, in addition to the above elements, also contain oxygen; and
3. Oils of which sulphur is an ingredient;—these also usually contain nitrogen.

Carbo-hydrogen Volatile Oils.

629. Oil of Turpentine—Camphene, $C_{20}H_{16}$.—This oil is procured by distilling common turpentine with water. Turpentine is obtained from several species of the pine, in the wood of which it is contained in large quantities. At the proper seasons of the year, incisions are made with an axe in the trunks of the trees, from which it gradually exudes, and is collected and preserved. This viscous substance is a solution of resin (common *rosin*) in oil of turpentine, and the latter is separated in the process of distillation.

Common *spirits of turpentine* is an impure camphene, containing a portion of resin in solution.

Exposed to the air, oil of turpentine evaporates rapidly, but at the same time oxygen is absorbed, and a portion of the oil is converted into resin, which remains in solution. It can therefore be preserved pure only in vessels from which the air is perfectly excluded. When pure, it is a limpid liquid, of specific gravity 0·87, which boils at 315°.

QUESTIONS.—Do the essential oils exist ready formed in the plant? 628. Into what three classes may they be divided? 629. How is oil of turpentine procured? From what is it obtained? What are the ordinary spirits of turpentine?

Oil of turpentine is not soluble in water, but dissolves readily in alcohol and ether, and in other essential oils. Mixed with three or four times its volume of strong alcohol, it constitutes the ordinary *burning fluid.*

When left for some time in contact with water, a portion of the two unite, forming a solid *hydrate of camphene,* which has the formula, $C_{20}H_{16},6HO$. Other hydrates may also be prepared, containing, severally, 4, 2, and 1 equivalents of water.

Many substances, as caoutchouc, insoluble in alcohol and ether, dissolve readily in this oil, which renders it an important substance in the arts. It is also largely used in mixing paints, in the manufacture of varnishes, and in the practice of medicine.

630. *Hydrochlorate of camphene,* or *artificial camphor,* $C_{20}H_{16},HCl$, is prepared by passing a current of dry hydrochloric acid gas through pure oil of turpentine. It makes its appearance as a white crystaline solid, which melts at about 302°, and may be sublimed, without change, at about 338°. In many of its properties it closely resembles common camphor.

The portion which remains liquid has the same composition as the crystals, but it cannot be solidified, so far as is known, by any temperature.

Hydriodic and hydrobromic acids form with oil of turpentine compounds similar to those formed by the hydrochloric.

Sulphuric and nitric acids act readily upon oil of turpentine, forming several interesting compounds.

631. Oil of Lemons, $C_{10}H_8$.—This oil is obtained by pressure from the yellow part of the peel of the fruit, and is isomeric with the preceding, having the same ultimate composition, but its equivalent being only one-half that of oil of turpentine. By distillation it yields two oils, which differ in their boiling points.

Oil of lemons has a specific gravity of 0·847, and it boils at 338°.

By the action of hydrochloric acid, it forms two camphors, similar to those formed from oil of turpentine, one being solid and the other liquid.

Orange peel yields an oil isomeric with the above, but having a different specific gravity, and a different boiling point.

Oil of elemi, oil of juniper, oil of pepper, &c., have a similar composition to the above, and they are all isomeric.

QUESTIONS.—What is the effect when oil of turpentine is left in contact with water? What use is made of this oil? 630. What is artificial camphor? How is it formed? 631. Describe the oil of lemons. What is said of oil of elemi, oil of juniper, &c.?

Oxygenated Volatile Oils.

632. Oil of Bitter Almonds.—Benzoic Series of Compounds. —Bitter almond oil, $C_{14}H_6O_2$, is obtained from the kernels of bitter almonds by distillation with water after the fixed oil contained in the seed has been expressed. It does not pre-exist in the seed, but is formed from the *amygdaline* of the seed, in a manner soon to be explained. Its density is a little above that of water, and it boils at 356°. It is quite colorless, and has a pungent taste and fragrant odor, and is poisonous.

After the fixed oil has been expressed, the pulp is allowed to stand for some hours, and a kind of fermentation takes place in the amygdaline by virtue of the presence of another proximate principle contained in the almond, called *synaptase*, or *emulsine*. During this fermentation there are produced, besides the oil, hydrocyanic acid and glucose.

633. *Amygdaline*, $C_{40}H_{27}NO_{22}$, may be obtained in a free state. It is a crystaline, colorless powder, very soluble in water and alcohol. *Synaptase* also may be isolated;—it is a yellowish-white solid.

634. Chlorinated Oil of Bitter Almonds—Chloride of Benzyle, $C_{14}H_5ClO_2$. —This substance is obtained by passing a current of dry chlorine through oil of bitter almonds, and expelling the excess of chlorine by heat. It is a colorless liquid, heavier than water, and has a peculiar, disagreeable odor. It is formed from the oil by the substitution of an atom of chlorine for one of hydrogen. Iodine and bromine form similar compounds.

When this chlorinated oil is treated with gaseous ammonia, there are formed *benzamide*, $C_{14}H_5O_2,NH_2$, and hydrochlorate of ammonia. Thus,

$$C_{14}H_5ClO_2 + 2NH_3 = C_{14}H_5O_2,NH_2 + NH_3,HCl$$

Benzamide is a white crystaline solid, and may be entirely purified from the sal-ammoniac by washing it with water. By remaining long in contact with water, benzamide is converted into hydrobenzamide, $C_{42}H_{18}N_2,6HO$; and this, by solution in alcohol and boiling, is reconverted into bitter almond oil and ammonia.

Benzoic Acid.—When exposed to the atmosphere, oil of bitter almonds rapidly absorbs oxygen, and is converted into *benzoic acid*, $C_{14}H_6O_4 = C_{14}H_5O_3,HO$. This acid is also formed when

---

QUESTIONS.—632. Describe bitter almond oil. From what proximate principle contained in the kernel is the oil formed? 633. Describe amygdaline. 634. Describe the chlorinated oil. Describe benzoic acid.

the oil is boiled with a solution of potassa. It is a beautiful, crystaline solid, which melts at 248°, and boils at about 463°. It may, however, be sublimed at a temperature considerably below its boiling point.

Benzoic acid is found in considerable quantity in the resinous substance usually called *gum benzoin*, a product of the *laurus benzoin*. From this compound it may be procured by direct sublimation, in the following manner. Place a small quantity of the resin, coarsely powdered, upon a plate of metal on a stand, and put over it a glass receiver, having suspended in it a small twig of mint, or other substance, as shown in the figure, and apply the heat of a lamp beneath it. In a short time the leaves will be covered with delicate crystals of the acid.

Benzoic acid with bases forms numerous salts called *benzoates*, but none of them are of sufficient importance to require description here.

635. *Benzoic Ether*, $C_4H_5O,C_{14}H_5O_3$, is formed by distilling a mixture of 1 part of benzoic acid, 2 parts of alcohol, and 6 parts of hydrochloric acid. It is a colorless liquid, heavier than water, and boiling at 410°. By substituting methylic instead of wine alcohol in the mixture, *benzoic-methylic ether* is formed, which boils at 226°.

636. *Benzoine*, $C_{28}H_{12}O_4$, is formed when crude oil of bitter almonds is shaken with an alcoholic solution of potassa, and is separated by crystalization. The crystals melt at 248°. It is isomeric with the oil. It may be sublimed without change; but by passing its vapor through a red-hot tube the oil is reproduced.

By heating benzoine with nitric acid, it loses 2 equivalents of its hydrogen, and a new compound is formed, called *benzile*. By the addition of oxygen, *benzilic acid* is formed.

637. *Benzone*, $C_{26}H_{10}O_2$, is a compound formed when benzoate of lime is distilled. It is a solid, little soluble in water, but very soluble in alcohol and ether.

638. *Benzene*, $C_{12}H_6$, at ordinary temperatures, is a liquid of specific gravity 0·85, which boils at 187°. It is formed by distilling a mixture of 3 parts of dry recently-slaked lime and 1 part of benzoic acid. It is

---

QUESTIONS.—Describe the mode of subliming a small quantity of benzoic acid. 635. Describe benzoic ether. 636. Describe benzoine. 637. Benzone. 638. Benzene.

of a sweet taste and agreeable odor, and crystalizes when cooled to 32°. It has also been called *benzol* and *phene;* and is produced by the decomposition of many organic substances by heat.

639. Oil of Spiræa Ulmaria.—Salicine Series of Compounds.—Flowers of the plant called meadow-sweet (*spiræa ulmaria*), when distilled with water, yield an acid essential oil, $C_{14}H_6O_4$, which, like the oil of bitter almonds, is of special interest because of its intimate relations with numerous other compounds. It does not pre-exist in the flowers, but is formed in the process of distillation. It is a liquid more dense than water, and boils at 385°. In composition it is isomeric with benzoic acid, and is sometimes called *salicylous acid.*

640. Salicine, $C_{26}H_{18}O_{14}$, is a substance obtained from the bark of certain species of willow (*salix*), and from many other plants, by digesting them with water, and then precipitating the gummy matter also contained in the solution by boiling with oxide of lead. After precipitating the lead by means of sulphuric acid, and sulphide of barium, salicine is obtained by evaporation in small, white, acicular crystals. Its taste is bitter, and it is very soluble in hot water and alcohol, but not in ether. Cold sulphuric acid dissolves it, forming a deep red solution.

A mixture of equal parts of salicine and bichromate of potash, digested for some time with 6 or 8 times their combined weight of dilute sulphuric acid, and afterwards distilled, yields salicylous acid, or oil of spiræa ulmaria, described above.

Treated with dilute sulphuric or hydrochloric acids, or with nitric acid, various other compounds are produced, which cannot be here described.

641. *Salicylic Acid,* $C_{14}H_6O_6 = C_{14}H_5O_5,HO$.—This acid is formed when salicylous acid is heated with an excess of caustic potash. The alkali is then separated by a mineral acid, and the salicylic acid is obtained in colorless crystals, which are soluble in boiling water and in alcohol and ether.

When salicylic acid is distilled with an excess of lime, carbonic acid and a new compound called *phenol* or *phenylic* alcohol, $C_{12}H_6O_2$, are formed. Thus,

$$C_{14}H_6O_6 = C_{12}H_6O_2 + 2CO_2.$$

QUESTIONS.—639. Describe the mode of procuring oil of *spiræa ulmaria.* 640. Describe salicine. How is salicylous acid, or the oil last mentioned, obtained from salicine? 641. Describe salicylic acid. What is phenol or phenylic alcohol?

Phenol is sometimes classed with the acids, and called *carbolic acid.* It resembles the alcohols in many of its properties.

642. By distilling a mixture of 2 parts of absolute alcohol, 1 part of sulphuric, and $1\frac{1}{2}$ parts of salicylic acid, *salicylic ether*, $C_4H_5O,C_{14}H_5O_5$, is obtained. It is a dense liquid, boiling at 437°, and possesses acid properties, readily combining with bases and forming proper salts. By substituting methylic alcohol in the above mixture, and using 2 parts of salicylic acid, we obtain *salicylic methylic ether*, $C_2H_3O,C_{14}H_5O_5$, which is of special interest as being identical with the chief constituent of *oil of winter-green (gaultheria procumbens)*, a substance well known, as obtained by distilling with water the leaves and berries of this plant.

The oil is an aromatic liquid, heavier than water, having its boiling point at 435°.

643. Oil of Cinnamon.—Cinnamic Series of Compounds.— Oil of cinnamon is prepared by distilling with water the cinnamon of commerce, which is the prepared bark of two or more species of trees, found in Ceylon and other Eastern countries. It possesses the peculiar taste and odor of the bark. Its composition when pure seems to be $C_{18}H_8O_2$, but when exposed to the air it absorbs oxygen, and perhaps undergoes other changes not understood, so that different chemists have arrived at different results as regards its true composition.

644. *Cinnamic Acid*, $C_{18}H_7O_3,HO$, is always found in the oil after it has been kept for a time exposed to the air, and is also contained with benzoic acid in the *balsams of Tolu* and *Peru*. It is a crystaline solid, which melts at 264°, and boils at about 570°.

Cinnamic acid forms ethers with common and methylic alcohols, analogous to those formed by salicylic acid.

When vapor of cinnamic acid is made to pass through a glass tube heated to dull redness, it is decomposed, and a carbo-hydrogen, $C_{16}H_8$, is formed, called *cinnamene*. It is a colorless liquid, of an agreeable, penetrating odor. This compound may also be obtained from the resinous substance called *styrax*, and has in consequence sometimes received the name *styrole*.

By treament with different reagents, still other compounds belonging to this series may be obtained.

645. Oil of Aniseed.—Anisic Series of Compounds.— Aniseed (the seeds of the *anisem sativum*), when distilled with water,

---

QUESTIONS.—642. Describe salicylic ether. What is said of the oil of winter-green? 643. How is oil of cinnamon procured? 644. Describe cinnamic acid. 645. Describe aniseed oil.

yields an essential oil from which crystals are deposited at a low temperature, having the composition, $C_{20}H_{12}O_2$. Treated with reagents, it forms a series of compounds known as the anisic series. Among them are *anisic acid*, $C_{16}H_7O_5,HO$, which, with wine and methylic alcohols, yields ethers, and *anisene*, $C_{14}H_8$. These sustain the same relation to each other, and to the oil, as cinnamon sustains to cinnamic acid and the oil of cinnamon.

646. *Oil of Cumin*, obtained by distilling with water the seeds of the *cuminum cyminum*, treated in a similar manner with reagents, forms a series of compounds, analogous to the above. Another series, called the *eugenic series*, is formed from the *oil of cloves* or *oil of pimento*.

647. *Oil of Peppermint*, $C_{20}H_{20}O_2$, is contained in the leaves and stem of the plant, from which it is separated by distillation with water. This oil is used in large quantities by confectioners; and in some of the western states the plant is extensively cultivated, to be distilled for the oil. Dissolved in alcohol, it forms the well-known *essence of peppermint*. This is the general method of preparing what are called essences.

Oil of peppermint is liquid at ordinary temperatures; but, when cooled gradually, it deposites a white crystaline *stearopten* (627), resembling camphor, which has the same composition as the oil.

There are very many other oxygenated volatile oils, but as they possess no properties giving them peculiar interest, they cannot be here described.

Sulphuretted Volatile Oils.

648. Oil of Black Mustard, $C_8H_5NS_2$.—This oil does not exist, as such, in the seed, but is formed from substances contained in them, after the fixed oil has been expressed, as in the case of bitter almond oil. The active substances in the mustard-seed, producing the oil when the bruised seed is digested with water, are called *myrosine* and *myronic* acid, both of which have been isolated; but their composition has not been determined.

Oil of mustard is a colorless liquid, boiling at 293°, and forming a most pungent, irritating vapor. Treated with reagents, it forms several other compounds.

QUESTIONS.—646. From what is oil of cumin obtained? 647. Describe oil of peppermint. 648. What is the composition of oil of black mustard?

649. **Oil of Garlic**, C_6H_5S.—This oil is procured by distilling garlic with water, and redistilling the product first obtained in a salt-water bath, and then rectifying it with potassium. It is a colorless liquid, lighter than water, and of an offensive odor. It has been called *sulphide of alyl*, being composed of sulphur and the carbo-hydrogen, C_6H_5, called *alyl*.

Camphors.

650. The camphors are allied both to the essential oils and to the resins. There is a large number of them, if we include the stearoptens (627), deposited at low temperatures from the vola tile oils; but only common or Japan camphor, and Borneo camphor, with a few of the compounds formed from them, will be here described

651. **Common, or Japan Camphor**, $C_{20}H_{16}O_2$.—This substance, which is always seen as a white crystaline solid, is obtained by distilling with water the roots and wood of the *laurus camphora*, a tree found in the island of Japan, and other parts of the East (see figure). It is a little lighter than water, and is readily soluble in alcohol, and ether, and every variety of ardent spirits. Water dissolves about one-thousandth of its weight, which is sufficient to communicate to it something of its peculiar, pungont, but agreeable odor. At 347° the solid melts, and at 410° it boils. In the open air, at ordinary temperatures, it gradually evaporates. When some lumps of it are contained in a close glass vessel, of somewhat larger capacity than is merely sufficient to receive it, the vapor that forms is gradually condensed in small crystals upon the sides of the

Laurus Camphora.

QUESTIONS.—649. What is the composition of oil of garlic? 650. To what are the camphors allied? 651. From what is common camphor procured? Describe its properties.

glass; and if the vessel stand in a place so that the light can fall only on one side, the chief deposition will be on the side exposed to the light.

When subjected to the action of chlorine, a part of the hydrogen of the camphor is replaced by chlorine, forming *chlorinated camphor*, $C_{20}H_{10}Cl_6O_2$.

By the action of heated nitric acid upon camphor, *camphoric acid*, $C_{20}H_{16}O_8 = C_{20}H_{14}O_6,2HO$, is formed, which is bibasic, and forms with alcohol a coupled acid (609) and a compound ether. By the action of a mixture of potassa and lime, at an elevated temperature, upon vapor of camphor, another acid is formed, called the *campholic acid*, the composition of which is $C_{20}H_{17}O_3,HO$.

Distilling this acid with phosphoric acid, we obtain the carbo-hydrogen, $C_{18}H_{16}$, called *campholene*, which is liquid, and boils at 275°.

Borneo Camphor, $C_{20}H_{18}O_2$. — This compound comes chiefly from the island of Borneo, and hence its name. In some of its properties it closely resembles the preceding, but its odor and taste are different. It melts at 383°, and boils at 419°. In composition it differs from Japan camphor only by containing two more equivalents of hydrogen. By heating it with nitric acid, these two equivalents are separated from it, and common camphor formed.

Coumarine.

652. By this name a crystaline, odoriferous substance is known, which is extracted from the *Tonka bean*, but is also contained in several plants, as the sweet vernal grass (*anthoxanthum odoratum*). It is procured by digesting the bruised beans in strong alcohol, which dissolves the odoriferous principle. Coumarine is the basis of the perfume called *vanilla*.

By the action of reagents it forms *coumaric acid*, $C_{18}H_7O_5,HO$.

QUESTIONS.—What effect is produced when common camphor is subjected to the action of chlorine? How is camphoric acid formed? What is said of Borneo camphor? 652. From what is coumarine obtained? In what perfume is it used?

FIXED OILS AND FATS.

653. The fixed oils are so called because, unlike the essential or volatile oils, they are incapable of being volatilized without change; consequently, when spread upon paper they produce a permanent stain. The fats are essentially the same as the oils, except that they are usually solid at ordinary temperatures, while the oils are generally liquid. But the distinction is unimportant.

Fatty substances—a phrase which may include both the fixed oils and fats—are found both in the animal and vegetable kingdoms; and their ultimate elements are always carbon, hydrogen, and oxygen.

The fatty substance of a plant is usually found in the seeds or pericarp, occasionally upon the surface of the leaves or bark. When contained in the seed or pericarp, it exists in peculiar cells, which require to be broken before it can be separated; and usually heat is applied to render it more liquid, when the separation is effected by severe pressure. Animal fats are separated from the tissues containing it either by pressure or by the protracted action of heat.

Most oils absorb oxygen from the atmosphere, by which their consistency is changed; but by some the absorption is much more rapid than by others. Such as absorb oxygen rapidly—as linseed oil—are called *siccative*, or *drying oils*, and are used in painting. When mixed with the coloring substance, or *pigment*, and spread upon the surface of wood, or other material, by means of a brush, oxygen is absorbed and a resinous coating formed, by which the coloring substance is firmly retained. By heating a drying oil nearly to the point at which decomposition begins to take place, the tendency to absorb oxygen is increased. This effect is still greater if, before heating, some highly oxydized body, as oxide of lead or manganese, is mixed with the oil.

Unctuous oils are such as do not absorb oxygen from the air, or do it but slowly. *Rancidity* in oils is usually occasioned by the absorption of oxygen. Some oils, while they absorb oxygen also give off carbonic acid or hydrogen.

654. Proximate Principles of the Fats. — All the fats and fixed oils are composed of several proximate principles, which

---

QUESTIONS.—653. Why are the fixed oils so called? Are they found both in animal and vegetable bodies? In what part of the plant are they usually found? What are siccative, or drying oils? What are unctuous oils? 654. What are the chief proximate principles of the fats?

are usually combined in definite proportions. The most common of these are *glycerine, stearine, margarine*, and *oleine*, of which nearly all the animal fats and oils are almost wholly composed. In several fatty substances, as butter, we find, in addition, small quantities of other proximate principles, as *butyrine, caprine*, and *caproine.* In some instances still other principles are found, which will be described in their proper places.

The proportion of stearine, margarine, and oleine in the different fats is exceedingly variable, and occasionally one or another of them may be wanting; in the more solid fats stearine and margarine are more abundant, while oleine is the chief constituent of the oils and more fusible fats. All of them contain glycerine except *spermaceti*, in which, instead of this principle, another peculiar compound is found, called *ethal*, a substance in composition allied to the alcohols.

The fats and fixed oils generally are capable of combining with the caustic alkalies to form *soaps*, and are therefore said to be *saponifiable.* The process consists in digesting the fatty substance with a solution of a fixed alkali, as potassa, or soda, when the stearine, margarine, oleine, &c., disappear, and corresponding acids, called *stearic, margaric, oleic*, &c., *acids*, are formed, which unite with the alkali, glycerine being at the same time set free.

These acids probably do not exist as such in the fats, or even in the proximate principles from which they are derived, but are formed during the process of saponification, as will appear more clearly hereafter. The presence of water is necessary to the process, a small portion of it, as we shall see, being required to produce the compounds formed.

Glycerine.

655. Glycerine, $C_6H_8O_6 = C_6H_7O_5HO$, was discovered by Scheele, and called the "sweet principle of fat." It is best obtained by boiling, for some time, a mixture of equal parts of olive oil and oxide of lead diffused in water, filtering the liquid which remains, and then passing through it a current of sulphuretted hydrogen, to separate all the lead. Lastly, it is heated for a few moments to the boiling point, and all the water carefully evaporated in a vacuum.

QUESTIONS.—In what fats do stearine and margarine usually abound? In what is oleine the chief constituent? What is said of spermaceti? What is meant by the saponification of a fat? What are formed in the process of saponification? 655. How is glycerine obtained?

As thus prepared, glycerine is an oily liquid, very soluble in water and alcohol, but insoluble in ether; quite inodorous, but having a sweet taste. Its specific gravity is about 1·27. Its name is derived from the Greek, *glukus*, sweet, in allusion to its taste.

With sulphuric and phosphoric acid it forms coupled acids (607) similar to those formed by these acids with the alcohols, which are called respectively the *sulphoglyceric* and the *phosphoglyceric* acids.

Glycerine is not volatile, but is decomposed when heated, yielding a peculiar, acrid volatile principle, called *acroleine*, $C_6H_4O_2$. It is this substance which constitutes the acrid, irritating fumes always perceived when any fatty substance is decomposed by heat, in such circumstances that the product of the decomposition cannot be immediately consumed, as in the manufacture of gas from oil.

Stearine and Stearic Acid.

656. *Stearine*, $C_{142}H_{140}O_{16}$. — Stearine (Greek, *stear*, tallow) forms the chief constituent of tallow, and is a white crystaline solid, much resembling spermaceti. It exists in nearly all fats, and in many of the oils, both animal and vegetable; and is found in greater proportion as the point of congelation of the fat or oil is more elevated. The best method to obtain it is to heat moderately some mutton suet with 8 or 10 times its volume of camphene, in which it will be dissolved; and, on cooling, the stearine will be deposited in white pearly crystals. These are entirely insoluble in water, and melt at 140°.

Stearine, by saponification with potash, forms stearate of potash, and hydrated glycerine is set free; and the stearate, by subsequent decomposition by a mineral acid, yields *stearic acid*, $C_{68}H_{68}O_7 = C_{68}H_{66}O_5,2HO$, which is obtained as a white, crystaline solid.

Stearic acid is the substance of which "stearine" and "adamantine" candles are formed; and therefore constitutes an important article of commerce. It melts at 167°, and solidifies at about 158°.

QUESTIONS.—Describe the properties of glycerine. What is acroleine? 656. What is the derivation of the word stearine? From what is this substance obtained? How is stearic acid formed? Describe it. What use is made of it?

657. Stearic acid for the manufacture of candles is prepared by two different processes. One method is to digest the natural fat, as tallow, with lime-water, aided by heat, by which an insoluble lime-soap is formed: and this is then decomposed by dilute sulphuric acid, the lime of course all being separated, as a sulphate. The fatty acids now rising to the top are washed to separate the glycerine, and compressed to remove the oleic and most of the margaric acids; and the cakes of nearly pure stearic acid are ready for use.

By the other mode dilute sulphuric acid is used, by which nearly the same effect is produced, this acid combining with the glycerine to form sulphoglyceric acid (655), which is washed away with the water. The fatty acids now obtained are distilled, at a high temperature, in an apparatus in which a partial vacuum is kept up, and through which a current of steam is continually passing. By pressure most of the oleic acid is now separated, and the mixed stearic and margaric acids obtained.

The composition of stearine corresponds to 2 equivalents of anhydrous stearic acid, and 1 equivalent of hydrated glycerine. Thus,

$$C_{142}H_{140}O_{16} = 2(C_{68}H_{66}O_5) + C_6H_7O_5,HO.$$

658. *Stearic ether* is formed by treating wine alcohol with stearic acid, and passing through the solution a current of hydrochloric acid; and by a similar process with methylic alcohol *stearic methylic ether* is formed Both are solid at temperatures below about 85° or 90°.

Margarine and Margaric Acid.

659. *Margarine* and *margaric acid* have the same composition, respectively, as stearine and stearic acid, of which they are isomeric modifications. The melting point of margarine is 118°, and that of margaric acid 140°.

660. *Margarone*, $C_{66}H_{66}O_2$, is a white pearly substance, formed by distilling a mixture of 4 parts of margaric acid and 1 part of lime. The following equation illustrates the chemical changes produced in the acid to form the margarone:

$$C_{68}H_{68}O_8 + 2CaO = C_{66}H_{66}O_2 + 2CaO,CO_2 + 2HO.$$

Oleine and Oleic Acid.

661. *Oleine* is the chief ingredient of most of the fixed oils, and is found also in many of the fats. It is difficult to obtain it pure, and therefore its formula cannot be given with certainty.

QUESTIONS.—657. What two methods are given for preparing stearic acid for the manufacture of candles? 658. Describe stearic ether. 659. What is said of the relation of margarine and margaric acid to stearine and stearic acid respectively? 660. What is margarone? 661. What is said of oleine?

It is best prepared from olive oil. Oleine is a yellowish liquid, of an oily consistency, which requires a temperature of about zero for congelation.

Oleic acid, $C_{36}H_{33}O_3,HO$, is an oily liquid, having a density of about 0·808, and solidifies at about 32°. It is formed by saponification of oleine, and subsequent decomposition of the soap by hydrochloric acid.

By distillation in close vessels oleic acid forms *sebacic acid*, $C_{10}H_8O_3,HO$; which, distilled with alcohol and hydrochloric acid, forms *sebacic ether*.

Action of Nitric Acid upon the Fatty Acids.

662. Nitric acid acts energetically upon the fatty acids, producing a variety of other acids, some of which are volatile, and others fixed at moderate temperatures. The former are the *formic*, *acetic*, *propylic* or *acetonic*, *butyric*, *valerianic*, *caproic*, *œnanthylic*, *caprylic*, *pelargonic*, and *capric acids*; all of which are homologous, and have the general formula, $C_{2n}H_{2n}O_4 = C_{2n}H_{2n-1}O_3,HO$.

At the same time the following non-volatile acids are formed, and may be separated by the proper means, viz., the *succinic*, *adipic*, *pimelic*, *suberic*, and *sebacic*. These also are homologous, and have the general formula, $C_{2n}H_{2(n-1)}O_8 = C_{2n}H_{2(n-2)}O_6,2HO$.

Most of these acids, it will be observed, are also obtained from other sources. The *pelargonic* is obtained from the rose-geranium (*pelargonium roseum*), the *succinic*, from amber (Latin, *succinum*), and the *suberic*, from cork, which is the prepared bark of the *quercus suber*.

Other Proximate Principles of the Fats, as Butyrine, Palmatine, &c.

663. *Butter* contains several proximate principles, besides those above described, as *butyrine*, *caprine*, and *caproine*, but it is difficult to separate them.

By digesting butter with an alkali, decomposing the compound formed by tartaric acid, and distilling, *butyric acid*, $C_8H_8O_4 =$

QUESTIONS.—Describe oleic acid. 662. What is said of the action of nitric acid upon the fatty acids? 663. What is said of butter? How is butyric acid obtained?

$C_8H_7O_3,HO$, is obtained, mixed with other substances, from which it is easily separated.

Butyric acid may also be formed from sugar, by mixing with a solution of it a little curds of milk and powdered chalk, and allowing it to stand for a time in a place where it shall be kept at a temperature of about 90°. A peculiar fermentation takes place, called the *butyric fermentation*, and the butyric acid, as it forms, combines with the lime, and forms butyrate of lime, which is decomposed by hydrochloric acid, and the butyric acid separated by careful distillation.

Butyric acid is liquid at ordinary temperatures, and boils at about 327°. Its density is about 0·963, and it is soluble in both water and alcohol.

With alcohol and sulphuric acid butyric acid forms *butyric ether*, $C_4H_5O,C_8H_7O_3$. This substance, dissolved in 5 or 6 times its volume of alcohol, forms *pine-apple oil*, which is used by confectioners for its pine-apple flavor.

Butyrone, C_7H_7O, and the compound, $C_8H_8O_2$, sometimes called *butyral*, are formed by distilling butyrate of lime. Butyral is the proper aldehyde (547) of butyric acid, and may be called *butyric aldehyde*.

Caproic, caprylic, and *capric acids*, are also obtained from the decomposition of butter. Each of them forms a vinic and a methylic ether.

664. *Castor oil* (obtained from the plant called *ricinus communis*) by saponification forms *ricinoleic acid*, $C_{38}H_{36}O_6 = C_{38}H_{35}O_5,HO$; and this acid, distilled with solution of potassa, yields the compound, $C_{16}H_{18}O_2$, which is evidently homologous with the alcohols, and has been placed in our list (547) as *caprylic alcohol*.

Palmatine, and *palmitic acid*, $C_{64}H_{64}O_8 = C_{64}H_{62}O_6,2HO$, are obtained from palm oil, which is imported largely into this country from the coast of Africa, and used in the manufacture of soap.

665. *Spermaceti*—called also *cetine*, when pure—is a beautiful white substance, found mixed with oil in cavities of the heads of certain species of whales. The oil is separated from it by pressure; and the hard, white, crystaline substance thus obtained

---

QUESTIONS.—How may butyric acid be formed from sugar? Describe its properties. How is butyric ether formed? What is butyrone? What is said of butyric aldehyde? 664. How is caprylic alcohol procured? 665. What is ethal? From what is it obtained?

melts at about 120°, and may be sublimed unchanged, in close vessels, at about 680°.

Spermaceti contains no glycerine; but in its place a peculiar principle called *ethal*, $C_{32}H_{34}O_2$, is found, which may be separated as a colorless, crystaline solid.

Ethal, in many of its properties, as has been stated, closely resembles the alcohols; and various bodies are derived from it similar to those derived from the alcohols. Thus, when treated with sulphuric acid, it forms a coupled acid, called the *sulphethalic acid*, which corresponds to the sulphovinic acid; and, distilled with perchloride of phosphorus, it yields a liquid ether, $C_{32}H_{33}Cl$, corresponding exactly with the hydrochloric ether, C_4H_5Cl, previously described.

This substance yields an acid, called the *ethalic*, $C_{32}H_{32}O_4 = C_{32}H_{31}O_3, HO$, corresponding to acetic acid in the common alcohol series.

666. Wax.—Many substances, mostly of vegetable origin, are known as wax, of which *bees'-wax* is the proper type. This, as is well known, is obtained from honey-comb, by heating it with water; the wax melts and swims upon the surface, while the impurities it contains are dissolved in the water, or settle to the bottom. It appears to be a compound of two principles, *cerine* and *myricine*, which may be separated by boiling alcohol. Common bees'-wax is of a yellow color, but is whitened by exposing it in thin layers to the action of the atmosphere and of light.

Bayberry tallow, or *myrtle wax*, is a fat obtained from the fruit of the common bayberry (*myrica cerifera*). It is obtained from the berries by steeping them in hot water, and is found in other vegetables. Its composition is essentially the same as common tallow, but it contains other principles in small quantity, among which is myricine. It melts at 117°, but is very hard when cold, and may be formed into candles.

667. *Myricine*, $C_{92}H_{92}O_4$, subjected to the action of a boiling alkaline solution, is converted into *palmitic acid*, $C_{32}H_{32}O_4$, and

---

QUESTIONS.—What is said of the properties of ethal? 666. From what is bees'-wax obtained? What two principles are mentioned as contained in it? From what is myrtle wax procured? What use is made of it? 667. What is melissic alcohol?

a compound called *melissine*, $C_{60}H_{62}O_2$, which, being analogous in composition to the alcohols, may be called *melissic alcohol.* Treated with potassa or lime, melissic alcohol forms *melissic acid*, $C_{60}H_{60}O_4 = C_{60}H_{59}O_3,HO$ (547).

Cerotine, $C_{54}H_{56}O_2$, and *cerotic acid*, $C_{54}H_{54}O_4 = C_{54}H_{53}O_3,HO$, are other compounds obtained from wax. The former substance in composition ranks with the alcohols, and may be called *cerotic alcohol.*

Soaps and Plasters.

668. Soaps.—Frequent allusion has already been made to the action of the alkalies upon the fats and oils;—when these are boiled together, union takes place between them, and a well-known and very important substance is formed, called *soap.* The acids contained in the fat or oil, as the margaric, stearic, oleic, &c., described above, combine with the alkalies, forming proper salts, which exist in the soap together with other substances.

The soaps formed by potassa, soda, and ammonia only are soluble in water. Potash soaps are generally soft, while those made with soda are hard; but their consistency depends also upon the nature of the fat used. All soaps contain a large proportion of water in their composition.

Potash or soda for soap-making should be in the caustic state, in order to act readily upon fatty substances.

Soft soaps are made entirely of potash and tallow or oil, and often other animal matters. For the coarser kinds, very impure fats are used, without even separating them from the animal tissues in which they are contained. The common yellow hard soaps contain a portion of common rosin.

Toilet soaps are often perfumed with the essential oils.

669. The mode in which soaps of all kinds operate to produce their cleansing effects is easily understood. All soaps are always more or less alkaline, as the alkalies contained in them are not entirely neutralized by the oily acids with which they are combined; they are therefore ever ready to combine with more fatty or oily matter with which they may come in contact, as that constantly given off in the insensible perspiration from

---

QUESTIONS.—What is cerotic alcohol? 668. How are soaps formed? What soaps are soluble in water? What are insoluble? What are soft, and what hard soaps? 669. Explain the action of soaps.

the skin. Caustic alkali always has a soapy feeling to the fingers, from the circumstance that it attacks the cuticle, and forms with it a soapy compound, at the same time absorbing a portion of water from the atmosphere.

670. *Plasters.*—The *lead-plaster*, or *diachylon*, used in surgery, is a kind of metallic soap, which is made by boiling olive oil and oxide of lead together, with a little water. The oleic and margaric acids contained in the oil unite with the oxide of lead, in the same manner as they combine with the alkalies (oxides of potassium and sodium) in the formation of soaps. There are two kinds of diachylon, the yellow and the brown the former of which is made with litharge, and the latter with red lead; but their properties are essentially the same. Both are quite hard at ordinary temperatures, but melt with a moderate heat.

RESINOUS SUBSTANCES.

671. Resins are solid substances contained in many plants and trees, usually in a state of solution in some essential oil. When exposed to the air the oil evaporates, and the solid resin is obtained. By friction they usually become negatively electrical. They are all insoluble in water, but most of them are soluble in alcohol, ether, and the essential oils.

Most of them act as weak acids, and are capable of combining with the alkalies and other metallic oxides.

Common resin, or *rosin* (French, *colophane*), is obtained from different species of pine, in which it exists combined with oil of turpentine. The substance called turpentine is obtained from the trunk of the tree by incisions made in it while growing; and this, by distillation, yields the oil of turpentine, the resin remaining behind in the boiler.

There are, in commerce, several different kinds of turpentine, as *Venice turpentine*, *Strasbourg turpentine*, *Canada balsam*, &c.

When the tree is felled, and the turpentine extracted by heat, it is partially decomposed, and acquires a dark color, and an offensive burnt odor, and is called *tar*. By heating tar, so as to expel all the oil of turpentine contained in it, the common *pitch* of commerce is obtained.

From the different turpentines, by the action of reagents, many important substances may be formed, as the *pimaric*, *silvic*, and *pinic acids*.

672. *Lac*, or *gum lac*, as it is often called, is procured from a species of tree called *ficus*, by punctures made in the bark by

---

QUESTIONS.—670. How is *lead-plaster* or *diachylon* formed? 671. What are resins? What is common resin or rosin? How is *tar* obtained? 672. From what is *lac* procured?

insects. It is soluble in alcohol and heated oil of turpentine, but is quite insoluble in water. Its composition seems to be quite complex, there being contained in the common lac of commerce a mixture of as many as five different resins.

Lac is much used in the preparation of varnish and the manufacture of *sealing-wax.*

Copal is the resin of the *rhus copallinum* and *eleocarpus copaliferus*, trees found only in warm climates. It is of a light yellow color, and very hard, and without odor or taste. Its specific gravity is about 1·11. It is believed to be a mixture of several different resins, which, by proper processes, may be separated from each other.

This resin is quite insoluble in common alcohol or ether, but is softened, though not dissolved, by some of the essential oils. Its complete solution is effected by carefully fusing it, and then treating it with boiling alcohol or oil of turpentine.

Copal varnish is made by fusing the resin in a deep copper vessel, to prevent it from igniting, and then pouring in hot linseed oil, and oil of turpentine, to give the proper consistency.

Mastic and *sandarac* are other resins used in the preparation of varnishes.

673. *Amber* (*electron* of the ancient Greeks, and *succinum* of the Romans,) is always found as a fossil, but it appears to be the resin of some ancient tree that has become extinct. It is found on the shores of the Baltic sea, on the Yorkshire coast of England, and in some of the United States, as in New Jersey.

It is found in masses seldom weighing more than a few ounces, and in small grains. Its color is usually some shade of yellow, often inclining to red. It has a specific gravity of 1·07 to 1·09, and is insoluble in water, and is acted on but slightly by alcohol, ether, or the oils.

It is capable of being worked in the lathe, and is often seen fashioned into ornaments of various forms. It is used also in the preparation of varnishes.

674. *Caoutchouc.*—This substance, sometimes called *gum elastic*, or *India-rubber*, is prepared from the milky juice which exudes

---

QUESTIONS.—What use is made of *lac?* From what is copal procured? What use is made of it? 673. Where is amber found? Give a description of it. 674. From what is gum elastic obtained?

from certain trees, as the *ficus elastica*, when incisions are made in them. The white milky juice which exudes is received on masses of clay, and dried by exposure to the heat and smoke of fires kindled for the purpose.

As usually seen, caoutchouc is a solid of a dark color, and specific gravity a little less than water. At 32°, and lower temperatures, it is very hard, and has little elasticity; but at 60° or 70°, it is exceedingly flexible and elastic. It is quite insoluble in water, and alcohol, but dissolves slightly in pure ether and some of the essential oils. Chloroform dissolves it readily, as does also a solution of sulphur in oil of turpentine.

By distillation it yields several peculiar products.

Vulcanized India-rubber is formed by heating caoutchouc with a portion of sulphur, which becomes incorporated with it, increases its elasticity, and renders it less liable to be affected by changes of temperature. It is used for many important purposes.

675. *Gutta Percha* is a substance, in many of its properties, closely resembling the preceding; it is prepared in the same way from the milky juice of a tree found only in tropical climates.

Like India-rubber, it is quite insoluble in water and alcohol, but is dissolved in small quantities by pure ether and some of the essential oils, and more largely by chloroform and sulphide of carbon. Its specific gravity is about 0·97.

Gutta percha is less elastic than caoutchouc, and at ordinary temperatures is quite hard, but becomes soft at 200° to 212°, and may be formed into any desired shape. It is becoming of considerable importance in the arts.

VEGETABLE ACIDS NOT INCLUDED IN THE PRECEDING GROUPS.

676. These acids are all found ready formed in plants, a circumstance in which they differ from most of those heretofore described. Sometimes they occur in a free state, but generally they are found in combination with bases.

QUESTIONS.—Describe the properties of gum elastic. How is vulcanized India-rubber prepared? 675. From what is gutta percha obtained? Describe it. 676. What is said of the acids of this group?

677. Oxalic Acid, $C_4H_2O_8 = C_4O_6,2HO$.—Oxalic acid, in combination with bases, is found in many plants, especially in certain species of the sorrel (*oxalis*), and also in certain minerals. It may likewise be prepared artificially, by digesting starch, or sugar, with nitric acid, and by other processes.

To prepare it artificially, add, in successive portions, 24 parts of starch to 144 of nitric acid of commerce, diluted with 10 parts of water, and heat gently till the nitrous vapors cease. When the action is over, set the whole aside to crystalize.

Pure oxalic acid is a crystaline solid, in external appearance not unlike Epsom salt, for which it has sometimes been mistaken. It is very soluble in water, exceedingly sour to the taste, and poisonous. Its composition in crystals is $C_4H_2O_8 + 2HO$.

678. Salts of Oxalic Acid.—Oxalic acid is bibasic, and, as in the case of other bibasic acids, both equivalents of its basic water may be replaced at the same time by a fixed base, or only one of them, producing the two series of salts of the forms $2(RO)C_4O_6$ and RO,C_4O_6,HO;—RO being used to indicate any base.

679. *Oxalate of lime*, $2CaO,C_4O_6$, is found in many plants, in the cells of which it may often be detected in small crystals by the microscope. The crystals when thus found, or if formed by other means, always contain 2 equivalents of water of crystalization. It is very insoluble. With potash it forms three salts, called the *neutral oxalate*, $2KO,C_4O_6$, the *binoxalate*, KO,C_4O_6,HO, and the *quadroxalate*, the latter containing twice as much acid as the next preceding. The binoxalate is often used in solution to remove stains of iron-rust, under the name of *salt of sorrel*.

Oxalic acid forms with the alcohols, both coupled acids (609) and compound ethers.

680. Tartaric Acid, $C_8H_6O_{12} = C_8H_4O_{10},2HO$.—Tartaric acid, in combination with potash, exists in many fruits, especially in grapes and in pine-apples. When the expressed juice of the grape is fermented, as in the manufacture of wine, this salt, in an impure state, is precipitated upon the inside of the cask, as *argol*, or *tartar*. From this the pure acid is obtained, which is a white solid, very soluble in water, and of an agreeable acid taste.

QUESTIONS.—677. In what is oxalic acid found? How may it be prepared artificially? Describe its properties. 678. What is said of the salts of oxalic acid? 679. In what is oxalate of lime found? 680. In what is tartaric acid found?

Tartaric acid is bibasic, and forms two series of salts, according as one only or both equivalents of its basic water may be replaced by the base.

There are only two or three important salts of this acid, among which the acid tartrate of potash, or *cream of tartar*, takes the first place. Its composition is $KO,C_8H_4O_{10},HO$. It is prepared entirely from the impure tartrate, or *argol*, forming the settlings of wine-casks by repeated crystalizations and filterings.

By saturating a solution of cream of tartar with soda, or carbonate of soda, a double tartrate of potash and soda is formed, called *Rochelle salt*.

681. *Tartar-emetic* is a double tartrate of potash and oxide of antimony, which is formed by boiling equal parts of cream of tartar and the oxide in 6 parts of water. It is much used in medicine.

By the action of heat upon tartaric acid its characters are changed, and it is converted into other acids, which are generally isomeric with itself, as the *metatartaric*, *isotartaric*, and *paratartaric* or *racemic acids*.

With the alcohols tartaric acid also forms coupled acids and compound ethers.

682. Citric Acid, $C_{12}H_8O_{14} = C_{12}H_5O_{11} + 3HO$.—Citric acid is obtained chiefly from the lemon (*citron*), but is found in other fruits, as the orange, currant, gooseberry, strawberry, &c. When pure it forms crystals, which are very soluble in water, and have an agreeable sour taste. It is used in calico-printing, and for medicinal and domestic purposes.

To prepare citric acid, lemon-juice is first saturated with lime, and then the citrate of lime, so formed, mixed with several times its weight of warm water, is decomposed by sulphuric acid. The clear liquid is then drawn off and evaporated until the crystals of citric acid are deposited as the solution cools.

Citric acid is tribasic, and forms three series of salts, according as one, two, or three equivalents of its basic water may be replaced by the fixed base.

By heat this acid is decomposed, forming carbonic acid and other products, among which is *aconitic acid*, C_4HO_3,HO, so called because first obtained from the *aconitum napellus*. By distillation it also yields the *itaconitic* and *citraconic acids*.

Citric acid with alcohol forms two different coupled acids, and a compound ether.

QUESTIONS.—What is said of the salts of tartaric acid? What is *cream of tartar? Rochelle salt?* 681. Describe *tartar-emetic*. What acids isomeric with tartaric acid are mentioned? 682. From what is citric acid obtained? Describe its properties.

683. Malic Acid, $C_8H_6O_{10} = C_8H_4O_8,2HO$.—Malic acid is found in many vegetables, both free and in combination with bases, especially in many fruits before coming to maturity, as the apple (*malum*), and the plum. It is abundant in the fruit of the mountain ash (*sorbus aucuparia*), and in the stalk of the rhubarb or pie-plant.

Pure malic acid forms crystals which are very soluble in water, and which melt at about 181°. With alcohol it forms a coupled acid and an ether, in the same manner as other bibasic acids.

By the influence of heat, carefully managed, it may be converted into two other acids, which are isomeric with each other, called the *maleic*, and the *paramaleic* or *fumaric acids*. The latter is also found in some plants, and in Iceland moss.

684. Tannic Acid, $C_{16}H_8O_{12} = C_{16}H_5O_9,3HO$. — Tannic acid (called also *tannin*), occurs in the bark and leaves of many trees, as the oak, chesnut, and hemlock, but is especially abundant in *nut-galls*, which are excrescences that form upon the leaves of several species of the oak.

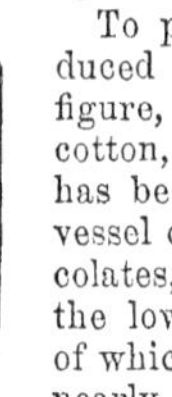

Prep. of Tannic Acid.

To prepare it, nut-galls, in coarse powder, are introduced into a funnel, of the form A, represented in the figure, the mouth having been loosely filled with a little cotton, and pouring over them some sulphuric ether that has been previously washed. The funnel is placed in a vessel of the form B, into which the liquid gradually percolates, and separates spontaneously into two portions;—the lower being a solution of the acid in water (a little of which was contained in the ether), with the lighter and nearly pure ether above it. The acid solution can be easily separated; and by evaporation it yields the tannic acid as a solid mass.

This acid is soluble in water, and has a peculiar astringent taste. It is a feeble acid, but forms salts with bases. With salts of the peroxide of iron, it forms a deep blue or black precipitate, which is the basis of writing-ink.* It forms an insoluble

* To prepare an excellent black writing-ink, pour 6 pints of boiling rain-water upon 6 ounces of the best nut-galls in powder, and add 4 ounces of gum-Arabic, and let the whole stand two days in a wooden or earthen vessel. Then strain the liquid and mix with it 4 ounces of clean copperas, and let it stand one or two months, stirring it frequently, and pour off the part free from sediment for use.

QUESTIONS.—683. Describe malic acid. 684. In what does tannic acid or tannin occur? Describe the mode of preparing it.

and very important compound with gelatine, which is the basis of *leather*. Raw hides, after the hair is removed, are soaked for a time in a decoction of bark which contains this substance, and are thus changed into leather.

685. Gallic Acid, $C_{14}H_6O_{10} = C_{14}H_3O_7,3HO$.—This acid is usually associated with the preceding, and is always formed when a solution of that acid is left for some time to the action of the atmosphere, or is boiled for a time with sulphuric acid. In the latter case, the reaction is attended by the formation of grape-sugar. This acid is readily obtained in crystals, which are quite insoluble in cold, but very soluble in hot water.

By the action of heat two other acids are formed from gallic acid, called the *pyrogallic* and *metagallic acids*.

ORGANIC ALKALIES, OR ALKALOIDS.

686. We have seen above, that many organic compounds are acids; so also there are others which are properly alkalies, as they readily, like potash, soda, &c., combine with acids which they neutralize, forming true salts.

All the organic alkalies contain nitrogen and hydrogen, and in some sulphur is found. Some of them exist ready formed in plants, but others are produced by their destructive distillation. Nearly all of them are poisonous.

687. Morphia, or Morphine, $C_{34}H_{18}NO_6$.—This substance is an essential ingredient of *opium*, which is the dried juice of certain species of the poppy, cultivated largely in different parts of Asia. To separate the morphia, the opium is digested several days in water, and the solution precipitated by ammonia, which is added cautiously in small portions. The impure morphia thus obtained is then further purified by dissolving it in boiling alcohol, from which it crystalizes on cooling.

The morphia in opium seems to be in combination with a peculiar acid called *meconic acid* (from *mecone*, a poppy).

QUESTIONS.—What is *leather?* 685. Describe gallic acid. 686. What are the organic alkalies? What is said of their composition? 687. From what is morphia obtained? Describe the method of separating it.

Pure morphia is a crystaline solid, but slightly soluble in water, and of a bitter taste. It combines readily with acids, forming salts, the most important of which are the *sulphate, acetate,* and *hydrochlorate.*

These salts are the compounds of this substance, generally used in medical practice under the name of *morphine*, and not the pure alkaloid, which is but slightly soluble in water, and therefore inert.

Narcotine and *codeine* are other compounds, of a similar character, procured from opium.

688. Quinia, or Quinine, $C_{38}H_{24}N_2O_4$.—Quinia is obtained only from the bark of certain species of a tree called *cinchona*, which grows chiefly in South America. In commerce it is called Peruvian bark, and is extensively used in medicine.

To prepare quinia, the bark, in powder, is digested in water, and from the solution obtained, the alkali is precipitated either by lime or ammonia. The quinia may then be dissolved in alcohol, and crystalized.

Quinia is a crystaline solid, slightly soluble in water, and intensely bitter. It combines readily with acids, as the sulphuric and hydrochloric; and the salts formed are extensively used in medicine, especially in certain fevers. With the sulphuric acid it forms two salts, a neutral and an acid sulphate.

689. Cinchonia, or Cinchonine, $C_{38}H_{24}N_2O_2$.—Cinchonia always accompanies quinia in Peruvian bark, and is obtained from it by a similar process. It differs in composition from quinia, in containing 2 atoms less of oxygen; but in most properties, the two substances are much alike. Alone, it is but slightly soluble in water, but the salts it forms with acids dissolve more readily, and are used in medicine.

Quinia and cinchonia exist together in the bark, the former being most abundant in that which is of a pale color, while the latter (cinchonia) occurs chiefly in the red bark.

QUESTIONS.—What are some of the properties of morphine? What other alkaloids are procured from opium? 688. From what is quinia prepared? Describe the mode of separating it from the bark? 689. In what does cinchonia differ from quinia? What is said of the color of the bark containing these alkaloids?

690. *Strychnia*, or *strychnine*, $C_{42}H_{22}N_2O_4$, is a vegetable alkali obtained from *nux vomica*, and other plants. It is the poisonous principle of the famous *Upas*, of the island of Java, of which so many fables are told. It forms an extensive series of salts with the acids, and is one of the most violent poisons known.

Brucia, or *brucine*, is a similar alkaline substance, obtained from the same source.

691. *Isatine*, $C_{16}H_5NO_4$, is derived from indigo by heating it with dilute nitric acid. It is a crystaline solid, of an orange-red color, soluble in hot water and in alcohol. From this substance *isatinic acid* is derived, and several other compounds.

Theine, and *Caffeine*, obtained from tea and coffee, appear to be the same substance. It is found also in the fruit of some other plants, and is contained in larger quantity in tea than in coffee. It is not certain that these articles, so extensively used among civilized nations, owe their peculiar properties to this principle.

Nicotine is an alkaloid obtained chiefly from the tobacco-leaf. *Piperine* is extracted from black pepper;—it is a crystaline solid, and acts as a feeble base.

Picrotoxine, from the coculus Indicus, *cantharadine*, from cantharides, *asparagine*, from the roots of asparagus and other plants, and *phloridzine*, from the fresh bark of the apple, pear, plum, and cherry trees, are compounds of a similar character.

ALKALOIDS OF THE ETHERS, OR CONJUGATED AMMONIAS.

692. The alkaloids of the ethers, like the ethers themselves, and the alcohols, are not found in nature, but are produced only by artificial means. They closely resemble ammonia in most of their properties, and indeed may be considered as ammonia in which one or more atoms of hydrogen has been replaced by equivalent quantities of the compound radicals, ethyle, C_4H_5, methyle, C_2H_3, acetyle, C_4H_3, &c. A few of these only can be described; and in doing so we shall find it advantageous occasionally to represent ethyle, C_4H_5, by Et; methyle, C_2H_3, by Me; amyle, $C_{10}H_{11}$, by Ayl; acetyle, C_4H_3, by Ae, &c.

QUESTIONS.—690. What is said of strychnia? From what is it procured? What other alkaloid accompanies it? 691. From what is isatine obtained? What is said of theine and caffeine? What other alkaloids are mentioned? 692. Are the alkaloids of the ethers found in nature? What do they resemble? From what may they be considered as formed?

693. Ethylamine, $C_4H_7N = NH_2C_4H_5 = NH_2Et$.—Ethylamine is formed by various processes, as by the action of strong aqua ammoniæ upon hydrobromic ether (bromide of ethyle), and then distilling from potash or lime. The ammonia and ether are put together into a glass tube, which is sealed hermetically, and immersed fifteen minutes in boiling water;—the reactions are as follows, viz.:

$$C_4H_5Br + NH_3 = C_4H_7N,HBr = NH_3,C_4H_5,Br.$$

If, as intimated above, we consider ethylamine a conjugated ammonia, that is, ammonia in which 1 atom of hydrogen is replaced by ethyle, then it is evident the compound, $NH_3C_4H_5 = NH_3Et$, is ammonium in which 1 atom of hydrogen is replaced by ethyle. It is called *ethylammonium*. The compound, NH_3, C_4H_5, Br, may therefore be called bromide of ethylammonium; and this when distilled with lime or an alkali yields ethylamine, precisely as sal-ammoniac, NH_4Cl, distilled with lime, yields ammonia (224).

Ethylamine is a colorless, limpid liquid, having a density of about 0·696, and boiling at about 68°. It has a pungent odor like ammonia, and alkaline reaction, forming salts with acids. With hydrochloric acid it forms dense white fumes, in the same manner as ammonia.

694. By the proper modes, 2, 3, and even 4 atoms of hydrogen in ammonia and ammonium may be replaced, producing the compounds, *biethylamine*, $NH(C_4H_5)_2 = NHEt_2$, *triethylamine*, $N(C_4H_5)_3 = NEt_3$, and *tetrethylamine*, $N(C_4H_5)_4 = NEt_4$. The latter, called also *tetrethylammonium*, like its prototype, ammonium, can only be obtained in combination, as an iodide, bromide, or chloride.

695. Methylamine, $C_2H_5N = NH_2C_2H_3 = NH_2Me$.—This compound is prepared in the same manner as ethylamine, simply substituting in the process a methylic ether. It is a colorless gas, having a strong ammoniacal odor, and alkaline reaction. By a cold a little below zero, it is condensed to a limpid, mobile liquid. In the gaseous state, it is more soluble in water than

---

QUESTIONS.—693. What is the composition of ethylamine? What is ethylammonium? Describe the properties of ethylamine. 694. What is biethylamine? Triethylamine? Tetrethylamine? 695. How is methylamine formed? In what does it differ from ethylamine?

any other known gas, water at 70° taking up 1000 times its own volume of the gas.

696. *Bimethylamine*, $NH(C_2H_3)_2 = NHMe_2$, *trimethylamine*, $N(C_2H_3)_3 = NMe_3$, and *tetramethylamine*, $N(C_2H_3)_4 = NMe_4$, sustain the same relations in the methylic compounds, as those of corresponding name, above described, in the ethyle series.

697. **Amylamine,** $C_{10}H_{13}N = NH_2C_{10}H_{11} = NH_2Ayl$. — Amylamine is prepared by a process altogether similar to those above given for preparing the corresponding compounds in the ethyle and methyle series. It is a limpid liquid, which boils at a little above 200°, and has a density of 0·750. It has the odor of ammonia, and is strongly basic, forming salts with the acids.

Biamylamine, &c., corresponding to ethyle and methyle compounds, above described, are known.

698. **Aniline,** $C_{12}H_7N = NH_2,C_{12}H_5$, is obtained from phenol, and also by the action of nitric acid upon indigo. It is an oily liquid, having a density of 1·028, and is distinctly alkaline, forming salts with the acids. It may be considered as ammonia in which 1 equivalent of the hydrogen is replaced by the group $C_{12}H_5$, called phenyle. The alkaloid might therefore be named *phenylamine*.

699. We may even have compounds answering to ammonia in which the three atoms of hydrogen have been replaced successively by different groups, or compound radicals, as in *ethylmethylphenylamine*, $N,C_2H_3,C_4H_3,C_{12}H_5 = N,MeEt,Pyl$.

700. In all these compounds, it plainly appears that the groups, or compound radicals, ethyle, C_4H_5, methyle, C_2H_3, &c., are equivalent to H, and replace it, in ammonia, NH_3, and ammonium, NH_4. Now, phosphorus, P, is equivalent in combination to N; and without experiment we might expect P to replace N in ammonia, as it does in the compound called phosphuretted hydrogen, PH_3. So a series of compounds has been discovered, answering to ammonia, with its hydrogen replaced by the groups, ethyle, methyle, &c., and the nitrogen by phosphorus. Thus we have the compounds, $P(C_2H_3)_3 = PMe_3$, $P(C_4H_5)_3 = PEt_3$, &c.

QUESTIONS.—696. What is bimethylamine? 697. Describe amylamine. 698. Describe aniline. Why may we name it phenylamine? 699. What is the composition of ethylmethylphenylamine? 700. May the nitrogen in these compounds be replaced by phosphorus?

701. The nitrogen in these compounds may even be replaced by a metal, as antimony, bismuth, or tin, as in the following examples.

Stibethyle, $Sb,C_{12}H_{15}=Sb(C_4H_5)_3=SbEt_3$.—Stibethyle is prepared by the action of hydriodic or hydrobromic ether upon antimonide of potassium; and afterwards distilling carefully in an atmosphere of carbonic acid. It is a transparent, colorless liquid which boils at 317°, and has a density of 1·33. A drop of it exposed in the open air, first emits dense, white fumes, and then takes fire spontaneously. Stibethyle combines readily with oxygen, sulphur, chlorine, and other elements, forming both binary compounds and proper salts.

702. *Stibmethyle,* $Sb(C_2H_3)_3 = SbMe_3$, is a corresponding compound, containing antimony and methyle; as is also bismethyle, $Bi(C_4H_5)_3 = BiEt_3$, of bismuth and ethyle. These latter, and many others of similar character, though formed on the type of ammonia, have less the characters of ammonia, and react with chemical agents more like simple metals. They are therefore sometimes called conjugate metals.

All the preceding compounds have the same molecular type as ammonia and ammonium; and it is worthy of notice that those having the type of ammonium, like ammonium itself, have been obtained only in combination.

703. There are other compounds similar in their general characters to the above, but the molecular type of ammonia does not appear in them. Of this character are such compounds as the following, viz., *bisethyle,* $Bi,C_4H_5 = Bi,Et$; *zincethyle,* $Zn,C_4H_5 = Zn,Et$; *stannethyle,* $Sn,C_4H_5 = Sn,Et$; *hydrargethyle,* $Hg_2C_4H_5$, &c. Each of these (and many others) combine with oxygen, sulphur, chlorine, &c., forming binary compounds; and these again uniting with other compounds form salts.

704. Cacodyle, $As,C_4H_6 = As,Me_2$. —This substance (named from the Greek, *kakos,* evil, and *ule,* principle,) serves as the basis of a long series of compounds, in which it performs the part of a quasi-metal. By distilling a mixture of equal parts of acetate of potash and arsenious acid, we obtain a liquid substance long

---

QUESTIONS.—701. May the nitrogen be replaced by certain metals? What is the composition of stibethyle? Describe its properties. 702. Describe stibmethyle. 704. From what does cacodyle receive its name?

known as *Cadet's fuming liquid.* Its composition is As,Me_2O, which corresponds to oxide of cacodyle, but it is more frequently called *alcarsine.* When pure it is a colorless liquid, which is highly corrosive and poisonous.

According to Regnault, gaseous alcarsine has a density of 7·8; by calculation we make it 7·824. Its equivalent, as given above, represents 2 volumes, and its density is calculated as follows:

4 vols. carbon vapor weigh		(·836 × 4)	3·344
12 " hydrogen	"	(·069 × 12)	·828
1 " arsenic vapor	"		10·370
1 " oxygen	"		1·106
Thus giving for 2 vols. alcarsine vapor,			15·648
The weight of 1 vol., or its density, therefore is			7·824

705. Alcarsine, by digestion in hydrochloric acid, is converted into chloride of cacodyle, $AsMe_2,Cl$, which is decomposed by a metal, as iron or zinc, and yields cacodyle, As,C_4H_6, which is a colorless, transparent liquid, having a density a little above that of water, and boiling at about 338°. In the open air it takes fire spontaneously; and even when covered by a film of water it gradually absorbs oxygen from the air, being converted first into *alcarsine*, As,Me_2O, and then into *alcargen*, or *cacodylic acid*, As,Me_2,O_3.

Cacodyle forms important binary compounds with oxygen, chlorine, sulphur, &c.; and from these many interesting salts are derived, not here described.

ORGANIC COLORING-MATTERS.

706. Color is a property which pertains to every substance; but it is proposed to treat here of certain compounds noted for their color, and the practical uses made of them in the art of coloring.

Infinite variety exists in the colors of organic substances; but

QUESTIONS.—How is Cadet's fuming liquid obtained? What is it composed of? What is it sometimes called? What is the density of alcarsine vapor? 705. What is formed when alcarsine is digested in hydrochloric acid? How is cacodyle obtained from its chloride? Describe cacodyle. What is said of the compounds of cacodyle? 706. What are to be treated of under the head of organic coloring-matters? What is said of the great variety of colors presented in nature?

the prevailing tints are red, yellow, blue, and green, or mixtures of these colors.

707. The *art of dyeing* consists in attaching the different coloring-matters to the fabrics to be colored. This is accomplished in various modes, as by adding some substance that forms an insoluble compound with the coloring-matter, or one that has the property of causing the coloring-matter to attach itself permanently to the fibre of the cloth. A substance used for this last purpose is called a *mordant.*

As a general rule, substances of an animal origin, as wool and silk, are more easily colored than those composed of vegetable matter, as cotton or flax.

Many coloring-matters combine readily with certain of the metallic oxides, as those of aluminum, tin, &c., forming insoluble compounds, called *lakes*, which are much used for painting in oils.

Nearly all colors are injuriously affected by the action of light; or, in popular language, are said to fade. This, it is believed, is generally occasioned by the absorption of oxygen, which causes a decomposition of the coloring substance.

708. *Indigo*, $C_{16}H_5NO_2$.—This is a beautiful blue coloring-substance, extracted by a kind of fermenting process, from the leaves of several different American and Asiatic plants, as the *isatis tinctoria, polygonum tinctorium, gymnema tingens,* &c.

The leaves of the plant are digested for eight or ten days in water, during which a yellow coloring-matter appears, and is held in solution by the water. This is then drawn off, and, by standing, absorbs oxygen from the air, and gradually becomes blue, and is deposited as a thick sediment. This is collected, and pressed in the form of square blocks, and constitutes the indigo of commerce.

Indigo of commerce is far from being pure;—it is a mixture of indigo with other matters. Pure indigo — sometimes called *indigotine*—may be obtained from it by sublimation with a gentle heat, and by other means.

Pure indigo is an insipid, inodorous substance, insoluble in water, but soluble in small quantity in alcohol. In strong sul-

---

QUESTIONS.—707. In what does the art of dyeing consist? What is a mordant? What are *lakes?* What occasions the fading of colors? 708. From what is indigo obtained? Describe the process of preparing it. What is said of the action of sulphuric acid upon indigo?

phuric acid it dissolves readily, with the formation of several new coupled acids, as the *sulphidigotic*, *sulphindylic*, &c. Alkalies act readily upon it, producing many important compounds, some of which have been described.

Deoxydizing bodies, as protosulphate of iron, protochloride of tin, &c., have the effect to destroy entirely the color of the above indigo—called often *indigo blue*—and produce another compound, called *indigo white*. This unites readily with bases, and forms several soluble compounds which serve admirably for coloring. The solutions, though yellowish at first, gradually become blue by exposure to the air; and articles impregnated with the solution undergo the same change.

Solution of indigo in sulphuric acid is much used for coloring the *Saxon blue*.

709. *Litmus*, *archil*, or *turnsol*, is a coloring substance resembling indigo, obtained from several species of lichen, as the *leconora parella*, and the *rocella tinctoria*. It is seen in small cubical masses, which are partially soluble in water, and the solution communicates a beautiful blue to substances immersed in it. This substance is much used in chemical investigations for detecting the presence of acids and bases; the former of which change its blue color to red, and the latter again restore the blue. It is a compound of several principles, as *lecanorine*, *orcine*, *roccelinine*, &c.

710. *Madder* is a red coloring-substance, obtained from the roots of a plant called *rubia tinctoria*. It contains several principles, the most important of which is a red crystaline compound which has received the name *alizarine*, $C_{30}H_8O_8$. Some of the other principles obtained from it are *madder purple*, or *purpurine*, $C_{28}H_{10}O_{15}$, *madder red*, $C_{28}H_9O_9$, and *xanthine*.

By means of different mordants, madder is made to produce various shades of red, purple, brown, and orange. The beautiful crimson, called *Turkey-red*, is produced by it by means of a complicated process which has not been fully explained.

QUESTIONS.—What is the effect of deoxydizing bodies upon indigo? 709. From what is litmus procured? What use is made of it? 710. From what is madder procured?

711. *Cochineal.*—This is a dried insect, which was named *coctus cacti* by Linnæus. It is a native of Mexico, and feeds upon a species of the cactus.

As seen in commerce, it is of a dark brown, inclining to red, but when macerated in water, it yields a very beautiful red coloring-matter, which is much used in dyeing. The beautiful red pigment called *carmine* is prepared from cochineal by boiling in clean water, and adding a little alum;—as the solution cools, the carmine is precipitated, and is collected and dried.

712. *Brezeline* is a crystaline solid, of an orange color, obtained from *Brazil-wood*, which is soluble in water and alcohol. With mordants, it gives a beautiful red. The same substance is contained in the African wood called *cam-wood.*

Hematoxyline, $C_{16}H_7O_6$.—This coloring-matter is procured from the wood of the tree called *hæmotoxylon Campeachianum*, and frequently, in commerce, *log-wood.* It is separated from the wood by digestion in water, and may be obtained in crystals. This substance, with salts of iron, gives a permanent black, and with other mordants, different shades of purple or red.

713. *Quercitrine*, $C_{16}H_8O_9,HO$, is a yellow coloring-matter, contained in the bark of the *quercus tinctoria*, and probably in other vegetable substances used in coloring yellow. A fine yellow is also produced by the *turmeric root*, and by the root of the common barberry (*berberis vulgaris*).

Chlorophyle.—This name has been applied to the green coloring-matter of the leaves of plants. It is contained only in those parts of plants which are exposed to the light, and this agent may therefore be supposed to have an important influence in its formation. It is a substance of a waxy nature, and is soluble in alcohol, but not in water.

The above are the most important coloring-substances obtained from vegetable bodies; by their combination with each other, and by the use of different mordants, in different modes, all the endless variety of tints are produced.

THE AMIDES AND NITRILES.

714. Amides.—An amide is a nitrogenized compound answering to a salt of ammonia less one (or more) atoms of water (or its elements). Thus, oxamide (731), $C_2O_2,NH_2=C_2O_3,HO,NH_3-2HO$;—that is, it answers to oxalate of ammonia less 2 atoms of water. So benzamide, $C_{14}H_7NO_2 = C_{14}H_5O_2,NH_2 = C_{14}H_5O_3,HO,NH_3$—

QUESTIONS.—711. What is cochineal? 712. From what is brezeline obtained? What is hematoxyline? 713. What is quercitrine? Describe chlorophyle. 714. Describe an amide.

2HO;—that is, it answers to benzoate of ammonia less an atom of water, or, less 2 atoms of water, if we have regard to the basic water of the salt.

It is also characteristic of the amides, that they are capable of resuming the water lost so as to regenerate the ammoniacal salt; or if a fixed alkali is used, a salt of this alkali is formed, and ammonia set free. Thus, when benzamide is treated with a boiling solution of potassa, benzoate of potassa is formed, and ammonia, being gaseous, escapes.

715. The amides are formed by several different modes, as by the simple distillation of a salt of ammonia, by the action of aqua ammoniæ upon the compound ethers, by the action of ammonia upon certain chlorides, &c. Oxamide, for instance, is formed by the distillation of oxalate of ammonia, but it may also be prepared by the action of aqua ammoniæ upon oxalic ether (616), alcohol being also at the same time reproduced from the ether. Thus,

$$C_4H_5O_2,C_2O_3 + NH_3 = C_4H_6O_2 + C_2O_2,NH_2.$$

Benzamide is usually prepared by the process last mentioned from the chloride of benzyle. Thus,

$$C_{14}H_5ClO_2 + 2NH_3 = C_{14}H_5O_2, + NH_4Cl.$$

It is solid, and capable of being crystalized;—at a temperature of about 240° it melts, and at a still higher temperature may be sublimed.

Oxamide also, mentioned above, is solid, and without odor or taste. At a high temperature it may be sublimed, but some of it will be decomposed.

716. *Malamide*, $C_8H_8N_2O_6 = C_8H_4O_6,2NH_2$, is the amide of malic acid (683), and is identical in composition with *asparagine*, a substance found in the young shoots of asparagus, in liquorice root, and the root of marsh-mallow, and in several other plants. It may be obtained in crystals, which require about 60 times their weight of water for solution. By the action of acids, it is converted into ammonia and *aspartic acid*, $C_8H_7NO_8$ =

QUESTIONS.—How may the ammoniacal salt be regenerated? 715. What are some of the modes by which the amides are formed? Describe the mode of preparing benzamide. 716. What is malamide, or asparagine?

$C_8H_5NO_6,2HO$. This acid is also to be considered as an amide, formed from the acid malate of ammonia by the loss of an atom of water. Thus,

$$C_8H_4O_8,HO,NH_3—HO = C_8H_5O_8,NH_2 = C_8H_5NO_6,2HO.$$

The above examples are given as illustrative of the mode in which bodies of this class are formed, and of their relation to the ammoniacal salts; many are known that cannot be here mentioned. Most of them are neutral in their reactions, but they may be acids—as aspartic acid—or bases. Those that are acid, are usually formed from acid salts of ammonia.

717. Nitriles.—Many of the amides are capable of parting with still another atom of water, being then converted into compounds called *nitriles*. There are not many known; and they are not important.

Acetonitrile, C_4H_3N, is formed from *acetamide*, $C_4H_3O_2,NH_2$, by distilling the latter with anhydrous phosphoric acid, which causes it to part with 2 atoms of water. Thus,

$$C_4H_3O_2,NH_2 — 2HO = C_4H_3N.$$

Butyronitrile, C_8H_7N, is an oily liquid, which boils at about 245°. It is prepared by distilling butyrate of ammonia, or butyramide, with anhydrous phosphoric acid, which separates an atom of water. *Valero-nitrile*, $C_{10}H_9N$, is obtained in a similar manner from valeramide.

CYANOGEN, AND ITS COMPOUNDS.

Symbol, C_2N, or Cy; *Equivalent* (12 + 14 =)26; *Density*, 1·82.

718. History.—Cyanogen has already been mentioned (315). It was discovered by Gay Lussac, in 1814, and in 1817 was recognized by Berzelius as a compound radical (536). In many of its properties it resembles chlorine, bromine, and iodine, combining like these with other simple substances, and forming compounds called *cyanides*.

Like the elements above mentioned, also, it combines with oxygen, forming oxacids, and with hydrogen forming a hydracid;

QUESTIONS.—Describe the mode of preparing aspartic acid. Are there acid and basic as well as neutral amides? 717. How are the nitriles formed? 718. When was cyanogen discovered?

and in many other respects, it reacts precisely as a simple substance.

719. Preparation. — Cyanogen may be prepared by several modes. Its elements do not combine directly, but a cyanide is first formed, and this being decomposed yields the pure cyanogen. The best method to procure a small quantity, is to heat gently in a small retort cyanide of mercury, and collect the cyanogen over mercury. At the same time a brown mass will be formed, and remain in the retort, which has received the name of *para-cyanogen.* It is isomeric with cyanogen.

720. Properties.—Cyanogen is a colorless gas, of a density 1·82, and has a peculiar, but not disagreeable odor, resembling that of wild-cherry water. By a temperature of —4°, or by a pressure of 4 atmospheres, at the ordinary summer temperature, it is converted into a liquid.

The following method serves well to prepare a small quantity in the liquid form. Introduce into a strong glass tube, bent as in the figure, a little cyanide of mercury, and seal it hermetically. It is then to be held horizontally, and the heat of a lamp applied at the extremity, *a*, containing the cyanide of silver. The other extremity, *b*, is to be kept cool; and in a short time the liquid cyanogen will be seen to collect in it.

Preparation of Cyanogen.

In the liquid form, cyanogen is colorless and limpid, and has a density of about 0·9. By long keeping, it undergoes a change, forming *para-cyanogen.*

Compounds of Cyanogen and Oxygen.

721. Cyanogen and oxygen form no less than three isomeric compounds, all of which are acids.

722. Cyanic Acid, CyO,HO. — Cyanic acid is always formed when an alkaline cyanide, as cyanide of potassium, is exposed to

QUESTIONS.—What simple substances does cyanogen resemble in many of its properties? 719. How is cyanogen prepared? Describe para-cyanogen. 720. What are the properties of cyanogen? May it be obtained as a liquid? May it be preserved in the liquid form? 721. How many compounds of cyanogen and oxygen are known? 722. Describe cyanic acid.

the air at a red heat; oxygen is absorbed, and the acid, as it is formed, combines with the alkali. But it is best prepared by heating gently cyanuric acid, a compound to be hereafter described.

It is a clear transparent liquid, with a penetrating odor not unlike that of acetic acid, and is very corrosive to the flesh.

Cyanates.—Cyanate of potash is prepared by heating the yellow prussiate of potash (soon to be described) with peroxide of manganese, and digesting the mass in alcohol. It is a white crystaline solid. *Cyanate of ammonia*, which is isomeric with *urea*, is formed by bringing together hydrocyanic acid gas and ammonia, and by the action of cyanate of potash upon sulphate of ammonia. It is also contained in the urine of animals, and is therefore called urea.

723. **Cyanic Ether,** C_4H_5O,CyO.—Cyanic ether is formed by distilling a mixture of cyanate of potassa, and the sulphovinate (607) of the same base. It is a liquid less dense than water, and has a pungent, irritating odor. Dissolved in ammonia, it forms a crystaline compound, $C_6H_8N_2O_2$, which has been called *ammoniacal cyanic ether*. When this substance is treated with water, carbonic acid is given off, and a new substance formed, called *cyamethene*, having the composition, $C_{10}H_{12}N_2O_2$.

Treated with solution of caustic potash, cyanic ether forms 2 equivalents of carbonate of potash, and is transformed into a new compound, called *ethylamine*, $C_4H_7N = NH_2,C_4H_5$, which in its properties closely resembles the vegetable alkaloids (693).

724. **Fulminic Acid,** $2CyO,2HO$.—Fulminic acid has not as yet been obtained in a separate state, nor even in combination with water. It is formed by pouring alcohol into a strongly acid solution of nitrate of mercury or silver, and applying a little heat, if necessary, so as to produce brisk ebullition. The acid, as it forms, combines with the oxide of the metal used, and the metallic salt formed is at once precipitated. By digesting a solution of the metallic fulminate with another base, as potassa, the fulminate is decomposed, and the acid transferred to the new base.

This acid receives its name from the tendency of its salts to explode violently by heat or friction, or other causes.

Fulminic acid is bibasic, and, like other acids of this character, forms two series of salts, the neutral, which contains 2 equivalents of fixed base to each equivalent of the acid, and the acid salts which, for each equivalent of acid, contain 1 equivalent of fixed base and 1 equivalent of water.

QUESTIONS.—What is *urea?* 723. How is cyanic ether formed? 724. Has fulminic acid been obtained in a separate state?

725. *The Fulminates.*—There are several fulminates known, but those of mercury and silver are the most important. The former, *fulminate of mercury*, is prepared by dissolving 200 grains of mercury in an ounce of strong nitric acid by the aid of heat, and, when cold, pouring into it 4 or 6 ounces of alcohol. The mixture should be contained in a large glass or earthen vessel, and a little heat applied, if the action does not commence in a few minutes after adding the alcohol. When the action has once commenced, it goes on violently, without the further aid of heat, attended by the evolution of white fumes, consisting of aldehyde, acetic, formic, and nitrous acids, and their corresponding vinic ethers, and the deposition of the solid fulminate in small crystals.

The powder thus formed should be immediately well washed and dried, and may then be preserved for any length of time, if kept in the dark, but is acted upon slightly by light.

Fulminate of mercury is a brown crystaline powder, without odor, but having a styptic, metallic taste. By slight friction with any hard substance, or by a blow, or by the contact of the minutest quantity of nitric or sulphuric acid, it explodes with great violence;—even when moist, it will often explode with a slight blow, and should therefore never be touched with anything harder than wood or paper.

726. Fulminate of mercury is much used for filling percussion caps, for exploding fire-arms; and for this purpose it is mixed with a little less than half its weight of nitre, and some solution of resin in alcohol, to cause it to adhere to the capsule, and to prevent the action of the air.

727. *Fulminate of Silver* is prepared very nearly in the same manner as the above. A dime is to be dissolved in about 2 ounces of common nitric acid, and the solution diluted with 2 ounces of distilled water, and then 2 ounces of alcohol added, and the whole mixed well by shaking. By application of heat, rapid ebullition will soon commence, when the heat should be removed; and the fulminate will be gradually deposited, and should be immediately washed and dried. It is a beautiful white solid, and should be handled with the utmost care, as it explodes most violently, even when damp, by the slightest friction, or by the mere contact of sulphuric acid.

QUESTIONS.—725. Describe the mode of preparing fulminate of mercury. Describe its properties. 726. What use is made of it? 727. How is fulminate of silver prepared? What is its character?

It is very improperly made the basis of a small toy called a *torpedo*, which consists of a little of the salt mixed with some gravel, and done up in paper. It explodes merely by being thrown upon a hard substance.

Fulminates of copper and zinc may be prepared by digesting the fulminates of mercury or silver with these metals.

728. *Cyanuric Acid*, 3CyO,3HO.—Cyanuric acid is a white crystaline solid, with little taste or odor, and may be sublimed; but a part of it is by the operation converted into cyanic acid. It is best formed by heating urea, and digesting the mass in strong sulphuric acid, adding a few drops of nitric acid until the solution becomes clear. To the whole then add an equal volume of water, and as it cools the acid will be deposited in smal crystals.

It is a tribasic acid, as the above formula indicates, and, like other tribasic acids, forms with bases three series of salts.

Cyanidide is an isomeric modification of this acid.

729. **Cyanuric Ether**, $3C_4H_5O,3CyO$, is prepared by distilling carefully a mixture of the cyanurate and sulphovinate of potassa. It is solid at ordinary temperatures, but melts at about 185°, and boils at 529°.

Boiled for some time with an alcoholic solution of potash it is decomposed, yielding alcohol, ammonia, and carbonic acid, according to the following equation:

$$3C_4H_5O,3CyO + 12HO = 3C_4H_6O_2 + 3NH_3 + 6CO_2.$$

Compound of Cyanogen and Hydrogen.

730. Cyanogen and hydrogen form one compound only; which, however, is exceedingly important.

Hydrocyanic Acid (Prussic Acid), C_2N,H, or HCy.—Hydrocyanic acid is prepared by various processes, one of which is to decompose cyanide of mercury by concentrated hydrochloric acid, by the aid of heat. The mixed gases are passed through a tube containing first pieces of carbonate of lime, to separate any free hydrochloric acid, and then fused chloride of calcium to absorb all the moisture; and the hydrocyanic acid, mixed now

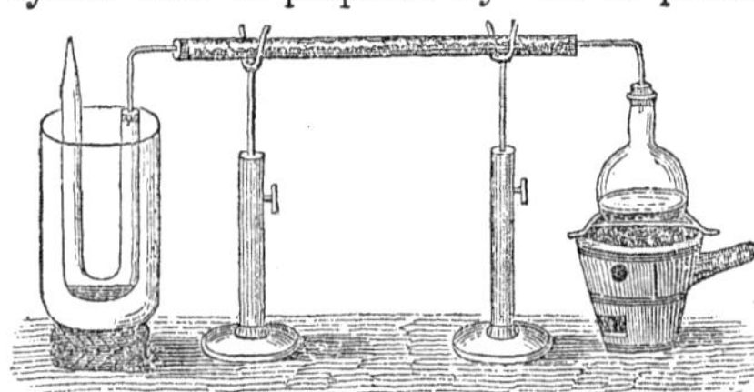

Preparation of Hydrocyanic Acid.

QUESTIONS.—728. Describe cyanuric acid. 729. Describe cyanuric ether. 730. How is hydrocyanic acid prepared?

with only carbonic acid, is condensed in a bent tube, surrounded by a freezing mixture, the carbonic acid escaping into the air.

Pure hydrocyanic acid is a limpid, colorless liquid, of a strong odor, similar to that of peach-blossoms. Its specific gravity is about 0·70. Its point of ebullition is 80°, and at 5° it congeals. When a drop of it is placed on a piece of glass, a part of it becomes solid, because the cold produced by the evaporation of a portion is so great as to freeze the remainder. It unites with water and alcohol in every proportion.

Hydrocyanic acid is a powerful poison, producing, in poisonous doses, insensibility and convulsions, which are speedily followed by death. A single drop of it placed on the tongue of a dog causes death in the course of a very few seconds; and small animals, when confined in its vapor, are rapidly destroyed.

The pure acid decomposes spontaneously when long kept, especially if exposed to the light; but if largely diluted with water it may be preserved, and is in this state used in medicine. It is contained in water distilled from the blossoms and leaves of the peach and other allied fruit trees.

Sulphocyanates or Sulphocyanides.

731. The *sulphocyanates* or *sulphocyanides* constitute a class of compounds corresponding to the cyanates with the oxygen replaced by sulphur. Thus the composition of cyanate of potassa is KO,CyO, while that of the sulphocyanate is $KCyS_2 = KS,CyS$.

732. Sulphocyanate of Potassium, $K,CyS_2 = KS,CyS$.—This is properly a sulphur salt. It is obtained by making a mixture of 46 parts of ferrocyanide of potassium, 17 parts of pearlash, and 16 of sulphur, and melting them together; and when the whole has cooled, dissolving out the sulphocyanate by boiling alcohol.

It is a solid, and crystalizes in long prisms, which are without color, and deliquesce in the open air. It is used as a delicate test of iron.

733. Sulphocyanic Acid, $HCyS_2 = HS,CyS$.—This acid, called also *hydrosulphocyanic acid*, is prepared by distilling sulphocyanate of potassa with phosphoric acid. It is a colorless liquid, of a sour taste, and forms with peroxide of iron salts of a deep red color.

QUESTIONS.—What are the properties of hydrocyanic acid? Is it used in medical practice? 731. What is said of the sulphocyanates? 732. How is sulphocyanate of potassium formed? 733. Sulphocyanic acid?

734. The following compounds are derived mostly from the sulphocyanates:

Mellon, C_6N_4, is a yellow insoluble compound, obtained by distilling sulphocyanate of potassium in an atmosphere of chlorine. It is capable of combining with metals, forming compounds which are called mellonides. *Melam*, $C_{12}H_{11}N_9$, is a grayish-white powder, procured by decomposing, by heat, the sulphocyanate of ammonia. *Melamine*, $C_6H_6N_6$, results from the action of solution of potash, at a boiling heat, upon melam. It is a transparent solid, scarcely soluble in water, if cold, but very soluble in boiling water. When dry it may be sublimed. At the same time with the compound just described is formed *ammeline*, $C_6H_5N_5O_2$, a basic substance, capable of combining with the nitric and other acids, to form salts. It crystalizes in fine silky needles. *Ammelide*, $C_{12}H_9N_9O_6$, is formed by boiling ammeline in some acid;—it is a white compound, insoluble in water and alcohol, and possesses little interest.

Compounds of Cyanogen and the Metals.

735. Cyanogen combines readily with nearly all the metals; but a few only of the more important of the compounds thus formed can be here noticed.

736. **Cyanide of Potassium**, K,C_2N, or KCy.—This compound is best prepared by heating in a covered crucible 8 parts of yellow prussiate of potash, and adding three parts of pearlash, both being first well dried. On the large scale for the manufacture of prussiate of potash above mentioned, it is prepared by heating together in an air-furnace animal charcoal and pearlash. This cyanide thus formed, acting afterwards upon iron, or oxide of iron present, produces the yellow prussiate soon to be described.

Cyanide of potassium is largely used at the present time in the processes of gilding and plating by means of galvanism.

737. *Cyanide of Mercury*, $HgCy$, is prepared by pouring a hot solution of nitrate of silver into a hot concentrated solution of cyanide of potassium, and purifying, by recrystalization, the solid cyanide of mercury which is deposited when the mixed solution cools. It has already been mentioned as affording a ready means for procuring cyanogen.

738. *Cyanide of Silver*, $AgCy$, is precipitated as a solid compound by mixing hydrocyanic acid with solution of any salt of silver. It is inso-

---

QUESTIONS.—734. What other compounds are mentioned as derived from the sulphocyanates? 735. What is said of cyanogen in relation to the metals? 736. How is cyanide of potassium formed? What use is made of it? 737. What is said of cyanide of mercury? 738. Cyanide of silver?

luble in water, but is soluble in aqua ammonia, and in solution of nitrate of silver.

By modes altogether similar to the above, the cyanides of gold, zinc, cobalt, nickel, lead, copper, palladium, &c., may be procured, but the compounds are not here described.

Double Cyanides.—Polycyanides.

739. Nearly all the metallic cyanides are remarkable for thei. tendency to combine with each other, so as to form double cyanides. This is especially true of the cyanides of iron, but many others of the class show the same tendency.

740. Hydroferrocyanic Acid, $FeCy,2HCy=H_2Cy_3Fe$.—This acid is best prepared by mixing a saturated solution of yellow prussiate of potash with strong hydrochloric acid, free from contact with the air, and then adding a small quantity of sulphuric ether. The acid is at once precipitated in small white crystals, which are to be collected and dried in a vacuum, after being washed with ether.

This acid may be preserved any length of time free from contact with the air and moisture; but in the air it is gradually decomposed, leaving a residue of Prussian blue. It acts readily upon the metals and metallic oxides, producing ferrocyanides.

741. Ferrocyanide of Potassium.—Yellow Prussiate of Potassa—$2KCy,FeCy,3HO=K_2Cy_3Fe,3HO$.—This compound is prepared, on the large scale, by heating potash with animal matter, as horns, hoofs, leather, woollen rags, hair, and other animal offal, by which cyanide of potassium is formed; which, being digested with warm water, containing fragments of iron or smithy scales, acts upon the iron to produce the compound in question. After filtering, the yellow Prussiate, in solution, is set aside to crystalize. Instead of animal substances, as here described, common charcoal alone may be used with the carbonate of potash, and bringing in contact with it when intensely heated atmospheric nitrogen, prepared by passing a current of air through burning coke.

QUESTIONS.—739. What is said of nearly all the metallic cyanides? 740. How is hydroferrocyanic acid prepared? 741. Describe ferrocyanide of potassium.

Ferrocyanide of potassium is a crystaline solid, of a lemon-yellow color, very soluble in water, but insoluble in alcohol and ether. It is much used in the arts as a coloring substance, and as an important reagent in the laboratory of the chemist.

Many other ferrocyanides, as those of sodium, barium, calcium, &c., must be passed by unnoticed.

742. Hydroferridcyanic Acid, $3HCy,Fe_2Cy_3 = H_3Cy_6,Fe_2$.—This compound is prepared by passing a current of hydro-sulphuric acid gas through a newly prepared solution of ferridcyanide of lead. It is obtained in crystals, which are soluble in water, and have an acid, astringent taste.

It may be considered as a compound of 3 equivalents of hydrocyanic acid with sesquicyanide of iron, as indicated by the first of the above formulæ, or as a compound of 3 equivalents of hydrogen with the assumed radical Cy_6Fe_2, as indicated by the second formula.

743. Ferridcyanide of Potassium.—Red Prussiate of Potash, $3KCy,Fe_2Cy_3 = K_3Cy_6Fe_2$.—Red prussiate of potash is prepared by passing a current of chlorine through a solution of the yellow prussiate, until it ceases to give a blue precipitate with solution of the persalts of iron, and evaporating the solution until crystalization takes place.

The action of the chlorine is to separate from 2 equivalents of the ferrocyanide one equivalent of potassium, thus forming chloride of potassium and the compound in question. Thus $2(2KCy,FeCy) + Cl = 3KCy,Fe_2Cy_3 + KCl$, the water always contained in the crystal of the ferrocyanide being omitted.

The crystals belong to the trimetric system, are of a red color, and readily burn when held in the flame of a candle. They are partially soluble in cold and very soluble in boiling water, but insoluble in alcohol.

Sometimes, during the process, the solution becomes green, by the formation of the magnetic cyanide of iron ($FeCy,Fe_2Cy_3$), which, however, disappears by boiling with a little caustic potassa.

Ferridcyanide of potassium forms characteristic colors with solutions of several of the metals, producing with them precipitates in which the

---

QUESTIONS.—What use is made of ferrocyanide of potassium? 742. How may hydroferridcyanic acid be procured? Of what is it composed? 743. Describe the red Prussiate of potash. What is said of the colors produced by this compound with metallic solutions?

potassium of the ferridcyanide is replaced by an equal number of equivalents of the new metal. Thus, with solution of the protosalts of iron it gives a blue, with those of uranium a reddish-brown, with those of nickel and bismuth a brownish-yellow, and with those of silver and zinc an orange yellow.

Similar compounds of sodium, barium, calcium, and magnesium are known.

744. Ferridcyanides of Iron.—Prussian Blues.—There are several different compounds known by these names. The one most common is prepared by precipitating an acid solution of sulphate of iron with solution of the yellow prussiate of potash. The compound thus formed is of a deep blue color, uncrystalizable, and insoluble in water and alcohol. When properly dried, it has a peculiar metallic, coppery lustre, and is much used by painters; but the color is liable to fade, and is at once destroyed by the action of the caustic alkalies which decompose it.

The production of the proper blue precipitate in this case requires the presence of the air, from which oxygen is largely absorbed. When a salt of the peroxide is made use of, this is not required.

The Prussian blue of commerce is very soluble in solution of oxalic acid, and a good blue writing-ink is prepared from it.

Prussian blue receives its name from the fact that it was first made at Berlin, Prussia, about the year 1710. It is sometimes called Berlin blue.

The compound just described as the ordinary Prussian blue of commerce, is probably a mixture of several different cyanides of iron.

Turnbull's or *Paris blue*, $3FeCy,Fe_2Cy_3$, is a definite compound; it is best prepared by mixing a solution of the red Prussiate of potash with solution of a protosalt of iron, the 3 equivalents of potassium in the Prussiate being replaced by 3 equivalents of iron. Its color is less intense but more clear than that of the ordinary Prussian blue. Its formula may be written $Fe_3Cy_6Fe_2$.

By precipitating solution of pernitrate or perchloride of iron with a solution of yellow Prussiate of potash, a definite compound is also formed sometimes called *neutral Prussian blue*, or *sesqui-ferricyanide of iron*, or *biferridcyanide of iron*. Its composition is $3FeCy,2Fe_2Cy_3$. This probably constitutes the chief part of the common Prussian blue of commerce.

QUESTIONS.—744. What is said of the Prussian blues? How may a blue writing ink be formed from Prussian blue? May other Prussian blues be formed?

Another compound, called *basic Prussian blue*, is formed by precipitating a protosalt of iron with solution of ferrocyanide of potassium, and exposing the white compound thus obtained for some time to the action of the air for the absorption of oxygen.

Several other similar cyanic compounds of iron are known, called *soluble Prussian blues*, but they cannot here be particularly described.

745. Cobaltocyanides.—Several other metals, as cobalt, nickel, and mercury, enter into combination with cyanogen, forming compounds in which these metals perform essentially the same office as the iron in those of the above list.

746. *Hydrocobaltocyanic Acid*, $3HCy,Co_2Cy_3 = H_3Cy_6Co_2$, is obtained by decomposing cobaltocyanide of lead by hydrosulphuric acid, separating the sulphide of lead by filtration, and then evaporating so as to crystalize. The crystals thus procured are colorless and fibrous, and very insoluble in water. The solution has a very sour taste, and acts readily upon the alkaline carbonates with effervescence.

It will be seen that this compound corresponds to the hydroferridcyanic acid of the iron series of cyanic compounds. Like that acid, too, it is tribasic.

747. *Cobalticyanide of Potassium*, $3KCy,Co_2Cy_3 = K_3Cy_6Co_2$, is formed by dissolving cobalt, or its carbonate, or cyanide, in solution of cyanide of potassium containing excess of hydrocyanic acid, and evaporating so as to crystalize. It will be noticed that the compound is altogether analogous in composition to the ferridcyanide of potassium, with which it is also isomorphous. The crystals are of a yellowish color, and are very soluble in water, forming a colorless solution.

Very many other metallic polycyanides are known, as the *platinocyanides, aurocyanides, argentocyanides, mercuriocyanides*, &c., but it would extend our work too much to describe them.

ALBUMINOUS, OR PROTEINE COMPOUNDS.

748. These compounds, three in number, are *fibrine, albumen*, and *caseine*. They are found both in vegetable and animal bodies, and, like sugar, starch, and woody-fibre, are capable, in certain circumstances, of being converted into each other. They are believed to be compounds of a proximate principle called

QUESTIONS.—745. What other metals form cyanides similar in their properties to those of iron? 746. Describe hydrocobaltocyanic acid. 747. Describe cobaltocyanide of potassium. Are there other metallic polycyanides? 748. What are the albuminous or proteine compounds?

proteine, (*proteuo*, I am first), and having the composition, $C_{40}H_{31}N_5O_{12}$. Its real composition, however, cannot be considered as settled.

To obtain proteine, albumen or caseine is to be digested successively in water, alcohol, and ether, in order to separate everything that is soluble in these liquids; and the mass that remains is to be soaked for a time in diluted hydrochloric acid, and then dissolved in a weak solution of caustic potash. From this, acetic acid, cautiously added, precipitates the proteine.

As thus obtained, proteine is a white, inodorous solid, which is insoluble in water or alcohol, and is capable of forming different compounds both with acids and bases. From its acid compounds it is readily precipitated by tannic acid, with which it forms an insoluble compound, by ferrocyanide of potassium, and by the alkalies.

In the open air it rapidly absorbs moisture, and becomes a gelatinous mass; and by heat is decomposed, exhibiting the phenomena usually attending the combustion of nitrogenized bodies. It leaves no residue after combustion.

749. Fibrine.—Gluten.—Fibrine receives its name from the circumstance that it enters largely into the composition of the muscular fibre of the animal system.

It is best obtained by whipping a quantity of fresh-drawn blood with a bunch of twigs, until it coagulates, and then carefully washing with water the stringy mass thus obtained. Subsequently it is to be washed with alcohol and ether, to remove all fatty matter adhering to it, and dried.

When thus obtained, fibrine is a yellowish opake mass, which is quite insoluble in water, alcohol, or ether. When long digested in water, at a very high temperature, a small proportion is dissolved, but slight decomposition also takes place. It is not soluble in the acids, but dissolves readily in dilute solutions of the alkalies.

Fibrine, obtained in the above manner from venous and arterial blood, does not appear to be exactly the same in all its properties;

QUESTIONS.—Is the composition of proteine known with certainty? How may it be obtained? What are its properties? 749. From what does fibrine derive its name? How may it be procured? Is it the same, whether obtained from venous or arterial blood?

as that from venous blood, when triturated with solution of nitrate of potash, at a high temperature, it becomes soluble, and very much resembles albumen, while that from arterial blood is not so affected.

Fibrine may also be obtained from the flesh of animals, of which it seems to form the essential part. When procured from this source, it resembles that from venous blood.

When the juice of various roots, as beets, turnips, carrots, &c., is allowed to stand for a time, a portion of it coagulates very much like blood, and may be separated from the uncoagulated part. It has been called vegetable fibrine, and seems to be the same as fibrine of animal origin.

750. *Gluten* is a substance obtained from the cereal grains It may be procured readily from wheat flour by kneading (554) it for some time in a large quantity of cold water, to wash away all the starch and other products, and then boiling for a time in alcohol. A soft, adhesive, elastic mass is thus obtained, which appears to be identical with fibrine. It is to this substance that wheat flour (which contains 20 or 25 per cent. of it) owes its adhesive, plastic nature; and it is this substance also, in a state of decomposition — mixed, probably, with starch in the same state—which constitutes *yeast.*

751. Albumen.—This compound is found nearly pure in the whites of eggs, from which it derives its name. Like fibrine, it exists in two states—as a liquid, in the whites of eggs, serum of the blood, humors of the eye, &c.—and, as a solid, in the brain and nerves of animals, and in the seeds of plants. The latter, called *vegetable albumen,* is considered as altogether identical with that of animal origin.

Liquid albumen may be coagulated, or solidified in various ways, as by heating to a temperature of 140°, or more, or by the action of chemical reagents, as alcohol, tannic acid, corrosive sublimate, and creosote; but when once coagulated it cannot be redissolved. As it forms an insoluble, and therefore inert compound with corrosive sublimate, it is considered a good antidote in cases of poisoning with the latter substance.

QUESTIONS.—What is said of the juice of certain roots, as those of beets, turnips, &c.? 750. What is gluten? What is yeast? 751. Describe albumen.

Vegetable albumen is found in the leaves and stalks of plants, and in the seeds, and in some cases in the roots. As obtained from plants, it is sometimes white, but is often colored. It is contained in considerable quantity in wood, and has great influence in causing its decay. It is therefore found that wood which has been saturated with a solution of corrosive sublimate is rendered more lasting when subjected to the influence of the atmosphere and moisture (501).

752. Caseine.—Legumine.—Caseine in many of its properties closely resembles albumen. It is most readily obtained from milk; and derives its name from *caseum*, curd. A very little sulphuric acid stirred in some skimmed milk causes it to coagulate in a short time; and the curds so formed being well washed with water, and digested for a time with carbonate of baryta to separate the acid, afford the pure caseine. When dried it is but slightly soluble in water, but very soluble in solutions of the alkalies, or alkaline carbonates.

Legumine, believed to be identical with the caseine from milk, is procured from peas, beans, and other similar seeds, by bruising them in water, and straining through a fine sieve; the liquid which passes through is a solution of this substance, and contains some starch, which settles by standing. The supernatant liquid, being poured off, possesses all the characters of skimmed milk, and from it the legumine, or vegetable caseine, may be procured in the same manner as from milk.

Caseine, in milk, is held in solution by free alkali present;—it is not coagulated by heat, like albumen, but coagulates readily by the action of *rennet*, or by acids, except the phosphoric.

The three important compounds above described very closely resemble each other in many of their properties, as well as in composition; but they are at the same time, in other properties, essentially different. Fibrine coagulates spontaneously by standing, but the other two undergo this change only by the application of certain agents. Albumen is not coagulated by acids, nor by rennet, but coagulates by heat, by contact of alcohol, and some of the metallic salts, and other compounds. Caseine does not coagulate by heat, nor by alcohol, but is readily coagulated by most of the acids, and by rennet. Probably what is considered as liquid caseine is only the substance held in solution by an alkali; and the coagulation of it by an acid results from the neutralization of the alkali by the acid.

QUESTIONS.—In what parts of plants is vegetable albumen contained? 752. Describe caseine. Describe legumine. How is caseine coagulated? How are fibrine and albumen coagulated?

Each of these substances appears always to contain a small quantity of sulphur; and fibrine and albumen contain, in addition, a little phosphorus. Decomposed by heating with acids or other reagents, they afford very complex results.

When kept for a time they undergo spontaneous decompostion, and in certain circumstances also induce peculiar changes in other organic compounds, as in the production of the alcoholic and other fermentations.

Being so nearly the same in composition, if not absolutely identical, it is altogether probable that, both in vegetables and animals, frequent transformations of one of these substances into another take place, by means and processes not yet understood.

CHEMICAL PHENOMENA OF VEGETATION.

753. *Germination* is the process by which a new plant originates from seed. A seed consists essentially of two parts, the *germ* of the future plant, endowed with a principle of vitality, and the *cotyledons*, or *seed-lobes*, both of which are enveloped in a common covering or cuticle. In the germ two parts, the *radicle* and *plumula*, may be distinguished, the former of which is destined to descend into the earth and constitute the root, the latter to rise into the air and form the stem of the plant. The office of the seed-lobes is to afford nourishment to the young plant until its organization is so far advanced that it may draw materials for its growth from extraneous sources.

The conditions necessary to germination are threefold; namely, moisture, a certain temperature, and the presence of oxygen gas. The necessity of moisture to this process has been proved by extensive observation. It is well known that the concurrence of other conditions cannot enable seeds to germinate provided they are kept quite dry.

A certain degree of warmth is not less essential than moisture. Germination cannot take place at a very low temperature; and a high temperature, as that of boiling water, destroys the vitality of the seeds. The most favorable temperature for the germination of most seeds is between 60° and 80°.

In the process of incipient germination a kind of saccharine fermentation (556) takes place, by which nutriment is supplied for the young

---

QUESTIONS.—753. What are the essential parts of a seed? Describe the process of germination in a seed? What are the conditions required for healthy germination?

plant during the early stages of its growth. This seems to be occasioned by the influence of the active principle *diastase*, which is not pre-existent in the seed, but is formed by the action of the air and moisture on the substances contained in it. When the process of germination is over, the plant is found provided with the necessary organs for procuring its nutriment from the atmosphere and soil; and there remains of the seed only its ligneous husk, which sometimes perishes in the ground, but at others rises to the surface and performs for a time the functions of leaves.

754. *Respiration of Plants.* — *Assimilation of Carbon.* — We have already noticed (213) the beautiful provision, by which the two great classes of organized bodies mutually compensate for the change each produces in the constituents of the atmosphere, and continue that proportion of them which is conducive to the healthful existence of both. Animals, by respiration, are constantly giving off carbonic acid, while plants, by the action of light upon their leaves, absorb carbonic acid and exhale oxygen.

It will be seen, therefore, that a process is ever going on in plants—at least when under the influence of light—not unlike the respiration of animals, in which the leaves perform the office of lungs. But the changes produced in the atmosphere being the reverse of those occasioned by the respiration of animals, the two processes serve to support each other; and the proportional quantity of oxygen and carbonic acid in the atmosphere is preserved ever the same.

The carbon obtained by plants by this respiratory process, and the water which is absorbed by the roots and leaves, furnish the elements of woody matter, and other substances, as sugar, starch, gum, &c., which contain oxygen and hydrogen in the proportion to form water. But the actions that really take place are usually much more complex than this would seem to indicate. The absorption of carbonic acid, and the liberation of oxygen, take place only in the light; in the dark, an opposite process sets in—oxygen is absorbed and carbonic acid is given off. But it has been well established that, on the whole, the quantity of carbonic acid absorbed is much greater that that evolved, whilst the quantity of oxygen evolved is much greater than that absorbed. Very many plants, it is believed, receive a large part of the carbon they contain from the atmosphere, while some few probably receive the whole. Others derive a part from the soil.

Now it is not difficult to trace some of the further changes which no doubt take place in the plant. All the softer parts of most plants con-

---

QUESTIONS.—754. What is meant by the respiration of plants? What different effects are produced upon the atmosphere by the respiration of plants and animals? Does the absorption of carbonic acid take place only in the light?

tain abundance of starch; even in the tubes and cells of ordinary wood it is found. This is so closely allied to sugar, gum, and woody-fibre, and the transformations of them from one to another so easy, that we may suppose them to be constantly taking place.

The other numerous secondary products, as the vegetable acids, oils, coloring-matters, &c., which characterize plants, may be formed from starch during the inverse respiratory action just alluded to, in which carbonic acid is given off, and oxygen absorbed from the atmosphere. During the day the assimilating power of the plant is in action, and carbonic acid is rapidly absorbed, and oxygen evolved; but during the night, while the plant is in repose, this nutritious action ceases, and a different process commences, during which the various substances natural to the plant are elaborated, attended by the absorption of oxygen from the atmosphere, and the evolution of carbonic acid and water. This change is well illustrated by certain plants, as the *cacalia ficoides*, the leaves of which are bitter in the evening, but in the morning are sour, like those of sorrel. During the night oxygen has been absorbed, and an acid generated from materials that were combined the evening previous, in a different mode.

Nitrogen is an essential ingredient in many of the products of plants;—it is believed to be obtained chiefly from the soil.

755. *Inorganic Constituents of Plants.*—Besides the substances which form the organic matter of plants, various inorganic bodies, as alkaline and earthy salts, are usually found contained in them, and no doubt serve an important purpose. If a plant is made to vegetate in a soil containing in it small quantities of several salts, we find that, while it continues in health, it seems to exercise a remarkable discretionary power, absorbing some, and rejecting others. Those absorbed most freely are such as are required for its proper growth, and are not given up to the water in which the plant may be immersed; while some are absorbed and again given off, and others still are entirely rejected, as being injurious. Most plants contain a small quantity of some salt of potassa, which exists as a carbonate in the ashes resulting from their combustion; but plants that grow near the sea, or springs of salt water, usually contain soda instead of potassa. Silica, lime, magnesia, &c., are also often contained in plants.

As these inorganic substances are necessary for the proper growth of plants, and all plants do not require the same substance, it is plain that a particular plant can be expected to flourish only in soils containing the substances it requires.

QUESTIONS.—Is nitrogen assimilated by plants? From what is it derived? 755. What are the inorganic constituents of plants? Do different plants absorb very different substances from the soil?

756. *Nature and Use of Manures.*—Every substance is called a manure which, applied to a soil, increases its productiveness. In a particular case it may be a substance which is required in the proposed crop, but of which the soil is deficient; or it may be a substance designed merely to give the soil a proper texture, so that it may more easily be penetrated by the rootlets of the plants, or to increase its adhesiveness and enable it to retain longer the moisture that falls upon it. It is in the manner last mentioned, probably, that most mineral manures, as lime, marl, &c., usually operate.

From the preceding remarks, the great advantage of *rotation of crops* may readily be seen. The mineral constituents of a soil are derived from the disintegration of its subjacent rocks; and some of them being contained only in small quantities, by continued succession of the same crop, may be entirely removed, and the soil become impoverished. By substituting another crop, which requires little or none of the material which has been exhausted, an abundant harvest may be obtained; and the soil, by the gradual decomposition of its subsoil, recover its former constitution.

757. The great value of organic manures is supposed to depend almost entirely upon the nitrogen they supply to the soil. The most of the nitrogen in plants is contained in the seed, tubers, &c., those parts which are selected by man for food, or for medicinal purposes, in consequence of the active principles they contain; and but little is found in the stem, leaves, root, &c., which are rejected as useless. The residue of a former season, therefore, may manure the land abundantly, so far as carbon is concerned, but there will be a deficiency of nitrogen, an essential element of the future desired crop. This deficiency is supplied by the decaying animal and vegetable substances used as manures; the value of which will therefore be very nearly in proportion to the nitrogen they supply. If mere ammoniacal salts are used, or perishable animal substances, their whole benefit is imparted to the crop immediately succeeding their application; but organic substances, which decay but slowly, yield a more permanent benefit: their nitrogen is gradually evolved, and though little benefit at first appears, the soil is at length found to be essentially improved.

758. In healthy vegetation, light serves a most important purpose; indeed, without it, no plant could come to perfection. A plant which grows in darkness, as in a cellar, however rich may be the soil in which

---

QUESTIONS.—756. What are manures? Explain the different modes in which manures operate to produce their effects? 758. What is said of the importance of light to healthy vegetation?

it stands, remains soft, its color pale, and its woody fibre unformed. When brought to the light, it perhaps increases in volume less rapidly, but the healthy action of the organ at once commences; the green color appears, and all the parts of the plant begin gradually to advance to maturity.

COMPOSITION OF THE ANIMAL TISSUES.

759. By the animal tissues we understand all the various solid found in the system, not excepting the bones and teeth.

The Muscles.—The muscles are the organs of motion to the animal, and they act solely by their contractions, which take place at the will of the animal. Their structure is always fibrous, and in the mammalia they are of a red color, and receive from the circulation a large supply of blood. The chief substances contained in them are fibrine, albumen, and gelatine. The two latter substances are contained chiefly in the membranes which envelope the fibres.

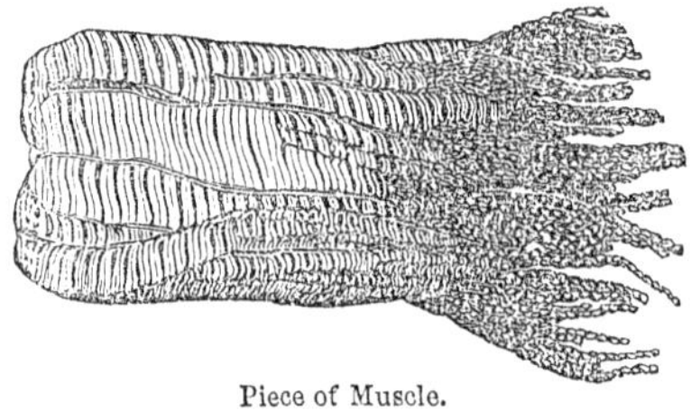
Piece of Muscle.

The accompanying figure represents a piece of muscle, and shows its fibrous structure. When taken from the animal and dried at a moderate temperature, it contracts greatly by the loss of water, and its weight is eventually reduced to one-fourth.

760. *The Cartilages, Skin, and Membranes.*—The *cartilages* appear to have essentially the same composition as the skin and the membranes generally. They are all composed of *gelatine* and *chondrine*, both of which are entirely dissolved by long continued boiling, and form with water a transparent jelly. Their exact composition has not been fully determined. Chondrine is most abundant in the cartilages.

The common *glue* of commerce is dried gelatine, and is prepared by boiling cuttings of parchment, or the skins, ears, and

---

QUESTIONS.—759. What are the muscles of animals? What are they chiefly composed of? 760. What is said of the cartilages, &c.? What is glue?

hoofs of animals, and evaporating the solution. Its use is well known. *Isinglass*, which is a very pure variety of gelatine, is prepared from the sounds of fish of the genus *acipenser*, especially from the sturgeon. The animal jelly of the confectioners is made from the feet of calves, the tendinous and ligamentous parts of which yield a large quantity of gelatine.

Gelatine is insoluble in alcohol, but is dissolved readily by most of the diluted acids, which form an excellent solvent for it.

Gelatine manifests little tendency to unite with metallic oxides. Corrosive sublimate and acetate of lead do not occasion any precipitate in a solution of gelatine, and the salts of tin and silver affect it very slightly.

By boiling gelatine with dilute sulphuric acid, and precipitating the acid with chalk, a sweet crystaline compound is obtained called *glycocoll* (*glukus*, sweet, and *kolla*, glue), or *gelatine sugar*, the composition of which is $C_4H_5NO_4$.

The action of tannin or tannic acid (684) upon gelatine is peculiar and important, resulting in the production, in certain cases, of the well known and most useful substance, *leather*.

Leather is usually formed by subjecting the skins of animals for some time to an infusion of bark which contains a large proportion of tannic acid; and the chief chemical change which takes place is believed to consist in a union of this acid with the gelatine of the skin. This is the common leather of which shoes are made: other varieties of this useful manufacture are prepared by different modes. Glove-leather, for instance, is prepared by impregnating the skin, after having been deprived of all its fatty matter by soaking in a weak alkali, with solution of common salt and alum, from which chloride of aluminum is formed, and unites with the gelatine of the skin.

The Brain and Nerves.—The substance of the brain, nerves, and spinal marrow differs from that of all other animal textures. The white and gray portions differ essentially in their nature, but are composed chiefly of water, albumen, a fatty matter, and traces of phosphates and other salts. Only about one-fifth of the whole is solid matter. The fatty matter is peculiar, and is quite unlike that of other parts of the system. The phosphorus in the brain is said to amount to 3 or 4 per cent. of all the solid matter contained in it.

761. *The Bones.* —The bones of animals consist of earthy matter, which is chiefly phosphate and carbonate of lime, and animal matter, which is essentially the same as cartilage. These

QUESTIONS.—What is said of the action of tannic acid upon gelatine? How is leather prepared? 761. What is said of the composition of the bones?

may easily be separated from each other. To separate the solid or earthy matter, it is only necessary to digest the bone for a time in dilute hydrochloric acid, in which both the phosphate and carbonate of lime are soluble, so that they will be entirely removed, leaving the cartilage soft and flexible, but retaining perfectly the form of the bone. The cartilage may be removed by heating the bone some time in the open air, so as to burn away all the organic matter; or by digesting it in solution of an alkali, or in water at a high temperature, under pressure. In some cases of disease, as rickets, to which children and youth are chiefly liable, there is a deficiency of earthy matter in the bones; and they are, in consequence, weak, and incapable of affording the necessary support to the system.

Horn differs from bone in containing only a trace of earth. It consists chiefly of gelatine and a cartilaginous substance like coagulated albumen. The composition of the nails and hoofs of animals is similar to that of horn; and the cuticle belongs to the same class of substances.

762. *The hair* seems to have essentially the same composition as horn, but the color is occasioned by an oil, which is soluble to ether, and may therefore be removed by it. It contains sulphur, and is therefore blackened by nitrate of silver, or other metallic salt.

When horn or hair is heated, it is first fused, and then swells up, giving off carbonate of ammonia, and carburetted hydrogen, the last of which takes fire, and burns with a brilliant flame.

Shells are composed of a mixture of phosphate and carbonate of lime. The shells of the crustacea, as lobsters, crabs, &c., usually contain 4 or 6 per cent. of phosphate of lime, and 50 or 60 per cent. of the carbonate, the rest being animal matter. The shells of the molusca, as the oyster and clam, are nearly pure carbonate of lime, containing only a mere trace of animal matter.

THE BLOOD.—PHENOMENA OF RESPIRATION AND DIGESTION.

763. The support of life in animals is attended by many chemical phenomena, which are constantly taking place in every part of the body, but especially in the blood, the lungs, the digestive system, and in the glands. The substances taken into the mouth are modified in the process of digestion, and distributed to every part

QUESTIONS.—How does horn differ in composition from bones? 762. What is said of hair? 763. What is said of the chemical changes which take place in the system?

by the circulation of the blood, where they come in contact with oxygen received from the air in the lungs. Among these different principles, received in the processes of nutrition and respiration, important chemical changes are taking place, by which the animal heat is kept up, and the waste of all the various tissues supplied, and, indeed, all the animal functions supported.

The blood of the different orders of animals is not the same, but we propose here to speak of that of the higher orders, or vertebrated animals, in which it is always of a red color.

In these, the blood which is brought from the lungs, and propelled by the heart through the arteries to every part of the system, is of a bright red or scarlet color, and is called *arterial blood;* but as it is returned to the heart by the veins, to be again sent to the lungs, it is of a dark red or purple, and is called *venous blood.* The blood is therefore of two kinds; but, chemically, they seem to be essentially the same.

The Blood.

764. *Composition of the Blood.*—The blood is easily distinguished from all the other fluids by its color. Its taste is slightly saline, its odor peculiar, and to the touch it seems somewhat unctuous. Its specific gravity is variable, but usually about 1·05; and in man its temperature is about 98° or 100°. While flowing in its vessels, or when recently drawn, it appears to the naked eye as a uniform homogeneous liquid; but if examined with a microscope of sufficient power, numerous particles of a discoid form are seen floating in a nearly colorless liquid. Most of these discs are red, and in the blood of man are about $\frac{1}{3200}$th of an inch in diameter, and one-fourth as thick; but there are also others which are without color. The figure on next page represents these discs in human blood, as seen under the microscope,—*a* and *b* colorless discs. The figure at the right represents these discs as they are sometimes seen united in rolls like pieces of coin.

QUESTIONS.—Is the blood the same in all animals? What two kinds of blood are there? 764. What is said of the composition of the blood? How does it appear when viewed by the microscope? What is the form of these corpuscles? Their size?

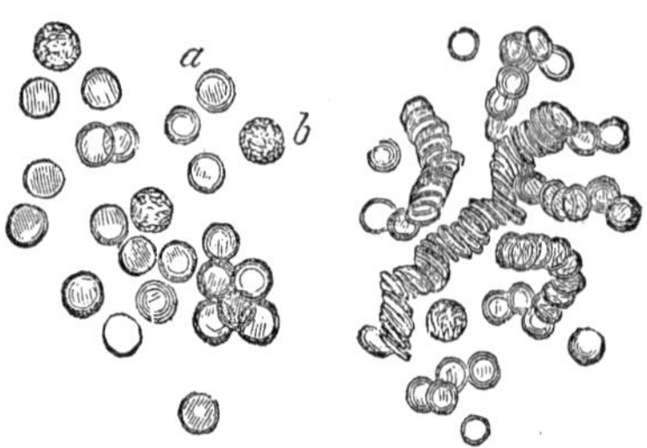

Discs in Human Blood.

These discs vary considerably in size in the blood of different animals, being considerably smaller in the blood of the common domestic animals than in man, but larger and of an oval form in oviparous vertebrates. They are larger in the blood of the elephant (see figure in the margin) than in that of any other mammalian species.

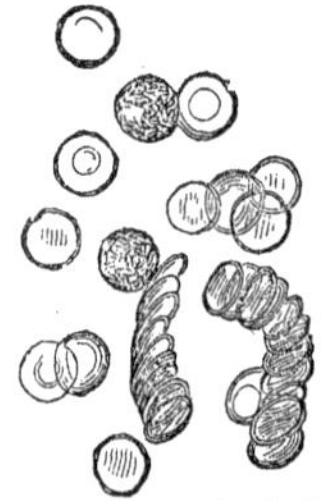
Discs in the Blood of the Elephant.

765. When the blood is withdrawn from the system, these discs or globules contract into a solid mass or *coagulum*, which, by standing, entirely separates from the liquid *serum*. This is a thin yellowish liquid, of sp. gr. about 1·03, and coagulates when heated to about 140°. It has a slight alkaline reaction, owing to the presence of soda, and contains a considerable quantity of albumen.

In the living body the blood also contains fibrine in solution, which, however, soon separates from the coagulum after its extraction from the system. Various salts also are found, as common salt, phosphates of lime, magnesia, and ammonia, and lactates of soda and magnesia.

The relative proportion of the ingredients of the blood must necessarily vary, independently of disease, even in the same individual, according as the nutrition is scanty or abundant. Slight variations are also occasioned by difference of age and sex.

QUESTIONS.—765. What change occurs in the blood when withdrawn from the system? What substances are mentioned as contained in the blood?

The following table contains the results of several analyses of blood:—

Water	90·5
Albumen	8·0
Chlorides of sodium and potassium	·6
Lactate of soda and extractive matter	·4
Soda and phosphate of soda	·41
Loss	·09
	100·00

The *coloring-matter* of the blood is confined entirely to the discs or corpuscles spoken of above, and has been called *hematosine.* In many of its properties it closely resembles albumen. It is composed of carbon, hydrogen, nitrogen, oxygen, and iron, the latter of which constitutes between six and seven per cent.

766. *Coagulation of the Blood.*—The coagulation of the blood, to which allusion has already been made (765), consists in the agglutination of the fibrine it contains, by which the red globules, and other suspended particles, are inclosed, forming the clot. The time required for the coagulation of the blood depends much upon temperature, being promoted by heat and retarded by cold.

The process is also influenced by exposure to the air. If atmospheric air be excluded, as by filling a bottle completely with recently-drawn blood, and closing the orifice with a good stopper, coagulation is retarded. It is singular, however, that if blood be confined within the exhausted receiver of an air-pump, the coagulation is accelerated.

Some substances, introduced into the blood, either retard or entirely prevent its coagulation, as saturated solution of chloride of sodium, hydrochlorate of ammonia, nitre, and a solution of potassa. The coagulation, on the contrary, is promoted by alum and the sulphates of the oxides of zinc and copper. The blood of persons who have died a sudden, violent death, by some kinds of poison, or from mental emotion, is usually found in a fluid state.

Phenomena of Digestion.

767. Digestion is the process by which the food is fitted to nourish the system, and supply the constant waste that is required for the support of the powers of life. The blood is the agent by

QUESTIONS.—What is the coloring-matter of the blood called? 766. What is said of the coagulation of the blood? 767. What is digestion?

which the matter required for the support of the system is supplied to the various parts where it is needed; but it is from the food that it is first received.

768. *The Saliva.*—The saliva is a slightly viscid liquid, secreted by several glands about the mouth, called the salivary glands. It serves to keep the mouth constantly moist; and the sight, or even thought of food, causes it to flow rapidly in the mouth. Besides water, and a small quantity of several salts, it contains a peculiar principle called *ptyaline,* which may be precipitated from it by absolute alcohol.

The use of the saliva is to mix with the food during mastication, and form a soft, pulpy mass, suitable to be swallowed. Probably it aids also in the process of digestion, by fitting the food to be more readily acted on by the juices of the stomach.

The saliva is often much affected by disease, and occasions the bad taste in the mouth, especially on rising in the morning. In some diseases, as intermittent fever, it is often very acid.

769. *The Gastric Juice.*—The gastric juice is an acid fluid secreted by the coats of the stomach, and contains in solution several salts, mucus, albumen, and a nitrogenized substance called *pepsine.* It has the property of softening down and dissolving, more or less perfectly, all the various substancss which serve for food. When the food is introduced into the stomach, it is there intimately mixed with this juice, by the agency of which it is converted into a semi-fluid matter called *chyme.* That this change is owing to the solvent power of the gastric juice has been fully determined; and it seems to be well established as a further fact, that the gastric juice secreted in the stomachs of animals of any species is especially adapted for dissolving the particular kind of food upon which the species usually feed.

While the food is in the stomach, it is constantly kept in motion by a peculiar movement of the muscular walls of the intestine, called its peristaltic motion. This has the effect thoroughly to mix every part, and also to propel the mass slowly forward.

QUESTIONS.—768. Describe the saliva. What purpose does it serve? 769. By what is the gastric juice secreted? What purpose does it serve? What is *chyme?*

The chyme, as it passes from the stomach, appears as a homogeneous, semi-fluid mass; but it has recently been found that the solvent action of the gastric juice is chiefly limited to the nitrogenized part of the fluid, as the albumen, fibrine, &c.; the non-nitrogenized parts, as the starchy and fatty compounds, being only mechanically divided and mixed up with the mass.

From the stomach, the chyme passes to a large intestine called the duodenum, where it is subjected to the action of two other fluids, the *bile* and the *pancreatic juice*, which we will now proceed to consider.

770. *The Bile.*—The bile is a liquid secreted by the liver, and preserved in the gall-bladder, which, when the stomach is filled, is slightly pressed and made to discharge its contents through a small duct into the duodenum. It is of a yellow, or greenish-yellow color, and has a peculiar sickening odor, and a taste at first sweet and then bitter, but exceedingly nauseous. Its consistence is variable, being sometimes limpid, but more commonly viscid and ropy. It contains a peculiar principle called *cholesterine*, and two organic acids called the *cholic*, and *choleic*, both of which are in combination with soda.

It seems now to be very well determined that one, if not the chief, purpose served by the bile is to aid in the digestion of the fatty and oily substances taken for food. In connection with the pancreatic juice, it also aids in the separation of the chyle.

Biliary calculi are concretions which occasionally form in the gall-bladder, or duct leading from it. They differ much in composition, but not unfrequently they are composed, in great part, of cholesterine.

771. *The Pancreatic Juice.*—This is a secretion of a certain organ called the pancreas. It is a limpid, colorless, alkaline liquid, with a saltish taste, like that of the serum of the blood. It is slightly viscous, and coagulates by heat, and by the acids.

The pancreatic juice is poured into the duodenum at the same time with the bile. Both of these liquids act energetically upon fatty and amylaceous substances, and produce important changes

---

QUESTIONS.—What fluids are poured into the blood next after it leaves the stomach? 770. By what organ is the bile secreted? Describe it. What compounds are contained in it? What purpose is served by it? 771. What is the character of the pancreatic juice?

in them, by which they are prepared to enter into the circulating system.

We have seen that the substances taken for food, after having been acted upon in the stomach by the gastric juice, and reduced to a homogeneous mass called *chyme*, pass into the duodenum, where the mass receives the bile and the pancreatic juice; immediately a change takes place, and a milky liquid appears diffused through the mass, called *chyle*. This is the part of the food, which, by the digestive process, has been prepared for the nourishment of the system, the rest being thrown off as refuse matter; and, as the mass passes on through the intestinal canal, it is absorbed by numerous little tubes opening into its side, called *lacteals*, and conveyed by them to a common reservoir, the *thoracic duct*. From this it is discharged into the left subclavian vein, and becomes incorporated with the blood, forming a part of its substance, and fitting it for the proper continuance of its functions.

Phenomena of Respiration.

772. All organized bodies require constantly the presence of atmospheric air for the proper performance of their various functions, and their continued existence in the living state. All the larger land-animals are provided with lungs, in which, by the action of certain muscles, air is constantly inhaled and again expelled, and thus a continued circulation is kept up; but in some of the smaller ones, as insects, the breathing is by tubes called trachea, or is performed by the whole surface of their bodies. In fishes, certain organs called gills perform the office of lungs, and the air that is breathed is first absorbed by the water, and from this imparted to the animal. In every case the process by which the air is made to perform its office upon the living being is called *respiration*.

The important chemical changes that attend the respiration of animals are, the constant absorption of oxygen, and the formation and evolution of carbonic acid.

QUESTIONS.—What is the *chyle?* What purpose does it serve? How is it conveyed to the blood? 772. Do all organized bodies require the constant presence of air? What purpose is served by the lungs of animals? What important changes attend the respiration of animals?

To show the presence of carbonic acid in the air expelled from the lungs, it is only necessary to apply a small tube to the mouth, and blow, a few seconds, into some recently-prepared lime-water;—the carbonic acid in the air from the lungs will unite with the lime, forming carbonate of lime, which gives the solution a milky appearance. It is true, there is a little carbonic acid in the air before it is taken into the lungs; but while there the quantity is very considerably increased. This is shown satisfactorily by blowing in the same manner by a hand-bellows into another portion of lime-water, when it will be found that a much longer time will be required to produce the milkiness alluded to (304).

Blowing through Lime-water.

773. The change in the blood from the venous to the arterial state, is effected in the living animal during the passage of the blood through the capillary vessels of the lungs, where it is exposed to the action of an extensive surface of atmospheric air, through the thin membranes which separate these from the air-vessels; and the arterial blood, in traversing the capillary system of the body, imparting nourishment to it, gradually assumes the dark-colored condition in which it is returned to the heart by the veins. The same change is produced when venous blood, just taken from the system, is brought in contact with atmospheric air, and is attended with the evolution of carbonic acid gas. It takes place more speedily when air is agitated with blood; it is still more rapid when pure oxygen is substituted for atmospheric air; and it does not occur at all when oxygen is entirely excluded. The quantity of carbonic acid developed very exactly corresponds with that of the oxygen which disappears.

The quantity of oxygen withdrawn from the atmosphere, and of carbonic acid disengaged, is variable in different individuals, and in the same individual at different times, depending very much upon the state of the system. In a state of health, anything that accelerates the respiration increases the amount of oxygen absorbed, and the quantity of carbonic acid exhaled; but in certain cases of disease, as in inflammatory fevers, though the respiration may be very rapid, little oxygen is withdrawn from the air, and the venous blood is found to be very florid.

QUESTIONS.—How may the presence of carbonic acid in the air exhaled from the lungs be shown? 773. How is the change in the blood from the venous to the arterial state effected? What change is again produced in the blood as it circulates through the system?

774. *Animal Heat.—Nutrition.*—It has long been known that there is a close connection between the function of respiration, and the development of heat in the animal system.

Thus, in all animals whose respiratory organs are small and imperfect, and which, therefore, consume but a comparatively minute quantity of oxygen, and generate little carbonic acid, the temperature of the blood varies with that of the medium in which they live. In warm-blooded animals, on the contrary, in which the respiratory apparatus is larger, and the chemical changes more complicated, the temperature is almost uniform; and those have the highest temperature whose lungs, in proportion to the size of their bodies, are largest, and which consume the greatest quantity of oxygen. The temperature of the same animal at different times is connected with the state of the respiration. When the blood circulates sluggishly, and the temperature is low, the quantity of oxygen consumed is comparatively small; but, on the contrary, a large quantity of that gas disappears when the circulation is brisk and the power of generating heat energetic. It has also been observed, that when an animal is placed in a very warm atmosphere, so as to require little heat to be generated within his own body, the consumption of oxygen is unusually small.

We have seen already that the digestion of the food is performed, in part, in the stomach, by the gastric juice dissolving the nitrogenized or albuminous portion, and partly in the duodenum, where the amylaceous and fatty parts are acted on by the bile and pancreatic juice. This indicates an important division of the substances taken for food into two kinds; and we find that the same distinction is also observed in the purposes they serve in the system. It has been determined that the albuminous or nitrogenized parts of the food, go to support the waste of the tissues, while the non-nitrogenized parts, as starch, sugar, and the fats and oils, are consumed in preserving the temperature

---

QUESTIONS.—774. Is there any immediate connection between the function of respiration of animals and the temperature of their systems? What facts are mentioned as illustrating and proving the point? What two purposes are served by the food? What are the elements of respiration and nutrition?

of the system. The latter may therefore be called the *elements of respiration*, and the former *elements of nutrition*.

775. The combustion of carbon and hydrogen in the open air, as is well known, is always attended by the evolution of heat; and the same effect is produced in the system when these substances unite with the oxygen introduced into the blood in the process of respiration. The only essential difference is, that in combustion the process is more rapid, and the heat produced proportionally more intense; but the absolute quantity of heat produced by the consumption of a given amount of carbon or hydrogen is, in all probability, always the same.

We here find the origin of the carbonic acid which is given off during respiration, and also by the insensible perspiration from the skin, while the water produced by the combustion of the hydrogen unites with the large quantity ever present in all parts of the body.

We see, therefore, that the oxygen of the air is tending to consume all living beings, as really as if they were in a burning fire; in fact, this consumption, which is a kind of combustion, is ever going on while respiration continues. To keep up the combustion, a constant supply of fuel is needed, which is found in the respiratory food, as before stated; and when these elements enter into combination within the animal system, heat is as necessarily developed as when the same thing takes place in the open air, in ordinary combustion.

776. *Use of the Fat in the Animal Economy.*—The fat and oils found in animals appear to be stores of respiratory food, laid up by a wise foresight of nature for time of need. Thus, it is well known that when food is abundant the fat accumulates, which is again gradually wasted away when the supply of food becomes deficient. When the supply of food is wholly withheld from an animal, the fat rapidly disappears, its carbon and hydrogen going to supply the demands of respiration; and when this has all been consumed, the substance of the muscles is attacked, which become lean and flaccid, and lose their contractile power; and at length the brain and nerves yield to the same influence,

QUESTIONS.—775. What always attends the combustion of carbon and hydrogen? Is the same effect produced in the animal system? 776. What purpose is served by the fat in the animal system? What becomes of the fat when the supply of food is short?

and death speedily closes the scene of suffering. In animals that lie torpid during the winter, nature has provided that in the summer season a large accumulation of fat is laid up to supply the demands of respiration during the time they lie torpid in their dens; and, on the approach of the warm weather of spring, they are consequently found lean and weak.

It is important to observe that the composition of the nitrogenized food consisting chiefly of the proteine compounds and gelatine, is the same a that of many of the tissues, the waste of which it supplies; and it is the opinion of many that these compounds are never produced in the anima' system, but are appropriated directly from the food, unchanged.

It is, however, known that some substances are formed in the system from the materials supplied by the food. Thus, it has been found that a lean goose, fed on Indian corn, will in a few days increase in weight several pounds, in consequence of the fat that will be formed in the system, which must have been contained in the corn, or formed from the substance of the corn by the organs of the animal. Now that the most of it has been formed in the latter mode is certain from the fact that all the oil or fatty matter contained in the corn will amount only to a small fraction of the fat found in the goose. So bees will form wax when fed on pure honey or pure sugar. It has been found by experiment that for twenty parts of honey consumed they will form about one part of wax.

SEVERAL ANIMAL SECRETIONS AND EXCRETIONS NOT BEFORE NOTICED.

777. *Milk.* — This well-known liquid is secreted by females of the class mammalia, for the support of their young. It is of a white color, and is a little heavier than water, usually having a density from 1·02 to 1·04. Its composition varies in different animals, and also in the same animal, according to the food that is taken. On an average, that of the cow contains, in an hundred parts, water 87·4 parts; butter, 4·0; milk-sugar, or *lactine*, and soluble salts, 5·0; caseine, albumen, and insoluble salts, 3·6.

When milk is examined by a microscope, it is found to be a transparent liquid, with numerous minute globules floating in it, which consist chiefly of fatty matter, and which, if the milk stand at rest a few hours, rise to the surface as *cream*. If, now, the cream is separated, and violently agitated for a time, the mem-

---

QUESTIONS.—777. What substances are contained in cow's milk. What is cream?

branes enveloping these fatty corpuscles are broken, and it collects into a mass of butter, which floats in the watery liquid.

Milk is not coagulated by heat, but this effect is readily produced by acids and by *rennet*, which is prepared by soaking the inner coat of a calf's stomach in water. The coagulum so formed is called *curd*, and when pressed and otherwise prepared, constitutes *cheese*.

778. When milk is allowed to stand some hours, at a temperature of between 60° and 70°, it sours; and, on examination, a peculiar acid is found in it called *lactic acid*, the composition of which is $C_6H_6O_6$. When highly concentrated, it is a thick, colorless liquid, of specific gravity 1·21, and has a very sour taste. It is soluble in alcohol and water, but cannot be crystalized.

Lactic acid may also be formed from sugar by mixing with it in solution some curds from milk, and some chalk to unite with it as it is produced, and allowing the whole to stand for some time at a temperature of about 80°. The chemical change that takes place appears to be very simple, as an atom of fruit-sugar contains exactly the ingredients of two atoms of lactic acid. Thus, $C_{12}H_{12}O_{12} = 2(C_6H_6O_6)$. It is this acid which is contained in the gastric juice.

Lactic acid is also formed when the juices of beets, carrots, &c., are allowed to ferment at high temperatures. At the same time a viscid, slimy substance is formed, from which circumstance this has been termed the *viscous fermentation*.

Milk may also be made to undergo the vinous or alcoholic fermentation. For this purpose it is exposed for some hours to a temperature of about 100°; and the alcohol, which, no doubt, is formed from the milk-sugar it contains, may subsequently be distilled from it. It has long been known that some barbarous nations are accustomed to prepare an intoxicating drink from milk, by some fermenting process.

779. *Lymph.*—Lymph is a watery fluid, secreted by many of the membranes; it lubricates all the cavities, and moistens the

QUESTIONS.—778. What acid is formed in the souring of milk? May it be formed from sugar? May milk be made to undergo the alcoholic fermentation? 779. What is lymph?

cellular tissues of the body. It much resembles the serum of the blood, and mixes readily with water. Sometimes, in diseases, this liquid, or a liquid analogous to it, is secreted in one or more organs, so abundantly as to constitute dropsy.

780. *Urine.*—This is a yellowish liquid, secreted by the kidneys, by which various substances, constantly accumulating in the blood, are separated from it during the healthy state of the system. This excretion is essential, as without it the substances thus thrown off would, in a short time, accumulate in such quantity as to destroy life. It has usually a specific gravity of about 1·02, and consists of water, holding in solution a small portion of *urea* and several different salts, as well as a little free acid, from which it acquires an acid reaction. A little mucus is also usually present, derived from the urinary passages; and in consequence of this it putrefies if kept for a time at a summer temperature.

The quantity of the urine is affected by various causes, especially by the nature and quantity of the food and the liquids received into the stomach; but on an average, a healthy person voids between thirty and forty ounces daily. The quality of this fluid is likewise influenced by the same circumstances, being sometimes in a very dilute state, and at others highly concentrated.

The urine of birds, insects, and reptiles, is solid, and consists chiefly of urate of ammonia. *Guano*, which has been so largely imported within a few years, to be used as a manure, is composed chiefly of the urine and other excrements of birds, which have been collecting for ages in the places where they are found. It is imported chiefly, if not entirely, from islands on the coast of Africa, and the west coast of South America.

781. *Urea*, $C_2H_4N_2O_2 = CyO,NH_3 + HO$.—This substance is always found in healthy urine, and may also be prepared artificially. Its crystals, when pure, are transparent and colorless, of a slight pearly lustre, and have commonly the form of a four-sided prism. It leaves a sensation of coldness on the tongue, like

QUESTIONS.—780. What purpose is served in the animal system by the secretion of the urine? What is guano? 781. From what is urea obtained?

nitre, and its smell is faint and peculiar, but not urinous. Its specific gravity is about 1·35.

Water, at 60°, dissolves more than its own weight of urea, and boiling water takes up an unlimited quantity. It requires for solution about five times its weight of alcohol at 60°, and rather less than its own weight at a boiling temperature. The aqueous solution of pure urea is very permanent, but if the other constituents of urine are present, it putrefies with rapidity, and, by a heat of 250°, it is melted and resolved into carbonate of ammonia, and other products.

Though urea has not any distinct alkaline properties, it unites with the nitric and oxalic acids, forming sparingly soluble compounds, which crystalize in scales of a pearly lustre.

Urea is prepared artificially as follows. Four parts of dry Prussiate of potash are intimately mixed with 2 parts of peroxide of manganese, and heated to dull redness in an iron vessel. When the mass has become cold it is digested in cold water, and frequently stirred, until all that is soluble is taken up by the water. This is now decanted, and mixed with 3 parts of sulphate of ammonia previously dissolved in water. Sulphate of potash and cyanate of ammonia are at once formed; and on the application of a moderate heat the latter compound is converted into urea, which is only an isomeric modification.

To separate the urea, the whole solution is evaporated to dryness, and the urea taken up by strong alcohol.

782. *Uric Acid*, $C_{10}H_4N_4O_6$, is found, in combination with bases, in the urine of many animals, especially that of birds of prey, and of serpents. It is a white, tasteless, and inodorous solid, but sparingly soluble in either cold or hot water. *Urate of ammonia* constitutes a large part of the best guano.

When uric acid is boiled with peroxide of lead a substance called *allantoine* is formed, which has the composition, $C_8H_6N_4O_6$, carbonic acid being also evolved. It is a solid, and capable of being crystallized. Boiled with the strong acids, it forms urea and *allantoic acid*, $C_6H_4N_2O_6$. *Alloxan, alloxantine, alloxanic,* and *dialuric acids*, are other closely allied substances.

783. Besides the above substances, there is found in the urine of herbivorous animals, as the cow and horse, another important

---

QUESTIONS.—Describe the artificial mode of preparing urea. 782. What is uric acid? 783. From what is hippuric acid obtained?

acid, called the *hippuric acid* (from *hippos*, horse). Its composition is $C_{18}H_9NO_6 = C_{18}H_8NO_5 + HO$.

To prepare this acid, the fresh urine may be concentrated by a gentle heat,—taking care not to cause it to boil,—until its volume is diminished one-half; and hydrochloric acid then added until an acid reaction is produced. By standing for a time, impure hippuric acid will make its appearance in the form of long, slender crystals. These are then to be redissolved in boiling water, mixed with some milk of lime and animal charcoal, and the resulting solution of hippurate of lime carefully filtered before cooling. To decompose this, hydrochloric acid is to be added until an acid reaction appears; and, on cooling, the pure hippuric acid will be obtained in beautiful crystals. These are only slightly insoluble in cold, but very soluble in boiling water.

Some of the relations of this acid are curious and important. When boiled for some time with a strong acid, it is converted into benzoic acid and glycocoll (760), 2 atoms of water being absorbed in the process. Thus,

$$C_{18}H_9NO_6 + 2HO = C_{14}H_6O_4 + C_4H_5NO_4.$$

By boiling with peroxide of lead, hippuric acid is converted into benzamide (634), carbonic acid and water being formed at the same time.

The action of nitrous acid upon the hippuric, results in the production of a new acid, called the *benzoglycollic*, which has the formula, $C_{36}H_{16}O_{16}$; water and nitrogen being eliminated. Glycocoll acted on by nitrous acid forms *glycocollic acid*, $C_8H_8O_{12}$.

QUESTIONS.—Describe the process of obtaining hippuric acid. When it is boiled with a strong acid, what compounds are formed from it?

APPENDIX.

Tables of Weights and Measures, Proportion of Alcohol in Spirits of different Specific Gravities, etc.

WEIGHTS.

English Imperial Standard Troy Weight.

24 Grains = 1 Pennyweight.
20 Pennyweights = 1 Ounce = 480 Grains.
12 Ounces = 1 Pound = 5760 Grains.

Apothecaries' Weight.

20 Grains = 1 Scruple ℈.
3 Scruples = 1 Drachm ʒ.
8 Drachms = 1 Ounce ℥ = 480 Grains.
12 Ounces = 1 Pound = 5760 Grains.

The Grain, Ounce, and Pound are the same in both the above weights.

Avoidupois Weight.

1 Drachm = 27·34 Grains.
16 Drachms = 1 Ounce = 437·5 Grains.
16 Ounces = 1 Pound = 7000 Grains.
28 Pounds = 1 Quarter.
4 Quarters = 1 Cwt. or 112 lbs.
20 Cwt. = 1 Ton.

French Decimal Weights.

Gramme	= 15·443 Grains.	Gramme	=		15·443 Grains.
Decigramme	= 1·544 "	Decagramme	=	10 Grammes	= 154·43 "
Centigramme	= 0·154 "	Hectogramme	=	100 "	= 1544·3 "
Milligramme	= 0·015 "	Kilogramme	=	1000 "	= 15443· "

The Kilogramme = 2·206 lbs. Avoid., and 2·68 lbs. Troy.

Troy weight is used in all cases in weighing the precious metals, the ounce being generally made the unit. Physicians, in making their prescriptions, and druggists, in compounding their medicines, make use of the subdivisions of the pound called Apothecaries' weight; but drugs are always bought and sold by Avoirdupois weight.

In scientific researches, as in chemical analysis, and the determination of specific gravities, either Troy weight or the French decimal system is used. When Troy weight is used, the grain is made the unit.

MEASURES OF CAPACITY.

Wine or Apothecaries' Measure.

60 Drops = 1 Drachm ʒ.
8 Drachms = 1 Ounce ℥.
16 Ounces = 1 Pint.
8 Pints = 1 Gallon = 231 Cubic Inches.

The cubic inch of pure water at 62° weighs 252·458 grains, or 16·283 grammes.

French Decimal Measure.

Litre = 61·025 Cubic Inches.
Decilitre = 6·102 "
Centilitre = 0·610 "
Millilitre = 0·061 "

The Litre = 1·056 quart wine measure.

MEASURES OF LENGTH.

French Decimal System.

Metre = 39·371 Inches.
Decimetre = 3·937 "
Centimetre = ·393 "
Millimetre = ·039 "

The French Metre is one ten millionth part of the distance, upon the surface of the earth, from the equator to either pole.

The Gramme (the unit of weight) is the weight of a cubic centimetre of pure water, at its greatest density.

TABLE showing the Specific Gravity of Liquids, at the Temperature of 55° Fahr., corresponding to the Degrees of Baumé's Hydrometer.

FOR LIQUIDS LIGHTER THAN WATER.

Deg.	*Sp. Gr.*	*Deg.*	*Sp. Gr.*	*Deg.*	*Sp. Gr.*	*Deg.*	*Sp. Gr.*	*Deg.*	*Sp. Gr.*
10 =	1·000	17 =	·949	23 =	·909	29 =	·874	35 =	·842
11	·990	18	·942	24	·903	30	·867	36	·837
12	·985	19	·935	25	·897	31	·861	37	·832
13	·977	20	·928	26	·892	32	·856	38	·827
14	·970	21	·922	27	·886	33	·852	39	·822
15	·963	22	·915	28	·880	34	·847	40	·817
16	·955								

FOR LIQUIDS HEAVIER THAN WATER.

Deg.	*Sp. Gr.*	*Deg.*	*Sp. Gr.*	*Deg.*	*Sp. Gr.*	*Deg.*	*Sp. Gr.*	*Deg.*	*Sp. Gr.*
0 =	1·000	15 =	1·114	30 =	1·261	45 =	1·455	60 =	1·717
3	1·020	18	1·140	33	1·295	48	1·500	63	1·779
6	1·040	21	1·170	36	1·333	51	1·547	66	1·848
9	1·064	24	1·200	39	1·373	54	1·594	69	1·920
12	1·089	27	1·230	42	1·414	57	1·659	72	2·000

TABLE showing the Quantity of Absolute Alcohol in (100 *parts of*) *Spirits of different Specific Gravities, at* 60° *Fahr., the spirits being supposed to contain nothing but pure alcohol and pure water.*

ALCOHOL.	SP. GRAVITY.	ALCOHOL.	SP. GRAVITY.	ALCOHOL.	SP. GRAVITY.
100	0·796	66	0·881	32	0·955
99	0.798	65	0·883	31	0·957
98	0·801	64	0·886	30	0·958
97	0·804	63	0·889	29	0·960
96	0·807	62	0·891	28	0·962
95	0·809	61	0·893	27	0·963
94	0·812	60	0·896	26	0·965
93	0·815	59	0·898	25	0·967
92	0·817	58	0·900	24	0·968
91	0·820	57	0·902	23	0·970
90	0·822	56	0·904	22	0·972
89	0·825	55	0·906	21	0·973
88	0·827	54	0·908	20	0·974
87	0·830	53	0·910	19	0·975
86	0·832	52	0·912	18	0·977
85	0·835	51	0·915	17	0·978
84	0·838	50	0·917	16	0·979
83	0 840	49	0·920	15	0·981
82	0·843	48	0·922	14	0·982
81	0·846	47	0·924	13	0·984
80	0·848	46	0·926	12	0·986
79	0·851	45	0·928	11	0·987
78	0·853	44	0·930	10	0·988
77	0·855	43	0·933	9	0·989
76	0·857	42	0·935	8	0·990
75	0·860	41	0·937	7	0·991
74	0·863	40	0·939	6	0·992
73	0·865	39	0·941	5	0·993
72	0·867	38	0 943	4	0·994
71	0·870	37	0·945	3	0·996
70	0·872	36	0·947	2	0·998
69	0·874	35	0·949	1	0·999
68	0·875	34	0·951	0	
67	0·879	33	0·953		

INDEX.

THE END.

www.ingramcontent.com/pod-product-compliance
Lightning Source LLC
LaVergne TN
LVHW021306110826
845150LV00003B/496

* 9 7 8 1 4 2 5 5 5 8 9 4 9 *